Rainer Feistel and
Werner Ebeling

**Physics of Self-Organization
and Evolution**

Related Titles

Dressler, F.

Self-Organization in Sensor and Actor Networks

386 pages
2007
Hardcover
ISBN: 978-0-470-02820-9

Nayfeh, A. H., Balachandran, B.

Applied Nonlinear Dynamics
Analytical, Computational, and Experimental Methods

700 pages
1995
Hardcover
ISBN: 978-0-471-59348-5

Rainer Feistel and Werner Ebeling

Physics of Self-Organization and Evolution

WILEY-VCH Verlag GmbH & Co. KGaA

The Authors

Dr. Rainer Feistel
Leibniz-Institut für Ostseeforschung
Rostock, Germany
rainer.feistel@io-warnemuende.de

Prof. Dr. Werner Ebeling
Institut für Physik
der Humboldt-Universität zu Berlin
Germany

Cover Picture
Photograph by Gregor Rehder, Waterspouts observed over the Baltic Sea on 18 August 2010 by r/v "Maria S. Merian"
Design by Susanne Feistel

All books published by **Wiley-VCH** are carefully produced. Nevertheless, authors, editors, and publisher do not warrant the information contained in these books, including this book, to be free of errors. Readers are advised to keep in mind that statements, data, illustrations, procedural details or other items may inadvertently be inaccurate.

Library of Congress Card No.: applied for

British Library Cataloguing-in-Publication Data
A catalogue record for this book is available from the British Library.

Bibliographic information published by the Deutsche Nationalbibliothek
The Deutsche Nationalbibliothek lists this publication in the Deutsche Nationalbibliografie; detailed bibliographic data are available on the Internet at http://dnb.d-nb.de.

© 2011 Wiley-VCH Verlag & Co. KGaA, Boschstr. 12, 69469 Weinheim, Germany

All rights reserved (including those of translation into other languages). No part of this book may be reproduced in any form – by photoprinting, microfilm, or any other means – nor transmitted or translated into a machine language without written permission from the publishers. Registered names, trademarks, etc. used in this book, even when not specifically marked as such, are not to be considered unprotected by law.

Cover Design Grafik-Design Schulz, Fußgönheim
Typesetting Thomson Digital, Noida, India
Printing and Binding Fabulous Printers Pte Ltd

Printed in Singapore
Printed on acid-free paper

Print ISBN: 978-3-527-40963-1
ePDF ISBN: 978-3-527-63681-5
ePub ISBN: 978-3-527-63680-8
mobi ISBN: 978-3-527-63682-2
oBook ISBN: 978-3-527-63679-2

Contents

Preface *IX*

1	**Introduction to the Field of Self-Organization** *1*	
1.1	Basic Concepts *1*	
1.2	History of Evolution as a Short Story *6*	
1.3	Structure, Self-organization, and Complexity *14*	
1.4	Entropy, Equilibrium, and Nonequilibrium *17*	
1.5	Dynamics, Stability, and Instability *25*	
1.6	Self-Organization of Information and Values *28*	
2	**Fundamental Laws of Equilibrium and Nonequilibrium Thermodynamics** *35*	
2.1	The Thermodynamic Way of Describing Nature – Basic Variables *35*	
2.2	Three Fundamental Laws and the Gibbs Relation of Thermodynamics *45*	
2.3	Thermodynamic Potentials, Inequalities, and Variational Principles *55*	
2.4	Irreversible Processes and Self-Organization *63*	
2.5	Irreversible Radiation Transport *70*	
2.6	Irreversible Processes and Fluctuations *76*	
2.7	Toward a Thermodynamics of Small Systems Far from Equilibrium *80*	
3	**Evolution of Earth and the Terrestrial Climate** *85*	
3.1	The Photon Mill *88*	
3.2	Black-Body Radiation Model of Earth *91*	
3.3	Local Seasonal Response *99*	
3.4	Atmospheric Cooling Rate *104*	
3.5	Black-Body Model with Atmosphere *106*

3.6	Humidity and Latent Heat	*110*
3.7	Greenhouse Effect	*119*
3.8	Spatial Structure of the Planet	*124*
3.9	Early Evolution of Earth	*149*
4	**Nonlinear Dynamics, Instabilities, and Fluctuations**	*163*
4.1	State Space, Dynamic Systems, and Graphs	*163*
4.2	Deterministic Dynamic Systems	*168*
4.3	Stochastic Models for Continuous Variables and Predictability	*177*
4.4	Graphs – Mathematical Models of Structures and Networks	*187*
4.5	Stochastic Models for Discrete Variables	*194*
4.6	Stochastic Processes on Networks	*200*
5	**Self-Reproduction, Multistability, and Information Transfer as Basic Mechanisms of Evolution**	*211*
5.1	The Role of Self-Reproduction and Multistability	*211*
5.2	Deterministic Models of Self-Reproduction and Bistability	*213*
5.3	Stochastic Theory of Birth-and-Death Processes	*218*
5.4	Stochastic Analysis of the Survival of the New	*222*
5.5	Survival of the New in Bistable Systems	*226*
5.6	Multistability, Information Storage, and Information Transfer	*230*
6	**Competition and Selection Processes**	*237*
6.1	Discussion of Basic Terms	*237*
6.2	Extremum Principles	*241*
6.3	Dynamical Models with Simple Competition	*244*
6.4	Stochastic of Simple Competition Processes	*253*
6.5	Competition in Species Networks	*264*
6.6	Selection and Coexistence	*278*
6.7	Hyperselection	*284*
6.8	Selection in Ecological Systems	*288*
6.9	Selection with Sexual Replication	*297*
6.10	Selection between Microreactors	*301*
6.11	Selection in Social Systems	*306*
7	**Models of Evolution Processes**	*311*
7.1	Sequence-Evolution Models	*314*
7.2	Evolution on Fitness Landscapes	*319*
7.3	Evolution on Smooth Fisher–Eigen Landscapes	*321*
7.4	Evolution on Random Fisher–Eigen Landscapes	*328*
7.5	Evolution on Lotka–Volterra Landscapes	*333*
7.6	Axiomatic Evolution Models	*340*
7.7	Boolean Behavior in the Positive Cone	*342*
7.8	Axiomatic Description of a Boolean Reaction System	*349*
7.9	Reducible, Linear, and Ideal Boolean Reaction Systems	*352*

7.10	Minor and Major of a Boolean Reaction System 355
7.11	Selection and Evolution in Boolean Reaction Systems 356

8	**Self-Organization of Information and Symbols** 363
8.1	Symbolic Information 364
8.2	Structural Information 368
8.3	Extracting Structural Information 371
8.4	Physical Properties of Symbols 375
8.5	Properties of the Ritualization Transition 381
8.6	Genetic Code 384
8.7	Sexual Recombination 390
8.8	Morphogenesis 392
8.9	Neuronal Networks 396
8.10	Spoken Language 402
8.11	Possession 405
8.12	Written Language 406
8.13	Money 409

9	**On the Origin of Life** 413
9.1	Catalytic Cascades in Underoccupied Networks 415
9.2	Formation of Spatial Compartments 418
9.3	Replicating Chain Molecules 421
9.4	Molecular Information Processing 428
9.5	Darwinian Evolution 433

10	**Conclusion and Outlook** 441
10.1	Basic Physical Concepts and Results 441
10.2	Quo Vadis Evolutio? 447

References   453

Index   501

Preface

The wall had developed a crack, and a strip of sunlight was dancing right through it.
Erich Kästner: The Schildburghers

At one time Latin was close to the universal language of what we now call "science". Then French took over followed by German and at long last English. Both the English and the Americans are infamous for being monolingual. If you don't publish in English, your work is likely to go unnoticed.
David Hull: Science and Language

The first German edition of this book was written in the years 1978–1980 in Rostock, Moscow, and Berlin, about 10 years before the fall of the Berlin wall, Figure 1.

In contrast to the general expectation of politicians, in particular of the various so-called east experts in the West, that the sudden implosion of the eastern political system was a completely unpredictable event, these authors believe, based on own experience, that already since the 1970s "something was in the air." Preceding the "perestroika" and the related social turnover after 1989, there was a developing spirit of enlightenment, similar to the "Aufklärung" in Germany and the "Siècle des Lumières" in France, before the revolutionary events happened in Europe in the eighteenth and nineteenth century. In retrospect, expressed in the language of physicists, the dawn of the phase transition was accompanied by internal fluctuations with increasing amplitudes.

When in 1970 Ilya Prigogine, Figure 2, presented his great lecture on self-organization in Moscow, this was appreciated by the audience as a kind of theoretical revolution. One of these authors was a witness of this outstanding event. The lecture room in the Moscow Academy of Science was overcrowded, the excitement was enormous. While most people supported the new ideas, a few also protested against them. The related emotional impression is well reflected in the text by the German poet Erich Kästner that we used as our motto above. Kästner (1899–1974) had a difficult life and was badly treated by the Nazi regime because of his protest against the militaristic German policy and the Second World War.

Figure 1 Crack in the Berlin wall, virtually foreseen already in 1954 by Erich Kästner, in the motto of this chapter. Photo taken in November 1989 in Berlin, courtesy of Jörg Hildebrandt.

The vivid resonance which the ideas on self-organization found among the public in particular in the former socialist countries resulted in the immediate translation to Russian of all books on self-organization available at that time. A similar response was observed after the lectures of Manfred Eigen and Ilya Prigogine at the end of the 1970s in Halle/Saale, presented during public conferences organized by the oldest German Scientific Academy, Leopoldina. As a matter of fact, Leopoldina strongly promoted the distribution of the ideas of self-organization in both parts of Germany. As well, Hermann Haken's books on Synergetics also played an important role for the emergence, propagation, and development of the new physical discipline. Nearly the same what had happened in Moscow in 1970 took place again in 1982 after an excellent lecture of Prigogine on "The Arrow of Time," given at the Pergamon Museum in East Berlin. In addition to physicists, chemists, mathematicians, and biologists, also many philosophers and journalists participated in the discussion, and Prigogine's ideas were transmitted on the radio to a broader public.

Figure 2 Ilya Prigogine during a lecture in the 1980s. (from the archive of the authors).

The original edition of this book was inspired by the feeling of this exciting atmosphere of the emergence of something "New" in those unique years. This way our first book was mainly a reaction to a widespread interest of the community in ideas of self-organization and evolution. These authors were very happy to get in close contact to scientific pioneers such as Ilya Prigogine, Herrmann Haken, and Manfred Eigen, as well as to various other experts in that field such as Conrad, Nicolis, Klimontovich, Romanovsky, Stratonovich, Velarde, and Volkenstein.

It was at the Heraeus Seminar on "Evolution and Physics – Concepts, Models, and Applications" in Bad Honnef in January 2008 when these authors decided to publish an English edition of the German "Physik der Selbstorganisation und Evolution" of 1982, almost exactly 30 years after the writing of the original manuscript had begun. During the meeting the impression prevailed that even after three decades the presentation given in the old book might still be worth to be made available to a wider public than to just the German readership. The quick first idea of an almost literal translation was soon dismissed; in the meantime, too much research was done and too many discoveries were made to be left completely unappreciated in the revision. The first and second German editions appeared in 1982 and 1986 in the Akademie-Verlag in Berlin. Rather than straightforwardly producing an English version of the original monograph, in discussions with the legal successor of the publisher we eventually decided to accept the challenge and largely rewrite the text in order to bring it to a modern state that reflects more recent developments both in the field and in the viewpoints of the authors. The following reasons motivated us to write an essentially new manuscript:

1) The revolutionary changes in the former GDR and in the other East European countries substantially changed our view on the evolution. On the other hand, it

occurred also that many of the essential statements were confirmed by the evidence of life. In particular, we have in mind a statement on page 209 of the first edition: "Refusal of innovation is virtually an optimal tactic, but ultimately it is a fatal strategy, as biological evolution has shown." Being used to read and write "between the lines" at that time, this was a critical formulation. Having in mind the experience from the last 22 years we may now better say "as was shown by biological and social evolution."

2) Our professional careers after "the wall had developed a crack" led one of us (R. F.) to Baltic Sea research and correspondingly to fields of oceanography and geosciences. The second author (W.E.) was founding speaker of a German collaborative research center on "Complex Nonlinear Processes – SFB 555" and later – already as a pensioner – in several new collaborations with foreign partners such as in Cracow, Madrid and Moscow. This experience as well as the contacts with new friends, colleagues, and collaborators widened our views and introduced new topics into our research which are in part reflected here.

3) In the last 30 years, the worldwide situation changed dramatically, new problems such as the warming climate, destruction of environment, and global financial crises have arisen which pose real dangers to the future of humankind. Similar to an earlier popular book (Ebeling and Feistel, 1994), also in this general book on evolution we felt some necessity to gather and review certain material regarding the past and the future of evolution on Earth, without claiming to be exhaustive in any way. We hope that some of our perspectives provide interesting alternatives to commonly discussed viewpoints.

The reader may recognize that this experience and the new tasks strongly influenced the manuscript and resulted in new views and several completely new topics. In particular, we put much more attention on the evolution of climate and environment, a topic that got a lot of actuality these days. Further we took into account many new results in the fields of nonlinear dynamics, chaos theory, and theory of self-organization. As a result, the emphasis of this edition differs from the original one in various respects. For instance, self-reproducing automata are no longer discussed; their theory did not develop as promisingly as it appeared to us in the past. Complexities and other similarity measures of sequences were a new and hot topic at that time; now the wealth of publications on genetic information and on language evolution issued since then is impossible to be reasonably reviewed in a single chapter. We decided to put more emphasis on the physical question how symbolic information carriers can emerge by mechanisms of self-organization than on techniques to describe and to analyze strings of letters or chains of molecules. On the other hand, elementary questions regarding the climate and the structure of our planet occupy much more space and attention in this book than it did before.

The authors worked many years on thermodynamics, theoretical physics, and dynamical systems. Accordingly, this book is written in the language of physics and makes use of standard tools from thermodynamics and mathematics. The target it aims at is beyond the traditional field of physics; it is the – still fascinating – question of how the fundamental laws of physics are related to the emergence of qualitative newness and the historicity of evolution. Evidently, the forms of life that exist on Earth, the basic laws that govern it, and the amazing diversity of functions and structures it produces including ourselves and the science we are engaged in, represent the main challenge and central problem of this book. Notwithstanding, here the focus is on physical conditions, theories and models that offer a better understanding for the emergence and observed perfection of life.

We like to convey our deep conviction that no mysteries are involved in the appearance and unfolding of life, neither divine beings, intelligent creators, nor extraterrestrial visitors, that is, neither systems that were never practically observed, nor ones that are even unobservable by definition of their human inventors. We share the view of the ingenious Lev Landau that his proverbial "gramophone spirit" is neither a necessary nor a particularly helpful theory for explaining the functioning of a gramophone. Rather, the laborious, meticulous and sometimes erratic search for consistent combinations of observational evidence, theoretical models, and causal explanations is the only promising and defendable scientific approach to reveal the secrets of life, as repeatedly outlined by authors such as Carl Sagan and Richard Dawkins. Accordingly, quite fundamentally, this book is not a place to find any kind of final answer or eternal truth. It is intended as an introduction to the research field for students, and as a review of selected aspects of self-organization and evolution for interested colleagues and other readers who share our curiosity regarding the physical basis of the miracle of life.

At the Rostock university, the research group on nonlinear irreversible processes was established with a special course read by Werner Ebeling already in 1973. Under the impression of the works of Prigogine, Eigen, and Haken, and by the experiments of Belousov and Zhabotinsky on oscillating reactions and chemical waves, as well as of Hartmut Linde on hydrodynamic surface waves, it was a fascinating experience to do research at the borderlines between physics, biology, and social science; quite in contrast to the classical scope of textbook physics. Easy access to the upcoming "personal computers" such as Sinclair ZX81 and ZX Spectrum permitted the investigation of numerical solutions of nonlinear differential equations as well as simulation experiments at so far unprecedented convenience and promptness. It was the emotional experience of passing a secret gate and entering a pristine new world of concepts, models, and theories which stimulated ideas and publications of the group.

Various meetings with other colleagues on workshops and conferences were always very inspiring and informative, in particular the *Conferences on Irreversible Processes* and *Self-Organization* held in Rostock in 1977 and in Berlin 1982, respectively (Figure 3), the first two in a series of four. While the 1982 book was in preparation, the group moved to the Humboldt University of Berlin and grew rapidly, in particular because of its scientific attractiveness to many students and postdocs. By coincidence, we all remember very well the end of the departure day of a workshop held by the group near Berlin when a specific historical "fluctuation" triggered a critical local instability with dramatic and enduring consequences: the Berlin Wall had developed a crack that gradually widened in an irreversible process of self-organization which eventually resulted in the end of the world-wide Cold War.

Figure 3 Ilya Prigogine at the II. Conference *Irreversible Processes and Self-Organization* in Berlin 1982. From left to right: Roman Ingarden, Yuri Klimontovich (deceased 2003), Rainer Feistel, Dagmar Ebeling, Ilya Prigogine (deceased 2003), Werner Ebeling. (photo from the archive of the authors).

For the research group, Figure 4, but as well for close friends such as Miguel Angel Jiménez Montaño (University of Veracruz at Xalapa, Mexico) and Yuri Mikhailovich Romanovsky (Moscow State University, Russia), Figure 5, with whom the authors frequently met and worked together since the 1970s, one of the most attractive meeting places is the cottage house of the Ebeling family in Born on the Darss Pensinsula.

When telling about the history of the relations between thermodynamics, statistical physics, and nonlinear science, with respect to historical events related to Berlin, Rostock, and Moscow as the towns closely connected with our personal education and careers, some prejudice may be excused by the reader. We cannot claim perfect accuracy of the historical details; this point we prefer to leave to professionals. We are

Figure 4 Autumn workshop 1978 at Ebeling's cottage house in Born, Darss Peninsula. From left to right: Horst Malchow, Jürn Schmelzer, Waldemar Richert, Lutz Schimansky-Geier, Matthias Artzt, Ulrike Feudel, Werner Ebeling, Ingrid Sonntag, Harald Engel, Reinhard Mahnke.

Figure 5 Science-family meeting in 2000 at Ebeling's cottage house in Born, Darss Peninsula. From left to right: Barbara Ebeling, Rainer Feistel, Sabine Feistel, Miguel Angel Jiménez-Montaño, Julia Hildebrandt, Werner Ebeling, Erik Hildebrandt, Yuri Mikhailovich Romanovsky, Stefan Feistel, Jörg Hildebrandt.

more interested in the origin of the ideas and their development than in biographical facts.

This book summarizes the knowledge, viewpoints, and opinions of the authors regarding the physics of self-organization. It is a textbook with some bias regarding the thermodynamics of evolutionary processes including the relations between nonlinear dynamics, statistical physics, and stochastic theory. It paints a physical picture of the origin of life, from the early universe and formation of the planet Earth up to competition and information processes characteristic for life in its various forms. In explaining the main ideas of our approach we prefer the inductive and sometimes the historical perspective. This way, this book is basically intended as an introduction for students of physics of higher semesters and for graduate students in physics and related research fields who search their path across the "jungle" of modern interdisciplinary research. The book's orientation is more on general understanding and illustrative examples than on theoretical or modeling details. Deeper knowledge is available from a wealth of scientific literature that is briefly reviewed in the text. In other words, in the first line our aim is to contribute to the education of students and young scientists, to explain what is holding together physics and natural sciences, and not so much to fill the readers' minds with too many equations. Nevertheless, we also include some original results of research from the group's working since the 1970s at the Rostock University and since 1980 at the Humboldt University Berlin.

This work is in part based on lectures held at universities in Berlin, Cracow, Moscow, Puebla, Riga, Saratov, Torun, and Xalapa on topics such as "Thermodynamics," "Dissipative Structures," "Statistical Physics," "Stochastic Theory," "Brownian Motion," "Nonlinear Dynamics," "Self-Organization," "Entropy and Information," "Physics of Evolutionary Processes," and on related subjects given by the authors in close cooperation with colleagues such as Heinz Ulbricht, Reinhard Mahnke, or Gerd

Röpke at the Rostock University, and Lutz Schimansky-Geier and Igor Sokolov at the Humboldt University Berlin between 1974 and 2010.

Many colleagues and friends inspired and improved this book, and assisted in various other ways: Kobus Agenbag, Chris van den Broeck, Dmitri S. Chernavsky, Michael Conrad (deceased), Olivia Diehr, Andreas Engel, Harald Engel, Ulrike Feudel, Jan Freund, Ewa Gudowska-Nowak, Eberhard Hagen, Hermann Haken, Peter Hänggi, Anna-Maria Hartmann, Ingrid Hartmann-Sonntag, Toralf Heene, Olaf Hellmuth, Kevin Hennessy, Hanspeter Herzel, Martin Heß, Günter Jost, Yuri L. Klimontovich (deceased), Jürgen Kurths, Hartmut Linde, Reinhard Mahnke, Horst Malchow, Trevor McDougall, Lutz Molgedey, Frank Moss, Günther Nausch, Gregoire Nicolis, Thorsten Pöschel, Ilya R. Prigogine (deceased), Gregor Rehder, Gerd Röpke, Yuri M. Romanovsky, Ruslan L. Stratonovich (deceased), Andrea Scharnhorst, Lutz Schimansky-Geier, Juern Schmelzer, Frank Schweitzer, Dirk Schories, Igor M. Sokolov, Franz Tauber, Heinz Ulbricht, Norbert Wasmund, Vladimir Vasiliev, Manuel G. Velarde, Mikhail V. Volkenstein (deceased), Yakob B. Zeldovich (deceased), and Christoph Zülicke.

We are particularly and deeply indebted to our families for their multiple assistance, stimulating cooperation, and inexhaustible patience, Barbara Ebeling, as well as Sabine, Susanne, Stefan, and Angela Feistel.

The present manuscript was mostly written in Warnemünde where one of these authors is now involved in Baltic Sea research, and in the village of Born located on the Darss Peninsula near Rostock where the second author enjoys new chances after retirement, and also partially in Berlin in close contact with the colleagues at the Humboldt University.

This book is dedicated to our former students, coworkers, and friends, who helped to work out our ideas in a harmonic but critical and altogether friendly atmosphere.

Berlin, Born, and Warnemünde, in February 2011.

1
Introduction to the Field of Self-Organization

Becoming is the transition from being to nothing and from nothing to being
  Georg Wilhelm Friedrich Hegel: Science of Logic

1.1
Basic Concepts

Our world is the result of a long process of evolution that took between 10 and 20 billions of years. Evolution is based on self-organization (SO); this insight we owe to great scientific results of the nineteenth and twentieth century which will be represented and discussed in this book. Our everyday experience and scientific investigations of natural and social processes have taught us that many complex systems have the ability of self-structuring and SO. The most remarkable examples for this statement are the evolution of life and society on our planet in the last 4 billion years. Although this conclusion seems to be obvious, the scientific interpretation of this process is very difficult and requires contributions from virtually all branches of science, including concepts of philosophers such as Hegel, developed already two centuries ago (Figure 1.1). The main aim of this book is to present in a concise form the most important contributions physics may provide to the solution of the conundrum of evolution.

Our key point is the concept of SO. To start with a kind of definition, under SO we understand an irreversible process, that is, a process away from thermodynamic equilibrium which through the cooperative effects of subsystems leads to higher complexity in spatial structures and temporal behavior of the system as a whole. Self-organization is the elementary step of evolution, while evolution consists of many such steps as shown schematically in Figure 1.2.

The importance of SO and evolution for all natural and social processes is quite evident and is subject of scientific investigations at least since the works of Kant, Hegel, Marx, and Darwin. In physics, several problems of evolution were studied first by Helmholtz, Clausius, Boltzmann, Einstein, Friedman, and Gamow. These

Figure 1.1 Memorial to the great philosopher Hegel, 1770–1831, who developed a deep understanding of evolutionary processes, at the Dorotheenstadt cemetery in Berlin. (photo by authors Dec. 2010).

great scientists created the first physical models for the evolution of the Universe. In the last 50 years, we observe the development of a new branch of physics, the physics of SO processes, which is mainly based on the work of Schrödinger, Turing, Prigogine, Eigen, Haken, and Klimontovich. As we mentioned already in the Preface to this book, many scientists from places all over the world, and in particular those in the former socialist countries, studied the new branch of science with great

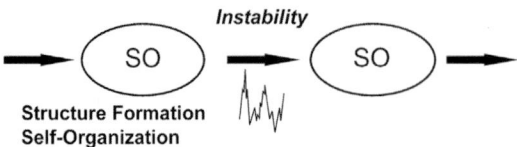

Figure 1.2 Evolution as a sequence of many subsequent SO steps. Each period of SO leads to some structure formation and after its ripening, eventually to an instability that drives the system away from its previous quasiequilibrium state and ultimately leads to another phase of SO.

excitement. The books of the pioneers were printed in large editions, were translated into many languages (see, e.g., Glansdorff and Prigogine, 1971, 1973; Nicolis and Prigogine, 1977, 1979; Haken and Graham, 1971; Haken, 1978) and several popular presentations found a wide public (Prigogine, Nicolis, and Babloyantz, 1972; Nicolis and Portnow, 1973; Ebeling, 1976a, 1979; Haken, 1981; Prigogine, 1980; Nicolis and Prigogine, 1987; Prigogine and Stengers, 1993; Ebeling and Feistel, 1994).

It was in particular the "magic year" of 1971 when the seminal publications on *dissipative structures* (Glansdorff and Prigogine, 1971), *molecular evolution* (Eigen, 1971), and *synergetics* (Haken and Graham, 1971) appeared almost simultaneously and elucidated the physical theory of SO from rather complementary perspectives. Those concepts by themselves represented an innovation, a "critical mutation" in the evolution of physical science; in an accelerated phase of social SO, new working groups were established in universities and research institutes, numerous subsequent articles and books were published, and conferences were held which propagated and amplified the new ideas and theories. The authors of this book were among those who were fascinated and inspired by this new physical picture of the world. The novel "excitation mode" included not only physicists but also mathematicians, chemists, biologists, social scientists, geoscientists, and even philosophers and politicians. Personally, the authors are aware of only one earlier, similar, self-organized development; that was when engineers and mathematicians developed the concepts of cybernetics, self-regulation, automata, learning systems, and roboters (Turing, 1937; Wiener, 1952; Steinbuch, 1961; von Neumann, 1966), which also spread into philosophy (Klaus, 1968) and inspired famous science-fiction authors such as Stanisław Lem, Isaac Asimov, or Karel Čapek, who once coined the term "roboter," derived from the Slavic word for "worker." The understanding of nonlinear feedback mechanisms was a key to the later theories of SO.

The summit of this development took perhaps place in the 1970s and 1980s of the last century, immediately before the dramatic social and political turnover in Eastern Europe and in many other parts of the world. The authors of this book had the opportunity to listen to the excellent lectures held by Ilya Prigogine, Hermann Haken, Manfred Eigen, and others in Moscow, Berlin, Halle, and elsewhere. We were much impressed by the excitement of many people, even beyond the field of science, about the ideas of SO. So our statement is that the reception of the ideas of SO contributed to the atmosphere of revolutionary changes that were dawning at the end of the 80s of the last century (see, e.g., Ebeling, 2010). And in fact the events that happened in the eastern societies appeared to be instructive examples for kinetic phase transitions and SO in complex systems, proceeding in real life (Haken, 1981). Critical fluctuations in the society became amplified to macroscopic structures and processes; randomly formed nuclei grew exponentially to mass demonstrations. Previous symmetries with respect to private and public ownership were suddenly broken. With the establishment of the different social structure, similar to Figure 1.2, fluctuations faded away and were finally slaved by the new master variables and dominating dynamical modes.

In this book, our topics are the scientific foundations rather than the social implications of SO. In fact, the theory of SO and evolution is based on many disciplines such as:

- thermodynamics of irreversible processes in open systems,
- nonlinear mechanics, electronics, and laser physics,
- chemical kinetics, far from equilibrium,
- nonlinear population dynamics and theoretical ecology,
- nonlinear systems theory, automata theory, cybernetics,
- theory of information, languages, sequences, complexity,
- probability theory, random noise, statistical methods, and so on.

This list shows already the interdisciplinary character of this field of science. So we may raise the question: What are the chances of physicists to contribute substantially to the science of SO and evolution? Gregoire Nicolis and Ilya Prigogine write in their book (Nicolis and Prigogine, 1977; here translated from the German edition):

In some sense we are in the position of a visitor from another planet who first encounters a house in a suburb and tries to understand its origin and function. Of course the house is not in contradiction to the laws of mechanics, otherwise it would collapse. But this is not the key question; of central interest is the technology available to the constructors, are the needs and requirements of the residents, and so on. The house cannot be understood without the culture in which it is embedded.

In order to develop our basic concepts, let us start with the notions of "elementary" and "complex." When we want to understand our world as a whole entity, the fundamental question arises: what is the relation between the laws for the elementary aspects and the complex ones? Several reasonable answers on this question are discussed (Simon, 1962; Anderson, 1972). The world outlook of physics relies on laws that physicists consider fundamental. The term fundamental in this context means that these laws cannot be reduced to more basic laws. They are the laws that regulate the attributes and dynamics of elemental particles and fields. Furthermore, they include the laws that formulate even more general rules of exclusion of certain processes, such as the first and second laws of thermodynamics and the Pauli principle. Summarizing we state (see also Ebeling and Feistel, 1994; Ebeling, 2006):

1) The fundamental laws of physics cannot be violated; they are valid without restrictions for complex systems as well.
2) Complex systems may have emergent attributes, the whole is more than the sum of its parts, and symmetry breaking plays a fundamental role.

The terms "emergent irreducible properties" and "emergent values" play a central role in our concept. As a first irreducible property of physical systems, we understand the *irreversibility of macroscopic processes*. This cannot be deduced in a simple way from the laws of motion of the microscopic constituents (Ruelle,

1993; Hoover, 2001; Lieb and Yngvason, 1999; Ebeling and Sokolov, 2005). As a second fundamental property of macroscopic systems, we understand the ability to show SO under certain conditions. We may define "SO as a process in which individual subunits achieve, through their co-operative interactions, states characterized by new, emergent properties transcending the properties of their constitutive parts."

Among the points we have to understand are the role of symmetry breaking, of order parameters and phase transitions, of fluctuations and kinetic transitions. Further we have to study the role of different time scales and of slaving of variables in processes of SO. A point of special interest for our picture is the role of *values*, which are indeed among the most relevant emergent properties. In Chapters 6–9 we will discuss the value of a species which means the fitness in the sense of Darwin. We will show that competition is always based on some kind of valuation. This way valuation is a central concept in the theory of evolution and is deeply connected with the origin of life. We will come back to this point at the end of this chapter and in Chapters 6–9.

Another key point is the *origin of information*. As we will see, it is quite easy to measure the amount of information, but it is extremely hard to say something about its value. This is a point which needs a careful discussion. Looking back on our list of key points, we see that it is already quite long and we have a lot of things to discuss.

At the end of this introductory paragraph, let us make the statement that we should not be "overly anthropomorphic" in our views on evolution. Just in order to illustrate this point of view, let us make a longer quotation from a wonderful science fiction book written by the Polish author Stanislaw Lem who spent most of his lifetime in the Polish town Cracow. Lem describes in his science fiction novel what the researcher Giese found on a newly detected planet Solaris:

Giese was an unemotional man, but then in the study of Solaris, emotion is a hindrance to the explorer. Imagination and premature theorizing are positive disadvantages in approaching a planet where – as has become clear – anything is possible. It is almost certain that the unlikely descriptions of the 'plasmatic' metamorphoses of the ocean are faithful accounts of the phenomena observed, although these descriptions are unverifiable, since the ocean seldom repeats itself. The freakish character and gigantic scale of these phenomena go too far outside the experience of man to be grasped by anybody observing them for the first time, and who would consider analogous occurrences as 'sports of nature,' accidental manifestations of blind forces, if he saw them on a reduced scale, say in a mud-volcanoon Earth.

Genius and mediocrity alike are dumbfounded by the teeming diversity of the oceanic formations of Solaris; no man has ever become genuinely conversant with them. Giese was by no means a mediocrity, nor was he a genius. He was a scholarly classifier, the type whose compulsive application to their workutterly divorces them from the pressuresof everyday life. Giese devised a plain descriptive terminology, supplemented by terms of his own invention, and although these were inadequate, and sometimes clumsy, it has to be admitted that no semantic system is as yet available to illustrate the behavior of the ocean. The "tree-mountains," "extensors," "fungoids," "mimoids," "symmetriads" and "asymmetriads," "vertebrids" and "agilus" are artificial, linguistically awkward terms, but they do give some impression of Solaris to anyone who has only seen the planet in blurred photographs and incomplete films. The fact is that in spite of his cautious nature the scrupulous Giese more than once jumped to premature conclusions. Even when on their guard, humanbeings inevitably theorize. Giese, who thought himself immune to temptation, decided that the "extensors" came into the category of basic forms. He compared them to accumulations of gigantic waves, similar to the tidal movements of our Terran oceans. In the first edition of his work, we find them originally named as 'tides.' This geocentrism might be considered amusing if it did not underline the dilemma in which he found himself.

We hope after this long quotation, our idea is becoming clear now: In order to understand evolution, we need more than just systematics; we need some fresh and unconventional look.

Our guidelines are the ideas of the great pioneers. Among them was, for example, Robert Mayer, who was the first who understood the *Sun as the driving force of the evolution* on our Earth. Further we mention the contributions of Alexander von Humboldt, who was one of the first researchers who took a *global view* on the processes on Earth as well as the contributions of Rudolf Clausius, Hermann von Helmholtz, and Ludwig Boltzmann, who looked first at the Universe from the point of view of physics. The problems about the history of the Universe which these researchers posed were to some extent solved in the twentieth century by other pioneers as Albert Einstein, Alexander Friedman, Edwin Hubble, and George Gamov. These contributions paved the way to the science of SO processes which started as far as we see with Erwin Schrödinger's (1944) Faustian question "What is life?" and Ilya Prigogine's (1947) dissertation "Etude thermodynamique des phenomenes irreversibles."

1.2
History of Evolution as a Short Story

Our Earth is very special and – as far as we are aware – is the only place where life is embedded in the Universe. The Universe is the system of stars and galaxies, so the evolution of the Earth is part of the evolution of the Universe. This book is mainly devoted to the story of the evolution on our planet even though we are not experts in astrophysical theories. Therefore we will only briefly discuss the evolution of the metagalaxy, which provides the general frame for the evolution of our Earth. At present, most experts seem to accept the hypothesis that the evolution of the metagalaxis started with a catastrophic event at about 10–20 billion years ago, which was a kind of explosion, the so-called *Hot Big Bang*. We follow this hypothesis since it seems to explain most, but not all, of the known facts about the Universe surrounding us.

It does not make much sense to ask what was before. Any history has to start with some moment which is given by records, with at least some data beyond speculations. In our case, this is in the first line the red shift interpreted due to an expansion of our cosmos, observed first by Edwin P. Hubble in 1929. The second fact is the so-called background radiation, which was first predicted theoretically in 1946 by George Gamov on the basis of an assumed cosmic expansion, and nearly 20 years later in 1964 observed by Arno Penzias and Robert Wilson. The third fact is given by the relative abundances of protons and neutrons (3:1) and the estimated abundances of the elements, in particular the ratio between hydrogen and helium in the Universe. These three observations as well as other ones are interpreted now as connected with some singular event which happened more than 10 billion years ago. By the way, the experiments with the Large Hadron Collider in CERN and elsewhere are approaching now temperatures of $T \approx 10^{13}$ K and generate extremely dense matter (Aad *et al.*, 2010; Aamodt *et al.*, 2010). This makes it possible to check several predictions of the Big Bang model experimentally in near future.

So let us take the Big Bang event similar to the opening of a box by the ancient Greek women Pandora which, according to the Greek mythology recorded by Homer and Hesiod, was created by Hephaistos, one of the 12 Greek Gods residing in the Olympus. When Pandora opened her box, many sins escaped and only hope remained. In our case, the box of Pandora released a relativistic, optically transparent quantum gas of extremely high density and high temperature which started to expand. Modern researchers do not believe in Greek mythology, but strangely enough they believe in thermodynamics and in relativistic theory. So let us assume following the standard assumption that the relativistic quantum gas observed the relation between temperature T and density ϱ, which is valid for ideal adiabatically expanding gases (another very strong assumption),

$$T \approx \text{const}\, \varrho^{1/3} \tag{1.1}$$

Since the density varies with some scaling distance $R(t)$ in the form

$$\varrho(t) \approx \text{const}\, R(t)^{-3} \tag{1.2}$$

we find

$$T \approx \text{const}\, R(t)^{-1} \tag{1.3}$$

In other words, we expect that the temperature is falling with the reciprocal scaling distance $R(t)$. The solution of Einstein's general relativistic equations for an expanding (radiation) cosmos found by Friedman provides us for the initial stage of some quantum gas of massless particles with the following time dependence (Dautcourt, 1976; Neugebauer, 1980; Zeldovich, 1983; Greene, 2004; Hoyng, 2006):

$$R(t) \sim \sqrt{t},\ \varrho(t) \sim \frac{1}{t^2},\ T(t) \sim \frac{1}{\sqrt{t}} \tag{1.4}$$

Introducing here some known facts, such as the knowledge that nowadays, after more than 10 billion years, the radiation has a temperature of about 3 K, we find as an estimate that the temperature decreased since the Big Bang approximately, according to the rule of thumb

$$T(t)[K] \sim \frac{10^{10}}{\sqrt{t[s]}} \tag{1.5}$$

Of course this is a very rough estimate based on several serious assumptions, in particular:

- the whole process is assumed to be adiabatic in the thermodynamic sense and
- the matter in the Universe is ultrarelativistic and radiation-dominated.

Let us briefly sketch now the scenario of what happened after the Big Bang, in the form of a short story consisting of 12 parts (Feistel and Ebeling, 1989; Ebeling, Engel, and Feistel, 1990a). So we divide the cosmic evolution into 12 epochs. The story is about the expansion of matter, which was very hot and dense at the "beginning" and cooled down during the later adiabatic expansion process.

1st *Epoch: Physical vacuum and space–time field foam*

There is not much known about this early epoch, sometimes called the Planck era, which ends with the formation of what we know as space and time at the so-called Planck time $t \approx 10^{-43}$ s.

In the Higgs-field hypothesis, all elementary interaction forces were still unified and their carrier bosons were massless similar to photons today.

2nd *Epoch: Mining the vacuum*

The epoch of mining the vacuum is the story about the formation of the primary soup which is a fluid form of matter with a high density and a very high temperature

$$T \approx 10^{32} \text{ K}$$

This period extends up to a time

$$t \approx 10^{-33} \text{ s}$$

At the beginning of this epoch, the Universe expanded extremely rapidly; this expansion was named "inflation." In the currently widely accepted inflation theory, it is assumed that the early Universe expanded within $\Delta t \approx 10^{-35}$ s by a factor of at least 10^{30} in its diameter, much faster than light (Greene, 2004). Driven by temporarily repulsive gravity, this process is thought of as a sudden phase transition of a subcooled Universe, similar to the explosive growth of a supercritical nucleus, during which fundamental symmetries between elementary particles were broken, the Universe was flattened, and initial quantum fluctuations were frozen in, and gave rise to the presently visible lumpy structure of galaxies and their clusters. The perhaps most convincing observational evidence for this theory is the fact that the angular correlation of temperature fluctuations of the cosmic background radiation that was measured by the COBE and WMAP satellites has a complicated shape but is perfectly consistent with related theoretical predictions carried out for acoustic oscillations of a dense quantum gas (Smoot, 2006). Thus, the present background radiation appears as a frozen-in image of the dense universe prior to the inflation event.

The primary soup, which was left at the end of the second epoch, consisted of quarks, antiquarks, leptons, photons, and other particles. At the end of the second epoch, the soup had cooled down to

$$T \approx 10^{28} \text{ K}$$

These are the starting conditions for the third epoch.

3rd *Epoch: Quark–gluon soup*

In this epoch, the Universe is a kind of quark–gluon plasma of high density, perhaps more similar to a "soup" than to a gas. The high temperature supported a state wherein the constituents of atomic nucleons – quarks and gluons – existed unbound. In an experimental effort to recreate such conditions, several researchers collided gold ions using the relativistic heavy ion collider (RHIC) and in recent experiments at CERN using the large hadron collider (LHC) they collide lead ions. The results were analyzed by the two groups – the ALICE collaboration and the ATLAS collaboration. Both groups concluded that the quark–gluon plasma is like a strongly interacting liquid (Aad *et al.*, 2010; Aamodt *et al.*, 2010). However many questions remain still open in this field; further experiments have to be carried out and analyzed. But let us return to our story of the primary soup.

At the end of this third epoch, the soup had cooled down to

$$T \approx 10^{15} \text{ K} \approx 1000 \text{ GeV}$$

The corresponding time is about

$$t \approx 10^{-12} \text{ s}$$

At the characteristic energy of this epoch which is in the order of 1000 GeV, one observes the breaking of the electroweak symmetry, according to the Weinberg-Salam theory. This change of symmetry leads to a change in the composition of matter due to quark annihilation, which leads to the next epoch.

4th *Epoch: Quark annihilation*

At the beginning of the time interval,

$$10^{-11} \text{ s} < t < 10^{-6} \text{ s}$$

quarks were still dominant in the Universe. However, near the end of this epoch the temperature went down to

$$T \approx 10^{13} \text{ K} \approx 1 \text{ GeV}$$

This corresponds to a particle energy at which the annihilation of quarks becomes possible, including the reactions between quarks and antiquarks

$$q + \tilde{q} \to 2\gamma, \quad q + \tilde{q} \to e + \tilde{e}$$

In this epoch, nearly all quarks were annihilated except for a small number of surplus quarks.

5th *Epoch: Formation of a nucleon–lepton–photon plasma*
The time interval

$$10^{-6}\,\text{s} < t < 10^{-3}\,\text{s}$$

is the epoch of the formation of nucleons. Due to the attracting chromodynamic forces between the quarks, the remaining quarks could form nucleons – either protons or neutrons

$$u + u + d \to p, \quad u + d + d \to n$$

Beside nucleons, in the fifth epoch, the metagalaxy was filled with electrons, positrons, photons, and neutrinos. Due to further cooling down according to $T \propto 1/\sqrt{t}$, the temperature was about 30 MeV at the end of the fifth epoch.

6th *Epoch: Neutrino decoupling*
We consider now the time interval

$$10^{-3}\,\text{s} < t < 1\,\text{s}$$

Due to further expansion and decrease in temperatures, the mass density and the temperature approached the values

$$\varrho \sim 10^8\,\text{g/cm}^3; \quad T \sim 10^{10}\,\text{K}$$

At such conditions, the neutrinos became uncoupled from the other particles. As we know from many experiments, in the dense matter on our Earth neutrinos have very large mean free pathways and may fly over very large distances. This started already in the sixth epoch; this was the source of an ocean of neutrinos which are filling our Universe.

7th *Epoch: Breaking the neutron–proton symmetry*
The age of the Universe is now already about 1 s and the temperature around 1 MeV corresponding to $T \sim 10^{10}$ K. So far the number of protons and neutrons was nearly equal. However, due to temperatures below 1 MeV, from now on a certain part of the neutrons changed into protons which have a smaller mass. Finally the relative abundances were 75% for the protons and 25% for the neutrons. These are the abundances of protons and neutrons which we observe today, and this is one of the correct predictions of the Big Bang model.

8th *Epoch: Synthesis of helium and other nuclei, fixation of element abundances*
We speak now about the time interval

$$10^2\,\text{s} < t < 10^4\,\text{s}$$

and temperatures approaching

$$T \approx 10^9 \text{ K} \approx 100 \text{ eV}$$

Then the following reactions become possible:

$$2\text{p} + \text{n} \rightarrow \text{He}^3, \quad \text{D} + \text{D} \rightarrow \text{He}^4$$

The abundances of the chemical elements became fixed due to the insufficient plasma temperature for other nuclear fusion reactions. Note that the heavier elements beyond helium, which we find now on Earth, were formed only in later epochs.

9th *Epoch: Atom formation and photon decoupling*

Together with electrons, protons, and nuclei of He produced in the previous epoch then formed a typical highly ionized plasma, which is intransparent for photons. So, in this phase photons had a rather short mean free path. However, when the temperature was further lowered to $T \approx 10,000-100,000 \text{ K} < 1 \text{ eV}$, the formation of H atoms and He atoms became possible. When the temperature approached 1000 K, nearly all electrons were caught by nuclei and atoms formed a neutral gas. Since neutral gases are transparent for light, we observe from now on a decoupling of the photon gas from the heavy matter. This resulted in an independent evolution of the two subsystems – heavy matter and the ocean of radiation. In the forthcoming phase, the radiation was further cooling down. Due to the low density of the gas and the long free path, some of the photons were able to travel for long times and over extremely large distances without significant scattering events. Further cooling down independently of the heavy matter, at the end the photons reached the level of 2.7 K observed today and formed the ocean of background radiation discovered by Penzias and Wilson. The story of heavy matter was much more complicated and requires the opening of a new chapter of our story.

10th *Epoch: Self-structuring of heavy matter*

About a million years after the Big Bang, the metagalaxy consisted of three extended and nearly independent subsystems:

1) the neutrino ocean,
2) the photon ocean, and
3) the system of heavy matter.

It is impossible to describe here in detail the complicated processes which led to the basic structures which are the constituents of our metagalaxy, the formation of stars and planets. The self-structuring of heavy matter and the formation of stars and planets is based on the action of gravitational forces. Starting from an initially uniform gas of hydrogen and helium atoms with a temperature below 1000 K, due to the attractive character of gravity, clusters of matter were formed. This way the homogeneity and isotropy of the distribution of heavy matter was lost. The new symmetry breaking is due to gravitational instabilities which tend to form condensed droplets of matter, similar to what we know from van der Waals gases. In the cosmos, the long-range gravitational forces between massive objects take over the role of the attractive short-range van der Waals forces between molecules. The dense droplets of heavy matter are heated up again due to an adiabatic compression caused by the gravitational forces (van Dokkum, 2011). In the interior of very dense clusters of matter, nuclear fusion was ignited again. Stars were born which started to radiate with a surface temperature of about

$$T \approx 10^4 \text{ K}$$

Some of the stars were accompanied by smaller clusters, the planets with surface temperatures

$$T \approx 10^2 - 10^3 \text{ K}$$

This two-temperature system was imbedded into the big sea of background photons with a temperature now below

$$T \approx 10 \text{ K}$$

Between the three systems of different temperatures in the metagalaxy, a new mechanism comes into action which works like a Carnot machine and is driving from now on the process of evolution.

11th *Epoch: The photon mill*

The so-called photon mill is the most important mechanism responsible for the SO of the terrestrial (and may be other) biosystems. The idea demonstrated in Figure 1.3 is that the steady flow of photons from the

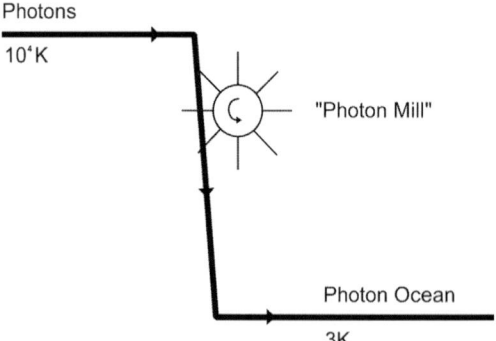

Figure 1.3 The steady photon flow from the temperature level of $T \sim 10{,}000$ K corresponding to the surface of stars, possibly passing through an intermediate station on planets and ending finally at the level $T \sim 3$ K corresponding to cosmic background radiation, is the driving force for the terrestrial evolution.

temperature level $T \sim 10{,}000$ K, corresponding to the surface of stars, passing through a possible intermediate station on planets with $T \sim 100\text{--}1000$ K, and finally ending at the thermal level of the sea of cosmic background radiation, is the thermodynamic driving force of evolution. We will come back to this fundamental mechanism several times in this book.

12th *Epoch: Self-organization on Earth*

Above we have given a short story sketching the complicated processes which led to the basic structures which are the constituents of our metagalaxy, the formation of stars and planets. We are interested here mainly in what happened on the particular planet Earth which was formed about 4–5 billion years ago. The first process of SO that occurred on our planet was the formation of geological structures. From the thermodynamic point of view, the driving force responsible for the formation of geological structures in the crust is based on the flow of heat coming from the hot interior of the Earth and flowing outward. This process occurs on a very long time scale. In the atmosphere, we have a more complicated interplay between heat flows from the interior and radiation flows from the surrounding, giving rise to structures such as climate and weather. The driving flows of heat are demonstrated in Figure 1.4. It will be discussed in detail in Chapter 3. The physical effects which finally led to the evolution of the Earth's crust and to geological structures also influence the climatic processes by hydrodynamic instabilities such as the Benard and Marangoni instabilities (Nepomnyashchy, Velarde, and Colinet, 2002), as will be discussed later.

The third and most complicated process of SO on Earth is the evolution of biological organisms. The processes which eventually led to the evolution of life on Earth are in the first line driven by the photon mill shown in Figure 1.3 but are also strongly influenced by geological and climatic factors. To model some of these processes, which are understood only in part so far, is the main topic of this book.

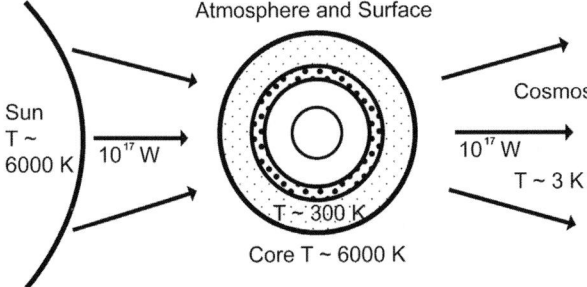

Figure 1.4 The evolution of geological structures is driven by a second mechanism connected with the flow of heat from the interior of the Earth to the surface crust, and from there to the atmosphere and to the ocean of cosmic background radiation, as will be discussed in detail in Chapter 3.

1.3
Structure, Self-organization, and Complexity

As a result of the SO in the 12th epoch of the evolution of the universe, many structures of different kinds were formed. Taking for a moment a more formal view, in some respect all the structures in nature and in society around us are relations between elements. In nature we study atoms made of electrons and nuclei, gases, liquids, and solid states made of atoms, and biological beings that are complex entities made of many atoms or molecules. The relations between objects in the realm of life, that is, biological or social objects, are more complicated; their relations may be characterized with the term "networks," at least on the formal side. In order to deal with evolutionary processes, we have to go beyond physics and study more general types of structures. Typically, biological, ecological, and social networks consist of many ties, connections, group attachments, meetings, and other events or activities that relate one individual to another. From the abstract point of view, bioecological and socioeconomic networks are also "structures." As we will describe later in detail, it is possible to give an abstract presentation of such structures by means of mathematical tools. The mathematical idea of structure stands in close relation to the terms *element, set, relation, and operation*.

In this context, the particular nature of the elements does not play any role with respect to the structure model. In contrast, the nature of the relations between the elements determines the specificity of a structure. In that sense, structure means the manifold of interactions between the elements. In the most general context, we speak about SO if the structures formed out by the elements are not determined by external influences but by the internal interactions and relations. As we will show later, there is a deep difference between SO in equilibrium and in nonequilibrium systems.

In other contexts, the nature of the elements is relevant. So, in the general case, the nature of the elements and of the relations is relevant. Later we will see that another essential factor is the dynamics of the elements, which may lead to even more complicated structures, to real complexity. So far, we follow first a more abstract and general approach. By using such a general level, we create methodologically the opportunity of respecifying the definition of both the elements and the relations for any application area one might think of.

The abstract idea of a "structure" is of great importance in our life – both in reality and in science. This is particularly obvious when social structures are analyzed. What does "structure" mean in the original sense? On the one hand, we have our conventional understanding of structure – understanding of structure in our real life. Alternatively, we may formulate a precise structure model in terms of mathematics, in system theory, and in the structural part of the theory of SO (Laue, 1970; Görke, 1970; Casti, 1979; Ebeling, Freund, and Schweitzer, 1998; Helbing, Herrmann, and Schreckenberg, 2000; Pyka and Küppers, 2002; Pyka and Scharnhorst, 2009):

Under a structure we understand the composition of elements and the set of relations respectively operations which connect the elements.

The German philosopher, Kröber (1967), writes about the idea of structure in real systems:

Each system consists of elements that are arranged in a certain way and are linked to each other by relations. We understand by the "structure of a system" the kind of arrangement and connections of their elements ... Of what kind the elements are we do not consider in this respect. If we speak about structure, we are not interested in the elements of the structure. We only consider the manifold of relations. In this respect the structure of a system is a well-defined connection between the elements of the system. These elements, which are arranged in a determined manner and connected by determined connections can be necessary or random, universal or unique, relevant or irrelevant.

With the famous book *The Elements of Mathematics*, the scientists group Bourbaki gives an example for systematically constructing mathematics as the science of such "structures." In the following section, we will give a short description of important concepts of the mathematical theory of structures. Especially we will give a summary of the theory of relations, graphs, and matrices to the extent we need it here, with the purpose to apply this abstract theory to socioeconomic networks. In ecological and in socioeconomic systems, the elements of structure are individuals or groups of individuals in different institutional and organizational forms. The socioeconomic connections between these elements are relations in the sense of this abstract theory of structure. Their description can be given graphically by a system of vertices, which model the elements (individuals, groups, firms) and of edges, which describe the relations (connections). Vertices and edges can be weighted. Edges can have a direction. This way we can include quantitative aspects. The decisive aspect of SO of networks is the formation of new connections, which generate new structures. In the following, we want to show that the instruments of the mathematical theory of structure in connection with the ideas and concepts of the theory of SO can contribute to the description of the connections of different groups of the complex socioeconomic network of the global world.

Socioeconomic systems are complex systems which on the one side consist of many connections between the elements. Therefore complexity is an additional idea to be defined. Together with Freund and Schweitzer, one of the authors proposed the definition (Ebeling, Freund, and Schweitzer, 1998):

As complex we describe holistic structures consisting of many components, which are connected by many (hierarchically ordered) relations respectively operations.

The complexity of a structure can be seen in the number of equal respectively distinct elements, in the number of equal respectively distinct relations and operations, as well as in the number of hierarchical levels. In the stricter sense, complexity requires that the number of elements becomes very large (infinite).

Here we are especially interested in the origin of complex structures. We are interested in the development of order (information) for the corresponding basic situation. At the end, we have to answer the question which parameter relations (order parameters) determine the qualitative behavior of the system. As already mentioned, pioneering work in the investigation of self-organizing systems was done by Ilya Prigogine and his coworkers (1970, 1977). Further important work in this field has been done by Eigen (1971) and together with Peter Schuster (Eigen and Schuster, 1977, 1978) by developing the hypercycle model. The mechanism of SO is clearly worked out by Prigogine. He also gives a stringent physical and mathematical formulation for these processes, in particular with respect to the energetic and entropic aspects. The investigation of such systems shows that the formation of order in complex systems can be associated with physical processes that occur far away from equilibrium conditions (Ebeling and Feistel, 1982). We underline that biological processes, just as socioeconomic processes, can be investigated with the help of the theory of SO because they obey the valid physical and chemical laws. However, processes which include real life (biological and socioeconomic systems) also obey additional rules and laws that are not determined by physics alone. This is evident already from the very general character of the structures we consider here. As discussed above, we formulate the idea of structure mathematically and refrain from the subjective side of this idea. As mentioned before, we need an abstraction in order to describe real systems with the help of graph theory. This will be explained later in more detail in Section 4.4.

By formulating our ideas in a mathematical language, we have the advantage to have access to the vast potential of formal tools and methods available in this field. An important basic approach is to generally start from the theory of sets. As pointed out above, the connection between element and set is the first and most important aspect of a structure. Furthermore we introduce relations and operations. This idea of structure reflects abstract properties of a system. Due to the abstract character, most of the ideas and results can be translated to other systems and comparisons are possible. In general, we distinguish between spatial, temporal, causal, and functional structures.

To illustrate the structures, we will use graphs which represent the elements and their connections by geometrical symbols (Harary, Norman, and Cartwright, 1965; Laue, 1970; Gardner and Ahsby, 1970; Casti, 1979; Ebeling and Feistel, 1982; Jain and Krishna, 2003; Hartmann-Sonntag, Scharnhorst, and Ebeling, 2009). Just to give an example of a complex network, we present an economic network with four levels in Figure 1.5. This graph represents the flow of materials and outcomes in a production process. The economic picture just serves as a schematic example; in later sections we will discuss mostly examples from biology and ecology.

As our example nicely shows, in economics the relations between economic agents, in particular the flow of commodities, can be represented in a network form. Similarly, information flows, for example, price signals between market participants, are exchanged. The structure of the network then describes a situation with local interaction (not every agent is informed about all other agents but the agents do not act isolatedly either) (Pyka and Scharnhorst, 2009). Socioeconomic networks can also describe a variety of different actions of agents that influence other agents (Saviotti, 1996; Saviotti and Mani, 1995; Saviotti and Nooteboom, 2000). The diffusion of

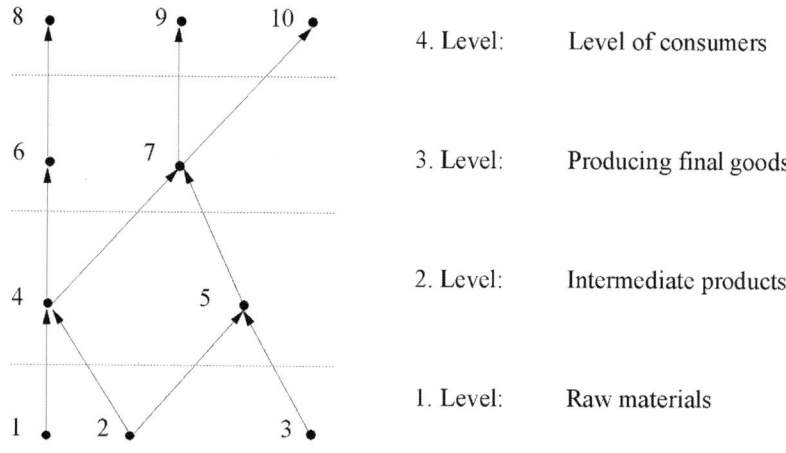

Figure 1.5 Schematic graph of a socioeconomic network with four levels: (1) raw material such as airts, fossils, plants, and so on; (2) intermediate products such as steel, coal, corn, and so on; (3) producing final goods; and (4) level of consumers.

technologies over firms can be described as a network of actions from the formation of a company (with a certain technology) over knowledge transfer between companies (in form of imitation or merging) to the exit of companies due to technological competition. We will come back to such a network interpretation in later sections for examples taken from chemistry, biology, ecology, and economy. We are especially interested here in the origin of complex structures, and in the development of order (information). In the end, we have to answer the question which parameter relations (order parameters) determine the qualitative behavior of the system.

1.4
Entropy, Equilibrium, and Nonequilibrium

Entropy is one of the central terms of this book. This term plays a central role in the modern science of our days. Its significance reaches from physics to probability theory, the theory of information, computer science, economics, psychology, just to give a few examples. What does entropy mean? If we consult Wikipedia first, we read there: "In physics, entropy, symbolized by S, from the Greek μετατροπή (metatropi) meaning 'transformation', is a measure of the unavailability of a system's energy to do work." We consider this as a good and quite useful definition. However, this is not the only way to define entropy. There exist now many publications on the topic of entropy; there is a special journal *Entropy* and on the Internet one may follow an ongoing and never-ending discussion on what entropy actually is. Entropy is something that has a Janus face; one can look at it from many perspectives and it might always look quite different. The term entropy was coined in 1865 by

Rudolf Clausius (1865) who played the most important role in the history of the two fundamental laws of thermodynamics. On page 390 of his seminal paper, he wrote:

*Sucht man für S einen bezeichnenden Namen, so könnte man, ähnlich wie von der Größe U gesagt ist, sie sey der Wärme- und Werkinhalt des Körpers, von der Größe S sagen, sie sey der Verwandlungsinhalt des Körpers. Da ich es aber für besser halte, die Namen derartiger für die Wissenschaft wichtiger Größen aus den alten Sprachen zu entnehmen, damit sie unverändert in allen Sprachen angewandt werden können, so schlage ich vor, die Größe S nach dem griechischen Wort η τροπη, die Verwandlung, die **Entropie** des Körpers zu nennen. Das Wort Entropie habe ich absichtlich dem Worte **Energie** möglichst ähnlich gebildet, denn die beiden Größen, welche durch diese Worte benannt werden sollen, sind ihrer physikalischen Bedeutungen nach einander so verwandt, dass eine gewisse Gleichartigkeit in der Benennung mir zweckmäßig zu seyn scheint.*

(In English: If one looks for a term to denote S, one could, similar to what is said about the quantity U that it is the heat and work content of the body, say about the quantity S that it is the transformation content of the body. Since I prefer to borrow the name for quantities so important for science from the old languages to permit their unaltered use in all languages, I suggest to name the quantity S the *entropy* of the body, after the Greek word η τροπη, the transformation. Intentionally I have formed the word *entropy* as similar as possible to the word *energy* because the two quantities denoted by those words are so closely related with respect to their physical meaning that a certain similarity in their nomenclature appears appropriate to me.)

Let us continue with a few additional historical remarks about the work of Clausius and the history of the second law (Ebeling, 2008). After studying physics in Berlin, Clausius taught for some years as a teacher at the Friedrich-Werdersches Gymnasium in Berlin and was a member of the seminar of Professor Magnus at the Berlin University. When at this time Clausius' fellow Hermann Helmholtz published his work about the first law, Professor Magnus asked Clausius to give a report about the essential news in that work. We do not know why Magnus asked Clausius rather than Helmholtz directly. Anyhow, this was a very good idea, since the extremely careful Clausius began with the roots of that science. More precisely, he studied first the nearly forgotten works of Clapeyron and Carnot. This way Clausius was able to find the link between Carnot and Helmholtz and could proceed to a higher level. Clausius' report on Helmholtz's fundamental work, given at Magnus' colloquium, was the beginning of a deep involvement with thermodynamical problems.

Building on the work of Helmholtz and Carnot, Rudolf Clausius developed, and published 1850 in Poggendorff's Annalen, a completely new law, the second law of thermodynamics. Clausius was fully aware of the impact of his discovery. The title of his paper explicitly mentions "laws." His formulation of the second law, the first of several later formulations, that *heat cannot pass spontaneously from a cooler to a hotter body* is expressing its essence already very clearly. Unlike Carnot and following Joule,

Clausius interpreted the transfer of heat as the transformation of different kinds of energy, in which the total energy is conserved. To generate work, heat must be transferred from a reservoir at a high temperature to one at a lower temperature. Clausius introduced here the concept of an ideal cycle of a reversible heat engine. In 1851, William Thomson (Lord Kelvin) formulated independently of Clausius another version of the second law. Thomson stated that it is impossible to create work by merely cooling down a thermal reservoir. The central idea in the papers of Clausius and Thomson was an exclusion principle: "Not all processes which are possible according to the law of the conservation of energy can be realized in nature." In other words, the second law of thermodynamics is a selection principle of nature. Although it took some time before Clausius' and Thomson's work was fully acknowledged, it was fundamental for the further development of physics, and in particular for the science of SO. In later works, Clausius arrived at more general formulations of the second law. The form valid today was reported by him at a meeting of the "Züricher Naturforschende Versammlung" in 1865. There, for the very first time, he used the term "entropy" and introduced the quotient of the quantity of heat absorbed by a body and the temperature of the body as the change of entropy. We will discuss the details of the thermodynamics theory in Chapter 2 but give already here a preliminary version of the two basic laws of thermodynamics.

First Law of Thermodynamics: There exists a fundamental extensive thermodynamic variable, the energy E. Energy can neither be created nor be destroyed. In can only be transferred or transformed. Energy is conserved in isolated systems.

Here, *extensive* means that the energy of a system is the sum of the partial energies of arbitrary subdivisions of that system. The energy production inside a system is zero. Any process is connected with a transfer or with a transformation of energy. Energy transfer may have different forms such as heat and work, and possibly chemical energy. This is expressed by the formulae

$$dE = d_e E + d_i E \tag{1.6}$$

where the internal change is

$$d_i E = 0 \tag{1.7}$$

and the exchange part is

$$d_e E = d'Q + d'A \tag{1.8}$$

Here we introduced an infinitesimal heat transfer $d'Q$ and an infinitesimal work transfer by $d'A$. The infinitesimal change of energy of a system equals the sum of the infinitesimal transfers of heat and work. The basic SI unit of energy is $1\,J = 1\,N\,m$, corresponding to the work needed to move a body 1 m against a force of 1 N.

Work may be expressed in terms of a bilinear form, in the simplest case just by $(-p\,dV)$, that is, pressure multiplied with change of volume. The hypothesis that the infinitesimal heat can also be expressed in a bilinear form led Clausius to the entropy concept. He assumed that d'Q may be written as the product of an intensive quantity and an extensive quantity. *Intensive* quantities are those which take equal values for arbitrary subdivisions of a given homogeneous system. The only intensive quantity which is related to heat is temperature T, and the related conjugate extensive quantity Clausius denoted by S. This way he introduced entropy as an extensive quantity that is conjugate to the temperature, $d'Q = T dS$. This equation may alternatively be written by division by the temperature; this takes us back to our first equation which was invented by Clausius:

$$dS = \frac{d'Q}{T} \qquad (1.9)$$

This says the differential of the state variable entropy is given by the infinitesimal heat, $d'Q$, divided by the temperature T. In more mathematical terms we may say, the temperature T is an *integrating factor* of the infinitesimal heat. To make more transparent what this means, we write the first law, Equation 1.6, in the form of a *Pfaffian differential equation*

$$d'Q = dE + p\, dV \qquad (1.10)$$

In general, no smooth mathematical function "heat," $Q(E, V)$, exists as an integral of Equation 1.10 with the properties $\partial Q/\partial E = 1$, $\partial Q/\partial V = p(E, V)$ for any given function $p(E, V)$. It is known that for Pfaffian forms of two independent variables such as Equation 1.10, there exists always a function, $\beta(E, V)$, termed the integrating factor, such that Equation 1.10 turns integrable after multiplication with β, and that the infinitesimal quantity $\beta \times d'Q$ is an exact differential (Stepanow, 1982). Clausius could prove that the reciprocal temperature, $1/T(E, V)$, is such a factor, Equation 1.9, and that a function $S(E, V)$ exists which satisfies the two equations

$$\frac{\partial S}{\partial E} = \frac{1}{T}, \quad \frac{\partial S}{\partial V} = \frac{p}{T} \qquad (1.11)$$

For clarity, we must add that Equations 1.9 and 1.11 are strictly true only for reversible processes; in that case the change of entropy, dS, in the system equals the flux of entropy through the boundary, as explained below. The basic SI unit of entropy is 1 J/K. One can easily see that entropy is in general not conservative, that is, Equation 1.9 is only a special case rather than being universally valid. Let as consider for example two bodies of different temperatures being in contact. Empirically we know that there will be a heat flow from the hotter body to the cooler one. When the heat flows down a gradient of temperature, it produces entropy. The opposite flow against a gradient of temperature is never observed. A generalization of this observation leads us to the following formulation of the second law.:

Second Law of Thermodynamics: Thermodynamic systems possess an extensive state variable termed entropy. Entropy can be produced, but never destroyed. The change of entropy in reversible processes is given by the exchanged heat divided by the temperature, Equation 1.9. During irreversible processes, entropy is produced in the interior of the system. In isolated systems entropy can never decrease.

We only mention here the deep relation between irreversibility and chaos (Schuster, 1984, 1995; Ruelle, 1993; Lanius, 1994a, 1994b; Lieb and Yngvason, 1999; Landa, 2001).

By splitting the entropy change into a part due to exchange and a part due to internal production, we arrive at the following mathematical formulation of the second law (details will be explained in Chapter 2)

$$dS = d_e S + d_i S \tag{1.12}$$

where the internal production part cannot be negative:

$$d_i S \geq 0 \tag{1.13}$$

and the exchange part is related to the heat transfer, (1.9),

$$d_e S = \frac{d'Q}{T} \tag{1.14}$$

or to other forms of entropy fluxes.

The next step in understanding the role of entropy is connected with the work of Ludwig Boltzmann and Max Planck. Both these great scientists were fascinated by the entropy concept. Ludwig Boltzmann found first a connection between entropy and the distribution function of molecules in a gas. Max Planck, who originally was not an enthusiast of the atomistic view, became also interested in the statistical foundations of entropy at the end of the nineteenth century. In fact, he was the first who introduced the "Boltzmann constant" when he explicitly wrote down the famous formula

$$S = k_B \log W \tag{1.15}$$

which connects the entropy with the number of microscopic states W that are consistent with a given macroscopic state (Klimontovich, 1982). This way it became clear that entropy is not only a thermodynamic quantity, but it is also related to the probability of states, and is a measure of disorder. Disordered states have the highest probability: thermodynamic equilibrium corresponds to states of maximum disorder (Figure 1.6). Entropy is like the head of Janus; there is still another meaning of entropy, which later has to be discussed: according to Hartley and Shannon, entropy is (up to a constant) a measure of information corresponding to the uncertainty removed by exploring the actual state of the system (Shannon, 1951; Jaglom and Jaglom, 1984; Hoover, 2001).

Following the great pioneers, we consider here *entropy* as the central statistical concept: entropy is a measure of uncertainty of the microscopic state. Later we will see that there is also a fundamental connection to the ideas of dynamic instability: in particular we will explain in Chapter 4 that the instability of the microscopic trajectories generates uncertainty of the thermodynamic state, that is, of entropy and of other macroscopic properties.

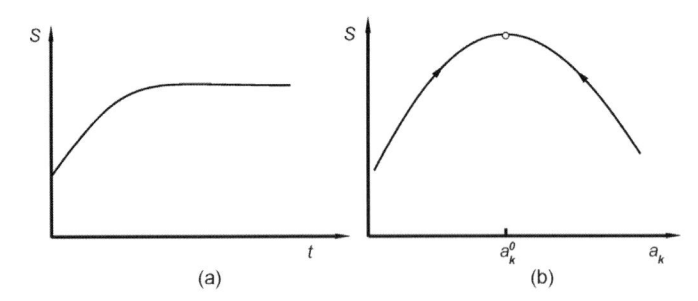

Figure 1.6 The tendency of entropy to approach the state of the highest entropy, corresponding to thermodynamic equilibrium and maximum disorder. (a) Entropy increases as a function of time. (b) Entropy as a function of parameters; the state with maximum entropy corresponds to equilibrium.

From our new point of view, structures may be classified into the two large categories of *equilibrium structures* and *nonequilibrium structures* (Klimontovich, 1982, 1991, 1995).

The theory of SO shows how the emergence of structures is associated with a decrease in entropy (Glansdorff and Prigogine, 1971; Nicolis and Prigogine, 1978; Ebeling and Feistel, 1982). The concept of "structure" is closely related to the term *information (entropy)*. Information is already a more difficult problem which we will discuss later (Ebeling, Freund, and Schweitzer, 1998). Now we are going to explain several of these aspects in more detail.

In order to prepare states which (at given energy) have an entropy lower than the entropy of the equilibrium (which is maximal in comparison to all other states of the system at the same energy),

$$S(E, X) < S_{eq}(E, X) \tag{1.16}$$

we need boundary conditions which permit a lowering of the entropy. Let us split the entropy change into two parts – the external change and the change due to internal irreversible processes, (1.12),

$$dS = d_e S + d_i S \tag{1.17}$$

The second law requires that the second term is always positive or zero. This implies that in order to have an increase in order by decrease of entropy, we need to export entropy

$$dS = d_e S + d_i S < 0 \tag{1.18}$$

This is the thermodynamic condition for SO. To express this condition in a different way, we have to satisfy the inequality

$$-d_e S > d_i S > 0 \tag{1.19}$$

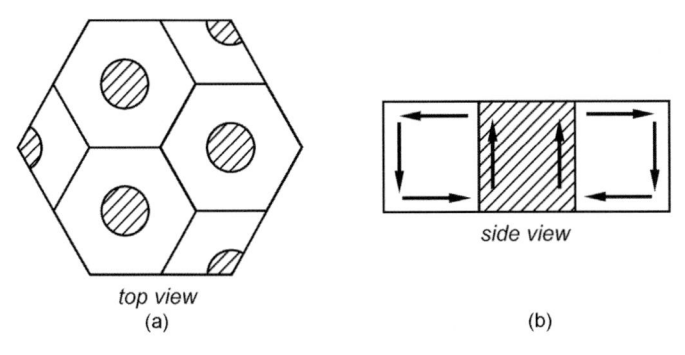

Figure 1.7 Benard cells in a liquid layer heated from below. (a) View at a Benard system from above; the shaded parts show the upwelling regions, the blank parts show the downwelling regions. (b) Vertical cross section showing one cell with ascending liquid in the center.

In other words, to lower the system's entropy we have to export more entropy than is produced by internal irreversible processes. Let us consider several examples. First we consider a liquid layer in a gravity field heated from below. Let T_1, T_2 be the temperatures below and above, respectively, and Q be the heat flow per unit of the surface and per unit time. Naturally we have $T_1 > T_2$ and therefore for the entropy production,

$$-\frac{d_e S}{dt} > \frac{d_i S}{dt} = \frac{Q}{T_2} - \frac{Q}{T_1} > 0 \tag{1.20}$$

We have shown above that the entropy export in a Benard system is positive. This means SO is possible. However, our criterion is only a necessary rather than a sufficient condition. The experiment displayed in Figure 1.7 shows that beyond some critical value of the temperature difference, the liquid starts to organize itself; it forms roll cells (Ebeling, 1976a, 1978a; Nepomnyashchy, Velarde, and Colinet, 2002). Self-structuring due to thermal convection is the most important form of dissipative structures in geophysics (see Chapter 3).

As another example from chemistry, we show a photograph of the Belousov–Zhabotinsky reaction, one of the classical examples of SO (Agladse et al., 1989; Linde and Zirkel, 1991; Linde and Engel, 1991), see Figure 1.8.

The Belousov–Zhabotinsky reaction is a rather complex redox reaction with the central part

$$Ce^{3+} \leftrightarrow Ce^{4+}$$

which occurs in the presence of sulfuric acid, malonic acid, and potassium bromate (Agladse et al., 1989; Linde and Zirkel, 1991). In the presence of a redox indicator (here, ferroin), the reaction can be visualized by observation of the color (red means more 3+ ions and blue means the 4+ ions dominate).

Figure 1.8 Snapshots of a Belousov–Zhabotinsky reaction showing spiral structures. (courtesy of Hartmut Linde).

As a final example for a structure formation due to SO, we show in Figures 1.9 and 1.10 the flow structure at an interface due to Marangoni effects (Linde, Schwarz, and Gröger, 1967; Nepomnyashchy, Velarde, and Colinet, 2002; Bestehorn, 2006).

Solitons are an example for localized structures, in our example moving fronts, which are relatively stable against perturbations, collisions, and so on (Nepomnyashchy, Velarde, and Colinet, 2002).

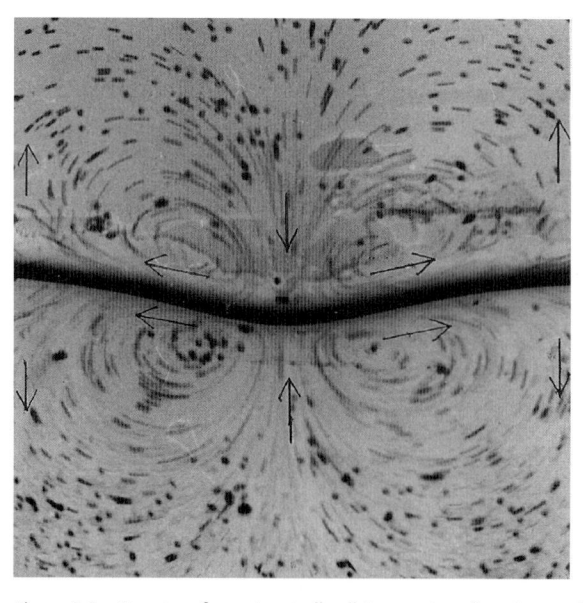

Figure 1.9 Structure formation (roll cells) at an interface due to the Marangoni instability. The arrows show the directions of the hydrodynamic flows. (courtesy of Hartmut Linde).

0.3 s before ---------->   head on collision ---------->   0.6 s after

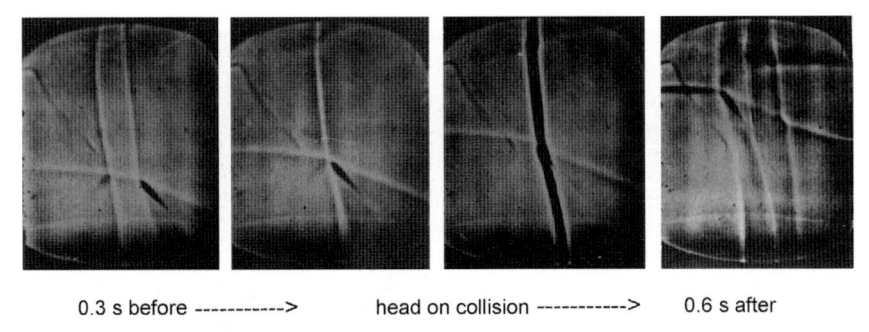

Figure 1.10 Soliton structures including collisions which are due to the Marangoni effect at an interface. (courtesy of Hartmut Linde).

1.5
Dynamics, Stability, and Instability

In the previous section, we demonstrated several examples for the dynamics of SO. Here we will explain in brief several concepts which are basic for the approach of physicists to the theoretical modeling of SO phenomena. More details and applications will be discussed later and the formal part will be developed in particular in Chapter 4. At first we introduce the notion of state space, dynamical models, and trajectories. In fact this is a view going back to the great pioneers of modern science Nicolaus Copernicus (1471–1543), Galileo Galilei (1564–1642), Johannes Kepler (1571–1630), and Isaak Newton (1643–1727). In the sixteenth and seventeenth century, these pioneers developed the idea of orbits for the description of planets which follow some rules or laws; in particular, they could show that the motion of all bodies is determined by gravitational forces. The very essence of this view appears in Newton's law in the form of differential equations. In the modern version which was basically formed by the pioneers of theoretical mechanics, Lagrange, Jacobi, and Hamilton, the geometrical description of all mechanical motions was developed as an extremely elegant mathematical tool. This beautiful theoretical construction was completed by Poincaré (1854–1912) by introducing the ideas of stability and instability of trajectories and by Helmholtz and Rayleigh by introducing the idea of self-sustained motions. All these ideas played an important role for the later concepts of SO which will be explained in detail in this book. Here we will briefly consider several formal aspects, notions, and mathematical tools, along with several physical concepts which led to the modern nonlinear dynamics on which the understanding of SO and evolution is grounded.

One can say that the geometrical interpretation of the trajectories of a dynamical system as orbits in an appropriate state space (phase space) appears to be one of the most important instruments for the investigation of dynamical processes. In fact, these concepts go back to the classical mechanics. Generalizing the concepts of mechanics, Poincare, Lyapunov, Barkhausen, Duffing, van der Pol, Andronov, Witt, Chaikin, Birkhoff, Hopf, and others laid the basis for the modern theory of dynamical systems (Eckmann and Ruelle, 1985). Parallel research was started in the biosciences by Lotka, Volterra, Rashevsky, and others. In the second half of the last century, dynamical systems theory was developed quite independently of physics and biosciences and found applications in most branches of science. In particular, the theory of dynamical systems is also the heart of the science of synergetics founded by Haken starting from laser physics (Haken, 1970, 1978, 1981, 1988) and the theory of SO developed by Prigogine, Glansdorff, and Nicolis, mostly backing on the concepts of irreversible thermodynamics (Glansdorff and Prigogine, 1971; Nicolis and Prigogine, 1977).

We will merely touch the surface here and will discuss only a few basic notions and examples. Let us consider a dynamical system, assuming that its state at the time t is given by a set of time-dependent parameters (coordinates). We consider this set as a vector and write

$$\mathbf{x}(t) = [x(t_1), x(t_2), \ldots, x(t_n)]$$

The set of state vectors, $\mathbf{x}(t)$, spans a vector space, $\mathbf{X}$, which will be termed the *state space* or phase space of the system. Let us assume now that the state at a time t_0, given by $\mathbf{x}(t_0)$, and the state at a later time $t > t_0$, given by $\mathbf{x}(t)$, are connected in a causal way. We assume that the current state is a function or a more general mapping of the initial state and certain additional parameters which we denote by the vector

$$\mathbf{u}(t) = [u_1, u_2, \ldots, u_m]$$

The set of all possible parameter values $\mathbf{u}$ forms the *control space* of the system. The parameters $\mathbf{u}$ take into account the influence of the environment and possibly also actions of external control. We assume that the connection is given by a map $\mathbf{T}$,

$$\mathbf{x}(t) = \mathbf{T}(\mathbf{x}(t_0), \mathbf{u}, t - t_0) \tag{1.21}$$

The set $\{\mathbf{X}, \mathbf{T}\}$ is referred to as the *dynamical model* of the system (Anishchenko et al., 2002; Landa, 2001; Ebeling and Sokolov, 2005). The choice of the phase space is not unique; there are many possibilities to make a choice for the phase space, depending on the model's details and convenience of representation. The existence of a dynamical map, $\mathbf{T}$, which defines the state at the time t, by means of the state at an earlier time, expresses the causality of the dynamical process under consideration. In the real world, we observe two different cases of causal relations between $\mathbf{x}(t_0)$ and $\mathbf{x}(t)$. If the relation between the two quantities is unambiguous, that is, if the future state is given by the map $\mathbf{T}$ in a unique way by the initial one, we speak about deterministic models. In contrast, if there are several possibilities for the future state $\mathbf{x}(t)$ depending on random effects, we speak about stochastic models.

A typical example for deterministic-type models is given by Onsager's relaxation equations which we will discuss in more detail in Chapter 2. They are a set of first-order differential equations for certain variables numbered $i = 1, \ldots, n$ describing the deviations from a stationary state and read

$$\frac{dx_i}{dt} = -\sum_j \lambda_{ij} x_j \tag{1.22}$$

In this equation there appears a set of relaxation variables, x_i, and a matrix, λ_{ij}, which defines their mutual relations. Newton's equations describing the movements of planets are another example. Further we may think of time series obtained by recording the outputs of some measuring instruments in certain time intervals, and by a stroboscopic observation of continuous trajectories or by periodic reports. There are other processes such as the annual reports on the production of a country, which are intrinsically discrete in time. Examples of models with discrete time will be considered in the later chapters.

Let us discuss here only the simplest case of the relaxation equations. These equations possess stationary states $\mathbf{x}^{(s)}$ given by the zeros of the right-hand sides, which lead to a vanishing time derivative of Equation 1.22. There are two kinds of stationary states – stable states and unstable states (Landa, 2001). Deviations from a stable state are followed by a relaxation to the original state.

Deviations from unstable states are amplified, as shown in Figure 1.11. The interplay between stability and instability belongs to the basic features of SO processes. Typical examples of structures created by instabilities we have already seen in Figures 1.7–1.10. In the case of Benard and Marangoni structures shown in Figures 1.7 and 1.9, the uniform state of the liquid at rest is unstable with respect to small hydrodynamic perturbations, which start to grow and bring the system to a new nonequilibrium structure. In the case of Figure 1.8, chemical instabilities give rise to spiral structures. We may consider these examples as concrete realizations of the general schema drawn in Figure 1.2.

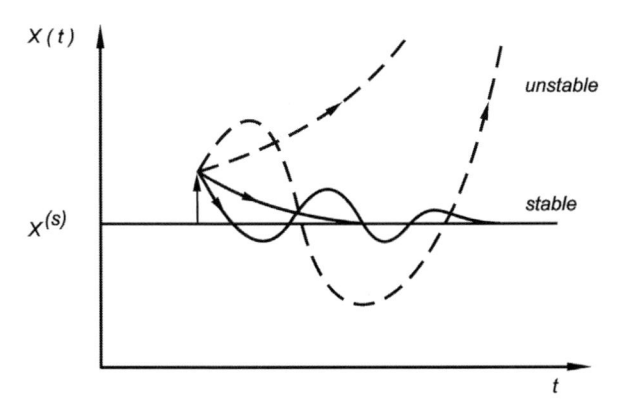

Figure 1.11 Stable and unstable states of a linear relaxation equation. The deviations from a stable state decay (solid curves), but the deviations from an unstable state start to grow (dashed curves).

1.6
Self-Organization of Information and Values

We know that the existence of all living beings is intimately connected with information processing. This we consider as the central aspect of life. We define *a living system as a natural, ordered and information-processing macroscopic system with an evolutionary history*. This may be used even as a criterion for decisions. Let us imagine that we are traveling on a spaceship far from our home-planet Earth. Suddenly we meet another unknown object moving in space, sending signals and doing maneuvers. The foreign spaceship hinders our further flight and takes an attitude which our captain considers as dangerous. He is asking us, the group of scientists on the ship: How should we behave? After some discussion, the scientists recommend to destroy the object if it is not a living being. But now the captain asks the scientist how to distinguish between a living and a nonliving item. After another discussion, the scientists group says: Evidently the foreign object is information processing, so check whether it is the result of a natural evolutionary process. If the answer is yes then we have to respect it, and if it is made just by another intelligence then you are possibly allowed to destroy it. Then the captain says, you are not really helpful, what I need is a more operational criterion, but the scientists answer that there is no other way to decide.

We consider information processing as a special, high form of SO. Information is an emergent property. We see several important problems here:

1) What is the general relation between SO and information and how may this relation be expressed in a quantitative way?
2) What is the origin of information processing in evolution? How did information processing emerge in the process of SO of biomolecules?

Both these problems seem to be unsolved so far. Key papers which reflect the main tendencies and the directions of search for final solutions were written by Eigen, Haken, and Volkenstein. How did information emerge by SO? Genuine information is symbolic information, needing a source that creates signals or symbols, a carrier to store or transport it, and finally a receiver that knows about the meaning of the message and transforms it into the structure or function the text is a blueprint for. This way symbolic information is always related to an ultimate purpose.

Information-processing systems exist only in the context of life and its descendants: animal behavior, human sociology, science, techniques, and so on. To our knowledge, the very first such system in evolution history was the genetic expression machinery of early life, where DNA triplets were used as symbols for amino acid sequences of proteins to be mounted. However the details, how life appeared, which way symbolic information developed out of nonsymbolic, native one, are hidden behind the veils of an ancient history. Other, later examples for the SO of information are much easier to study, and this was done in a very elucidating manner by Sir Julian Huxley in the beginning of the last century in behavior biology. The evolutionary process of the transition from use activities of animals to signal activities he discovered is termed ritualization." In our concept the transition to "ritualization" or "symbolization" is a central point. A more detailed view at this transition process reveals rather general features which we consider as a universal way to information processing.

When a process or a structure becomes ritualized (symbolized), its original full form of appearance is successively reduced to a representation by symbols, together with a building-up of its processing machinery, which is still capable of reacting to the symbol as if the complete original were still there. At the end of the transition, the physical properties of the symbolic representation are no longer dependent on the physical properties of its origin, and this new symmetry (termed coding invariance) makes drift and diversification easily possible because of the newly achieved neutral stability. Neutrally stable states are those which do not exhibit restoring forces in response to modification. If symbols are arranged as linear chains in space or time, the information system is regarded as a language (such as genetic information); if not, it is termed a signal system (such as traffic signs or the hormone system).

Let us make now a few remarks on the relation of *chemical information processing and life*: the scenario we propose is mainly based upon the fundamental ideas of Eigen, who first introduced the concept of SO far from equilibrium into the problem of the origin of life, and of Oparin, who proposed the first scientific approach to the problem by underlining the importance of insulated self-assembling droplets (coacervates). Our aim is to formulate an own-standing hypothetical staircase to life, trying to merge Eigens and Oparins ideas to a common scenario under mainly physical rather than chemical or biological aspects. Although many researchers discuss special aspects such as particular molecule types and environmental conditions in detail, we believe those to be only specific circumstances, while the main questions are posed by the succession of steps of SO, of symmetry-breaking, of information processing, the onset of most primitive forms of Darwinian selection.

This concept will be briefly surveyed, since it differs in a few details from the well-known Eigen–Schuster concept. Beside some mathematical points (referring, e.g., to the mathematical formulation of the quasispecies concept and to stochastic effects), the main difference is our emphasis on a very early formation of individual spatial compartments, similar to Oparin's approach. Further, instead of hypercycles, some hypothetical RNA-replicase cycles play the dominant role in our scenario. This argument was based on an estimate of the probabilities for a spontaneous generation of catalytic feedback structures. The theory of random catalytic networks (Chapters 2, 4, 6) suggests that very simple structures such as paths, branching systems, semicycles, and small cycles have much higher probabilities to come into existence spontaneously than larger cyclic structures. Especially the genetic expression machinery, often assumed to be the very first evolving structure which must have been appeared just by chance, is considered here as a result of a long chain of preceding chemical evolution steps (Ebeling and Feistel, 1982).

In our model, six qualitative stages are assumed and characterized in detail in Chapters 8 and 9:

1) *Physicochemical SO*: Spontaneous formation of polypeptides and polynucleotides, local increase in the concentration in compartments (coacervates, microspheres, pores, etc.), catalytic assistance of the synthesis in networks and

cascades, first self-reproduction of polynucleotides (RNA) assisted by catalytic proteins (replicases), competition and selection between compartments and between replicative cycles inside compartments.

2) *Formation of protocells and of a molecular language*: Genesis of self-reproductive units consisting of RNA and proteins, division of labor between RNA and protein is more and more developing, generation of elements of a molecular language. DNA takes over the role of the memory and RNA is specializing in transcription functions. The building-block principle for the synthesis of proteins is worked out.

3) *Genetic code, ritualization, and division of work*: The full coding is successively replaced by a kind of stenography, the first triplet code arises and the direct chemical meaning of a complete sequence is replaced by a symbolic notation (ritualization), the development of the genetic code and the division of work led to the first living "minimum organism" in the sense of Kaplan (1978).

4) *Cellular organization*: The triplet code generates more and more complex structures such as membranes that are stabilizing the compartments. Spatial separation and nonlinear advantage leads to a freezing of the code. Controlled division of cells occurs.

5) *Genesis of autarkic systems*: A sharp selection pressure generates systems that are able to use primary sources of energy by photosynthesis. Heterotrophy and food-webs appear. Division of labor inside the cells is leading to compartments and especially to the formation of nuclei. Sexual reproduction is invented.

6) *Morphogenesis*: Cell associations are formed. The metabolic products take over regulating functions; division of labor between cells creates multicellular organisms. Certain cell groups, neurons, specialize on information processing. The basic mechanisms which were first based on direct chemical or stereochemical relations are replaced by the use of symbols. The corresponding second ritualization transition (symbolization transition) leads to complex nervous systems.

We are convinced that any further progress in the understanding of the origin of life on Earth will depend on the clarification of the points mentioned above.

There is another point which needs a better understanding, namely the problem of *combinatorial explosion*. One of the basic problems of understanding the SO of complexity is connected with a choice among an enormous number of combinatorial possibilities. For example, the number of different possibilities to arrange n monomers belonging to λ types (4 for DNA, 20 for proteins) is given by

$$N_n = \lambda^n$$

This is a very large number and the corresponding probability to find it by a random search

$$P_n = \lambda^{-n}$$

is extremely small. Evolution arguments shed new light on this important problem (Chapters 7 and 8, Feistel, 1990; Krug and Pohlmann, 1997; Ebeling, Freund, and Schweitzer, 1998).

We believe that all these problems are of high importance for any search or design of macromolecular systems with desired properties. In fact, to estimate the chances for a successful search and to envisage a good strategy we need the entropies. In this way, entropy research should be an immanent part of any research program in the field.

We conclude this section with a discussion of the concept of *values as emergent irreducible properties*. As we have pointed out above, we understand SO as a "process in which individual subunits achieve, through their cooperative interactions, states characterized by new, emergent properties transcending the properties of their constitutive parts." In this respect we would like to stress the role of values, which are indeed among the most relevant emergent properties. An example is the value of a species which means the fitness in the sense of Darwin. Competition is always based on some kind of valuation.

Evidently the concept of values was first introduced by Adam Smith in the eighteenth century in an economic context. The fundamental ideas of Adam Smith were worked out later by Ricardo, Marx, Schumpeter, and many other economists. In another social context, the idea of valuation was used at the turn of the eighteenth century by Malthus. Parallel to this development in the socioeconomic sciences, a similar value concept was developed in the biological sciences by Darwin and Wallace. Sewall Wright developed the idea of a fitness landscape (value landscape) which was subsequently worked out by many authors; in the last few years, many new results on the structure of landscapes were obtained by the group of Peter Schuster.

The concepts of values and fitness landscapes are rather abstract and qualitative. Our point of view is that values are an abstract nonphysical property of subsystems (species) in a certain dynamical context. Values express the essence of biological, ecological, economic, or social properties and relations with respect to the dynamics of the system (Feistel, 1986, 1990, 1991; Ebeling and Feistel, 1994; Ebeling, 2006). From the point of view of modeling and simulations, values are emergent properties. The valuation process is an essential element of the SO cycles of evolution. The existing theory has already anticipated several value concepts such as the value of energy (i.e., entropy), the information value, the selection value in biology and the exchange value in economy. All these value concepts have several features in common:

1) Values assigned to elements (subsystems) of a system or to the modes of their dynamics incorporate a certain entireness of the system; they cannot be understood by a mere view of the subsystem without its whole environment. In other words, here the whole is more than the sum of its parts. Values represent irreducible features of the system on its level of complexity.
2) Values are central for the structure and dynamics of the entire system; they determine the relations between the elements and their dynamical modes as well as the dynamics of these relations. Competition or selection between elements or dynamical modes are typical elements of the dynamics.

3) The dynamics of systems with valuation is irreversible; it is intrinsically connected with certain extremum principles for the time evolution of the values. These extremum principles may be very complex and can, in only a few cases, be expressed by scalar functions and total differentials.
4) The necessary physical condition for any form of valuation is the pumping with high-valued energy. Isolated systems show a general tendency to devaluation, which is caused already by the devaluation of the energy in the system, due to the second law.

Only in the simplest case, the value of a subsystem (species) with respect to the competition is a real number. Since real numbers form an ordered set, the species are ordered with respect to their values. In such systems, competition and valuation may be induced by the process of growth of species having high values and decay of species having low values (or the opposite). In many cases, the growth of "good" species is subject to certain limitations. A standard case studied in detail by Fisher (1930) and Eigen (1972) is the competition by average. Here all species better than the average over the total system, $V > \langle V \rangle$, grow and all others that are worse $V < \langle V \rangle$ decay. Because the occupation $N(t)$ is changing, this leads to an increase in the averaged value $\langle V \rangle$ in time. In some cases, the values are given by the distribution itself, for example,

$$V_i = -\log(N_i/N)$$

This valuation favors equal distribution and the averages correspond to entropies. It is interesting to note that in this view, the entropy appears as a special kind of valuation. In physics, according to Boltzmann's concept, entropy maximization is subject to the condition of constant energy, and the elements are defined by a partition of the phase space of the molecules. The lowering of entropy with respect to its maximum (at given energy) expresses, after Clausius, Helmholtz, and Ostwald, the work value of the energy in the body. The second law of thermodynamics expresses a general tendency to disorder (equipartition), corresponding to a devaluation in isolated systems.

Many competition situations in biology, ecology, and economy are associated with a struggle for common raw material or food. Here, the result of the competition can still be predicted on the basis of a set of real numbers (scalar values). In more general situations, the values are merely some property of the dynamical system rather than well-defined numbers. One may attribute three functions to valuation (Feistel, 1990, 1991):

1) Regulating functions, establishing stable attractor states among the winners of the competition,
2) Differentiating functions, awarding "better" competitors and eliminating the defeated ones, and
3) Stimulating functions, amplifying favorable fluctuations which destabilize the established parent regime.

Valuation is absolutely central to the origin of life; in order to survive, a living creature needs a standard of basic values as food, shelter, protection, comfort, and so on. In modern societies, there exists an exchange value – the money. A role similar to that of a currency is played by adenosine triphosphate (ATP), which is exchanged between the parts of an organism. The generation of values in biological and social societies is deeply connected with the origin of information.

The question of the value of information is a special and very important problem, which raised a never-ending discussion (see, e.g., Volkenstein,1986; Marijuan, 2003). We will explain our point of view in Chapters 5–9. In short, we see the value of information for a living system in the gain of some of the basic values due to the transfer of information to the system. We will explain examples in Section 5.6. Normally the basic values needed for survival are difficult to achieve, think only about the search for energy-rich molecules in the primary soup, for the recruitment of food, shelter, and protection in ecological systems and in early human societies. Here the information transfer by signals comes into play. It was basically information transfer which increased the access to the basic values and made the explosion of life on earth possible. This way, information value is not a new value per se, but it is the gain of basic values as food, shelter, protection, comfort, money, and so on by transfer of information.

We are concluding this chapter with these rather general remarks. As we will see, in nearly all problems which will be treated later, the real difficulties are in the details, but this is the subject of the following chapters.

2
Fundamental Laws of Equilibrium and Nonequilibrium Thermodynamics

Wir wollen also bei unserer Entwickelung von der Annahme ausgehen, dass die Wärme in der Bewegung der kleinsten Körper- und Aetherteilchen bestehe, und dass die Quantität der Wärme das Maass der lebendigen Kraft dieser Bewegung sei.
                         Rudolf Clausius: Mechanische Wärmetheorie

The phenomena of radiant heat ... should not be neglected in any complete system of thermodynamics.
              Josiah Willard Gibbs: Elementary Principles of Statistical Mechanics

2.1
The Thermodynamic Way of Describing Nature – Basic Variables

Why are the laws of equilibrium and nonequilibrium thermodynamics so important for the physics of self-organization and evolution? We note the following arguments:

1) Thermodynamic laws, in particular the first and the second law, tell us which processes are allowed and which are forbidden in nature. This way the thermodynamic laws determine the time arrow of processes (Mackey, 1992; Prigogine, 1967; Andrieux *et al.*, 2007; Hoffmann, 2008).
2) Processes of self-organization and evolution are described in terms of various physical quantities such as temperature, density, concentrations, specific heat, and so on. These quantities are internally connected by the existence of a few thermodynamic potentials such as entropy or free energy (Callen, 1960; Feistel, 2003, 2008a; Feistel and Wagner, 2005, 2006).
3) Determined by the thermodynamic laws of irreversible processes, flows of matter, heat, electricity, and so on play a central role for self-organizing systems (Prigogine, 1947; Prigogine, Nicolis, and Babloyantz, 1972).

For all these reasons, we start with an explanation of the thermodynamic way of describing nature. At the thermodynamic level, matter is described by concepts, such as *thermodynamic system, thermodynamic state, components, phases,* and so on, and by *thermodynamic variables* such as energy, entropy, temperature, and quantities (masses) of different kinds of matter (Callen, 1960). Following Schottky, a thermodynamic system is a finite macroscopic body with a well-defined border (surface) through which the system is exchanging matter, heat, and work with the surrounding. A classification of thermodynamic systems may be based on the types of admitted exchanges with the surrounding.

Open systems: Matter, heat, and work may be exchanged as well.
Closed systems: The system may exchange heat and work, but not matter.
Adiabatic systems: Work is exchanged, but heat and matter cannot penetrate the border.
Isolated systems: No macroscopic exchange with the surrounding is admitted.

In order to present at least one example of each kind of these systems, for an open system let us think of a glass of water as that is exchanging vapor with the surrounding, and as an example of a closed system a gas enclosed in a vessel within a heat bath with a moveable piston embedded. Further we consider as examples of an adiabatic system a gas enclosed in a vessel with a moveable piston which is not in thermal contact with a heat bath, or a gas enclosed in a Dewar vessel for an isolated system. At this moment, when speaking about the state of such a system we mean just a set of measurable macroscopic physical quantities which somehow characterize the macroscopic properties of the system. A thermodynamic system is regarded as *homogeneous* if measurements at different positions inside the system always yield the same results, otherwise the system is *inhomogeneous*. *Components* of a thermodynamic system are the chemical constituents (chemical elements and compounds) present in the system. These chemicals are denoted here by a Latin subscript i which is running from 1 to the total number of components, s. The *composition* of a system is the set of its chemical constituents such as (N_2, O_2, CO_2, H_2O) in the case of air.

The *phases* of a thermodynamic system are the relatively uniform, spatially separated parts. In a vessel containing water we may find, for example, vapor and liquid in different regions of the volume. Here, phases will be denoted by a Greek subscript α and will be numbered from 1 to σ. A quantity x is termed as a *thermodynamic variable* if it has the following properties:

- The quantity x is expressed by a real number or by a set of real numbers.
- There exists a prescription for the measurement of changes of the quantity x.
- The quantity x characterizes certain macroscopic properties of the system.

Here, the measurement recipe may not always be a direct or absolute one. In the case of the thermodynamic variables energy and entropy of a given state, thermodynamic measurements can only provide their respective differences when measurable amounts of heat or work are exchanged between the system and the outer world. The absolute values can either be arbitrarily fixed by suitable reference-state conditions (Feistel *et al.*, 2008b) or by additional theoretical arguments derived from, for example, statistical mechanics or the theory of relativity.

A variable is regarded as a *state variable* if it depends solely on the actual state but not on the history of the system. Thermodynamic state variables can be distinguished as conjugate pairs under three important aspects:

- *dependent* and *independent* variables,
- *extensive* and *intensive* variables,
- *internal* and *external* variables.

Out of the extended set of all thermodynamic variables associated with a given system, arbitrary values can be assigned (within reasonable limits such as positivity of density) only to a certain subset of those variables, referred to as the *independent* variables (or control parameters). All other variables, the *dependent* ones (or order parameters), are then adjusted by physical processes that proceed within the system. For stable states, the result of this relaxation process (i.e., the equilibrium state) is unique and independent of the system's history. The mathematical relation between dependent and independent variables can be formulated as an *equation of state* which is unique and specific for the particular system considered. The number of independent variables available in a system is determined by Gibbs' phase rule. For example, the pressure of a gas may be calculated from the volume and the temperature if an equation of state is available for that gas. In this special case of an equilibrium system with one phase and one component, there exist two independent variables (degrees of freedom); the values of two thermodynamic variables can arbitrarily be chosen to entirely specify the current state of that system, and in turn all the additional properties. Similar to the use of mechanical coordinates, the choice of the independent variables is arbitrary and may be guided by usefulness, uniqueness, and convenience.

In certain situations, the choice of the independent variables is not completely free. As an example, a vessel that contains the two-phase system of liquid water and vapor cannot possess arbitrary pairs of values for temperature and pressure; once the pressure is given, the temperature takes the boiling temperature of water related to that pressure, and if the temperature is given, the pressure will adjust to the vapor pressure of water at that temperature. In a temperature–pressure diagram, the set of liquid–vapor equilibria forms a one-dimensional curve (the *vapor–pressure curve*, $T(p)$) rather than a two-dimensional region, see Figure 2.1. The two-phase system has two degrees of freedom, such as temperature and volume, but temperature and pressure are not independent of each other. This problem appears with intensive variables such as temperature and pressure, which we consider now. A variable is termed *extensive* if its value for a system consisting of the parts 1 and 2 is the sum of the values of the parts

$$X = X_1 + X_2 \tag{2.1}$$

Examples are volume, V, and mass, M. *Intensive* variables, on the other hand, have the property to possess the same value in each volume element of a homogeneous system, independent of the size of that volume element

$$x = x_1 = x_2 \tag{2.2}$$

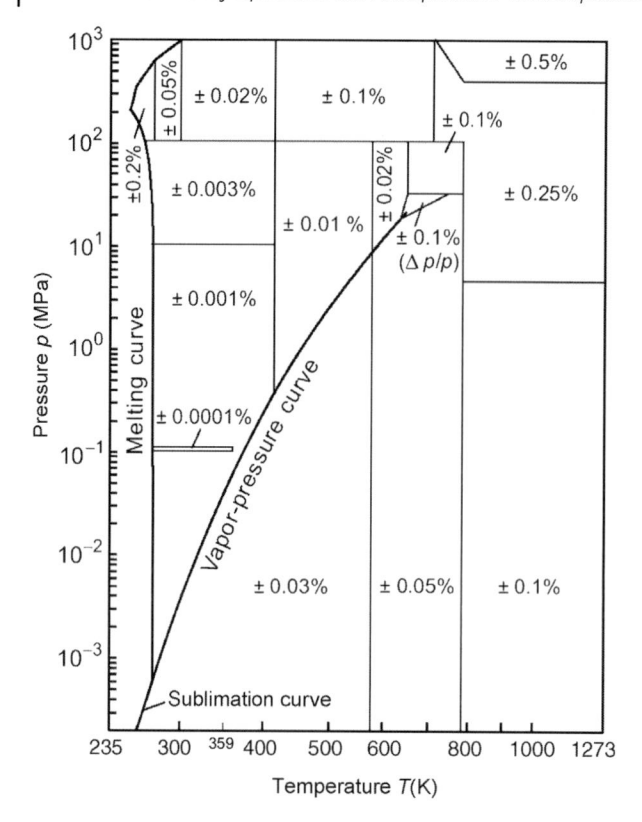

Figure 2.1 Phase diagram of water. Bold curves separate the different phases from one another, ice on the left of the melting curve, liquid water above, and vapor below the vapor–pressure curve. The sublimation curve separates ice and vapor. The uncertainties reported in different T-p regions refer to the currently most accurate density of water available from the IAPWS-95 equation of state (IAPWS, 2009a; Feistel et al., 2008b; IOC et al., 2010). Reproduction permitted by IAPWS.

Examples are *pressure*, p, and *chemical potential*, μ. In general, extensive variables will preferably be denoted by capital letters and intensive variables by lower-case letters. An exception is temperature, which is traditionally denoted by the capital letter T in spite of its intensive character. In geophysics, also absolute pressure is often denoted by P in contrast to the pressure p taken relative to atmospheric pressure, similar to t used for the temperature relative to the freezing point of water, in contrast to the absolute temperature T. If an equation of state is available in the implicit form

$$\Phi(x_1, ..., x_n, X_1, ..., X_m) = 0 \tag{2.3}$$

where x_i are intensive and X_i extensive variables, then this equation can equivalently be written as

$$\Phi(x_1, \ldots, x_n, \lambda X_1, \ldots, \lambda X_m) = 0 \tag{2.4}$$

for any positive number λ. This general property of equations of state is known as the scaling invariance (Baus and Tejero, 2008). It is merely expressing the empirical fact that the equation of state of a certain substance such as water does not depend on the total mass of water we are actually considering; the same equation holds equally for a test tube and for an ocean.

Internal variables are variables which depend only on the internal state, such as the positions and velocities of the molecules; examples are the temperature and the pressure. *External* variables depend on the state of the bodies in the surrounding; an example is the gravitational field. This distinction is not always clear and unambiguous; for example, volume is often considered an external parameter (Baus and Tejero, 2008), but fluid parcels in meteorology and oceanography are usually specified by their temperature and pressure, and the related density (and in turn its volume) of the parcel appears as an internal variable. In contrast, gravity is sometimes formally treated as an external parameter even though it describes the gravitational interaction between parts of the system (Neugebauer, 1980). Alternatively, if internal variables were rigorously defined, say, as those properties which can be expressed in terms of the masses, coordinates, and velocities of the system's particles, this definition would imply that volume and entropy are external variables, which appears somewhat counterintuitive. Here we touch already the fundamental problem of *emergent properties*. They are relevant, holistic properties of complex systems which cannot be reduced to the properties of the systems elemental constituents. Emergent properties appear in particular as a result of self-organization and evolution processes.

Let us go now into more detail. The quantity of matter contained in a system is given by the mass M and the quantities of the components by the individual masses, $M_1, M_2, \ldots, M_s$. The measurement of masses does not pose a problem, the SI units are kg or g. Rather than mass-based units, we may alternatively use molar masses m_i known from chemistry, such as approximately 1 g for one mole of atomic hydrogen, H, 2 g for hydrogen molecules, H_2, 16 g for oxygen, O_2, and 18 g for water, H_2O. According to Avogadro in 1811, Loschmidt in 1865, and Planck in 1899, the number of particles corresponding to the molar mass is the universal Avogadro constant,

$$N_A = 6.022\,141\,79(30) \times 10^{23}\,\text{mol}^{-1} \tag{2.5}$$

Here, the number in parentheses indicates the current measurement uncertainty of the last two digits. The value of N_A is intended to be fixed exactly after a redefinition of the SI unit for mass, the kilogram. Such a redefinition is necessary because the kilogram prototype stored in Paris is slowly but permanently losing mass (Brumfiel, 2010; Wynands and Göbel, 2010). The successive substitution of primary SI standards in the form of carefully prepared artifacts such as the kilogram prototype, or previously the meter prototype, by fundamental physical constants such as the speed of light improves the metrological stability, reproducibility, and comparability of measurements. This progress in reducing measurement uncertainties is highly relevant for many fields of science and technology, for example, for the reliable detection of tiny climatological changes in oceanography by comparison of measurements that were carried out more than a century apart by different persons with different devices and methods (Wielgosz and Calpini, 2010).

We introduce further the *mole numbers* of the sort i in a system by

$$\tilde{N}_i = \frac{M_i}{m_i} \tag{2.6}$$

and the particle numbers by

$$N_i = \tilde{N}_i N_A \tag{2.7}$$

The total particle number is denoted by N. Besides the extensive variables discussed so far, we introduce also the following intensive variables:

$n_i = N_i/V$ – particle densities,
$v_i = \tilde{N}_i/V$ – molar concentrations,
$\varrho_i = M_i/V$ – mass densities,
$x_i = N_i/N$ – mole fractions.

In the case that our system consists of several phases, we add to the variable a Greek superscript to denote the phase, for example, x_i^α would denote the mole fraction of the component i in the phase α.

Now we shall briefly discuss the mechanical variables V and p. The volume V of a system is measured by using either geometrical formulae, or the Archimedes principle. In order to measure the pressure, we introduce into the system at one end a small cylinder with a movable piston of the cross section A. If F is the force exerted by the body on the piston, the ratio $p = F/A$ is the pressure. The SI unit of the pressure is $1 \text{ N/m}^2 = 1$ Pa (pascal). In most cases, the pressure is independent of the position and the direction of the piston used for the measurement, that is, p is a scalar. However, there exist also more complicated cases where the pressure is a tensor and where the properties depend not only on volume but also on the shape of the system.

The central thermal variable of any thermodynamic system is the temperature, T. We can experience temperature immediately by our natural senses; when touching two objects, we can tell which one is warmer. The temperature of thermodynamic equilibrium systems is a well-defined quantity that takes the same value for two systems when they are in thermal equilibrium with one another. A scientific method for measuring temperatures was first introduced in 1592 by Galilei during a lecture when he presented a "thermoskop." By means of this measuring device, he reduced temperature measurements to length measurements. Using a similar principle, we introduce here absolute temperatures by means of a *gas thermometer*. It consists of a vessel with a moveable piston containing a gas with a very low density, n, and a possibility to measure the position of the piston, and in this way the volume. Then the relative temperature is defined by

$$T = T_t \cdot \lim_{n \to 0} \frac{V}{V_t} \tag{2.8}$$

Here T_t is a reference temperature and V_t the volume of the gas at that temperature. The gas in the device should have a very low temperature. In this way, the measurement of the temperature is reduced to a comparison of volumina of a dilute gas, or to a length measurement to determine the piston displacement. As the reference temperature we take the temperature of water at the liquid–solid–gas triple point, Figure 2.1, which is exact by definition of the International Temperature Scale of 1990, ITS-90,

$$T_t = 273.16 \, \text{K} \tag{2.9}$$

Our temperature definition is based on the empirical observation that dilute gases show a kind of universality: at low density, all gases behave in the same way, independent of the chemical composition. This property guarantees the universality of our temperature definition.

Also in this case, a redefinition of the SI unit kelvin (K) is in preparation because the triple point of water is actually not a "point" in the T–p diagram, rather, it is a tiny region as a result of the different isotopes of hydrogen and oxygen that are mixed in the current Vienna Standard Mean Ocean Water (VSMOW). If the isotope ratios in the different phases of water vary, the locus of triple point is slightly displaced within an estimated range of 14 μK (Nicholas, Dransfield, and White, 1996; White *et al.*, 2003a; Rudtsch and Fischer, 2008; Feistel *et al.*, 2008b). Moreover, recent measurements made by novel and more reliable experiments revealed systematic deviations between the latest temperature scale, ITS-90, and the thermodynamic temperature (Fischer *et al.*, 2011), see Figure 2.2. Thermodynamic temperature is the quantity that

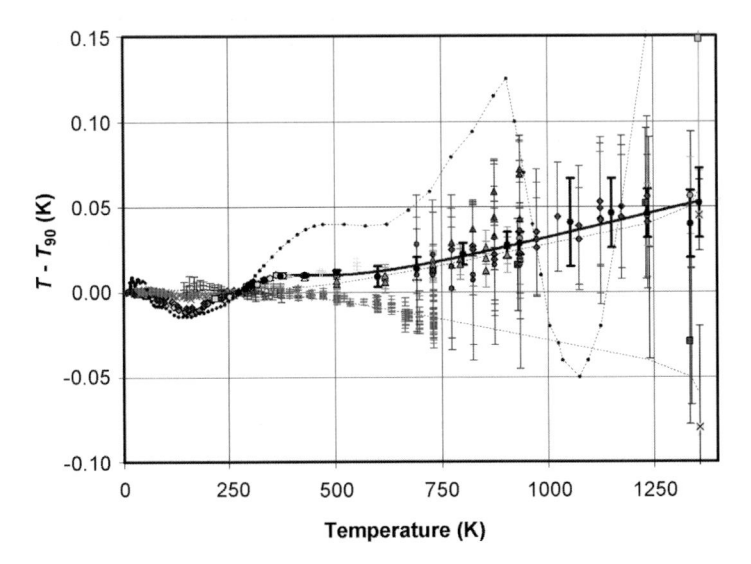

Figure 2.2 Deviation of the currently best estimate for the thermodynamic temperature, T, shown as the bold curve, from the temperature T_{90} expressed in the current International Temperature Scale of 1990 (ITS-90). The dotted curves represent the measurement uncertainty of ITS-90. The dotted lines connecting bigger dots show the deviation of the previous temperature scale of 1968, IPTS-68, from ITS-90. Additionally, results of various precision measurements are shown with error bars. Diagram adapted from Fischer *et al.* (2011), courtesy of Joachim Fischer.

appears in thermodynamic equations while temperature scales are practical techniques to actually determine the temperature of a given real system, that is, to calibrate a thermometer.

In the new SI definition of temperature, one kelvin will be defined in terms of the energy unit joule, $1\,J = 1\,N\,m$, by

$$1\,K = 1\,J/k_B \tag{2.10}$$

Here

$$k_B = 1.380\,6513(18) \times 10^{-23}\,J\,K^{-1} \tag{2.11}$$

is the Boltzmann constant to which an exact numerical value will be assigned by definition in the near future (Fellmuth, 2003; Fellmuth *et al.*, 2007; Quinn, 2007; Strehlow and Seidel, 2007; Seidel *et al.*, 2007; Jones, 2009; Wynands and Göbel, 2010; Fischer *et al.*, 2011). To measure the best value of k_B on the currently valid temperature scale, beside the gas thermometry for low temperatures as explained above, radiation thermometry is used for high temperatures, making use of the theoretical relation between temperature and energy available from Planck's (1906) law for black-body radiation (Hollandt *et al.*, 2007). A third fundamental method for temperature determination from alternative physical properties is noise thermometry which exploits the equation of Nyquist (1928) between temperature and thermal noise intensity (Edler and Engert, 2007).

Coming back to the definition of the state of a system, we may say that the thermodynamic state is described by a set of independent thermodynamic variables which completely characterize the thermodynamic system. There exists a distinct state of thermodynamic systems which has unique properties, the *equilibrium state*. A system is in the state of *thermodynamic equilibrium* if the following conditions are fulfilled:

1) The system is characterized by a unique set of extensive and intensive variables which do not spontaneously change in time except for irrelevant random fluctuations.
2) After isolation of the system from its environment, all the thermodynamic variables remain unaltered within the uncertainty of thermal fluctuations.

Any states which do not satisfy these conditions are nonequilibrium states. For example, if a gas is kept for a long time in a vessel under fixed environmental conditions, it will be in equilibrium and its subsequent isolation from the environment will not change its state. It must be mentioned that sluggish or metastable systems exist which relax to equilibrium so slowly that practically no change can be detected within reasonable observation times even though equilibrium may not yet be established. Examples are stably stratified solutions such as oceans where molecular salt diffusion or thermal conduction are extremely slow vertical transport processes, or distortions in cold crystalline solids such as inhomogeneities in the terrestrial crust which do not "heal" during billions of years and still offer to us structural information (see Chapter 8) on the history of their formation processes in the distant past. All our houses, devices, and books are in nonequilibrium states but their relaxation is hardly observed within a human lifetime, except that meticulous methods are applied as in the case of the kilogram prototype stored in Paris.

Summarizing the findings given above, we formulate now a first working principle focused on the equilibrium state and the temperature variable (Guggenheim, 1949). Following Fowler, this principle is referred to, as the *zeroth law of thermodynamics*:

Thermodynamic systems possess a special state, called thermodynamic equilibrium. For systems in thermodynamic equilibrium there exists a scalar intensive variable, called the temperature, which is uniquely defined. The temperature has equal values for any two systems in thermal equilibrium with each other. When two systems (1) and (2) are each separately in equilibrium with a third one (3), then the system (1) is also in equilibrium with the system (2).

The property described in the last part of our principle expresses the property of transitivity of the binary relation "in equilibrium with," and in particular of the temperature variable. A universal relative temperature scale can be defined as described here by means of low-density gas thermometers. There are other methods, such as definitions through the efficiency of Carnot machines, which however require more advanced concepts. Some practical methods were briefly described in the beginning of this section. The temperature measurement is fundamentally based on the property of transitivity.

At the first glimpse, the zeroth law may look like a self-evident triviality. But, imagine the polar ocean where the ice cover is in equilibrium with the seawater underneath with respect to freezing/melting, and the ice is also in sublimation equilibrium with the humid air above it. Seawater and air are not in contact with each other. When we make a hole in the ice and let the seawater get in contact with the air, then, as a result of the zeroth law, neither evaporation of water nor condensation of vapor will take place (Feistel et al., 2010b). Does that still sound trivial?

How do we practically compare the temperatures of two systems? We use a third system, known as thermometer. We bring the thermometer in contact with system (1) and wait for relaxation to equilibrium between the system and the thermometer. Then we repeat the experiment with system (2) and the same thermometer. If the second reading equals the first, we may conclude by means of the zeroth law that system (1) and system (2) have the same temperature, even though no direct comparison between the two was done. Typically, system (1) may be a heat bath of well-defined temperature in a calibration laboratory, and system (2) may be a patient in a clinic. Owing to the zeroth law, it is unnecessary to take ill patients to a calibration laboratory to measure their body temperature.

Transitivity of a binary relation is by no means a trivial property. In Chapter 4, we shall consider this aspect in more detail, including simple examples, and in Chapters 6 and 7 we shall see that the difference between irreversibility and evolutionary progress is basically a question of transitivity.

Systems that are not in equilibrium may spontaneously change their state in time without external interference. Sequences of states changing in time are called *processes*. We introduce the notations:

1) *Quasistatic processes* are sequences of equilibrium states. Processes which are sufficiently slow may be considered approximately as quasistatic.
2) *Reversible processes* may proceed in both directions. A film of a reversible process shown in backward time sequence will not contradict any law of nature. Especially we have to require that the time-reverse process does not violate the second law of thermodynamics.
3) All processes that do not belong to the classes (1) and (2) are *irreversible processes*.

Evidently all *real thermodynamic processes* belong to the third class. However the concepts of quasistatic and reversible processes are very useful as abstractions modeling limiting situations. Real thermodynamic processes show a fundamental asymmetry between future and past; they have the property of irreversibility. Let us discuss now in more detail our definition of irreversibility. We consider an arbitrary real process. In order to decide whether this process is reversible or irreversible we produce a film of the process. Let the sequence of observed states correspond to a time series

$$x(t_1), x(t_2), \ldots, x(t_n)$$

In the next step, we show our film in the opposite direction. This defines here the inverse process or backward process, corresponding to the time series

$$x(t_n), x(t_{n-1}), \ldots, x(t_1)$$

If this inverse process is in no contradiction to any known law of nature, the original process is regarded as a reversible process. In all other cases, we speak about irreversible processes. In order to illustrate this idea, let us consider two examples. First we consider the orbit of an electron around an atomic nucleus. In this case, there is no natural law that would forbid the backward motion. Forward and backward motion are both possible, both are allowed motions according to the laws of classical or quantum mechanics. Classical or quantum-mechanical motions are of reversible type. Our second example is a stone falling down to the ground where it comes to rest while converting kinetic energy into heat and producing entropy this way. The fictitious backward motion would correspond to a stone collecting heat from the surrounding and converting it to directed kinetic energy. This process would destroy entropy, which is impossible according to the second law. The second law postulates that macroscopic processes can produce entropy, but under no circumstances they may destroy entropy. The backward process associated with a macroscopic process would show a process that destroys entropy. In this way the backward process conjugate to any macroscopic process is forbidden by the second law. This proves that in the strict sense, any macroscopic processes are irreversible.

Reversible processes represent a limiting case of irreversible processes; the opposite, however, is not true. How can we describe the difference between reversibility and irreversibility in a more technical (mathematical) way? Formally, reversible processes can be described by mathematical groups, while irreversible processes belong to semigroups, lacking an inverse element (Prigogine et al., 1977; Alberti and Uhlmann, 1981). In the next section, we shall show that entropy and other thermodynamic quantities provide us with appropriate tools for finding characteristic functions, the so-called *Lyapunov functions*, as suitable measures for the quantitative description of irreversibility.

Let us introduce now in a quite preliminary way the two basic variables of any thermodynamic description: *energy* and *entropy*. The two quantities are extensive variables which characterize any thermodynamic system. We begin with the term energy.

It is extremely difficult or even impossible to give a strict definition of the term *energy*. In principle, energy is given "a priori," that is, it must be considered as an axiomatic element of thermodynamic systems. Due to the fundamental character of the energy, it is very difficult to avoid tautologies in its definition. However, the same difficulty was already discussed by Newton with respect to mass and force. Poincare discussed the difficulties to define the term *energy* in his lectures on thermodynamics in 1893. He said that in every special instance, it is clear to us what energy is, and we can give at least a provisional definition of it. However it seems to be impossible to give a general definition of energy. The same is true for entropy. So let us start with the following preliminary definition:

Energy is a universal extensive quantity characterizing a real thermodynamic system; it is conservative in isolated systems.

Correspondingly we can characterize entropy by the statement:

Entropy is a universal extensive quantity characterizing a real thermodynamic system; it is nondecreasing in isolated systems.

We see that energy characterizes the conservative aspects of a thermodynamic system and entropy the irreversible aspects. For isolated systems, entropy is a Lyapunov function. A more precise characterization will be given in the next section.

2.2
Three Fundamental Laws and the Gibbs Relation of Thermodynamics

Based on our preliminary statement on energy, we shall characterize now the central variable of thermodynamics, *energy*, in more detail. In addition to being the central variable of thermodynamics, energy is also the central variable of all branches of physics and natural sciences, a quantity which links all known real processes. Energy is subject to a conservation law, called the *first law of thermodynamics*:

Energy can neither be created nor be destroyed. It is conserved in isolated systems. Energy can be converted to other forms, or transferred to another system, without changing its total amount.

Being aware of the difficulties associated with any rigorous definition of energy, we restrict ourselves to a description of the properties of energy. The first property of energy is extensivity. Further we find that any finite system of physical or nonphysical origin contains a certain amount of energy. In this respect, energy has properties similar to mass. The deeper reason for this close relation may be Einstein's formula, $E = mc^2$, which relates the total energy E contained in a body to its total mass m, including the rest mass. Therefore, strictly speaking, only one of these quantities is independent. Another property of energy is that it accompanies any process in our world we know. Any process is connected with a transfer or with a transformation of energy. Energy transfer may take the different forms of heat, work or chemical energy. The SI unit of energy is $1\,\mathrm{J} = 1\,\mathrm{N\,m}$, corresponding to the work needed to move a body 1 m against a force of 1 N.

We denote infinitesimal heat transfer by $d'Q$ and the infinitesimal work transfer by $d'A$. If there is no other form of transfer, that is, if the system is closed, we find for the energy change

$$dE = d'Q + d'A \tag{2.12}$$

In other words, the infinitesimal change of energy of a system equals the sum of the infinitesimal transfers of heat and work. This is a mathematical expression of the principle given above: any change of energy must be due to a transfer, since creation or destruction of energy is excluded. If the system is open, that is, if exchange of matter by an amount of dN_i per sort i is admitted, we assume the energy balance

$$dE = d'Q + d'A + \sum_i \mu_i d N_i \tag{2.13}$$

Here, the so-called chemical potential μ_i denotes the amount of energy transported by transfer of a unit of particles of the chemical sort i. Here μ_i has the dimension of energy per particle or per mole. In the simplest case, the infinitesimal work takes the form

$$d'A = -p\,dV \tag{2.14}$$

In the case that there are also other forms of work, we find more contributions each having a bilinear form

$$d'A = \sum_k l_k \, d L_k \tag{2.15}$$

where l_k is an intensive and L_k is an extensive variable, and in particular we have $l_1 = -p$ and $L_1 = V$. Strictly speaking, this expression for the infinitesimal work is valid only for reversible forms of work. Later we shall come back to the irreversible case. In this way, the balance equation for the energy changes (2.12) takes the form

$$dE = d'Q + d'A = d'Q + \sum_k l_k d L_k \tag{2.16}$$

In this equation, there remains only one quantity which is not of the bilinear structure, namely the infinitesimal heat exchange $d'Q$. The hypothesis that bilinearity holds also for the infinitesimal heat leads us to the next fundamental quantity, entropy. We shall assume that $d'Q$ may be written as a product of an intensive quantity and an extensive quantity. The only intensive quantity which is related to heat is T, and the conjugate extensive quantity will be denoted by S. In this way, we introduce entropy as the extensive quantity which is conjugate to temperature:

$$d'Q = T dS \tag{2.17}$$

This equation may be interpreted also in a different way, by writing

$$dS = \frac{d'Q}{T} \tag{2.18}$$

The differential of the state variable entropy is given by the infinitesimal heat $d'Q$ divided by temperature T. In more mathematical terms, temperature T is an integrating factor of the infinitesimal heat.

The variable *entropy* was introduced in a lecture given on 24 April 1865 in Zürich, Switzerland, by Clausius. He borrowed the term from the Greek word for metamorphosis, "η τροπη" (Clausius, 1865, 1876). In turn, it was Gibbs (1873) who shortly after suggested to denote processes at constant entropy as "isentropic." The SI unit of entropy is 1 J/K. One can easily show that entropy is not conservative. For example, let as consider two bodies of different temperatures, $T_1 > T_2$, being in thermal contact. Empirically we know that there will be a heat flow from the hotter body 1 to the cooler one denoted by 2. Conservation of energy provided, we find that

$$d'Q_1 = T_1 dS_1 = d'Q_2 = T_2 dS_2 \tag{2.19}$$

Due to our assumption, $T_1 > T_2$, we get $dS_1 < dS_2$, that is, the heat flow down a gradient of temperature produces entropy. An opposite flow against a gradient of temperature was never observed. A generalization of these observations leads us to the *second law of thermodynamics*:

Thermodynamic systems possess the extensive state variable entropy. Entropy can be created but never be destroyed. In reversible processes, the change of entropy is given by the amount of exchanged heat, divided by the temperature. During irreversible processes entropy is produced in the interior of the system. In isolated systems entropy can never decrease.

Let us come back now to our relation (2.13) which reads after introducing entropy by Equation 2.17,

$$dE = TdS + \sum_k l_k dL_k + \sum_i \mu_i dN_i \qquad (2.20)$$

This equation is known as the *Gibbs fundamental relation*, which is valid for quasistatic and reversible processes. Since the Gibbs relation contains only differentials of state variables, it may be extended (with some restrictions) also to irreversible processes. Equation 2.20 represents a relation between the differentials dE, dS, dL_k, and dN_i in the form of a Pfaffian differential equation (Margenau and Murphy, 1943; Smirnow, 1968; Stepanow, 1982). For quasistatic or reversible processes, the Gibbs relation is an integrable differential equation, and its solution can be used to express one of those extensive variables as a dependent variable. In particular, we may write, for example,

$$E = E(S, L_k, N_i) \qquad (2.21)$$

or

$$S = S(E, L_k, N_i) \qquad (2.22)$$

In order to avoid misunderstandings, we state explicitly:

The Gibbs fundamental relation (2.20) should be interpreted as a relation between the extensive variables of a system in states of thermodynamic equilibrium.

The existence of such a fundamental intrinsic dependence between the extensive variables is the reason for most of the relations used in modern thermodynamics which usually are relations between measurable quantities. Because of the integrability of the Pfaffian form, Equation 2.20, its coefficients can be expressed as partial derivatives of the function (2.21)

$$T = \frac{\partial E}{\partial S}, \; l_k = \frac{\partial E}{\partial L_k}, \; \mu_i = \frac{\partial E}{\partial N_i} \qquad (2.23)$$

Hence, the intensive properties of the system can be computed from the function (2.21) by partial derivatives. For this reason, the function (2.21) is termed as *thermodynamic potential* or a *fundamental equation of state*. The seminal discovery that all equilibrium properties of a given systems can be obtained just by mathematical operations from a single function, once it is known, was reported by Gibbs (1873) in a surprisingly incidental way. In a footnote on page 310, he wrote (symbols of variables and equation numbers are suitably adapted here):

An equation giving E in terms of S and V, or more generally any finite equation between E, S, and V for a definite quantity of any fluid, may be considered as the fundamental thermodynamic equation of that fluid, as from it by aid of Equations 2.12, 2.14 and 2.17 may be derived all the thermodynamic properties of the fluid (so far as reversible processes are concerned).

It must be emphasized that in contrast to the fundamental equation of state, Equation 2.21, arbitrary relations between thermodynamic variables do not necessarily provide comprehensive information on the system. For example, the entropy of an ideal gas cannot be computed from the ideal gas equation, $pV = Nk_B T$. Contained in Equation 2.21, the complete information on the thermodynamic properties of a given system is preserved only by appropriate mathematical transformations of variables from one thermodynamic potential to another (Landau and Lifshitz, 1966; Alberty, 2001). We will return to this point in the next section.

When properties of certain substances are measured experimentally, such as the density of water at different temperatures, the resulting data points are subject to uncertainties of random and systematic origin. Random scatter is largely eliminated when a smooth correlation curve is fitted to the measurements, but systematic deviations are difficult to detect this way. If, say, all data points have one and the same constant erratic offset, a smooth correlation is still possible and will not reveal this problem. If various properties of a substance are known from different experiments or theoretical formulas, one can replace the separate fit of independent correlations, say, for a heat capacity formula, for the sound speed, for the freezing point, and so on, by a single combined regression procedure to determine a thermodynamic potential of that substance. Since this fit then requires mutual consistency between the different data sets in order to be compatible with the existence of the Gibbs fundamental relation, this method does not only reduce random errors of single readings, it also provides a sensitive indicator for systematic errors of certain data sets. This is an essential criterion for the assessment whether particular data sets are reliable or not (Feistel, 2003, 2008a, 2011b; Feistel and Wagner, 2005, 2006).

Since Gibbs' fundamental relation is so important, let us repeat again: Equation 2.16 expresses a balance between the energy change in the interior, dE (the left-hand side of the equation), and the transfer of energy in different forms through the boundary (the right-hand side). On the other hand, Equation 2.20 expresses the mutual dependence between the state variables at different equilibrium states, that is, a completely different physical aspect. For irreversible processes the Gibbs relation in the form of Equation 2.16 remains unchanged because the energy of a system can change only by transfer of heat, work, or matter. On the other hand, for irreversible processes the balance (2.20) has to be modified into an inequality since entropy is produced in excess to the heat exchange, that is

$$dE \leq TdS + \sum_k l_k dL_k + \sum_i \mu_i dN_i \tag{2.24}$$

The inequality sign is easily understood if we imagine an isolated system in an initial state away from equilibrium. Then, no fluxes may pass the boundary, the energy will remain constant, $dE = 0$, and the entropy inside will grow nevertheless, $dS \geq 0$. This change is also due to the fact that there may be a transfer of energy which enters the system in the form of work and returns from the system in the form of heat, that is, the system exports entropy that was not imported in advance. An example is an electric heater that imports electric power and exports the same amount of energy as heat, or a coffee pot that warms up (very slightly) when it is mechanically stirred with a spoon. In the following, for convenience we shall again denote by $d'A$ the jointly transferred work. Taking into account those contributions, the balance (2.24) reads

$$dE \leq T dS + d'A + \sum_i \mu_i dN_i \qquad (2.25)$$

For later applications, we formulate now the first and the second law in a notation due to Prigogine (1947). The balance of an arbitrary extensive variable, X, can always be written in the form

$$dX = d_e X + d_i X \qquad (2.26)$$

where the subscript "e" denotes the exchange with the surrounding and the subscript "i" the internal change. Then, the balances for energy and entropy read

$$dE = d_e E + d_i E \qquad (2.27)$$

$$dS = d_e S + d_i S \qquad (2.28)$$

In these terms, the first and the second law take the mathematical forms

$$dE = d_e E \quad d_i E = 0 \qquad (2.29)$$

and

$$dS \geq d_e S \quad d_i S \geq 0 \qquad (2.30)$$

respectively. With respect to the exchange of heat and work, we can write now

$$d_e E = d'Q + d'A + \sum_i \mu_i d_e N_i \qquad (2.31)$$

and

$$d_e S = \frac{d'Q}{T} \qquad (2.32)$$

These equations are equally valid for equilibrium and nonequilibrium situations and are particularly useful for our later considerations. The total change of particle numbers

$$dN_i = d_e N_i + d_i N_i$$

may need to be completed by rate equations for particle transformations, such as chemical or nuclear reactions, for example in the form

$$d_i N_i = f_i(T, V, N_k) dt \qquad (2.33)$$

Formally, the extensive variable V can also be split into external and internal changes in the form

$$dV = d_e V + d_i V, \quad d_i V = 0 \qquad (2.34)$$

because any expansion of the system's volume is always achieved at the cost of a volume loss of its environment, that is, volume can be exchanged but not produced.

We mention that the relation (2.32) for $d_e S$ is to be considered as a definition of exchanged heat. The investigations of De Groot and Mazur (1962) and Haase (1963) have shown that for open systems, alternative definitions of heat may be more useful. As to be seen so far, the "most natural" definition of heat exchange in open systems is

$$d_e S = \frac{d^* Q}{T} + \sum_i s_i d_e N_i \qquad (2.35)$$

The new quantity $d^* Q$ is termed "reduced heat," and s_i is the specific entropy carried by a particle of the kind i. The idea behind the definition (2.35) is that the entropy contribution which is solely due to a simple transfer of molecules should not be considered as proper heat. Another advantage of the reduced-heat notation is that it possesses several useful invariance properties (Haase, 1963; Keller, 1977). In practice, an unambiguous determination of the specific entropies s_i is difficult, as we shall discuss below. Therefore, as soon as particles are exchanged, the related change of the system's entropy is not always well-defined and may require additional specifications.

The first and the second law of thermodynamics formulated above in a very compact form represent a summary of several hundred years of physical research. They constitute the most general exclusion rules employed in physics.

The *third law of thermodynamics* is more technical but it also constitutes a summary of a long experience. We formulate this law in the following form:

Third law of thermodynamics:

Energy and entropy are extensive scalar quantities, being finite for finite systems and bounded from below. For real macroscopic systems, the inequalities $E > 0$ and $S > 0$ hold. In the limit of zero temperature, $T \to 0$, of single-phase systems, the derivatives of entropy with respect to extensive variables disappear asymptotically.

In the traditional formulations of Nernst (1918) and Planck (1954), the limiting value for the entropy at zero temperature is assumed to be zero. But already in 1914 Einstein suspected that systems with statistical disorder may violate that hypothetically universal law of vanishing entropy. This could be demonstrated for ice crystals by Giauque and Stout (1936) and Pauling (1935), compare Section 8.4. An excellent review of the history of the third law is given by Gutzow and Schmelzer (2011).

The famous statistical formula carved in the marble of Boltzmann's tombstone in Vienna (Figure 8.8)

$$S = k_B \ln W \qquad (2.36)$$

tells us that zero entropy corresponds to a unique microstate, $W = 1$, of a macroscopic system. If the ground state the system takes at $T = 0$ is unique, the related zero-point entropy disappears. If the ground state is degenerate, that is, if several different microstates, $W > 1$, exist at the same total energy, then the zero-point entropy is a positive constant.

In the case of energy, Einstein's famous equation provides us with a universal absolute value for the specific internal energy of any thermodynamic system, independent of its temperature or pressure

$$\frac{E}{M} = c^2 = 89\,875\,517\,873\,681\,764 \text{ J kg}^{-1} \qquad (2.37)$$

since the speed of light is exact by definition. Theoretically, if the rest mass M of the system were measured sufficiently accurately, the asymptotic value of the internal energy at $T = 0$ could be computed from it.

In the practice of thermodynamic measurements, the absolute values for energy and entropy are not relevant; they can neither be measured in thermodynamic experiments, nor does the knowledge of their exact values improve or degrade any properties computed from a thermodynamic potential. Because the Gibbs relation (2.20) is an integrable differential equation, the thermodynamic potential of a given system, Equation 2.21, can be determined in the form of a path integral

$$E(S, L_k, N_i) - E(S^{\text{ref}}, L_k^{\text{ref}}, N_i^{\text{ref}}) = \int_C \left\{ T dS + \sum_k l_k dL_k + \sum_i \mu_i dN_i \right\} \qquad (2.38)$$

along an arbitrary curve C in the phase space from an arbitrary reference point to the point of interest. The integrands, Equation 2.23, can be measured experimentally at a number of points in a certain region of the state space. Then, the integral (2.38) can be represented as a correlation valid in that region if the path C and in particular the reference point is located within that region. Hence, the zero point is not necessarily a preferred reference point for many important substances. Humid air or seawater, for instance, undergo various phase transitions including precipitation or condensation of certain constituents when they are cooled down from ambient conditions to the extremely low temperatures. For this reason, the International Association for the Properties of Water and Steam defines the reference states for pure water, seawater, or humid air in the vicinity of ambient conditions (IAPWS, 2008, 2009a, 2009c, 2010). For instance, energy and entropy of liquid water are set to zero at the triple point (Wagner and Pruß, 2002; Feistel et al., 2008b).

The four principles of thermodynamics and the Gibbs fundamental relation are general laws of nature. For the formulation of the zeroth principle, the third principle, and the Gibbs relation, we need the concept of temperature. That means that these principles are applicable to thermal systems only. The first and the second principle are much more general. Following the dissertation of *Max Planck*, the community of physicists claims that *the fundamental laws are valid for any macroscopic process in nature and society* (Ebeling, 2008; Hoffmann, 2008). Similarly, already Clausius (1865, p. 400) had conveniently formulated the first and the second law, respectively, in a compact form:

1) *Die Energie der Welt ist constant* (the energy of the world is constant)
2) *Die Entropie der Welt strebt einem Maximum zu* (the entropy of the world aspires to a maximum).

It is an intriguing idea that the laws of physics may provide universal hard bounds (similar to the light cone in the theory of relativity) for any possible process in nature and society (Greene, 2004). However, in a more conservative approach, care must be taken when the laws found for our terrestrial environment are extrapolated to extreme conditions such as those of the early universe or of black holes. In the theory of general relativity, energy conservation cannot rigorously be formulated for strong gravity fields because of the curvature of space and time (Landau and Lifschitz, 1967b, §85, §100; Davis, 2010). Entropy, on the other hand, is well-defined only for thermodynamic equilibrium states and certain states in its vicinity, such as those of local equilibrium. We possess no formula, say, for the entropy of an arbitrary set of moving mass points because entropy cannot be expressed as a function of their coordinates and momenta. What is the correct entropy of a galaxy, then? Formally considering a cluster of gravitationally interacting objects, say, as a statistical ensemble similar to a gas is problematic since the kinetic energies of those "particles" (i.e., the "temperature" of the cluster) do not obey the zeroth law or an equipartition theorem, as the example of the very different kinetic energies of the planets and moons in the solar system obviously demonstrates, regardless of millions of years of their mutual interaction (even though the objects do obey the virial theorem, Landau and Lifschitz, 1967a, §10). Despite the definition of temperature and entropy of black holes suggested by Bekenstein (1973) and Hawking (1974), the theoretical relation between entropy and gravity is in general unsettled (Neugebauer,1977, 1980, 1998; Muschik and v. Borzeszkowski, 2008). We are not aware of any formulation of the second law under conditions where the formulation of the first law is not possible.

From the first law in the form of the Gibbs relation, Equation 2.10, and the second law in the form of a maximum-entropy condition for equilibria, Equation 2.24, it can be concluded that the local intensive variables, Equation 2.23, must take the same values everywhere in a given volume. Even on the two sides of a phase boundary, or under the influence of a weak gravity field, equal temperature, pressure, and chemical potentials are required for equilibrium states (Landau and Lifshitz, 1966). In contrast, the conjugate local variables exhibit stratification; particle density and entropy density develop gradients in direction of the external gravity force. The two distinguished states of a fluid in a tank reactor, the equilibrium state and the well-stirred state, respectively, are indistinguishable without gravity but turn complementary otherwise (Feistel and Feistel, 2006). In the wind-mixed ocean, salinity and specific entropy are homogeneous while temperature, pressure, and chemical potentials exhibit spatial gradients. Because of those gradients, the usual ambient ocean state is a nonequilibrium state; heat flow and molecular diffusion are permanently producing entropy. Similar conditions prevail in the turbulent troposphere (Feistel et al., 2010b; Chapter 3).

The complete relaxation of a stratified ocean to its equilibrium state proceeds extremely slowly and is practically never observed, not even in the deep ocean where the water masses remain isolated from the atmosphere for thousands of years. At the theoretical equilibrium state under gravity, all thermodynamic properties of the ocean depend on just one parameter, the vertical coordinate, z (Landau and Lifschitz, 1974). If we measured the values of salinity, c, temperature, T, and pressure, p, at various locations in the ocean, and plot the related data points in a c–T–p space, those points would form a one-dimensional (1D) curve with the parametric representation $c(z)$, $T(z)$, $p(z)$. On the contrary, if the ocean is in an arbitrary initial state, the values of $c(\mathbf{r})$, $T(\mathbf{r})$, $p(\mathbf{r})$ at different locations $\mathbf{r}$ may scatter independently and may cover some three-dimensional (3D) region in the c–T–p space. We formulated the hypothesis that the second law tends to reduce the fractal dimension of the manifold occupied by a distributed system in the appropriate state space to a minimum value consistent with the particular boundary conditions (Feistel and Hagen, 1994; Feistel and Feistel, 2006). In fact it is observed that the properties in the deep ocean form a two-dimensional (2D) manifold, that is, between the three independent thermodynamic variables of seawater a constraint, $\Phi(c, T, p) = 0$, exists for attractor states of "aged" ocean waters. This relation represents a pre-equilibrium, quasi-steady-state that is approached by the system after a much shorter time than relaxation to complete equilibrium would take. In this respect, the real ocean as a whole is thermodynamically almost 2D, while deviations from the attractor surface, $\Phi(c(\mathbf{r}), T(\mathbf{r}), p(\mathbf{r})) \neq 0$, occur at locations $\mathbf{r}$ where the ocean gets in contact with the atmosphere and "young" water is formed (McDougall and Jackett, 2007). In a thermodynamic 2D ocean, the Jacobian of the spatial property distributions disappears identically,

$$(\nabla T \times \nabla p)\nabla c = 0 \tag{2.39}$$

as a result of the mutual dependence of the three functions, $c(\mathbf{r})$, $T(\mathbf{r})$, $p(\mathbf{r})$, on the attractor's 2D surface, $\Phi(c, T, p) = 0$. Due to Carathéodory's theorem, Equation 2.39 is a necessary and sufficient condition of vanishing *helicity* of the ocean and for the existence of *neutral surfaces* which play a key role for ocean mixing (Margenau and Murphy, 1943; Buchdahl, 1949; McDougall, 1987; McDougall and Jackett, 1988; Feistel and Hagen, 1994; Pogliani and Berberan-Santos, 2000; IOC et al., 2010).

There are more examples for the interesting interplay between gravity and thermodynamics. Gravitationally contracting spherical mass accumulations, such as stars or planets, embedded in an infinite empty space are in a thermodynamic nonequilibrium state (Landau and Lifshitz, 1966). From the outer atmosphere, random thermal motion results in a continual loss by those particles which move faster than the planet's escape velocity (Lammer and Bauer, 2003). Gravitational escape expands the atmosphere and cools down the celestial object, similar to the irreversible evaporation of a liquid phase into a subsaturated vapor phase. We will return to this phenomenon in Chapter 3. On the other hand, even an ideal gas in thermodynamic equilibrium may collapse as a result of density fluctuations and the Jeans instability as soon as gravity interaction between the particles is involved and the total mass is beyond a critical value (Jeans, 1902; Iqbal, Khan, and Masood, 2011). This behavior is inconsistent with the classical Sackur–Tetrode entropy formula of an ideal gas where density fluctuations always result in a lowering of entropy.

In spite of the central position of energy, entropy, and work in physics, there exist many different definitions and interpretations (Ebeling, 1992; Hoffmann, 2008). Below we will be concerned with several of those interpretations. We introduce mechanical energy, heat and work. Furthermore, we talk about the *Clausius entropy*, the *Boltzmann entropy*, the *Gibbs entropy*, the *Shannon entropy*, and the *Kolmogorov entropy*. When extending the physical categories of energy and entropy to other sciences, inconsistency or confusion may easily arise. In order to avoid any misinterpretation, one has to be very careful when talking about these generalized categories.

There should be no doubt, however, that energy and entropy are central quantities for the description of nature. But, due to their fundamentality a specific problem of philosophical character arises: it is extremely difficult or even impossible to avoid tautologies in their definition. In conclusion we may say that energy and entropy should be elements of an axiomatic of science.

As we mentioned above, the difficulty in defining fundamental quantities was already discussed by Poincaré with respect to energy. In his lectures on thermodynamics, Poincaré said in around 1893:

In every special instance it is clear what energy is and we can give at least a provisional definition of it; it is impossible however, to give a general definition of it. If one wants to express the (first) law in full generality, ..., one sees it dissolve before one's eyes, so to speak leaving only the words: There is something, that remains constant (in isolated systems).

We may transfer this statement to the definition of entropy in the following way:

In every special instance it is clear what entropy is and we can give at least a provisional definition of it; it is impossible however, to give a general definition of it. If one wants to express the second law in full generality, ..., one sees it dissolve before one's eyes, so to speak leaving only the words: There is something, that is non-decreasing in isolated systems.

In this way our suggested definition of entropy is with necessary precision:

Characterizing a real thermodynamic system, entropy is a fundamental and universal quantity that is non-decreasing over time in isolated systems.

2.3
Thermodynamic Potentials, Inequalities, and Variational Principles

As stated already by *Planck*, the most characteristic property of irreversible processes is the existence of so-called *Lyapunov functions*. This type of function was defined first by the Russian mathematician *Lyapunov* more than a century ago. A *Lyapunov function* is a nonnegative function with the following properties:

$$L(t) \geq 0 \quad \frac{dL(t)}{dt} \leq 0 \tag{2.40}$$

As a consequence of the two relations, Lyapunov functions are by definition never increasing in time. Our problem is now to find a Lyapunov function for an arbitrary macroscopic system.

Let us assume that the system initially ($t=0$) is in a nonequilibrium state, and that for $t>0$ we are able to isolate the given system from the surrounding. From the definition of thermodynamic equilibrium it follows that, after isolation, certain changes will occur. Within the natural uncertainties, the energy E will remain fixed but the entropy will monotonously increase due to the second law. Irreversible processes connected with a positive entropy production, $P > 0$, will drive the system finally to an equilibrium state located on the same energy surface. In thermodynamic equilibrium, the entropy S takes the maximal value,

$$\lim_{t \to \infty} S(t) = S_{eq}(E, \mathbf{X})$$

which is a function of the energy and certain other extensive variables, $\mathbf{X}$. The total production of entropy during the process of equilibration of the isolated system may be obtained by integration of the entropy production over time

$$\delta S(t) = \int_0^t dt\, P(t) \qquad (2.41)$$

Due to the condition of isolation, there is no exchange of entropy during the whole process and due to the nonnegativity of the entropy production we have

$$\frac{d_i S}{dt} = P(t) \geq 0 \qquad (2.42)$$

The total production of entropy, $\delta S(t)$, is a monotonous, nondecreasing function of time. The concrete value of $\delta S(t)$ depends on the actual path Γ from the initial to the final state and on the rate of the transition processes. However, the maximal value of this quantity, $\delta S(\infty)$, should observe some special conditions. In particular for the case that the transition occurs without any entropy exchange, this quantity should be identical with the total entropy difference between the initial state and the equilibrium state at time, $t \to \infty$

$$\delta S(t) = S_{eq}(E, \mathbf{X}) - S(E, t) \qquad (2.43)$$

This quantity is connected with the so-called lowering of entropy with respect to the initial value

$$\delta S = S_{eq}(E, \mathbf{X}) - S(E, t=0) \qquad (2.44)$$

which is simply the difference between the two entropy values. By changing parameter values infinitely slowly along some path γ, we may find a reversible transition and calculate the entropy change in a standard way, for example, by using the Gibbs fundamental relation (2.20). An important property of the quantity δS is that it is independent of the path γ from the initial to the equilibrium state. On the other hand, the entropy change along an irreversible path may depend on details of the microscopic trajectory. On average over many realizations (measurements), the relation

$$\delta S(\infty) = \int_0^\infty dt\, P(t) = \delta S \qquad (2.45)$$

should hold. This equality follows from the fact that S is a state function; its value should be independent of the path along which the state has been reached. Assuming for a moment that Equation 2.44 is violated, we were able to construct a cyclic process that contradicts the second law. The macroscopic quantity

$$\Delta S = \langle \delta S(\infty) \rangle \tag{2.46}$$

may be estimated by averaging the entropy production over many realizations of the irreversible relaxation from the initial state to equilibrium under the condition of strict isolation from the outside world. The result

$$\Delta S = \delta S \tag{2.47}$$

is surprising: it says that we can extract equilibrium information, δS, from nonequilibrium (finite-time) measurements of the entropy production. We shall come back to this point in the next section. Here let us proceed on the way of deriving Lyapunov functions. By using the relations given above we find for macroscopic systems the following Lyapunov function:

$$L(t) = \Delta S - \Delta S(t) \tag{2.48}$$

which for isolated systems yields (Klimontovich, 1982),

$$L(t) = S_{eq} - S(t)$$

Due to the inequalities

$$\frac{dL(t)}{dt} = -P(t) \leq 0$$

and

$$\Delta S \geq \Delta S(t) \tag{2.49}$$

we conclude that the function $L(t)$ has indeed the necessary properties (2.40) of a Lyapunov function.

So far we considered systems specified basically in terms of their entropy and volume. Let us now consider a system which is in contact with a heat bath of a given temperature T. Following Helmholtz, we define the thermodynamic function

$$F = E - TS \tag{2.50}$$

which is termed *free energy* in the German nomenclature (which we prefer here) and *Helmholtz energy* in the English literature. According to Gibbs' fundamental relation, Equation 2.20, the differential of the free energy is given by

$$dF = dE - TdS - SdT = \sum_k l_k dL_k + \sum_i \mu_i dN_i - SdT \tag{2.51}$$

We see that the proper variables for the free energy are the temperature, T, and the extensive variables L_k (note that $L_1 \equiv V$), and N_i where $i = 1, \ldots, s$. In other words, we have found the dependence of the free energy on the independent variables,

$$F = F(T, L_k, N_i)$$

The total differential (2.51) may also be read as a balance relation for the free energy change for a quasistatic transition between two neighboring states.

In the form of the specific free energy, the *Helmholtz function* $f(T, \varrho) = F(T, V, N)/M$ is a preferred thermodynamic potential for the quantitative description of fluids such as water or air (Lemmon et al., 2000; Wagner and Pruß, 2002; Feistel et al., 2010b, 2010a; Lemmon and Span, 2010; Feistel, 2011c). The definition (2.50) is a Legendre transform that establishes F as a potential function if expressed in the independent variables given in Equation 2.51. The main reasons for using Helmholtz functions are that F is strictly additive for mixtures of ideal gases, that F can be expressed statistically by means of the canonical distribution for the computation of virial coefficients, and that F is a unique, single-valued function for fluids on both sides of the liquid–vapor phase transition curve and in the entire critical region, Figure 2.1. The practical disadvantage is that at given temperature and pressure, density as the input variable must first be evaluated by numerically solving an implicit equation (Wagner and Kretzschmar, 2008; Feistel et al., 2010a).

Let us now consider the differential of free energy under more general conditions, admitting also dissipative processes without particle exchange. Then we find

$$dF = dE - SdT - TdS = d'A - d'Q - SdT - Td_eS - Td_iS$$

or, making use of $d'Q = Td_eS$

$$dF = d'A - SdT - Td_iS \qquad (2.52)$$

In situations where the temperature is fixed, $dT = 0$, and the exchange of work is excluded, $d'A = 0$, we get the inequality

$$dF = -Td_iS \leq 0, \quad \frac{dF}{dt} = -TP \leq 0 \qquad (2.53)$$

As a consequence of Equation 2.52, free energy F is a nonincreasing function for systems contained in a heat bath which excludes exchange of work, in particular at constant volume, V. Under these conditions free energy takes its minimum at the thermal equilibrium. Consequently, the Lyapunov function of the system is given by

$$L(t) = F(t) - F_{eq}$$

which possesses the necessary Lyapunov properties given in Equation 2.40.

Another important situation is an environment with given temperature, T, and pressure, p. Then, the potential function is the *free enthalpy* (Gibbs energy)

$$G = E + pV - TS \qquad (2.54)$$

with the total differential

$$dG = dE + p\,dV + V\,dp - S\,dT - T\,dS$$

and the balance relation

$$\begin{aligned}dG &= d'A + d'Q + V\,dp + p\,dV - S\,dT - T\,d_e S - T\,d_i S \\ &= V\,dp - S\,dT - T\,d_i S\end{aligned}$$

At given temperature and pressure we get

$$dG = -T\,d_i S \le 0, \quad \frac{dG}{dt} = -TP \le 0 \tag{2.55}$$

Consequently, free enthalpy G is a nonincreasing function for systems imbedded in an isobar and isothermal reservoir. If several equilibrium states are possible for the same values of T and p, such as in the vicinity of a phase transition, the state with the lower value of G is stable and any other states are metastable or unstable, such as a subcooled vapor or a superheated liquid. At thermal equilibrium, the minimum value of G_{eq} is taken. Therefore, the difference

$$L(t) = G(t) - G_{eq} \tag{2.56}$$

is a Lyapunov function possessing the necessary properties given in Equation 2.40.

In the form of the specific free enthalpy, the *Gibbs function* $g(T, p) = G(T, p, N)/M$ is a preferred thermodynamic potential for quantitative description of substances in regions away from critical points (Feistel, 1993, 2003, 2008a, 2010; Tillner-Roth, 1998; Feistel and Wagner, 2006; Feistel et al., 2010a). The definition (2.54) is a Legendre transform that establishes G as a potential function if expressed in temperature and pressure as the independent variables, and in additional composition variables if mixtures are considered. These input variables are often measured directly and routinely, such as air temperature, air pressure, and humidity in meteorology, or sea temperature, sea pressure, and salinity in oceanography. In the vicinity of phase transitions, however, Gibbs functions are multivalued which poses a serious difficulty for numerical implementations (Kittel, 1969; Stanley, 1971; Wagner and Kretzschmar, 2008; Feistel et al., 2010a).

Let us now consider the most general situation in which our system is neither isolated nor in a reservoir with fixed conditions during its course to equilibrium. The definition of entropy production is in general a nontrivial problem. In the special case, however, that the only irreversible process which is participating is production of heat by destruction of mechanical work, the definition of $P(t)$ is quite easy. Since $P(t)$ is then given by the heat production divided by the temperature, the calculation of $L(t)$ requires only the knowledge of the total mechanical energy that is dissipated.

Energy, entropy, and work are the central categories of thermodynamics and statistical physics (Zurek, 1990). The fundamental character of these phenomenological quantities requires our full attention. The entropy concept closes the gap between phenomenological theory and statistical physics. Energy and entropy are not independent; rather, they are connected in a sophisticated, deep way. Our point of view is based on a valoric interpretation of entropy (Schöpf, 1984; Ebeling and Volkenstein, 1990; Ebeling, 1993). This very clear interpretation will be taken as the basis for a reinterpretation of the various entropy concepts developed by Clausius, Boltzmann, Gibbs, Shannon, and Kolmogorov. As a key point, we consider the value of energy with respect to work. The discussion on this relation started already in the nineteenth century and is still continuing today. The valoric interpretation was given first by Clausius (Schöpf, 1984) and was worked out by Helmholtz and Ostwald. But then, due to a strong opposition from the side of Kirchhoff, Hertz, Planck, and others, the interpretation was nearly forgotten except for a few authors (Schöpf, 1984; Ebeling and Volkenstein, 1990; Ebeling, 1992). As a matter of fact, however, for Clausius himself the valoric interpretation of entropy was the key to the introduction of this new concept in 1864–1865. Many physicists do not know that the entropy concept commonly taught as the Clausius concept at universities is much closer to the reinterpretation given by Kirchhoff than to the original of Clausius (Schöpf, 1984). Here, we keep close to the original interpretation in terms of a value concept in combination with some more recent developments (Ebeling and Muschik, 1993).

According to Klimontovich, we now consider the so-called *entropy lowering* defined by

$$\delta S = S_{eq}(E, \mathbf{X}) - S(E, t = 0) = \int_0^t dt' \, P(t') \tag{2.57}$$

This quantity, the entropy difference between a given state and the equilibrium state at the same energy, was introduced by Klimontovich as a measure of organization contained in a nonequilibrium system (Klimontovich, 1982, 1991). Several examples were given such as the entropy lowering of oscillator systems, of turbulent flows in a tube, and of a nonequilibrium phonon gas in a crystal generated by a piezoelectric device (Ebeling and Klimontovich, 1984; Ebeling, Engel, and Feistel, 1990a). The basic idea of this concept is that a comparison of entropies makes sense only if states of the same energy are compared.

Assuming that the initial state 1 is a nonequilibrium state, we know from the definition of equilibrium that, after isolation, changes will necessarily occur. Under the condition of isolation, the energy E will remain fixed within natural fluctuations, but the entropy will monotonously increase due to the second law. Irreversible processes connected with positive entropy production, $P > 0$, will eventually drive the system to its equilibrium state located on the same energy surface in the phase space. In thermodynamic equilibrium, entropy takes the maximum equilibrium value which is a function of the energy and certain other extensive variables. The entropy change is a Lyapunov function and may be computed by integration of the entropy production over time. Equation 2.57 is a way to obtain the entropy lowering and the entropy of nonequilibrium states. We may consider the entropy lowering as a measure of order contained in the body in comparison with maximal disorder in equilibrium. Equation 2.57 suggests also the interpretation as a measure of distance from equilibrium.

So far the thermodynamic meaning of entropy was discussed, but entropy is like the faces of Janus; it permits alternative interpretations. With respect to statistical physics, the most important of those is the interpretation of entropy as a measure of uncertainty or disorder. This is based on the concepts developed in the pioneering works of Boltzmann, Planck, and Gibbs. These authors showed that in statistical mechanics, the entropy of a macrostate is defined as the logarithm of the thermodynamic probability, W

$$S = k_B \log W \tag{2.58}$$

As an interpretation of this concept, irreversible processes at constant energy are connected with an expansion of the occupied part of the energy shell in the phase space. This way, the relaxation on the energy shell may be interpreted as a monotonous increase in the effectively occupied phase volume. This process implies a devaluation of the energy (Ebeling and Volkenstein, 1990).

Let us come back now to the relation between free energy and work. The energetic basis of any human activity is *work*, a term which is also difficult to define. The first law of thermodynamics expresses the conservation of the energy of systems. Energy may take various forms such as heat or work that appear in processes of energy transfer between systems in contact. They may be of different value with respect to their ability to perform work. The (work) value of a specific form of energy is measured by the entropy of the system. As shown first by Helmholtz, free energy represents the amount of energy in a body at fixed volume and temperature which is available for performing work. Before going to explain this in more detail, we return for a moment to a system with fixed energy and with fixed other external extensive parameters. Then the capacity to do work (the work value) takes its minimum of zero at the thermodynamic equilibrium where entropy has its maximum value. Relying on this property, above we developed a special entropy concept based on the term "value." In the framework of this concept, we consider the entropy lowering as a measure of the "value" of the energy contained in the system. In terms of the phase space volume of statistical mechanics, this measure has the following meaning: it gives the relative part of the phase space in the energy shell which is occupied by the system. The second law of thermodynamics tells us that entropy can be produced in irreversible processes but never be destroyed. Since entropy is a measure of value of the energy, this leads to the formulation that the distance from equilibrium and the work value of energy in isolated systems cannot increase spontaneously. In other words, the Lyapunov functions introduced above express the tendency of devaluation of energy. In order to increase the value of energy in a system, one has to export entropy. In this way, we have shown that the meaning of the thermodynamic concept of entropy may well be expressed in terms of the distance from equilibrium, of the value of energy, or of the relative phase–space occupation, rather than of the most frequently used interpretation of entropy as a measure of disorder.

Now we come back to free energy. This term was introduced into thermodynamics by Helmholtz and may be interpreted in the way that the total energy consists of a free part which is available for work and a bound part which is not available for work. A related concept is that of *exergy* which is of much interest for technical or geophysical applications (Tailleux, 2009) but is not further considered here. Due to the relation

$$E = F + TS = E_f + E_b \tag{2.59}$$

the energy in a body consists of two parts

$$E_f = F \quad E_b = TS \tag{2.60}$$

Correspondingly, the first contribution to the energy may be interpreted as that part of energy, the "free" energy, which is available for work. The product of entropy and temperature, TS, may be interpreted as the bound part of the energy, which is not available for work. Similarly, in the definition of *enthalpy*, H, by

$$H = G + TS = H_f + H_b \tag{2.61}$$

the product TS also represents the bound part of enthalpy (G being the "free" enthalpy). It follows from the second law, as shown in the previous section, that under isochoric–isothermal conditions free energy, F, is a nonincreasing function of time

$$\frac{dF}{dt} \leq 0$$

and that under isobaric–isothermal conditions free enthalpy, G, is nonincreasing

$$\frac{dG}{dt} \leq 0$$

The tendency of F and G to irreversibly decrease at given temperature is in fact determined by the general tendency expressed by the second law to devaluate energy (or enthalpy) with respect to their ability to do work. The work W performed on a system during a finite-time transition from an initial nonequilibrium state to a final equilibrium state

$$W = \delta F = F_{ne} - F_{eq} \tag{2.62}$$

will on average exceed the free energy difference (Jarzynski, 1996):

$$\langle W \rangle \geq \delta F = F_{ne} - F_{eq} \tag{2.63}$$

Here the averaging is to be carried out over an ensemble of transitions (measurements). The difference

$$W_{diss} = \langle W \rangle - W \geq 0 \tag{2.64}$$

is just the dissipated work, W_{diss}, associated with the increase in entropy during the irreversible transition. In one article, Jarzynski (1996) discussed the above relations between free energy and work from a novel perspective. The new relations derived by Jarzynski (1996) as well as other related developments will be discussed in more detail in the next sections.

To conclude this section, let us come back to the splitting of the energy into two parts proposed by Helmholtz. In this context, the idea was developed that entropy may also be split in a way similar to energy into a bound part and into a free part (Ebeling, 1993)

$$S = S_b + S_f \tag{2.65}$$

The bound part reflects the entropy bound in the microscopic motion, the free part is the part of entropy available for information processing. The splitting of entropy into two parts reminds of the decomposition

$$S = Y + Z$$

given originally by Clausius in 1864 in his book *Mechanische Wärmetheorie I*. There, Clausius defines the part Y as the exchange value of heat and Z as the degradation.

However, since Clausius' argumentation is notoriously difficult, we do not exactly know what the great pioneer had in mind with his decomposition. Schöpf's deep analysis makes plausible that Clausius considered the valoric interpretation as the central point of his entropy concept (Schöpf, 1978, 1984). The relation of entropy and information, that is, the splitting of entropy into different contributions, was discussed already in the papers of Szilard and Brillouin and later by many other workers (Weber, Depew, and Smith, 1988). A relation similar to the one we have described for entropy itself should also hold for the entropy transfer. There exist many forms of entropy transfer, such as heat conduction, and entropy transferred by matter; information transfer appears to be a special form of entropy transfer (Ebeling, 1993). The important problem of entropy transport and entropy production by thermal radiation is separately considered in Section 2.5 and in Chapter 3. In Chapter 8, we shall return to the problem of the relation between entropy and information.

2.4
Irreversible Processes and Self-Organization

So far we have written all balances for extensive quantities such as E, S, F, G, and so on, without referring to time. All the equations we have written down were mostly concerned with differences between thermodynamic equilibrium states, corresponding to an infinitesimal time step of a process, or to the comparison between equivalent but infinitesimally different members of a statistical ensemble. Equilibrium thermodynamics does not include time as a parameter.

Here we consider the time evolution and irreversible processes more explicitly, making use of the formalism of balance equations. For the three most important extensive quantities energy, entropy and substance, we write

$$\frac{dE}{dt} = \frac{d_e E}{dt} = \Phi_E \tag{2.66}$$

$$\frac{dS}{dt} = \frac{d_e S}{dt} + \frac{d_i S}{dt} = \Phi_S + P \tag{2.67}$$

$$\frac{dN_k}{dt} = \frac{d_e N_k}{dt} + \frac{d_i N_k}{dt} = \Phi_k + Q_k \tag{2.68}$$

Here, Φ_E, Φ_S, and Φ_k denote the flow rates of energy, entropy, and matter into (positive values) or out of (negative values) the system, and P and Q_k denote the production rate for entropy and conversion rates between the substances, respectively. Recall that the production of energy is impossible due to the first law, and that the production of entropy is always nonnegative, $P \geq 0$, as required by the second law. Self-organization is related to states of matter out of equilibrium with reduced entropy, that is, more ordered states of higher valued energy than the corresponding equilibrium state of the given system at the same total energy. In this regard, we introduced in the previous section the term *entropy lowering*. In order to generate states that at given energy possess an entropy lower than the entropy of the equilibrium which is maximal in comparison to all states of the system at the same energy, that is,

$$S(E, \mathbf{X}) < S_{\text{eq}}(E, \mathbf{X}) \tag{2.69}$$

we need conditions which lower the entropy. In other words, we need processes which satisfy the condition

$$\frac{dS}{dt} = \frac{d_e S}{dt} + \frac{d_i S}{dt} < 0 \tag{2.70}$$

This is the fundamental thermodynamic condition for self-organization.

To say this in a different way, for self-organization we need processes that realize the inequality

$$-\frac{d_e S}{dt} > \frac{d_i S}{dt} > 0 \tag{2.71}$$

Equivalent to this condition is the writing

$$\Phi_S < 0, \quad (-\Phi_S) > P > 0 \tag{2.72}$$

To express this relation in words, what we need is *entropy export*. For self-organization, we have to find conditions where the system exports more entropy than it generates in the interior of the system.

Let us look in more detail now at the third balance equation, Equation 2.68. This third equation is related to the transformation of matter and is also restricted by certain laws. The total mass cannot be generated or destroyed due to the conservation of mass; however, the different kinds of substances may be converted into one another. This is expressed in part by the chemical laws of stoichiometry, for example, for the reaction

$$2H_2 + O_2 \rightarrow 2H_2O$$

we have the stoichiometric relation

$$-2Q(H_2) = -Q(O_2) = 2Q(H_2O)$$

where Q are the rate functions, Equation 2.68.

2.4 Irreversible Processes and Self-Organization

Let us assume that in the general case R chemical reactions may proceed which we enumerate with $r = 1, 2, \ldots, R$, and s chemical species, $A_1, \ldots, A_s$ are participating. Then, the reaction r may be formally represented in the form

$$\nu_{1r} A_1 + \nu_{2r} A_2 + \cdots \to \nu_{mr} A_m + \cdots \nu_{sr} A_s \quad r = 1, 2, \ldots, R.$$

where the coefficients, ν_{kr}, termed the *stoichiometric coefficients*, are either positive or negative integers, or zero. We may write the reaction symbolically as

$$\sum_{k=1}^{s} \nu_{kr} A_k = 0 \tag{2.73}$$

Here, the produced species carry a positive and the consumed species a negative sign of the related stoichiometric coefficient. Using this convention we may express the conversion of matter by chemical or by nuclear reactions in the following form:

$$Q_k = \frac{d_i n_k}{dt} = \sum_{r=1}^{R} \nu_{kr} \frac{d\xi_r}{dt} \tag{2.74}$$

Here, ξ_r is the so-called *reaction rate* of the reaction r which is defined in such a way that $\xi_r = 1$ means that ν_{kr} moles of species k are produced in the reaction r, $\xi_r = 2$ means that $2\nu_{kr}$ moles are produced, and so on. An axiomatic formalism for the description of chemical reactions was developed by Aris (1965, 1968).

Let us proceed now to describe irreversible processes in uniform systems, which is of particular simplicity. Introducing the entropy production per volume

$$\sigma = \frac{P}{V} = \frac{1}{V}\frac{d_i S}{dt} = \frac{1}{VT}\frac{d' A_{\text{diss}}}{dt} - \frac{1}{T}\sum_{k=1}^{s} \mu_k \frac{1}{V}\frac{d_i n_k}{dt} \tag{2.75}$$

where $d'A_{\text{diss}}$ is dissipated work, and the following definitions of the affinity, A_r, and the speed, w_r, of the reaction r

$$A_r = -\sum_{k=1}^{s} \nu_{kr} \mu_k, \quad w_r = \frac{1}{V}\frac{d\xi_r}{dt} \tag{2.76}$$

we get for the entropy production density

$$\sigma = \frac{1}{VT}\frac{d'A_{\text{diss}}}{dt} + \sum_{r=1}^{R} \frac{A_r}{T} w_r \tag{2.77}$$

In order to rewrite the first term, we now assume that the particles of species k are moving with the current density, J_k, under the influence of an external force, F_k. For the work which is dissipated in this process we assume Ohm's law

$$d'A_{\text{diss}} = V \sum_{k=1}^{s} J_k F_k \, dt \tag{2.78}$$

Using these relations, we get for the entropy production density the bilinear expression

$$\sigma = \sum_{k=1}^{s} \frac{\mathbf{F}_k}{T} \mathbf{J}_k + \sum_{r=1}^{R} \frac{A_r}{T} w_r \qquad (2.79)$$

Defining thermodynamic forces by $X = F/T$ or $X = A/T$, we may bring this expression to the following simple form:

$$\sigma = \sum_{\alpha=1}^{s+R} X_\alpha J_\alpha \qquad (2.80)$$

According to Onsager, we may assume that close to equilibrium the flows are linear functions of the forces:

$$J_\alpha = \sum_{\beta=1}^{s+R} L_{\alpha\beta} X_\beta \qquad (2.81)$$

This so-called linear Onsager relation leads to a quadratic form for the entropy production density in linear irreversible thermodynamics

$$\sigma = \sum_{\alpha=1}^{s+R} \sum_{\beta=1}^{s+R} L_{\alpha\beta} X_\alpha X_\beta \qquad (2.82)$$

Since the entropy production must be nonnegative, we infer for the quadratic form the inequality

$$\sum_{\alpha=1}^{s+R} \sum_{\beta=1}^{s+R} L_{\alpha\beta} X_\alpha X_\beta \geq 0 \qquad (2.83)$$

that is, the Onsager matrix **L** is always positive definite for consistency with the second law. We see that the entropy production does not depend on possibly existing antisymmetric parts of the matrix **L**; those are irrelevant for the basic thermodynamic quantity entropy production. In 1931, Onsager showed in a fundamental work for which he received the Nobel Prize in 1968, that the symmetry of the coupling matrix can be derived from the time symmetry of the microscopic dynamical equations (see Onsager's collected works edited by Hemmer, Holden, and Kjelstrup Ratkje, 1996). For anisotropic systems, the situation is more complicated, and we refer to standard textbooks (De Groot and Mazur, 1962). We note that the coupling matrix is symmetric,

$$L_{\beta\alpha} = L_{\alpha\beta} \qquad (2.84)$$

as long as magnetic fields or Coriolis forces are not involved.

Additional conditions follow from the general symmetry of the underlying system. As was first pointed out by Pierre Curie, the symmetry of a cause (the given forces) cannot be higher than that of the resulting effects. In practice, we have to do with scalars (reaction rates and affinities), with vectors (forces and flows), and possibly also with tensors (pressure and tension). The Curie principle forbids that in an isotropic medium, a scalar "force" can give rise to a vectorial flow or vice versa. The Curie principle was applied by Ilya Prigogine to the problems of irreversible thermodynamics (Prigogine, 1947). This leads to the Curie–Prigogine principle:

The Onsager coefficients for couplings between fluxes and forces of different tensorial rank are zero. In isotropic systems, these quantities are uncoupled.

For the linear range of irreversible processes, a very interesting evolution theorem was derived by Prigogine (1947). Let us consider a stationary state of a linear system with stationary values of fluxes, $J_\alpha^{(0)}$, forces, $X_\alpha^{(0)}$, and entropy production, $\sigma^{(0)}$. Looking at deviations from the stationary value we find

$$\delta\sigma = \sigma - \sigma^{(0)} = \sum_{\alpha,\beta=1}^{s+R} L_{\alpha\beta}\delta X_\alpha \delta X_\beta \tag{2.85}$$

From the positive definiteness of the matrix **L**, it then follows that

$$\delta\sigma \geq 0 \tag{2.86}$$

where the equal sign holds if and only if $\delta X_\alpha = 0$. From this inequality, we infer that the entropy production at the stationary state is minimum in comparison to the states in the vicinity.

This derivation leads to the *Theorem of Prigogine* (1947) for linear irreversible processes:

$$\frac{d\sigma}{dt} \leq 0 \tag{2.87}$$

The entropy production of linear irreversible processes is minimum at stationary states. The evolution of thermodynamic fluxes and forces toward the stationary state is associated with a decrease of the entropy production.

In its linear range of validity, the Prigogine principle is consistent with the second law. There are numerous experimental verifications, and there cannot be any doubt about the validity of this principle in the linear regime. There are several efforts to generalize the Prigogine theorem to nonlinear irreversible processes, in particular we mention the theorem by Glansdorff and Prigogine which was published in several variants since 1954 (Glansdorff and Prigogine, 1971). The *Glansdorff–Prigogine evolution criterion* has the following form:

$$\frac{d_X \sigma}{dt} \leq 0 \qquad d_X \sigma = \sum_\alpha J_\alpha \, d X_\alpha \tag{2.88}$$

Since only the forces are varied here, in contrast to the Prigogine theorem, mathematically Equation 2.88 is a Pfaffian form rather than an exact differential, and the properties of its integrals are subject to Carathéodory's theorem. As far as we are aware, all the generalizations of Prigogine's theorem suffer from several limitations and are restricted in their applicability. Several authors started extended and rather polemic discussions with respect to the Glansdorff–Prigogine criterion and tried to find counterexamples. In this regard, the critics partially succeeded; it is nowadays generally accepted that the Glansdorff–Prigogine criterion is of limited validity. Our point of view is that one should consider the problem as objectively as possible and without emotional polemics. Evidently the intense and enduring search for evolution principles that are generally valid for any nonlinear processes was in vain so far. The examples investigated in this book also come to the conclusion that only in exceptional cases, a "goal" of nonlinear and evolution processes can be found. Perhaps one cannot reasonably expect a universal optimization principle to exist which rules any kind of process in the world. All we might expect is that there are principles having a limited range of validity, valid under specific circumstances. This way we have to accept that the Glansdorff–Prigogine principle has a restricted application range. Although that range is not precisely known yet, it certainly includes at least the field of nonlinear chemical processes which satisfy the Gibbs relation for the chemical potentials.

We completely refrain here from discussing the complicated field of irreversible thermodynamics in anisotropic systems; in this regard, we refer the reader to several excellent textbooks and surveys such as those of De Groot and Mazur (1962), Luzzi, Vasconcellos, and Ramos (2000) or Muschik (2008, 2009).

At the end of this section, we want to discuss some tendencies or even empirical rules for the entropy production at very large distances from equilibrium. Following the work of Linde, who is one of the pioneers in the investigation of Marangoni instabilities in thin liquid layers (Linde and Schwarz, 1963; Linde and Winkler, 1964; Linde, 1978; Linde et al., 2001, 2005; Linde and Eckert, 2010; Nepomnyashchy, Velarde, and Colinet, 2002; Bestehorn, 2006), we consider structures at interfaces due to this type of instabilities, which we discussed already in Chapter 1.

Similar to the thermal convection instability of Benard cells, Marangoni instabilities occur in thin layers under the influence of surface tension and concentration gradients. Unwanted structures of this kind are often observed when fresh paint is drying quickly on a large smooth surface, such as on a car or on a table plate. Figure 2.3 shows roll cells of first, second, and third order, corresponding to an increasing driving force (Linde and Eckert, 2010). Here, as discussed already in Chapter 1, the driving force is the Marangoni stress, $(d\sigma/dc) \times (dc/dx)$, where σ is the interfacial tension, c is the concentration of the diffusing substance, and x is the coordinate perpendicular to the interface. The figure demonstrates that with increasing Marangoni stress, structures with different symmetries appear. The bifurcations between different symmetries occur at critical Rayleigh numbers, Ra. In some sense, this picture reminds of our general schematic presented at the beginning of this book, namely that evolution is a sequence of self-organization steps. Figure 2.4 shows an estimate of the entropy production as a function of the Rayleigh number up to very large values, covering three successive bifurcations. A surprising result of Linde is the following: at the first bifurcation the entropy production is larger than an extrapolation of the linear branch, but at subsequent bifurcations no uniform tendency is visible. Sometimes the entropy production curve is above the analytical

RC I   RC II   RC III

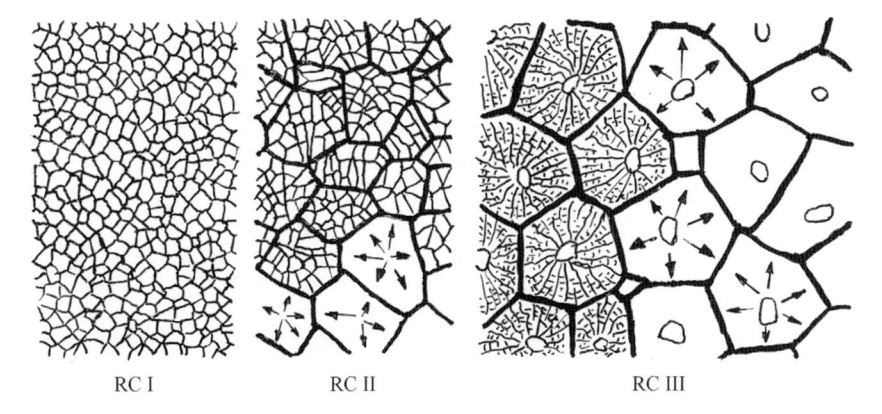

Figure 2.3 Roll cells created by Marangoni instabilities of first, second, and third order which correspond to increasing driving forces (here, Marangoni stress $(d\sigma/dc) \times (dc/dx)$, where σ is the interfacial tension, c concentration of the diffusing matter, and x is the coordinate perpendicular to the interface). The bifurcations between the structures with different symmetries occur at critical Rayleigh numbers Ra. (courtesy of Hartmut Linde).

continuation of the previous branch, and sometimes it is below. This finding supports the statement made above that beyond the first bifurcation there is no clear rule anymore for the increase or decrease in entropy production as the result of the particular new self-organization process. The Prigogine theorem of minimum entropy production for linear processes is no longer applicable then. Some specialists (in particular meteorologists, see e.g., Ozawa et al., 2003) suggest that even at very large distances from equilibrium, there exists still a universal tendency to maximize the entropy production. In contrast, we are not aware of any general, rigorous thermodynamic proof of such a hypothesis that is obviously inconsistent with experimental findings such as those shown in Figure 2.4.

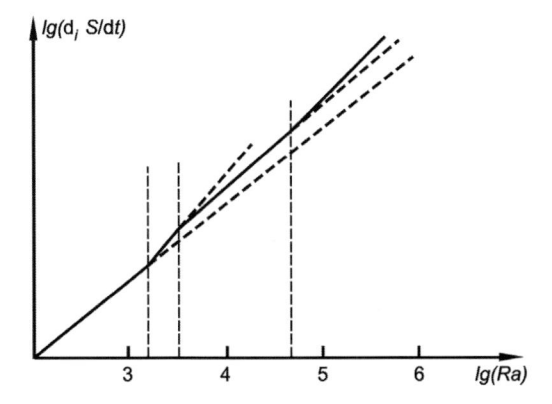

Figure 2.4 The behavior of the entropy production at subsequent Rayleigh instabilities according to Linde (1978) (courtesy of Hartmut Linde).

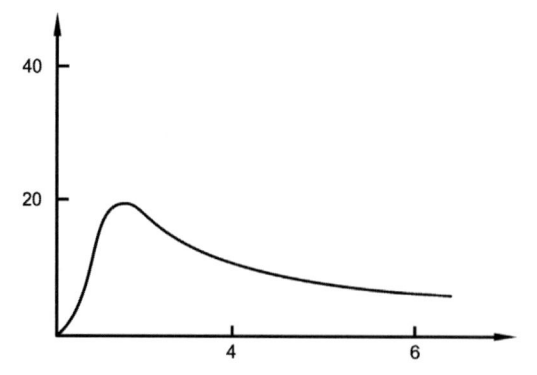

Figure 2.5 Entropy production of a fertilized amphibian egg. On the vertical axis, the entropy production is shown in calories per gram of the egg and per hour, and on the horizontal axis the number of days after the time of fertilization. (adapted from Lurie and Wagensberg, 1979).

Another interesting example for the temporal behavior of the entropy production of a living organism is shown in Figure 2.5. Adapted from Lurie and Wagensberg (1979), here the entropy production of a fertilized amphibian egg is displayed.

The remarkable result based on measurements of the heat production of the egg shows that the entropy production first increases over time and decreases later on. A plausible explanation could be that the organism does not care about entropy production in the first steps of individual development; it only cares about the formation of necessary structures and functions, and only in the last step it tries to bring the entropy production down. We believe that similar phenomena may be observed during the technological development of a new motor for a car or other machines. In the beginning, the engineers do not worry about losses of material and gasoline, and only after reaching the goal of successful functioning they try to optimize efficiency and to minimize entropy production.

2.5
Irreversible Radiation Transport

The radiation processes in the universe are of fundamental importance for evolution. Without the radiation from the Sun to Earth, no life on Earth was possible in the form we are familiar with. For this reason, we consider the irreversible radiation transfer here in a separate section.

First this problem was studied by Max Planck in 1906 in his book *Theorie der Wärmestrahlung*. However even now, more than a century after Max Planck's book, alternative expressions for the entropy flux related to thermal black-body radiation under nonequilibrium conditions are still controversially discussed in the scientific literature. In this section, we consider the radiative heat exchange between two planar black bodies of different temperatures. It will be shown (Feistel, 2011a) that the mathematical formula for the related nonequilibrium entropy flux can uniquely be inferred from the Stefan–Boltzmann law solely by exploiting the second law, without taking advantage of the statistical Planck distribution. Expressions for the entropy production at the emitting surfaces are derived and an effective temperature associated with the radiation between the two bodies is suggested.

Max Planck was the first who derived the entropy formula for the thermal black-body radiation. This problem is relevant not only for understanding the conditions of the terrestrial evolution but also for efficiency estimates of solar power production as well as for climatic entropy budgets on Earth (Emden, 1913; Fortak, 1979; Ebeling and Feistel, 1982; Yan, Gan, and Qi, 2004; Feistel and Feistel, 2006; Chen et al., 2010). By application of the second law to black-body radiation, Boltzmann (1884a, 1884b) theoretically confirmed Stefan's empirical emission law. A very detailed and thoughtful analysis of the radiation entropy is provided by Kabelac (1994) and Pelkowski (1995).

The entropy S of an equilibrium photon gas at black-body temperature T is related to its internal energy E in the form (Planck, 1906; Landau and Lifschitz, 1966)

$$S = \frac{4E}{3T} \qquad (2.89)$$

In a nonequilibrium process such as the solar or terrestrial radiation, the related photon gas is no longer isotropic, and the validity of a relation similar to Equation 2.1 between the fluxes of energy and entropy is no longer obvious. Some authors argue that the radiative entropy flux J_S should be related to the energy ("heat") flux J_Q in the same manner as in the case of heat conduction (Ozawa et al., 2003; Chen et al., 2010), that is, $J_S = J_Q/T$. Others suggest that the Planck factor of 4/3 must be maintained even in that case (Ebeling and Feistel, 1982; Stephens and O'Brien, 1993; Pelkowski, 1994, 1995; Weiss, 1996; Wright et al., 2001; Wright, 2007), that is, $J_S = 4J_Q/(3T)$. Somewhat surprisingly, upon integration of the Planck distribution, Essex (1984) obtained even a third version, $J_S = 4J_Q/(3\pi T)$, which may or may not be a simple calculation failure.

Because of its fundamental importance for the thermodynamics of the terrestrial climate, we shall consider in this section the radiative heat exchange between two black bodies of different temperatures as a tutorial example. It will turn out that the mathematical formula for the entropy flow is entirely determined by the second law, more precisely, that the related entropy balance is consistent with the second law if and only if the constant Planck factor of 4/3 rather than any other mathematical expression is used for the ratio TJ_S/J_Q. This result is somewhat exceptional since the second law usually imposes rather general constraints, such as positivity of the heat capacity or the compressibility (Planck, 1935; Landau and Lifshitz, 1980; IOC et al., 2010). The fact that a mathematical formula for a substance property can explicitly be derived from the second law is surprising and may be related to the special circumstance that photons form a rather peculiar kind of matter.

We imagine the standard situation of irreversible thermodynamics in which heat is transferred from a hot reservoir with the given temperature T_H to a colder one with

2 Fundamental Laws of Equilibrium and Nonequilibrium Thermodynamics

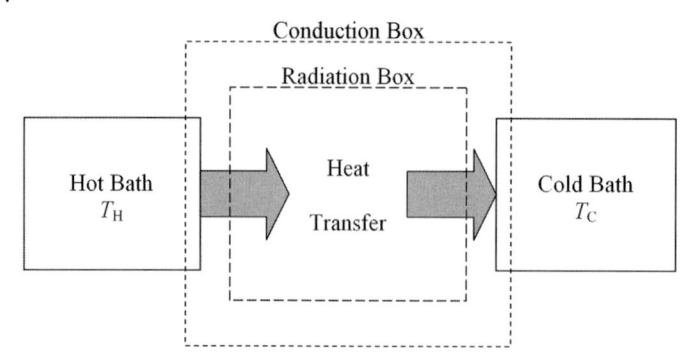

Figure 2.6 Heat transfer from a hot to a cold reservoir. The heat flow enters and exits the "conduction box" by heat conduction at the different temperatures T_H and T_C, respectively. Inside the conduction box, the heat is transferred by means of a certain carrier substance, such as a photon gas in the case of the "radiation box."

the given temperature $T_C < T_H$, as displayed in Figure 2.6. The resulting heat flow J_Q from one reservoir to the other will depend on the effective "thermal resistance" of the system between the heat baths, that is, on the details of the particular physical mechanism that represents the actual transfer process, such as by heat conduction, thermal convection, thermal radiation, and so on. We assume that the two baths are heat-conducting solids. We encapsulate the transfer system by some "conduction box" that includes the cross sections of two heat baths immediately below their surfaces, so that the box does not permit any other energy exchange than the entropy flux of J_Q/T_H which is entering the box and the entropy flux of J_Q/T_C which is exiting from it. In a stationary regime the entropy of the box is constant

$$\frac{dS}{dt} = 0 = \frac{d_e S}{dt} + P \tag{2.90}$$

such that the conduction box produces entropy at a rate of

$$P = -\frac{d_e S}{dt} = J_Q \left(\frac{1}{T_C} - \frac{1}{T_H} \right) \geq 0 \tag{2.91}$$

Here only the quantitative value of J_Q reflects the physical processes within the boxes that are driven by the external temperature difference, no matter what the nature of those heat transfer processes actually is.

Another box, the "radiation box," encloses the photon gas and is located immediately outside of the two reservoirs, Figure 2.6. It is used here to describe the amount of energy and entropy that some carrier substance imports, conveys, and exports. While it is clear that in a stationary regime the carrier must import and export energy at an equal rate of J_Q, it is less obvious what the temperatures, the entropy flux, and the entropy production are with respect to the transfer of J_Q between the heat baths and the carrier, and with respect to the particular physical properties of the carrier substance and process.

Between the hot and the cold body in the form of planar "infinite" plates we assume black-body thermal radiation. The energy emitted per surface area is $\sigma_{SB}T_H^4$ from the hot bath, and $\sigma_{SB}T_C^4$ from the cold bath, where σ_{SB} is the Stefan–Boltzmann constant. The resulting net flow of energy is

$$J_Q = \sigma_{SB}(T_H^4 - T_C^4) \tag{2.92}$$

The total entropy production within the conduction box caused by this flow is given by Equation 2.91.

The photon gas between the plates is homogeneous but not isotropic, that is, it is in a nonequilibrium state. There is no physical reason for any entropy production associated with the photon propagation inside the "radiation box," so the production P, Equation 2.91, must occur in the contact zones between the photon gas and the two reservoirs. Let the related entropy production rates be denoted here by P_H and P_C, respectively,

$$P = J_Q\left(\frac{1}{T_C} - \frac{1}{T_H}\right) = P_H(T_H, T_C) + P_C(T_H, T_C) \tag{2.93}$$

Because of the physical symmetry of the problem, the entropy production on the two surfaces can be described by the same but yet unknown mathematical function, P_S, with reversed arguments

$$P_H(T_H, T_C) = P_S(T_H, T_C), \; P_C(T_H, T_C) = P_S(T_C, T_H) \tag{2.94}$$

At the equilibrium state, this production must vanish for any temperature

$$P_S(T_H, T_C) = 0 \quad \text{if} \quad T_H = T_C \tag{2.95}$$

and because of the second law, it must otherwise be positive

$$P_S(T_H, T_C) > 0 \quad \text{if} \quad T_H \neq T_C \tag{2.96}$$

Note that the function P_S does not need to be symmetric in its parameters; the first argument, T_H, is the temperature of the emitted radiation while the second, T_C, is that of the absorbed radiation on the same surface S.

We formulate now the entropy balance. Let J_H be the entropy flux carried by the hot radiation, and J_C that of the cold radiation. Then, for the total entropy flux that enters the radiation box at the hot plate we infer the balance, as shown in Figure 2.7,

$$J_H = \frac{J_Q}{T_H} + P_H + J_C \tag{2.97}$$

and similarly for the entropy flux that exits the conduction box at the cold plate, as shown in Figure 2.7,

$$\frac{J_Q}{T_C} = J_H + P_C - J_C \tag{2.98}$$

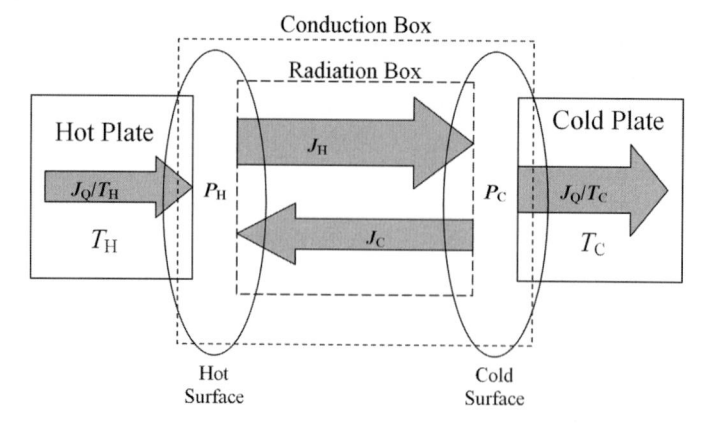

Figure 2.7 Stationary entropy fluxes of the radiation between two black bodies with the temperatures T_H and T_C, see Figure 2.6. At the hot surface, the entropy balance must be closed, consisting of conductive inflow J_Q/T_H, production P_H, radiative inflow J_C, and radiative outflow J_H, Equation 2.97. A similar balance must hold for the cold surface, Equation 2.98. The second law demands that entropy cannot be destroyed, that is, $P_H \geq 0$ and $P_C \geq 0$, for any pair of given temperatures T_H and T_C.

Thus, for P_S we get from Equations 2.97 and 2.94 the relation

$$P_S(T_H, T_C) = J_H - J_C - \frac{J_Q(T_H, T_C)}{T_H} \geq 0 \tag{2.99}$$

The second relationship which corresponds to Equation 2.98 follows from Equation 2.99 if the subscripts H and C are formally swapped, considering that $J_Q(T_H, T_C) = -J_Q(T_C, T_H)$

It is physically reasonable to assume that the radiative entropy fluxes J_H and J_C of the two independent parts of the ideal photon gas are described by the same but yet unknown function of their particular emitter temperatures, $J_H = J_R(T_H)$ and $J_C = J_R(T_C)$, with the property $J_R(0) = 0$. As we shall see, already under those rather general assumptions the second law in the form of the functional inequality (2.99),

$$J_R(T_H) - J_R(T_C) \geq \frac{\sigma_{SB}(T_H^4 - T_C^4)}{T_H} \tag{2.100}$$

uniquely defines an analytical solution for the radiative entropy flux function, $J_R(T)$. Exchanging T_H and T_C in (2.100) results in

$$J_R(T_H) - J_R(T_C) \leq \frac{\sigma_{SB}(T_H^4 - T_C^4)}{T_C} \tag{2.101}$$

We consider $T_H > T_C$ and combine the inequalities (2.100) and (2.101) in the form of lower and upper bounds,

$$\frac{\sigma_{SB}}{T_H} \leq \frac{J_R(T_H) - J_R(T_C)}{T_H^4 - T_C^4} \leq \frac{\sigma_{SB}}{T_C} \tag{2.102}$$

2.5 Irreversible Radiation Transport

In the limit $dT = (T_H - T_C) \to 0$, the finite difference turns into a derivative (l'Hospital's rule), and the interval width defined by the lower and upper bounds shrinks to zero. This results in the differential equation

$$\frac{1}{4T^3}\frac{dJ_R}{dT} = \frac{\sigma_{SB}}{T} \tag{2.103}$$

Upon integration we find the formula

$$J_R(T) = \frac{4}{3}\sigma_{SB}T^3 \tag{2.104}$$

for the radiative entropy flux emitted from a black body. This and only this mathematical expression for $J_R(T)$ is consistent with the second law. To see more easily that the solution (2.104) obeys the second law, we may write the inequality (2.100) in the form

$$f(x) \equiv 1 - 4x^3 + 3x^4 = (1-x)^2(1 + 2x + 3x^2) \geq 0 \tag{2.105}$$

For $x \equiv (T_C/T_H) > 0$, the function $f(x)$ has a minimum of zero at $x = 1$, that is, the entropy production vanishes at equilibrium and is positive otherwise.

We note that Equation 2.103 has the form of the local-equilibrium relation,

$$dE = TdS \tag{2.106}$$

where $dE = J_Q(T + dT, T)$ is the radiative energy flow at slightly different black-body temperatures, and $dS = J_R(T + dT) - J_R(T)$ is the related net entropy flow.

It is convenient to associate with the radiation box, Figure 2.6, a formal temperature, T_R, of the anisotropic photon gas, in the form

$$T_R = \frac{3}{4}\frac{(T_H^4 - T_C^4)}{(T_H^3 - T_C^3)} \tag{2.107}$$

On the temperature axis, T_R is located in the interval between T_H and T_C. This follows immediately from inserting Equation 2.104 into Equation 2.102, in the form

$$\frac{\sigma_{SB}}{T_H} \leq \frac{\sigma_{SB}}{T_R} \leq \frac{\sigma_{SB}}{T_C} \tag{2.108}$$

Consequently, the two entropy production rates, Equations. 2.94 and 2.106, when expressed in terms of T_R,

$$P_H = J_Q\left(\frac{1}{T_R} - \frac{1}{T_H}\right), \quad P_C = J_Q\left(\frac{1}{T_C} - \frac{1}{T_R}\right) \tag{2.109}$$

are nonnegative due to the second law. The temperature T_R associated with the radiation exchanged between a pair of black bodies coincides with the radiation temperature in the limiting case of equilibrium. The entropy production formulas, Equations. 2.99 and 2.109, equal the expression reported by Essex (1984), but in contrast to the conclusion drawn by that author, P_H and P_C take the universal bilinear form of fluxes times forces, Equation 2.109, if expressed in terms of the temperature

T_R assigned to the radiation field in between the two plates. In a more general formalism, a temperature T_R defined by the inequalities (2.109) is termed a "contact temperature" (Muschik and Berezowski, 2004; Muschik, 2008, 2009).

In terms of the contact temperature T_R, the net entropy flux within the radiation box takes the common form

$$J_S = J_H - J_C = J_R(T_H) - J_R(T_C) = \frac{J_Q}{T_R} \tag{2.110}$$

Provided the limiting case of emission into an empty space, $T_C \to 0$, the entropy flux emitted from a black body with the temperature T_H is $J_S = \frac{4}{3}\sigma_{SB}T_H^3 = \frac{\sigma_{SB}T_H^4}{T_R}$, the entropy production of the emitting surface is $P_S = \frac{1}{3}\sigma_{SB}T_H^3$, the energy flux associated with the conductive heat flow to sustain the body's temperature is $J_Q = \sigma_{SB}T_H^4$, and the temperature associated with the anisotropic photon gas is lower than that of the black body by 25%, $T_R = \frac{3}{4}T_H$. In this special limit of Equation 2.107, T_R is regarded as "flux temperature" by Kabelac (1994).

Here it is suggested to apply the term "contact temperature" to the more general case of a radiation field between two black bodies, Equation 2.107. This terminology has the advantage that at equilibrium the value of T_R coincides with the thermodynamic temperature, in contrast to Kabelac's original proposal, $T_R = 3T_H/4$, for which an asymptotic limiting value for the equilibrium case cannot be obtained.

The stationarity condition implies that the timescale for which the expressions (2.104) and (2.107) are derived here must significantly exceed the relaxation time of the radiation field, which at given black-body temperatures is in the order of L/c, where L is the distance between the plates and c is the speed of light.

2.6
Irreversible Processes and Fluctuations

The evolution of life on Earth started from fluctuations in small subsystems as we will show later. Therefore the theory of fluctuations is a fundamental ingredient of any theory of evolution. We start from the Boltzmann–Planck–Gibbs principle. We explained already earlier that in the pioneering work of Boltzmann, Planck, and Gibbs the entropy of a macrostate was expressed as the logarithm of the thermodynamic probability W

$$S = k_B \ln W \tag{2.111}$$

which is defined as the total number of equally probable microstates corresponding to the given macrostate. The factor, k_B, in front of the logarithm is the *Boltzmann constant*, Equation 2.11. In the simplest case of classical systems, the number of states with equal probability corresponds to the volume of the available phase space, $\Omega(A)$, divided by the smallest accessible phase volume h^3 (where h is Planck's constant). Therefore the entropy is given by

$$S_{BP} = k_B \ln(\Omega^*) \quad \Omega^*(A) = \Omega(A)/h^3 \tag{2.112}$$

Here, **A** is the set of all macroscopic conditions. In isolated systems at thermal equilibrium, the available part of the phase space is the volume of the energy shell enclosing the energy surface

$$H(q, p) = E \tag{2.113}$$

where $H(q, p)$ is the mechanical Hamiltonian in terms of coordinates, q, and momenta, p. If the system is isolated but not in equilibrium, only a certain part of the energy shell is available to the microstates. In the course of relaxation to equilibrium, the probability is spreading over the whole energy shell filling it finally with constant density.

Equilibrium means equal probability, and as we shall see, least information on the current distribution of microstates on the shell. In the nonequilibrium states, the energy shell shows regions with enhanced probability (attractor regions). We may define an effective volume of the occupied part of the energy shell by

$$S(E, t) = k_B \log(\Omega^*_{\text{eff}}(E, t)) \tag{2.114}$$

where the effective phase volume is

$$\Omega^*_{\text{eff}}(E, t) = \exp[S(E, t)/k_B]$$

In this way, the relaxation on the energy shell may be interpreted as a monotonous increase in the effectively occupied phase volume. This is accompanied by a devaluation of the energy.

The modern theoretical approach to fluctuations basically goes back to Einstein's insight about a deep connection between fluctuations and dissipation, both phenomena seemingly not related to one another. Einstein applied the Boltzmann–Planck principle discussed above to fluctuations. This was done in several early papers of Einstein published in 1902–1906. Let x be a fluctuating quantity of a thermodynamic system, $x = x(q, p)$. Under the assumption of thermodynamic equilibrium at fixed values of E, V, N, we will introduce a conditional entropy for the microscopic states, $S(x|E, V, N)$, where a certain value of x is realized. Geometrically, the manifold $x = $ const is some subset of the hypersurface $E = $ const. Therefore, this entropy is determined by the number of microscopic states, or in other words by the thermodynamic weight. The main idea in Einstein's approach consists in the calculation of the entropy difference δS for two states based on thermodynamic relations, that is, on Gibbs' fundamental equation. Additionally the distribution is normalized so that we need to know only relative changes of S rather than the full thermodynamic entropy, $S(E, V, N)$. Further we abbreviate $S(x|E, V, N) \equiv S(x)$ for simplicity. Following Einstein, we assume that the probability for the realization of certain value of x is proportional to the corresponding part, ω, of the phase space, that means

$$\omega(x) = \frac{\exp[S(x)/k_B]}{\int dx' \exp[S(x')/k_B]} \tag{2.115}$$

General properties of the probability distribution near equilibrium are as follows:

1) The thermodynamic potentials have extremum properties with respect to x.
2) From the second law it follows that the maximum, $S \to$ max, is reached in equilibrium, that is, if x reaches the equilibrium value x_0.

Therefore a series expansion at a stable equilibrium state gives

$$S(x) = S(x_0) - \frac{1}{2} k_B \beta (x-x_0)^2 + \cdots \tag{2.116}$$

with the coefficient

$$\beta = -\frac{1}{k_B} \left[\frac{\partial^2 S}{\partial x^2} \right] > 0 \tag{2.117}$$

This way we obtain that the fluctuations around equilibrium states are Gaussian,

$$\omega(x) = \text{const} \times \exp[-\beta(x-x_0)^2/2] \tag{2.118}$$

The standard deviations are determined by the second derivatives of the entropy

$$\left\langle (x-x_0)^2 \right\rangle = 1/\beta \tag{2.119}$$

In a similar way, we obtain Gaussian distributions for other thermodynamic embeddings where β is the positive second derivative of the corresponding thermodynamic potential.

According to Einstein's view, any macroscopic quantity x may be considered as a fluctuating variable which is determined by some probability distribution $\omega(x)$. The mean value is given by the first moment of the probability distribution

$$x_0 = \langle x \rangle = \int x \omega(x) dx \tag{2.120}$$

In a stationary state, we may shift the origin of the coordinate and assume $x_0 = 0$, without loss of generality

$$S(x=0) = \max$$

that is,

$$\left[\frac{\partial S}{\partial x} \right]_{x=0} = 0, \quad \left[\frac{\partial^2 S}{\partial x^2} \right]_{x=0} \leq 0 \tag{2.121}$$

According to Onsagers's view, the relaxation dynamics of the variable x is determined by the first derivative of the entropy, which is different from zero outside equilibrium. Starting from a deviation from the equilibrium (an entropy value below the maximum), the spontaneous irreversible processes should drive the entropy to increase,

$$\frac{d}{dt} S(x) = \frac{\partial S(x)}{\partial x} \times \frac{dx}{dt} \geq 0 \tag{2.122}$$

In this expression, two factors appear which were interpreted by Onsager in a quite ingenious way. According to Onsager's view the derivative

$$X = -\frac{\partial S(x)}{\partial x} \tag{2.123}$$

is the driving force of the relaxation to equilibrium. In irreversible thermodynamics, this term is termed the thermodynamic force in analogy to mechanics. This analogy implies that the (negative) entropy takes over the role of a potential. The second term

$$J = -\frac{dx}{dt} \tag{2.124}$$

is considered as the thermodynamic flux. We are repeating now in a simplified form some of Onsager's ideas which we explained already in Section 2.4. As pointed out there, Onsager empirically postulated a linear relation,

$$J = LX \tag{2.125}$$

The idea behind this assumption is that the thermodynamic force is the cause of the thermodynamic flux and both should disappear only simultaneously. The coefficient L is known as *Onsager's phenomenological coefficient*, or *Onsager's kinetic coefficient*. From the second law, it follows that the Onsager coefficients are strictly positive,

$$P = \frac{dS(x)}{dt} = JX = LX^2 \geq 0 \tag{2.126}$$

Onsager's postulate on a linear relation between thermodynamic forces and fluxes is the root for the development of the thermodynamics of linear dissipative system, also referred to as *linear irreversible thermodynamics*. A remarkable property is the bilinearity of entropy production as a sum of products of fluxes times forces,

$$P = JX \tag{2.127}$$

The universality of bilinearity was put in question for radiation processes by Essex (1984), but in appropriate variables the validity of the bilinearity principle could be confirmed also for this case (Pelkowski, 1995; Feistel, 2011a, see Section 2.5).

We find for the neighborhood of the equilibrium state the relations for the force

$$X = -\frac{\partial S(x)}{\partial x} = k_B \beta x \tag{2.128}$$

and for the flux

$$\frac{dx}{dt} = -k_B L \beta x \tag{2.129}$$

Using the previous equations, we finally obtain the linear relaxation dynamics

$$\frac{dx}{dt} = -\lambda x$$

with the abbreviation for the so-called relaxation coefficient of the quantity x,

$$\lambda = k_B L \beta \tag{2.130}$$

This linear kinetic equation describes the relaxation of a thermodynamic system brought initially out of equilibrium. Starting from the initial state $x(0)$, the dynamics of the variable $x(t)$ is given by an exponential law of decay,

$$x(t) = x(0) \exp[-\lambda t]$$

We see that $t_0 = \lambda^{-1}$ plays the role of the decay time of the initial deviation from equilibrium. This way we eventually arrived at a so-called fluctuation–dissipation relation. Onsager assumed that deviations from equilibrium and fluctuations around the equilibrium obey the same kinetics.

Now we generalize our result for the case of multiple thermodynamic variables. Taking the entropy in dependence on x_i, $i = 1, \ldots, n$, the entropy production reads

$$S(x_1, \ldots, x_n) = S_{\max} - \frac{1}{2} k_B \beta_{ij} x_i x_j + \cdots \tag{2.131}$$

Following Onsager's ideas we get the formulas

$$X_i = -\frac{\partial S(x_1, \ldots, x_n)}{\partial x_i} = k_B \beta_{ij} x_j \tag{2.132}$$

and

$$J_i = -\frac{dx_i}{dt} \tag{2.133}$$

The generalized linear Onsager ansatz reads

$$J_i = \sum_j L_{ij} X_j \tag{2.134}$$

Again, the second law requires positivity of the entropy production. This implies that

$$L_{ij} X_i X_j \geq 0 \tag{2.135}$$

that is, the matrix of Onsager coefficient is positive definite.

At the end of this chapter, we would like to add a few remarks on recent developments, mostly related to small systems.

2.7
Toward a Thermodynamics of Small Systems Far from Equilibrium

The extension of thermodynamics to situations far from equilibrium is extremely difficult. There are various approaches to solve this problem. We mention the early works of Onsager and Machlup (1953), Machlup and Onsager (1953), Prigogine

(1967), Glansdorff and Prigogine (1971), and Nicolis and Prigogine (1977). An advanced theory of nonlinear irreversible processes is due to the work of the late Rouslan Stratonovich (1994). We also mention the extended thermodynamics and other new approaches (Jou, Casas-Vazques, and Lebon, 1993; Ebeling and Muschik, 1993; Luzzi, Vasconcellos, and Ramos, 2000; Öttinger, 2005; Muschik, 2008).

In order to give a brief overview of relevant approaches, we first consider the method developed by Grmela and Öttinger, which deserves special attention (Grmela and Öttinger, 1997), named GENERIC (General Equations for the Non-Equilibrium Reversible-Irreversible Coupling), which seems to contain most of the theories mentioned above as special cases (Öttinger, 2005). The general time–evolution equation postulated by Grmela and Öttinger is of the following structure:

$$\frac{dx}{dt} = L\frac{\delta E}{\delta x} + M\frac{\delta S}{\delta x} \qquad (2.136)$$

Here x represents a set of independent variables; in many cases, x will depend on continuous position-dependent fields, such as mass, momentum, and energy densities. Further, E and S are the total energy and entropy expressed in terms of the state variables, and L and M are certain linear operators or matrices. The application of a linear operator may include integrations over continuous variables and then $\delta/\delta x$ typically implies functional rather than partial derivatives. There is no room here to go into the details of these more or less fruitful but not exhaustive methods. The whole field is still in development.

Let us discuss now several applications to small systems. As mentioned already and as will be shown later in more detail, the evolution of life started from fluctuations in some small subsystems of our Earth. This way the study of the thermodynamics of small systems is essential in our context. It is a hot topic of recent research mainly in connection with modern technologies of the manipulation of individual molecules.

In this context, a new theorem published by Jarzynski (1996) is of great interest. Here we present at least the basic idea. From the second law it follows, as shown in the previous sections, that under isochoric–isothermal conditions free energy is a nonincreasing function of time

$$\frac{dF}{dt} \leq 0 \qquad (2.137)$$

and that under isobaric–isothermal condition free enthalpy is nonincreasing,

$$\frac{dG}{dt} \leq 0 \qquad (2.138)$$

The tendency of F and G to decrease is in fact determined by the general tendency expressed by the second law to devaluate energy (or enthalpy) with respect to their capacity to do work. Repeating in part the arguments given in Section 2.3, let us now consider the work W performed on a system during a finite transition from an initial nonequilibrium state to a final equilibrium state. Then, as we have shown

$$W = \delta F = F_{ne} - F_{eq} \qquad (2.139)$$

is the work corresponding to a process of changing the parameters infinitely slowly along a path from the starting nonequilibrium point to the final equilibrium state. This relation is no longer true if the parameters change along the path at a finite rate. Under those conditions, the process is irreversible and the work W will depend on the microscopic initial conditions of the system and the reservoir, and will, on average, exceed the free energy difference (Jarzynski, 1996),

$$\langle W \rangle \geq \delta F = F_{ne} - F_{eq} \tag{2.140}$$

The averaging is to be carried out over an ensemble of transitions (measurements). The difference

$$W_{diss} = \langle W \rangle - W \geq 0 \tag{2.141}$$

is just the dissipated work W_{diss} associated with the increase in entropy during the irreversible transition. In his work, Jarzynski (1996) discussed the above relations between free energy and work from a new perspective. He derived a new relation in the form of the equality

$$\langle \exp[-\beta W] \rangle = \exp[-\beta W] \tag{2.142}$$

rather than the inequality (2.131), where $\beta \equiv 1/k_B T$. The nonequilibrium identity (2.142), proven by Jarzynski (1996) using different methods, is indeed surprising: it says that we can extract equilibrium information

$$\delta F = W = -k_B T \ln \langle \exp[-\beta W] \rangle \tag{2.143}$$

from nonequilibrium (finite-time) measurements. In recent times, these new relations found numerous applications (Seifert, 2005). A different but related approach results in the *Crooks theorem*, which gives a relation of probabilities for obtaining work from forward and reverse processes

$$\frac{P_F(W)}{P_R(-W)} = \exp\left[\frac{W - \delta G}{k_B}\right] \tag{2.144}$$

Obviously, this relation is determined by the lowering of free enthalpy.

Closely related to the previous relations is the Evans–Galavotti–Cohen fluctuation theorem (Evans, Cohen, and Morriss, 1993; Evans, Searles, and Mittag, 2001; Galavotti and Cohen, 1995a,b). This theorem may be considered as a simple consequence of a time-reversal symmetry. In reversible systems any initial data, with the exception of a set of zero volume in phase space, have the same statistical properties in the sense that all smooth observables admit a time average independent of the initial data and expressed as an integral with respect to a probability distribution on phase space, called the *natural stationary state*, or simply the *stationary state*. The theorem provides, asymptotically in the observation time, a quantitative and parameter-free relation between the stationary-state probability of observing a value of the average entropy production rate and its opposite. The fluctuation theorem acquires physical interest only in connection with the chaotic hypothesis. Under the latter general assumption, combined with time reversal, it predicts a *universal relation* between an entropy creation rate value and its opposite, accessible to simulations and possibly to laboratory experiments. The fluctuation theorem reads

$$\frac{P_t(\sigma)}{P_t(-\sigma)} = \exp\left[\frac{t\sigma}{k_B}\right] \qquad (2.145)$$

As said above, the physical interpretation of the theorem as a property of stationary states in nonequilibrium statistical mechanics is based on reversibility. The fluctuation theorem is supported by numerical experiments, which were used to determine the relation between the probabilities P_t to find positive or negative entropy production in numerical experiments (Evans, Cohen, and Morriss, 1993; Evans, Searles, and Mittag, 2001; Galavotti and Cohen, 1995a,b). The entropy production was obtained through measuring the heat production $\sigma = Q/(Tt)$ over a time interval t.

Another most advanced theory for fluctuations in small systems was derived by Gaspard (1998, 2007) and collaborators (Andrieux et al., 2008). The *Gaspard theorem* gives a relation between probabilities P of opposite fluctuations related to affinity, A, and current, J:

$$\frac{P_F(J)}{P_R(-J)} = \exp[A \times Jt] \qquad (2.146)$$

According to Gaspard (1998, 2007), the essence of his findings may be formulated in his *theorem of nonequilibrium temporal ordering*:

In nonequilibrium steady states, the typical paths are more ordered in time than their corresponding time reversals in the sense that their temporal disorder is smaller than the temporal disorder of the corresponding time-reversed paths.

This way thermodynamic entropy production can be related to the breaking of time-reversal symmetry in the statistical description of nonequilibrium systems, and this is most transparent in small systems.

3
Evolution of Earth and the Terrestrial Climate

And as five zones th' aetherial regions bind, Five, correspondent, are to Earth assign'd: The sun with rays, directly darting down, Fires all beneath, and fries the middle zone: The two beneath the distant poles, complain Of endless winter, and perpetual rain. Betwixt th' extreams, two happier climates hold The temper that partakes of hot, and cold.

Publius Ovidius Naso: Metamorphoses

All of human history has happened on that tiny pixel, which is our only home.

Carl Sagan: Pale Blue Dot

The rise of the concentration of industrially produced carbon dioxide (CO_2) in the terrestrial atmosphere is well documented by the famous Keeling Curve (Keeling and Whorf, 1998). Its potential impact on the climate in the form of the greenhouse effect and global warming was and is subject to countless scientific articles and conferences; the topic is almost omnipresent in the mass media, and repeatedly it triggers controversial political debates, in private circles as well as on international meetings. In this chapter, rather than attempting to review those studies or to assess the various arguments, we will focus on some general aspects that are most fundamental from the viewpoint of self-organization – the entropy balance of the planet Earth, the entropy production, the stability and sensitivity of the climate system, and the effect of fluctuations. In particular, we like to emphasize the predominant thermodynamic role of water for the functioning principles of the "steam engine Earth" (Ebeling and Feistel, 1982, 1994), and paint a picture of the dynamical, self-organized structure of our planet as a coupled system from the inner core to the top of the atmosphere.

The pivotal model of Arrhenius (1896) considered planet Earth as an absorber of solar irradiation and as a black body that emits thermal radiation according to the Stefan–Boltzmann law. Since the resulting surface temperature of the radiation balance is lower than the one we experience (Peixoto and Oort, 1992), Arrhenius concluded that a greenhouse effect by atmospheric carbon dioxide (CO_2) may explain the difference. Note that in the strict sense the common terminology is slightly misleading; the original greenhouse in the garden owes its warming effect mainly to the suppression of convective heat transport (wind), rather than blocking infrared emission as in the case of the climatic "greenhouse."

The simplified climate models discussed here are similar and have the form of globally averaged surface-radiation balances. In contrast to what one may expect, they show that classical Arrhenius-like systems suffer systematically from unrealistic, overestimated relaxation times. This problem can only be solved by more complex models which include the latent heat of water exchange between ocean and atmosphere. In those models, however, the greenhouse effect takes the form of the atmospheric water-vapor instability. Together with clouds, water vapor (rather than CO_2) is the strongest greenhouse gas. When the atmosphere gets a little warmer, it stores more vapor, which in turn increases the greenhouse effect, and so on. In a simple linear radiation-balance model, this instability amplifies the initial warm or cold temperature fluctuation. The positive feedback results in either the complete evaporation of all oceans on a Venus-like hot planet, or a complete freezing of all oceans on a Mars-like cold planet, that is, the model does not possess a stable stationary state with liquid oceans and moderate temperature. Hence, simple radiation–evaporation balance models of the Arrhenius type are dynamically unstable for the ocean. Our conclusion is that the climate attractor is chaotic rather than a randomly fluctuating stable steady state. Thus, the restoring forces which stabilize the terrestrial climate attractor may not primarily be those suggested by Arrhenius' popular greenhouse theory; on the contrary, they are still poorly understood (Lozán, Graßl, and Hupfer, 1998; Haigh, 2010). We bring forward the hypothesis that the bimodal hypsographic structure of the terrestrial crust and the related extended surface fractions of both sea and land are the key to the dynamic stability of the terrestrial climate.

Among the various instabilities that may occur in nonequilibium fluid systems, three seem to be most relevant for processes in the troposphere, the lowest part of the atmosphere (Ebeling and Feistel, 1983; Sempf et al., 2007). The convective instability permits fluctuations to trigger large-scale vertical air motion and the formation of low-pressure cells (cyclones, hurricanes, tornados, dust devils). Cyclones can also emerge from so-called baroclinic instabilities within front zones between air masses of very different temperatures and humidities (Pedlosky, 1987). Additionally, atmospheric jet streams can occur within front zones as a result of dynamic instabilities (Raethjen, 1961). Cyclones and jets have lifetimes of days or weeks – they are triggered by critical fluctuations and decay when their dissipation exceeds the available free-energy supply. Thus, weather processes fit well into the schematic of self-organization and evolution processes discussed in Chapter 1; dissipative structures form under suitable conditions and develop instabilities which amplify certain random fluctuations. The result is the appearance of another dissipative structure which in turn will prove sensitive to certain fluctuations, and so on. Such an unlimited series of subsequent self-organization steps possesses intrinsic randomness and historicity; the exact location, time, and form of the particular critical fluctuation may depend on tiny subscale events similar to the famous "butterfly effect." Minor details may determine where and when exactly, say, a tornado is born, and what route it takes, with all the consequences this may further imply. The dependence of macroscopic phenomena on irreversibly amplified random fluctuations is termed historicity. How would modern physics look like if, say, Edmond Halley had not published Newton's "Principia" (Bryson, 2003)?

Self-organized criticality (SOC) is another relevant mechanism for the formation of terrestrial structures. As an example for SOC, consider a traveling dune. On the crest, the wind is piling up sand from the windward side. Every now and then, an avalanche slides down the lee side. When the transport rate of sand is high enough to accumulate sand but inappropriate to sustain a continuous overflow, the system is toggling between two different dynamical regimes, sticking and slipping sand, crossing the critical point back and forth over and over again. Although the critical point itself is unstable, it acts as an attractor that keeps the system in its immediate vicinity such that specific fluctuations suffice to trigger the transition. Similar instabilities can be observed at active cliffs (Figure 3.1), when a violin is

Figure 3.1 Chalk cliff on Rügen Island, German coast of the Baltic Sea, as an example for self-organized criticality. Triggered by weather fluctuations, the slope of an active cliff is occasionally restored by violent landslides rather than by any permanent flow of material, smoothly creeping downhill. Photo taken in fall 2010.

played (Ebeling, 1976a) or when humans speak, sing, or cry (Herzel *et al.*, 1994; Tokuda *et al.*, 2007).

Under the wide range of pressure–temperature conditions typical for the terrestrial interior, mixtures of silicium and magnesium minerals possess a variety of solid and liquid thermodynamic equilibrium phases (Hirose, 2010). They have different densities and appear as qualitatively distinct, relatively homogeneous spheres separated by sharp interfaces ("discontinuities"), similar to atmosphere, hydrosphere, and lithosphere. Deep inside Earth, radioactive decay is producing heat that is carried away to the surface and from there emitted to the cosmic background. Since there are only sources and no sinks below the crust, the second law requires that the related radial component of the stationary temperature gradient must point inward, associated with outward heat conduction. Hence, similar to the density, also the temperature cannot decrease with greater depth. Thermal conduction across a thick layer of solid rock is extremely slow. Underneath, heat will accumulate and temperature will rise until a transition takes place to a state with sufficiently fast thermal transport, such as a liquid state with convection. If that convective transport is now faster than the heat supply from below, temperature will drop again, possibly even until the liquid returns to a solid state. As a result, the inner shells of our planet may be adjusted to temperatures near the melting point of the particular solid mineral. The dynamical attractor regime of the shell is not necessarily a stable stationary state, as we know from other examples of SOC.

Equipped with this toolbox of theoretical models, we combine available facts on the early evolution of the solar system and terrestrial planets in the form of a fictitious narrative. We develop the hypothesis that indirectly the Moon might have been an indispensible ingredient for the origin of life, and consequently, as an illustration for the fundamental historicity of evolution processes, also for the existence of humans and, ultimately, for the exceptional physical process of measurement, that is, the conscious scientific exploration of the world.

3.1
The Photon Mill

The possibility of terrestrial life results ... from the vast increase of entropy associated with the transformation of hot solar radiation into colder terrestrial radiation

Emden, 1913

In the 1970s, several global climatic energy balances were published (Svatkov, 1974; Budyko, 1978; Monin and Shishkov, 1979). This progress inspired the question for a related estimate of the entropy production that must necessarily accompany the multiple self-organization phenomena and sustained dissipative structures observed in our natural environment. The answer was the simple schematic (Ebeling and Feistel, 1982) that was later referred to as the "photon mill" (Feistel and Ebeling, 1989; Ebeling and Feistel, 1994, 2005), see Figure 3.2.

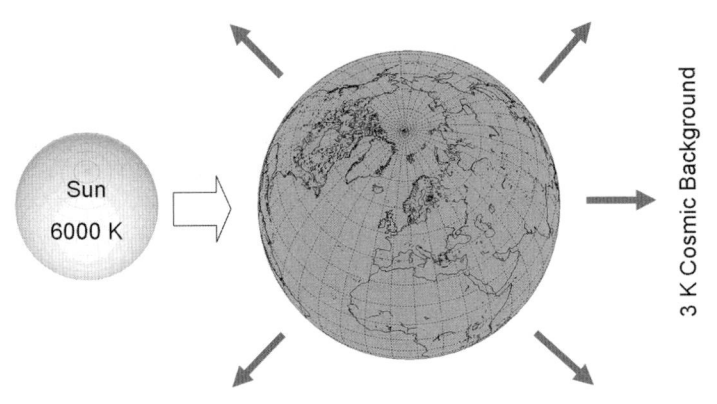

Figure 3.2 Schematic of the "Photon Mill" (Ebeling and Feistel, 1982) for the estimation of the terrestrial entropy export. With a mean surface temperature of about 300 K, Earth is located in the radiative thermal gradient between the hot Sun surface (6000 K) and the cold cosmic background (3 K).

From a cosmic point of view, the Earth is a thermally isolated system except for radiation. Located in the thermal gradient between the hot Sun surface ($T_{Sun} \approx$ 6000 K) and the cold cosmic background ($T_{CB} \approx$ 3 K), the terrestrial atmosphere and hydrosphere is passed through by a continuous flow of solar radiation energy described by the areic power density of $(1-\alpha)J_E \approx 230$ W m^{-2} (Svatkov, 1974) at a temperature of typically $T_{Earth} \approx$ 300 K. An estimated fraction of $\alpha \approx 30\%$, the albedo, of the incoming radiation, J_E, is immediately scattered back to space (Palle et al., 2004; Wielicki et al., 2005; Petty, 2006; Cockell, 2008; Haigh, 2010) and does not enter the terrestrial entropy balance. The energy flow is irreversible and implies a mean entropy production P per global surface area A of

$$\frac{P}{A} \approx \frac{4}{3}\left(\frac{1}{T_{Earth}} - \frac{1}{T_{Sun}}\right) \times (1-\alpha)J_E \approx 1 \text{ W m}^{-2} \text{ K} \tag{3.1}$$

Here, the Planck factor of 4/3 in the formula for the areic entropy production density, P/A, relates to the entropy/energy flux ratio of the thermal photon gas (Feistel, 2011a); we shall return to this – occasionally disputed – number further below. It was considered even more thoroughly in Section 2.5.

The actual solar intensity, I_E, the "solar constant," is higher than J_E by the geometrical factor of (surface area)/(cross section) = $4\pi R^2/(\pi R^2) = I_E/J_E = 4$. Budyko (1978) estimated a value of $(1-\alpha)J_E \approx 244$ W m^{-2} for the solar heating. The most accurate estimate for the mean total solar irradiation at the top of the atmosphere is $I_E = 1366$ W m^{-2} (Tobiska and Nusinov, 2006; Schmutz, 2010), corresponding to a value of $J_E = 341$ W m^{-2}. The flux is modulated seasonally by the elliptical orbit (about 3.5%, Spencer, 1971; Rozwadowska, 1991; Peixoto and Oort, 1992; Rozwadowska and Isemer, 1998; Grena, 2008) and by the 11-year sun-spot cycle (about 0.1%, Palle et al., 2004). The solar minimum observed in 2009 was lower than the previous two minima in 1986 and 1996 by $\Delta I_E = 0.3$ W m^{-2} (Fröhlich, 2009). This recent apparent downward trend on the decadal scale does not yet exceed the uncertainty of $U(I_E) = 0.4$ W m^{-2} estimated for the measurements (Schmutz et al., 2009; Schmutz,

2010). Only sun-spot minima can be used for the trend analysis; unfortunately, just two full solar cycles have passed since appropriately equipped satellites are taking their measurements. In contrast, models suggest an increase in solar radiation since 1900 by $\Delta I_E = 1.6$ W m^{-2} (Foster, 2004; Schöll et al., 2007) that occurred mainly in the period before the peak value of 1960. This value appears large as compared to the lowering during the Maunder minimum between 1645 and 1715 in the Little Ice Age of only $\Delta I_E = 0.9$ W m^{-2} that was estimated from beryllium isotope ratios found in ice cores (Steinhilber, Beer, and Fröhlich, 2009). A recent downward trend has been found in the sun-spot magnetic field strength (Penn and Livingston, 2010). The response of the climate system to small variations of the solar irradiance is highly nonlinear, non-Markovian, much more complex than just an immediate global warming or cooling, and is incompletely understood yet (Meehl et al., 2009; Haigh, 2010; Šesták, Hubík, and Mareš, 2010; Haigh et al., 2010).

The value of the total entropy production, Equation 3.1, represents the upper bound to any dissipative structures and self-organization processes that may occur on our planet at the present temperature. Practically, this limit is not very restrictive for evolution processes (Šesták, Hubík, and Mareš, 2010). An estimated fraction of nearly 1% of the solar energy input (recall: about 240 W m^{-2}) is transformed into mechanical energy of the atmosphere (almost 2 W m^{-2}, Peixoto and Oort, 1992) and is finally dissipated there (Wulf and Davis, 1952). This efficiency is far below that of the corresponding Carnot cycle in the order of magnitude of $1-(3\text{ K}/6000\text{ K}) = 99.95\%$. Note that in turn a small fraction of 1% of the atmospheric mechanical power would be sufficient to cover the total power consumption of the human society (about 20 mW m^{-2}, of which 4 mW m^{-2} is electrical power, Archer and Jacobson, 2005; Schiermeier et al., 2008; Lu et al., 2009). The entropy production relations are similar (Weiss, 1996). The biological annual primary production was estimated by Field et al. (1998) to be 105 Pg of carbon per year, with roughly equal contributions from land and oceans (Mackas, 2011). A more recent value of 123 ± 8 Pg of carbon per year was reported by Beer et al. (2010) for just the terrestrial part. Contrary to expectations, the primary production on land shrank by 0.55 Pg over the last decade (Zhao and Running, 2010). If we optimistically infer that globally about 2×10^{14} kg of carbon may be converted to organic matter by photosynthesis in the course of a year, and use a chemical binding energy of 10 kcal g^{-1} = 42 MJ kg^{-1} (Cummins and Wuycheck, 1971), we find a total ecological power consumption of 0.5 W m^{-2}, that is, just 0.2% of the solar energy supply but 25 times as much as the human demand. Note that these numbers involve substantial uncertainties; there are other estimates which claim that cyanobacteria alone contribute about 25%, or 450 TW, to the marine photosynthetic productivity, which would imply a total primary production of perhaps 7 W m^{-2} on the global average. We will assume here the more conservative value.

The average total power consumption per human, including food, transport, and industrial products, is about 2 kW; the solar irradiation perpendicular to an area of 10 m^2 would suffice to cover that need if we were able to exploit 100% of it. On the global average, about 100,000 m^2 of surface area are available per capita, including oceans, ice cover, deserts, jungles, and mountain ranges.

We conclude that only slightly more than 1% of the incoming solar energy is converted into mechanical and chemical "useful" energy; the rest is immediately dissipated in the form of heat. It must be emphasized that the related entropy produced directly or during the subsequent degradation processes, Equation 3.1, is continuously exported from the Earth to the cosmic background. The "cold bath" in combination with the hot "heat bath" are indispensible for the functioning of the "steam engine Earth." This important aspect may easily be overlooked when fictitious ideas such as the Dyson sphere (Dyson, 1959) are discussed.

It is intriguing to investigate in more detail where and how the global entropy production occurs. In a model similar to Equation 3.1, Fortak (1979) estimated a total terrestrial entropy export rate of $P = 600\,\mathrm{TW\,K^{-1}}$, corresponding to $1.2\,\mathrm{W\,m^{-2}\,K^{-1}}$ with respect to the terrestrial surface area $A = 5.1 \times 10^{14}\,\mathrm{m^2}$. From 36 months of observation by the ERBE satellite, Stephens and O'Brien (1993) calculated a mean terrestrial entropy production of $1.3\,\mathrm{W\,m^{-2}\,K^{-1}}$, with climatic fluctuations and annual variability in the order of 2%. Peixoto and Oort (1992) found an entropy export from the top of the atmosphere of $0.9\,\mathrm{W\,m^{-2}\,K^{-1}}$, and $0.6\,\mathrm{W\,m^{-2}\,K^{-1}}$ of entropy production inside the atmosphere. From a very detailed study of a gray atmosphere model, Pelkowski (1994, 1995) derived a value of $1.2\,\mathrm{W\,m^{-2}\,K^{-1}}$. Ozawa et al. (2003) also estimated an export of $0.9\,\mathrm{W\,m^{-2}\,K^{-1}}$ but argued that the factor of $4/3$ in Equation 3.1 is inappropriate for an anisotropic photon gas such as the solar radiation flux. Actually, the correctness of either the use of the Planck factor for the entropy flow of thermal radiation (Stephens and O'Brien, 1993; Pelkowski, 1994; Wright et al., 2001; Wright, 2007) or its neglect (Ozawa et al., 2003; Chen et al., 2010) is still subject to an ongoing controversial discussion in the literature. Here we will include the Planck factor of $4/3$ because it is necessary and sufficient for consistency with the second law (Feistel, 2011a), in particular for stationary anisotropic radiation, as considered in detail in Section 2.5.

Recently, Yan, Gan, and Qi (2004) showed that the world ocean exports about $0.6\,\mathrm{W\,m^{-2}\,K^{-1}}$ entropy on average, and Feistel and Feistel (2006) estimated the entropy export of the Baltic Sea to be about $0.3\,\mathrm{W\,m^{-2}\,K^{-1}}$. These figures are consistent with the global entropy balance of the photon mill and underline the essential role the ocean plays in this context. In the following sections, we will consider some selected, rather simplified thermodynamic models for certain key processes of the climate engine. Our analytical models are more of a qualitative, tutorial, and illustrative nature than of quantitative accuracy but they may assist to reveal the relative importance of particular branches of the complex spatial, temporal, and causal cascades of climatic interactions.

3.2
Black-Body Radiation Model of Earth

In the classical model of Arrhenius (1896), Earth's surface is considered as an absorber of solar irradiation in the optical spectrum and as a black body that emits infrared light. The resulting equilibrium temperature is lower than the one observed climatologically; from this deviation, the terrestrial greenhouse effect is estimated which has become commonplace in today's public opinion and significantly influences political and economic decisions.

The Arrhenius model was developed and is usually interpreted from the viewpoint of inhabitants of the continents, but actually the major fraction of the terrestrial surface is ocean (Budyko, 1969; Sellers, 1969). In this section, we will construct a simple tutorial thermodynamic model for the radiation balance of the ocean, ignoring the direct influence of the atmosphere in first place. For example, we may have in mind the subtropical latitudes where the humidity is low and the terrestrial radiation has its maximum (Peixoto and Oort, 1992; Trenberth and Solomon, 1994; Cockell, 2008). In obvious contrast to virtually self-evident expectations, we shall see that this approach in terms of Arrhenius-like simple models is physically essentially inappropriate to describe our observed climate, and conclude that the popular Arrhenius greenhouse concept must be considered more cautiously and carefully than this is occasionally the case in mass media or in various premature projects of "geo-engineering" to counteract global warming.

Our three-box model is similar to the classical thermodynamic heat engine, running between a hot and a cold heat bath, but we are actually not primarily interested in any kind of work it may perform. We describe this system in the form of a model for the photon mill, Figure 3.3. While the Earth is assumed here to be a black body, for some models that will be considered later, we imagine some kind of "water world" in which a motionless layer of seawater absorbs and emits the radiation. Real seawater is obviously not black in the visible spectrum but almost perfectly black in the infrared window of terrestrial long-wave emission.

We first consider some details of the radiation between the Sun, Earth, and the cosmic background. As a simple geometric model, we imagine that Sun and Earth are isothermal black spheres with the temperatures T_S and T_W, respectively, located inside a black hollow sphere, the cosmic background. For simplicity, we think of the latter as distant interstellar matter with the temperature T_C. The distance between Sun and Earth, R_{SE}, is large compared to their radii, R_S and R_E, respectively.

The total solar thermal emission is

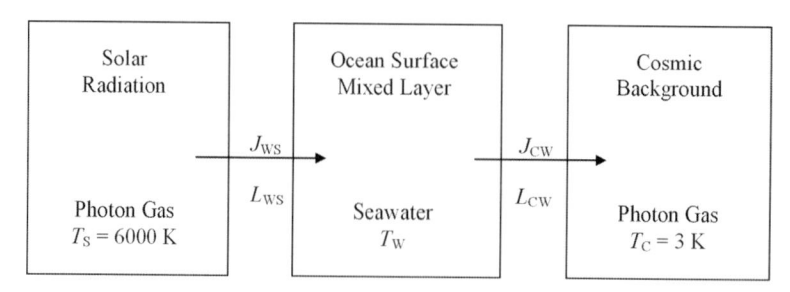

$$J_{Sun} = 4\pi R_S^2 \sigma_{SB} T_S^4 \tag{3.2}$$

Figure 3.3 Three-box model of the "heat engine ocean," driven by the thermal gradient between the solar surface and the cosmic background, the "photon mill." Here, the Js denote the heat flows and the Ls the heat transfer coefficients.

where $\sigma_{SB} = 5.6704 \times 10^{-8}$ W m^{-2} K^{-4} is the Stefan–Boltzmann constant. The irradiation per surface area at the distance of the Earth is

$$I_E = \frac{J_{Sun}}{4\pi R_{SE}^2} \qquad (3.3)$$

and is commonly termed the solar constant. It produces the currently best known value of $I_E = 1366$ W m^{-2} (Tobiska and Nusinov, 2006; Schmutz, 2010) if the estimates $T_S = 5778$ K, $R_S = 6.955 \times 10^8$ m, and $R_{SE} = 1.496 \times 10^{11}$ m are used. The fraction of the total solar irradiation absorbed by the cross section of the Earth modeled as a black body is

$$J_{ES} = \pi R_E^2 I_E \qquad (3.4)$$

Vice versa, the total terrestrial emission is, assuming a black body with the "water" temperature T_W, Figure 3.3,

$$J_{Earth} = 4\pi R_E^2 \sigma_{SB} T_W^4 \qquad (3.5)$$

The intensity at the Sun's distance is

$$I_S = \frac{J_{Earth}}{4\pi R_{SE}^2} \qquad (3.6)$$

and the fraction of the total terrestrial radiation that hits the Sun is

$$J_{SE} = \pi R_S^2 I_S \qquad (3.7)$$

Thus, the mean radiation energy balance between Sun and Earth reads per surface area, A, of the Earth (or of the sea surface provided it covers the entire globe),

$$j_{WS} = \frac{1}{4\pi R_E^2}(J_{ES} - J_{SE}) = \sigma_{WS}(T_S^4 - T_W^4) \qquad (3.8)$$

Here, the effective radiation constant is

$$\sigma_{WS} = \frac{R_S^2}{4 R_{SE}^2}\sigma_{SB} \qquad (3.9)$$

Note that the simple difference, Equation 3.8, must be replaced by a much more complicated expression if the black-body model for the energy exchange between Sun and Earth is replaced by spectrally resolved absorption and emission coefficients of a more realistic terrestrial atmosphere model. But even in that case, the flux j_{WS} must vanish if $T_S = T_W$, that is, j_{WS} must always be proportional to $(T_S - T_W)$, or more precisely, the ratio $j_{SW}/(T_S-T_W)$ must remain finite in the limit of thermodynamic equilibrium, $T_W \to T_S$, for any reasonable radiation model.

We repeat this consideration now for the exchange between the Earth and the cosmic background. The total terrestrial emission is given by Equation 3.5. If we neglect the small fraction of energy transferred from Earth to the Sun, Equation 3.7, in comparison to the entire solid angle (neglecting in the sky the tiny spot covered by the Sun), Equation 3.5 also describes the energy absorbed by the cosmic background,

$$J_{CE} = 4\pi R_E^2 \sigma_{SB} T_W^4 \tag{3.10}$$

To estimate the amount of energy which the Earth receives from the cosmic background, J_{EC}, we imagine that this radiation comes from a large hollow sphere with the Earth located inside it. Within the sphere, the radiation forms an isotropic and homogeneous equilibrium photon gas. The radiation passing through any selected surface element is the same regardless of where that element is located or how it is oriented. Thus, the emission (and immission) intensity on the fictitious hollow sphere is the same as the immission intensity on the terrestrial surface. The total background radiation captured by the black Earth is therefore

$$J_{EC} = 4\pi R_E^2 \sigma_{SB} T_C^4 \tag{3.11}$$

Thus, the radiation energy balance between Earth and the cosmic background per terrestrial surface area reads

$$j_{CW} = \frac{1}{4\pi R_E^2}(J_{CE} - J_{EC}) = \sigma_{CW}(T_W^4 - T_C^4) \tag{3.12}$$

Here, the effective radiation constant is

$$\sigma_{CW} = \sigma_{SB} \tag{3.13}$$

Equations 3.8 and 3.12 enable us to formulate the mean enthalpy budget per surface area of the terrestrial water box, Figure 3.3, in the photon-mill model, in the form,

$$\frac{dH_W}{dt} = j_{WS} - j_{CW} = \sigma_{WS}(T_S^4 - T_W^4) - \sigma_{CW}(T_W^4 - T_C^4) \tag{3.14}$$

The entropy production density related to this heat flow is caused by the absorption of solar and background irradiation on Earth, and of the terrestrial thermal radiation by the Sun and the cosmic background. With the radiation field between each pair of black bodies, a formal "flux temperature" or "contact temperature" can be associated (Feistel, 2011a, see Section 2.5), which in our case reads

$$T_{WS} = \frac{3(T_S^4 - T_W^4)}{4(T_S^3 - T_W^3)} \quad T_{CW} = \frac{3(T_W^4 - T_C^4)}{4(T_W^3 - T_C^3)} \tag{3.15}$$

for the exchange between Sun and Earth, T_{WS}, and between Earth and cosmos, T_{CW}, respectively. At the interface between the black body and the radiation field, entropy is produced. The related production on the Sun surface is, expressed per terrestrial surface area A

$$\frac{P_S}{A} = \frac{j_{WS}}{T_{WS}} - \frac{j_{WS}}{T_S} > 0 \tag{3.16}$$

on the terrestrial surface

$$\frac{P_W}{A} = \frac{j_{WS}}{T_W} - \frac{j_{WS}}{T_{WS}} + \frac{j_{CW}}{T_{CW}} - \frac{j_{CW}}{T_W} > 0 \qquad (3.17)$$

and in the cosmic background,

$$\frac{P_C}{A} = \frac{j_{CW}}{T_C} - \frac{j_{CW}}{T_{CW}} > 0 \qquad (3.18)$$

Note that the terrestrial entropy production reported in the beginning of this chapter, Equation 3.1, follows from Equation 3.17 for steady-state conditions, $j_{CW} = j_{WS}$, and from Equation 3.15, $T_{CW} \approx 3T_W/4$, in the obviously justified approximation of $T_S \gg T_W \gg T_C$.

As a function of the terrestrial surface temperature, T_W, the instationary total entropy production of the system Sun–Earth-background is

$$\frac{P}{A} = \frac{P_S + P_W + P_C}{A} = -\frac{j_{WS}}{T_S} + \frac{(j_{WS} - j_{CW})}{T_W} + \frac{j_{CW}}{T_C} = \sum_k j_k X_k \qquad (3.19)$$

The fluxes j_k are proportional to the nonequilibrium Onsager forces X_k which are defined by

$$X_{WS} = \frac{1}{T_W} - \frac{1}{T_S} \quad X_{CW} = \frac{1}{T_C} - \frac{1}{T_W} \qquad (3.20)$$

The Onsager coefficients,

$$L_{WS} = \frac{j_{WS}}{X_{WS}} = \sigma_{WS} T_S T_W \left(T_S^3 + T_S^2 T_W + T_S T_W^2 + T_W^3 \right) \qquad (3.21)$$

and similarly for L_{CW}, can be considered as independent of the Earth's temperature T_W only in some immediate vicinity of thermodynamic equilibrium, $T_S \approx T_W \approx T_C$

$$L_{WS} \approx 4\sigma_{WS} T_S^5 \qquad (3.22)$$

In this case, in agreement with Prigogine's theorem, the entropy production, Equation 3.19, has a minimum

$$\frac{1}{A}\frac{\partial P}{\partial T_W} = 0 = -\frac{2L_{WS}}{(T_W)^2}\left(\frac{1}{T_W} - \frac{1}{T_S}\right) + \frac{2L_{CW}}{(T_W)^2}\left(\frac{1}{T_C} - \frac{1}{T_W}\right) = 2\frac{j_{CW} - j_{WS}}{(T_W)^2} \qquad (3.23)$$

at the terrestrial steady-state temperature, T_W, at which the energy import equals the export, $j_{CW} = j_{WS}$.

In reality, the Sun's temperature is very different from that of the cosmic background, $T_S \gg T_C$. The related minimum of the entropy production with respect to the terrestrial surface temperature is given by the equation

$$\frac{1}{A}\frac{\partial P}{\partial T_W} = 0 = 2\frac{j_{CW} - j_{WS}}{(T_W)^2} + X_{WS}^2 \frac{\partial L_{WS}}{\partial T_W} + X_{CW}^2 \frac{\partial L_{CW}}{\partial T_W} \qquad (3.24)$$

and does no longer coincide with the steady state, $j_{CW} = j_{WS}$. Thus, even in the absence of kinetic phase transitions and dissipative structures, the Prigogine theorem of minimum entropy production is in general not applicable to situations far from equilibrium. This simple result is in contrast to some hypotheses regarding the existence of general thermodynamic extremum principles that govern the overall climatic regime of our planet (Paltridge, 1975, 2001; Nicolis and Nicolis, 1980; Grassl, 1981; O'Brien and Stephens, 1995; Nicolis, 1999, 2010; Pujol et al., 1999; Pujol & Fort, 2002; Lorenz, 2002; Kleidon et al., 2010). The most rigorous mathematical proof that the terrestrial radiation balance does not correspond to extrema of entropy production was given by Pelkowsky (1995, 1997).

What remains generally valid even in far-from-equilibrium situations is the vanishing force-related differential at the steady state (Glansdorff and Prigogine, 1971),

$$\frac{1}{A}\frac{\partial_X P}{\partial T_W} \equiv \sum_k j_k \frac{\partial X_k}{\partial T_W} = \frac{j_{CW} - j_{WS}}{(T_W)^2} \tag{3.25}$$

Upon integration of this expression, various functions of T_W can formally be derived that possess extrema at the steady state, but usually none of those is of any general physical relevance, except in the vicinity of the equilibrium.

The steady-state solution, $j_{CW} = j_{WS}$, of Equation 3.14 defines the terrestrial blackbody temperature of this model (Petty, 2006), of

$$T_W^0 = \sqrt[4]{\frac{\sigma_{WS} T_S^4 + \sigma_{CW} T_C^4}{\sigma_{WS} + \sigma_{CW}}} \approx T_S \sqrt{\frac{R_S}{2R_{SE}}} \approx 279\ \text{K} \tag{3.26}$$

This value is higher than the usual estimate of about 255 K (Peixoto and Oort, 1992) because we omitted the terrestrial albedo (typically, 30%) in the formula for σ_{WS}, Equation 3.9.

We estimate now the Lyapunov coefficient of the steady state (3.26), that is, the relaxation time of fluctuations toward this state. For this purpose, we need some additional knowledge on certain substance properties of our system, in particular, on how the temperature of the water box in Figure 3.3 depends on the entropy.

The thermodynamic potential of seawater is available in the form of its Gibbs function, g^{SW}, which is the specific Gibbs energy (also termed specific free enthalpy) expressed as a function of salinity, temperature, and pressure (Feistel, 2008a; IAPWS, 2008; IOC, SCOR, and IAPSO, 2010). Enthalpy is computed from its temperature derivative, in the form

$$H_W = M_W \left(g^{SW} - T \frac{\partial g^{SW}}{\partial T} \right) \tag{3.27}$$

(where M_W is the mass of seawater in the water box), and in turn the heat capacity from

$$C_p = \frac{\partial H_W}{\partial T} \tag{3.28}$$

3.2 Black-Body Radiation Model of Earth

From the enthalpy balance of the water box, Equation 3.14, we infer by means of Equation 3.28 the dynamic equation that is controlling the surface temperature,

$$\frac{dT_W}{dt} = \frac{1}{C_p}\frac{dH_W}{dt} = \frac{1}{C_p}(j_{WS} - j_{CW}) \tag{3.29}$$

Series expansion with respect to small deviations, $\delta T_W = T_W - T_W^0$, from the stationary solution yields, up to the linear term

$$\frac{d}{dt}\delta T_W = -\frac{\delta T_W}{\tau} \tag{3.30}$$

For a realistic example to estimate this decay time, we may assume conditions typical for the subtropical Atlantic. From the equation of state of seawater with 18 °C (off St. Helena Island) and salinity 35 g kg^{-1}, we compute the specific heat capacity, $c_p \approx 3995$ J kg^{-1} K^{-1}, and the density, $\varrho = 1021$ kg m^{-3}. We estimate for the oceanic mixed layer a thickness of $h = 200$ m and obtain for the heat capacity of the water column with a surface area of 1 m^2 the value, $C_p = c_p \varrho h = 8.2 \times 10^8$ J K^{-1} m^{-2}, and hence the thermal relaxation time of the ocean, Equation 3.30, of

$$\tau = \frac{C_p}{4(\sigma_{WS} + \sigma_{CW})(T_W^0)^3} \approx 5 \text{ years} \tag{3.31}$$

in agreement with the climatological value reported by Schwartz (2007). In its order of magnitude, this result is not very sensitive to the various simplifications and approximations we have made in this model. In reality, the typical phase shift of 2–3 months observed between the local seasonal signals of solar irradiation and sea-surface temperature (SST) (Prescott and Collins, 1951; Frankignoul and Hasselmann, 1977; Frankignoul, 1985; Hagen and Feistel, 2008; Stine, Huybers, and Fung, 2009) accounts for just 5% of the black-body value (Equation 3.31). But in reality, if the SST were controlled by Arrhenius' radiation balance, the oceans would not exhibit any relevant seasonal temperature variation.

To give an observational example, we know that the solar irradiation incident in the subtropical region varies by a factor of about 2 between winter and summer. This ratio results in a measured seasonal change of the ocean's surface temperature of typically 2–4 °C, or about 1% of its absolute temperature, see Figure 3.4. In turn, the related black-body emission of the ocean is proportional to T_W^4; because of the power law it varies seasonally by just 4 × 1% which is much too small to compensate the change of 100% in the insolation rate. The thermal inertia of the ocean, if expressed in terms of the radiative relaxation time, Equation 3.31, is inconsistent with the observed seasonal variability.

Before discussing this in more detail in the next section, we roughly estimate the amplitude of the seasonal cycle of the subtropical ocean for the hypothetical case that it was controlled by radiation only. We estimate the subtropical seasonal irradiation by

$$j_{WS} = \langle j_{WS} \rangle \times \left(1 + \frac{1}{3}\exp(i\omega t)\right) \tag{3.32}$$

Seasonal Air Temperature on St. Helena Island

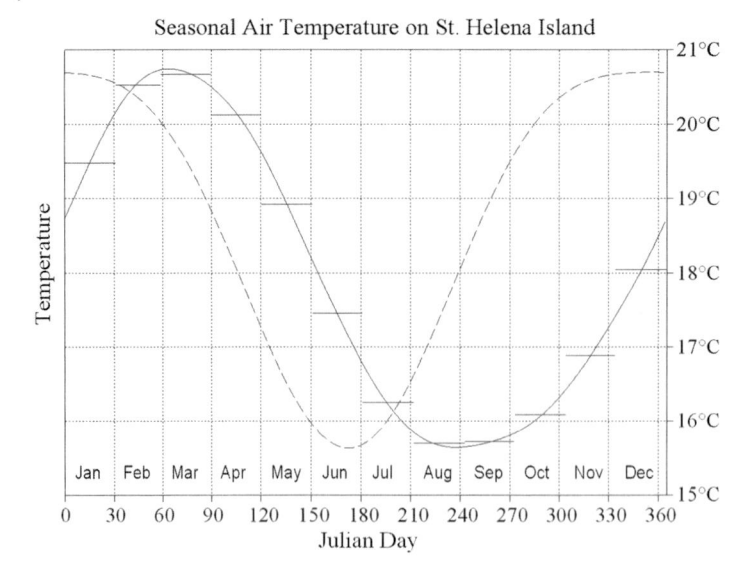

Figure 3.4 Climatological mean seasonal cycle of air-temperature 1893–1999 on St. Helena Island (Feistel, Hagen, and Grant, 2003), South Atlantic. Measured monthly mean values are indicated by horizontal bars. The daily resolved curve is generated mathematically by adjusting its monthly averages to the given data. For comparison, the dashed line shows the solar irradiation computed from Equation 3.40 with a suitably scaled amplitude.

and the outgoing radiation by

$$j_{CW} = \sigma_{CW} \left((T_W^0)^4 + 4(T_W^0)^3 \delta T_W \right) = \langle j_{WS} \rangle \left(1 + 4\frac{\delta T_W}{T_W^0} \right) \tag{3.33}$$

From Equation 3.29 we infer the temperature amplitude

$$|\delta T_W| = \frac{1}{3} \frac{\langle j_{WS} \rangle}{|i\omega C_p + 4\langle j_{WS} \rangle / T_W^0|} \tag{3.34}$$

From estimates for the mean irradiation, $\langle j_{WS} \rangle \approx 200 \text{ W m}^{-2}$, the terrestrial angular velocity, $\omega \approx 2 \times 10^{-7} \text{ s}^{-1}$, the heat capacity of the mixed layer, $C_p \approx 8 \times 10^8 \text{ J K}^{-1} \text{ m}^{-2}$, and the surface temperature, $T_W^0 \approx 300 \text{ K}$, we get a seasonal amplitude of

$$|\delta T| = \frac{200 \text{ K}}{|480i + 8|} \approx 0.4 \text{ K} \tag{3.35}$$

In the denominator, the part related to radiation amplitude is negligible in comparison to the heat capacity term. The resulting annual change of the ocean temperature significantly underestimates the observed values, such as 2.5 K shown in Figure 3.4. Hence, the seasonal amplitude of the real ocean heat loss is expected to be much higher than its radiation loss alone. In particular, even without any greenhouse effect, thermal radiation is insufficient to explain the rapid cooling observed in Austral fall, Figure 3.4, when the temperature drops by about 1.5 °C per month. With the same parameters as above, the required heat loss excess amounts to $C_p \Delta T / \Delta t \approx 460$ W m^{-2}, in addition to the compensation of the seasonal irradiation. That loss is higher than the mean irradiation by a factor of about 3, rather than by only 4% of thermal radiation increase in Austral summer, as estimated above.

3.3
Local Seasonal Response

While Equation 3.30 was developed as a model for the mean global radiation balance, it may also be understood as the related balance for a selected water parcel at the ocean surface if the parcel is not too small (such that exchange with neighboring parcels can be neglected on the time scale considered) and not too large (such that spatial homogeneity can reasonably be assumed). This may approximately hold for a part of the ocean mixed layer that extends over a few degrees in latitude and longitude and is observed on a time scale between several days and months.

For the astronomical variation of the solar irradiation at a certain position on the globe, detailed mathematical models are available (Monin, 1982; Peixoto and Oort, 1992; Rozwadowska and Isemer, 1998). For our present purpose of illustration, we consider a simpler approximate model with a circular orbit and a spherical shape of the Earth that is still sufficiently accurate but easier to apply. The radial unity vector of spherical coordinates, **n**,

$$\mathbf{n} = \begin{pmatrix} \cos\beta \sin\varphi \\ \sin\beta \\ \cos\beta \cos\varphi \end{pmatrix} \tag{3.36}$$

is perpendicular to the terrestrial surface at the latitude β and the longitude φ. To be transformed into a coordinate frame defined by fixed positions of Earth and Sun, the vector **n** is tilted by the Earth-axis inclination of $\psi = 23.45°$, and turned by the revolution angle α, by means of rotation matrices

$$\mathbf{N} = \begin{pmatrix} \cos\alpha & 0 & \sin\alpha \\ 0 & 1 & 0 \\ -\sin\alpha & 0 & \cos\alpha \end{pmatrix} \cdot \begin{pmatrix} 1 & 0 & 0 \\ 0 & \cos\psi & -\sin\psi \\ 0 & \sin\psi & \cos\psi \end{pmatrix} \cdot \mathbf{n} \tag{3.37}$$

The scalar product of **N** (pointing upward from the Earth's surface) with the unity vector pointing from the Sun to the Earth is the z-component of **N**, computed by evaluating Equation 3.37,

$$N_z = a \cos\varphi + b \sin\varphi + c \tag{3.38}$$

where the abbreviations are defined as

$$a \equiv \cos\alpha \cos\psi \cos\beta, \ b \equiv -\sin\alpha \cos\beta, \ c \equiv \cos\alpha \sin\psi \sin\beta \tag{3.39}$$

The local daily mean solar irradiation $q(\alpha, \beta)$, as a function of the season, α, given as an angle between 0 and 2π, and of the latitude, β, is proportional to the daytime integral over N_z

$$q(\alpha, \beta) = -\frac{q_0}{2\pi} \int_{\varphi_1}^{\varphi_2} N_z d\varphi = \frac{q_0}{2\pi} (2w - c(\varphi_2 - \varphi_1)) \qquad (3.40)$$

between the longitudes of sunrise, φ_1, and of sunset, φ_2. The terminator circle on the globe is given by horizontal incidence, that is, by the condition $N_z = 0$,

$$\sin\varphi_{1,2} = \frac{ac \mp bw}{a^2 + b^2} \qquad \cos\varphi_{1,2} = \frac{bc \pm aw}{a^2 + b^2} \qquad (3.41)$$

if $w = \sqrt{a^2 + b^2 - c^2}$ is a real number. Otherwise, there is either polar day ($N_z < 0$), $q = 2\pi|c|$, or polar night ($N_z > 0$), $q = 0$, but neither case applies to the subtropical ocean which we have preferably in mind here. The revolution angle $\alpha = 2\pi(j + 9\,\mathrm{d})/365\,\mathrm{d}$ depends on the Julian day number j, with a 10 d offset to the December solstice. Any leap-year adjustment is neglected for simplicity.

We decompose the radiation flux (3.40) into its seasonal spectral components,

$$q(\alpha, \beta) = \sum_m Q_m(\beta) \exp(im\alpha) \qquad (3.42)$$

that is, complex amplitudes of multiples $m = 0, \pm 1, \pm 2, \ldots$ of the fundamental frequency $\omega_0 = 2\pi/365\,\mathrm{d}$, as functions of the latitude, $Q_m(\beta)$,

$$Q_m(\beta) = \frac{1}{2\pi} \int_0^{2\pi} d\alpha\, q(\alpha, \beta) \exp(-im\alpha) \qquad (3.43)$$

Since q is a real function, the Fourier amplitudes and their complex conjugates satisfy the symmetry relation $Q_{-m}(\beta) = Q_m^*(\beta)$. The magnitudes of these functions are shown in Figure 3.5. Only the modes 0, 1, and 2 are relevant. For latitudes higher than 4°, the principal seasonal cycle ($m = \pm 1$) is dominating the temporal signal.

After those preparatory steps, we return to the enthalpy balance, Equation 3.29, for the case of local periodic forcing

$$\frac{dT_W}{dt} = \frac{1}{C_p}(j_{WS}(t) - j_{CW}) \qquad (3.44)$$

where the outgoing radiation is $j_{CW} \approx \sigma_{SB} T_W^4$, and the seasonal input of solar irradiation, Figure 3.3, is approximated by its fundamental annual harmonic as a function of the latitude, β, corresponding to Figure 3.5, using $q_0 = I_E$ in Equation 3.40, that is, omitting the albedo,

$$j_{WS}(t) = q(\alpha, \beta) = Q_0(\beta) + Q_1(\beta) \exp(i\omega_0 t) + Q_{-1}(\beta) \exp(-i\omega_0 t) + \cdots \qquad (3.45)$$

Seasonal Spectra of Solar Irradiation

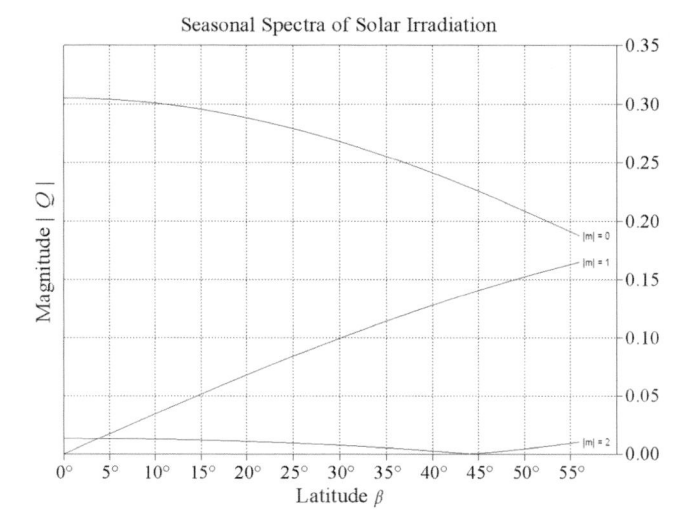

Figure 3.5 Meridional dependence of the magnitude of the Fourier components of local daily solar irradiation, $|Q_0(\beta)|$ and $|Q_m(\beta)| + |Q_{-m}(\beta)| = 2|Q_m(\beta)|$, for the annual mean value, $m = 0$, the annual cycle of 12 month period, $m = 1$, and the semiannual "double" cycle of 6 month period, $m = 2$, computed from Equation 3.43 for $q_0 = 1$, that is, in units of the solar constant, $I_E = 1366\,\mathrm{W\,m^{-2}}$.

The annual mean temperature is

$$T_W^0(\beta) = \sqrt[4]{\frac{Q_0(\beta)}{\sigma_{SB}}} \tag{3.46}$$

The value of 284 K resulting for the subtropics, $\beta = 30°$, is sufficiently realistic for our purpose. For the time-dependent seasonal anomaly, $\delta T_W(\beta, t) = T_W - T_W^0(\beta)$, we use the linearized form of Equation 3.44,

$$\frac{d}{dt}\delta T_W = -\frac{\delta T_W}{\tau(\beta)} + f(t) \tag{3.47}$$

where the relaxation time is

$$\tau(\beta) = \frac{C_p}{4\sigma_{SB}[T_W^0(\beta)]^3} \tag{3.48}$$

With $C_p = 8.2 \times 10^8\,\mathrm{J\,K^{-1}\,m^{-2}}$, Equation 3.31, the resulting value at $\beta = 30°$ is $\tau \approx 5$ years. The forcing function in Equation 3.47 is

$$f(t) = 2\mathrm{Re}\left\{\frac{Q_1(\beta)}{C_p}\exp(i\omega_0 t)\right\} \tag{3.49}$$

where we made use of the symmetry, $Q_{-1}(\beta) = Q_1^*(\beta)$, Equation 3.43. By means of Equation 3.47 we can estimate the amplitude and the phase for the seasonal cycle of the SST under the idealized model assumption that it is directly controlled by the photon mill, that is, exclusively by the local radiation balance between the ocean, the Sun, and the cosmic background. The time-dependent, quasistationary solution of Equation 3.47 is the exponential response

$$\delta T_W(t) = \int_0^\infty dt' \exp\left(-\frac{t'}{\tau}\right) f(t-t') \qquad (3.50)$$

This response function is a low-pass filter that effectively suppresses spectral components of f that run on shorter scales than the response time τ. Integration of Equation 3.49 results in the SST oscillation

$$\delta T_W(t) = 2\,\text{Re}\left\{\frac{Q_1(\beta)\tau}{C_p(1-i\omega_0\tau)} \exp(i\omega_0 t)\right\} = A(\beta)\cos(\omega_0 t + \varphi_0 + \varphi) \qquad (3.51)$$

Here, the seasonal temperature amplitude is

$$A(\beta) = \frac{2\tau(\beta)}{C_p} \frac{|Q_1(\beta)|}{\sqrt{1+[\omega_0\tau(\beta)]^2}} \qquad (3.52)$$

The initial phase φ_0 of the solar forcing signal at $t=0$ is given by

$$\tan\varphi_0 = \frac{\text{Im}\,Q_1(\beta)}{\text{Re}\,Q_1(\beta)} \qquad (3.53)$$

and the phase shift φ between the irradiation signal and the SST response follows from

$$\tan\varphi = \omega_0\tau(\beta) \qquad (3.54)$$

It is obvious that the phase lag will always be within the interval $0 \le \varphi < \pi/2$, corresponding to a positive delay of no more than about 91 days. In particular, if $\omega_0\tau \gg 1$, as in the case of Equation 3.48, the maximum delay of about 3 months will be rather independent of the actual value of the relaxation time τ. Between 20° and 60° latitude, the delay computed from Equation 3.54 varies only between 89 and 90 days. In contrast, Figure 3.6 shows the delay between solar irradiation and the principal mode of zonally averaged climatological SST data from the South Atlantic over the period 1985–2004. Between 9° and 30°, the delay decreases almost linearly from about 95 to 70 days. South of 30 °S, the influence of the west-wind belt and the Antarctic Circumpolar Current increasingly modify the subtropical conditions.

We conclude that the results from the local black-body response, Equation 3.54, are in agreement with the order of magnitude of 3 months time lag, but inappropriate to explain the observational details. In particular, the relaxation time of 5 years, Equation 3.48, is inconsistent with measured delay values that are significantly shorter than 90 days. Observed delays of more than 91 days cannot be explained in terms of a simple linear response to a given seasonal excitation, Equation 3.54. Similar time lags over 100 days were already reported by Prescott & Collins (1951).

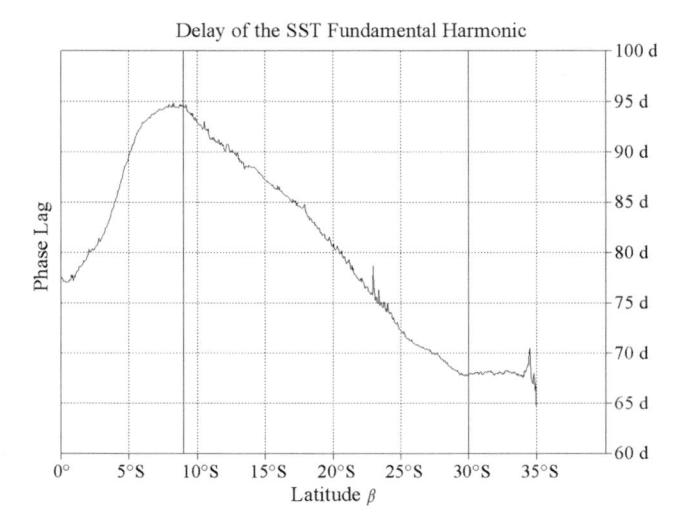

Figure 3.6 Phase shift in days between the seasonal fundamental harmonics of astronomical solar irradiation and zonally averaged sea-surface temperature (SST) of the South Atlantic (extracted from monthly Pathfinder v5 SST data 1985–2004 compiled by Kobus Agenbag, Capetown, 2006). In the subtropical range of latitudes, indicated by vertical lines, the delay shows a linear poleward decrease.

From regular measurements taken on St. Helena Island located at 15° 57′ S, 5° 42′ W in the subtropical South Atlantic, the monthly climatology is available since 1892 (Feistel, Hagen, and Grant, 2003). Although the island's air temperatures are similar to those of the surrounding ocean, there is probably some land influence in the form of higher daily maxima and a reduced seasonal delay. St. Helena is renown for its healthy and pleasant climate; the smooth climatological daily temperature curve shown in Figure 3.4 is reconstructed such that the monthly mean values derived from it coincide with those observed. The curve is delayed by 78 days behind the local sunshine cycle, and the summer–winter difference is about 5 °C. From the satellite SST data exploited for Figure 3.6 the same temperature range is found. By comparison, the temperature amplitude computed from Equation 3.52 amounts to $A(16°S) \approx 0.5$ °C only. Thus, also with respect to the seasonal temperature range, the results calculated from the local black-body radiation model of the ocean deviate significantly from the observation.

For comparison with the black-body model, the observed seasonal temperature cycle of St. Helena can approximately be expressed in terms of an empirical exponential response function, Equations 3.47 and 3.50, with respect to the local astronomical irradiation anomaly, $\delta j_{WS}(t)$, Equation 3.44, in the form

$$\delta T_W(t) \approx \frac{1}{C_{SH}} \int_0^\infty dt' \exp\left(-\frac{t'}{\tau_{SH}}\right) \delta j_{WS}(t-t') \tag{3.55}$$

The local relaxation time, τ_{SH}, Equation 3.54, would then be

$$\tau_{SH} = \frac{365 \, d}{2\pi} \tan\left(2\pi \frac{78 \, d}{365 \, d}\right) \approx 250 \, d \tag{3.56}$$

and the local effective heat capacity, Equation 3.52, would amount to

$$C_{SH} = \frac{2\tau_{SH}}{2.5\,K} \frac{|Q_1(16°)|}{\sqrt{1 + [\omega_0 \tau_{SH}]^2}} \approx 1.5 \times 10^8 \, J\, K^{-1}\, m^{-2} \quad (3.57)$$

Figure 3.4 shows that the observed delay is about 1.5 months in the austral autumn cooling and about 3.5 months in the austral spring, when the temperature is rising. This cannot be explained by a local linear response model such as Equation 3.55, without allowing for clouds, the rainy season, and so on.

We conclude that the real ocean's surface temperature, and in connection with it, to some extent, also the global climate, cannot be governed by a simple radiation balance between the surface, the Sun, and the cosmic background. While the corresponding energy balance, the Arrhenius model, is in reasonable agreement with observations, the related time constant estimated for the ocean is rather unrealistic. Beyond some greenhouse blocking, the atmosphere must be more fundamentally involved in the terrestrial part of the photon mill, which is already evident from the fact that the atmospheric cooling rate matches the global entropy export, as we shall see in the next section. We combine the black-body ocean with the cooling atmosphere model and raise the question again of whether this coupled system is a more suitable simple climate model than the classical Arrhenius model without atmosphere.

3.4
Atmospheric Cooling Rate

In reasonable approximation, the turbulent motion of the atmosphere can be modeled by an isentropic temperature profile as a function of the pressure. This idealized nonequilibrium state shows an almost realistic constant vertical temperature gradient and is in contrast to the isothermal equilibrium state of maximum entropy that is sometimes discussed (Dutton, 1973; Fortak, 1979).

In our model, the specific entropy, s, is assumed to be independent of the vertical coordinate, z, as a result of adiabatic vertical turbulent transport. Using the specific enthalpy, $h(s, P)$, as a thermodynamic potential function that has the property

$$\frac{1}{\varrho} = \left(\frac{\partial h}{\partial P}\right)_s \quad (3.58)$$

the hydrostatic equation, $\partial P/\partial z = -\varrho g_E$, with density ϱ, pressure P, and terrestrial gravity g_E can be integrated in the form (Landau and Lifschitz, 1966)

$$h(s, P) = h_0 - g_E z \quad (3.59)$$

The subscript "0" indicates surface values at $z = 0$. Expressed in terms of temperature T and pressure, the Gibbs function of air, $g(T, P)$, provides expressions for the specific entropy, $s = -(\partial g/\partial T)_P$ and the specific enthalpy, $h = g + Ts$ (Feistel et al., 2010b). In the ideal-gas limit, g reads

$$g(T, P) = g_{\text{ref}} + \int_{T_{\text{ref}}}^{T} \left(1 - \frac{T}{T'}\right) \cdot c_P(T') \, dT' + RT \ln \frac{P}{P_{\text{ref}}} \qquad (3.60)$$

where the subscript "ref" denotes some particular constant values at an arbitrarily specified reference state, and R is the specific gas constant of air. The enthalpy h of an ideal gas is degenerate with respect to the pressure. If the heat capacity in Equation 3.60 is independent of the temperature, the specific enthalpy takes the simple form (Kluge and Neugebauer, 1976)

$$h(T, p) = h_{\text{ref}} + (T - T_{\text{ref}}) c_p \qquad (3.61)$$

Thus, we obtain from Equation 3.59 the linear vertical temperature profile

$$T(z) = T_0 - \Gamma z \qquad (3.62)$$

where the surface temperature is $T_0 = T_{\text{ref}} + (h_0 - h_{\text{ref}})/c_p$ and the adiabatic lapse rate is $\Gamma = -dT/dz = g_E/c_p$ (Gill, 1982; Feistel et al., 2010b). This linear profile has only two parameters – the surface temperature and the vertical slope. Since the latter is a constant (depending on the humidity, assumed here as fixed) in this model, any change in the thermal state of the atmosphere results in a barotropic variation of the temperature, that is, $\Delta T(z) = \Delta T_0$ is independent of the altitude z. A radiative cooling rate, $K = -dT/dt$, with this property is in fact observed. For the subtropics, its typical value is $K \approx 2 \, \text{K} \, \text{d}^{-1}$ (Ridgway, Harshvardan, and Arking, 1991; Clough, Ianoco, and Moncet, 1992; Zhong and Haigh, 1995; Mlawer et al., 1997; Hartmann, Holton, and Fu, 2001). We shall relate that daily cooling rate now to the entropy of the atmosphere.

A hydrostatic and isentropic atmosphere satisfies the conditions

$$\frac{\partial P}{\partial z} = -\varrho g_E \qquad (3.63)$$

and

$$s(T(z), P(z)) = s(T_0, P_0) \qquad (3.64)$$

Per surface area, its total entropy S is given by the integral

$$S = \int_0^Z \varrho s \, dz = -\frac{s(T_0, P_0)}{g_E} \int_{P_0}^{0} dP = \frac{s(T_0, P_0)}{g_E} P_0 \qquad (3.65)$$

where we have assumed $P = 0$ at the top of the atmosphere for simplicity. From Equation 3.60, we obtain the specific entropy

$$s(T, P) = -\left(\frac{\partial g}{\partial T}\right)_P = s_{\text{ref}} + c_p \ln \frac{T}{T_{\text{ref}}} - R \ln \frac{P}{P_{\text{ref}}} \qquad (3.66)$$

As an aside, we note that this equation provides the isentropic vertical temperature–pressure profile $T(P)$ from the condition (3.64). Inserting Equation 3.66 into the expression for the total entropy of the air column, Equation 3.65, we get the equation we are looking for

$$S(T_0, P_0) = \frac{P_0}{g_E} s(T_0, P_0) = \frac{P_0}{g_E} \left\{ s_{\text{ref}} + c_p \ln \frac{T_0}{T_{\text{ref}}} - R \ln \frac{P_0}{P_{\text{ref}}} \right\} \qquad (3.67)$$

The atmospheric radiative cooling into space at constant surface pressure is, therefore, equivalent to an entropy export rate of

$$\frac{d_e S}{dt} = \left(\frac{\partial S}{\partial T_0} \right)_{P_0} \frac{dT_0}{dt} = -K \frac{P_0 c_p}{g_E T_0} = -\frac{K P_0}{\Gamma T_0} \qquad (3.68)$$

The actual value to be used for the adiabatic lapse rate depends on the humidity (Gill, 1982; Feistel et al., 2010b). As a suitable representative for the subtropical South Atlantic, St. Helena Island has a climatological lapse rate of $\Gamma = 7.7$ mK m^{-1} (Feistel, Hagen, and Grant, 2003). In the mean seasonal cycle on St. Helena, Figure 3.4, the air temperature T_0 varies between 16 and 23 °C, that is, by $\pm 1.3\%$, while the sea-level air pressure P_0 varies between 1016 and 1020 hPa, that is, by $\pm 0.2\%$. In the course of the year, the minimum temperature is observed in conjunction with the pressure maximum. Hence, the air density, P_0/RT_0, changes within about $\pm 1.5\%$ which is irrelevant for our estimate. With the rough approximations of $K \approx 2$ K d^{-1}, $P_0 \approx 10^5$ Pa, $\Gamma \approx 8$ mK m^{-1}, and $T_0 \approx 300$ K we calculate from Equation 3.68 the related entropy export rate of

$$\frac{d_e S}{dt} \approx -1 \text{ W m}^{-2} \text{ K}^{-1} \qquad (3.69)$$

With respect to the various crude estimates made in our model, this value is in excellent agreement with the photon-mill model, Equation 3.1, even though it is derived from completely different empirical facts and simplified theoretical relations. We conclude that the isentropic model atmosphere considered here with just one thermal degree of freedom, the surface temperature, may be helpful to illustrate the working principles of the heat engine "climate."

3.5
Black-Body Model with Atmosphere

In the literature on climate change, global power budgets similar to Figure 3.7 are frequently presented (Schneider, 1990; Peixoto and Oort, 1992; Cockell, 2008; Haigh, 2010). They differ from the earlier generation of heat balance schematics (Svatkov, 1974; Budyko, 1978) mainly by the greenhouse effect in the form of infrared radiation exchanged between the surface and the atmosphere to an amount that is similar to the total solar irradiation. Carbon dioxide contributes only a small fraction to this effect; most of it results from water vapor, liquid water, or ice crystals in the atmosphere, (Emden, 1913), which are almost perfectly "black" in the infrared spectrum.

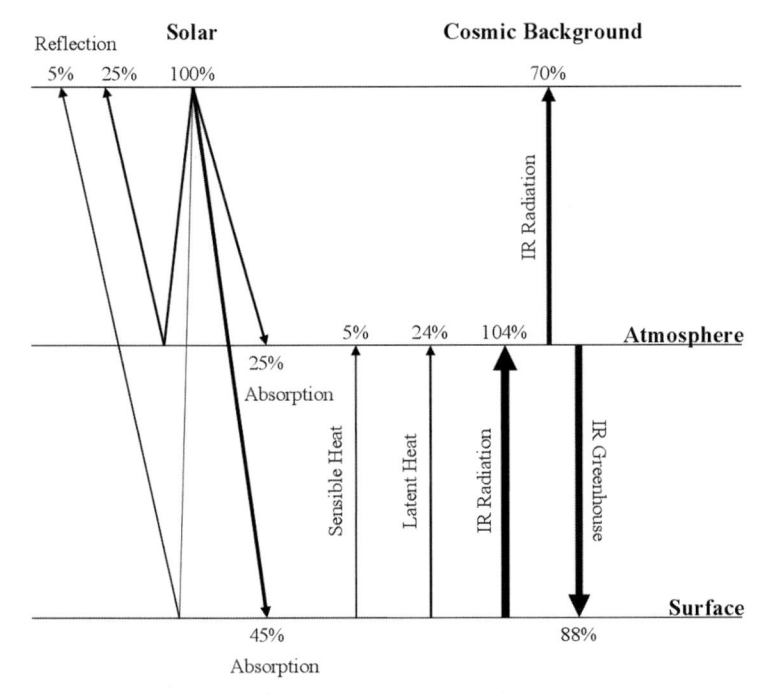

Figure 3.7 Schematic of the globally averaged climatic power budget, The energy exchanged between surface and atmosphere by the infrared (IR) greenhouse effect is more than twice the solar input at the surface, similar to the black-body model shown in Figure 2.7 in Section 2.5. If this averaged balance is applied to the ocean alone, a substantially larger latent heatflow must be considered. Adapted from Schneider (1990), Peixoto and Oort (1992), Cockell (2008), Trenberth, Fasullo, and Kiehl (2009), and Haigh (2010).

In this section, according to Figure 3.7 we shall tentatively assume that the infrared radiation is the dominating physical mechanism of energy exchange between the atmosphere and the surface, in particular, the ocean surface. Under this assumption, it appears obvious that a strengthened greenhouse effect will reduce the cooling of ocean currents which carry heat from the tropics to the high latitudes, such as the Gulfstream, with all the implications this may have for the moderate and polar climate zones.

As a generalization of the three-box model shown in Figure 3.3, we insert the atmosphere as a new box between the ocean and the cosmic background, Figure 3.8, consistent with equivalence of global entropy export and atmospheric cooling rate, Section 3.4. The thermal states of the atmosphere and of the oceanic mixed layer are represented by their surface temperatures T_A and T_W, respectively, which are the only dynamical variables of this model.

The temperatures are subject to energy balance equations, similar to Equation 3.29, in the form

$$C_W \frac{dT_W}{dt} = (1-a)\Phi - F(T_W, T_A) \quad (3.70)$$

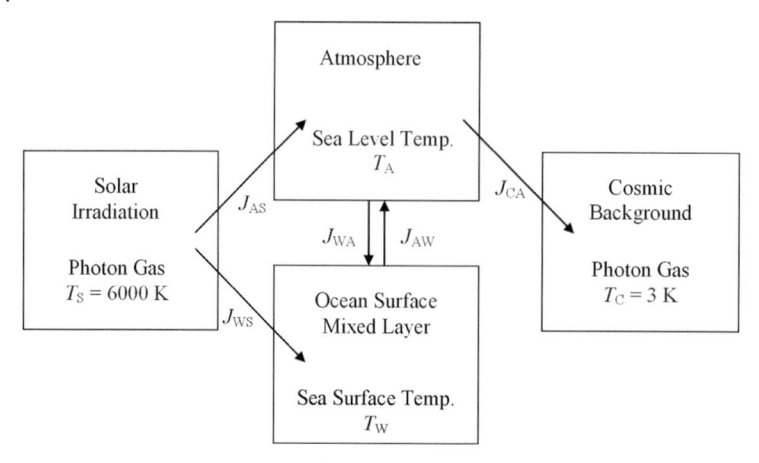

3 Evolution of Earth and the Terrestrial Climate

Figure 3.8 Four-box model of the photon mill. The planetary surface is split into an atmospheric and oceanic part with mutual thermal interaction. Only the atmosphere is exporting energy and entropy.

$$C_A \frac{dT_A}{dt} = a\Phi + F(T_W, T_A) - \sigma_A(T_A)^4 \tag{3.71}$$

Here, C_A and C_W are the heat capacities of the air and the water column, respectively, $\Phi = J_{AS} + J_{WS}$ is the nonreflected part of the astronomical solar irradiation, where J_{AS} and J_{WS} are the solar irradiation intensities absorbed by the atmosphere and the ocean, respectively, $a = J_{AS}/\Phi$ is the fraction absorbed by the atmosphere, σ_A is the effective Stefan–Boltzmann constant for the atmospheric long-wave radiation into the outer space, $J_{CA} = \sigma_A(T_A)^4$, and $F = J_{AW} - J_{WA}$ is the ocean–atmosphere energy exchange which, in first approximation, will be assumed later in this section to consist only of the greenhouse effect, Figure 3.7.

The unique steady-state solution for the atmospheric temperature is inferred from (3.70) and (3.71) as

$$T_A^0 = \sqrt[4]{\Phi/\sigma_A} \tag{3.72}$$

Using the black-body value for σ_A, the resulting temperature (Peixoto and Oort, 1992), $T_A^0 = 255$ K, is below the mean terrestrial surface temperature of 287 K (NCDC, 2010). This deviation was attributed to the greenhouse effect of the atmosphere by Arrhenius (1896). Here we argue that the atmosphere is vertically not isothermal, Equation 3.62, and that its effective radiation constant σ_A is smaller than the black-body value. In particular, if the air temperature T_A^0 is rising due to a stronger greenhouse effect, σ_A will be lowered in order to remain in balance with the solar constant, $\Phi = J_{CA}$, Equation 3.72.

The related stationary water temperature, T_W^0, is given implicitly by the solution of the equation

$$F(T_W^0, T_A^0) = (1-a)\Phi \tag{3.73}$$

It is helpful to transform the system (3.70), (3.71) to dimensionless temperature and time variables by means of the relations

$$\vartheta_A = (T_A - T_A^0)/T_0, \quad \vartheta_W = (T_W - T_W^0)/T_0, \quad \tau = t/t_0 \tag{3.74}$$

where the reducing constants are $T_0 = \sqrt[4]{\Phi/\sigma_A}$ and $t_0 = C_W T_0/\Phi$. We obtain the dimensionless equations

$$\frac{d\vartheta_W}{d\tau} = 1 - f(\vartheta_W, \vartheta_A) \tag{3.75}$$

$$\varepsilon \frac{d\vartheta_A}{d\tau} = f(\vartheta_W, \vartheta_A) - (1 + \vartheta_A)^4 \tag{3.76}$$

Here, the reduced air–sea interaction function is defined as

$$f(\vartheta_W, \vartheta_A) = a + F(T_W, T_A)/\Phi \tag{3.77}$$

and $\varepsilon = C_A/C_W$ is a small parameter in the order of magnitude of 0.02. From Equation 3.76 we infer that the air temperature, ϑ_A, is a fast variable that is a slave to the slower water temperature, ϑ_W. For small fluctuations, we can linearize Equations 3.75 and 3.76 in terms of the partial derivatives, f_W, f_A of f, in the form

$$\frac{d\vartheta_W}{d\tau} = -f_W \vartheta_W - f_A \vartheta_A \tag{3.78}$$

$$\varepsilon \frac{d\vartheta_A}{d\tau} = f_W \vartheta_W + f_A \vartheta_A - 4\vartheta_A \tag{3.79}$$

and find the quasi-steady state of Equation 3.79,

$$\vartheta_A = \vartheta_W \frac{f_W}{4 - f_A} \tag{3.80}$$

After inserting this solution into Equation 3.78, the slow mode is governed by the equation

$$\frac{d\vartheta_W}{d\tau} = -\vartheta_W \frac{4 f_W}{4 - f_A} \tag{3.81}$$

Transformed back to the original units, the related relaxation time of the ocean is

$$t_W = \frac{C_W}{F_W} \left[1 - \frac{F_A T_A^0}{4\Phi} \right] \tag{3.82}$$

The derivatives of F are abbreviated by $F_A = \partial F/\partial T_A$ and $F_W = \partial F/\partial T_W$, to be evaluated at the steady state.

Equation 3.82 is valid for arbitrary exchange rates F between the ocean and the atmosphere. In the case of a radiative balance, we may assume black-body thermal radiation laws

$$F(T_W, T_A) = \sigma(T_W)^4 - \sigma(T_A)^4 \tag{3.83}$$

Then, Equation 3.82 takes the form

$$t_W = \frac{C_W}{2\sigma(T_W^0)^3} \quad (3.84)$$

This result is by a factor of 2 greater than the relaxation time of the ocean without atmosphere, Equations 3.31 and 3.48. We conclude that the radiative greenhouse coupling is insufficient to explain the ocean's quick thermal response observed in the climate system. This implies that the net thermal radiation can only account for a minor fraction of the ocean–atmosphere energy transfer, in contrast to the schematic presented in Figure 3.7. Consequently, the greenhouse effect is not blocking the direct ocean cooling very strongly. Rather, it indirectly even intensifies the export of latent heat from the sea surface due to lower relative humidity of warmer air, resulting in an additional water-cooling effect.

As we stated already before, Arrhenius' radiation model provides reasonable estimates for the temperature but fails completely in explaining the time scales, even if an atmospheric greenhouse is included in the model. Another important player is obviously missing in this simple model. Therefore, we will add water to the atmosphere in the following section.

The model displayed in Figure 3.8 can be interpreted in two different versions. As a local model, it is understood as a column of ocean water with a column of air on top. Taking a trade wind speed of $10\,\mathrm{m\,s^{-1}}$, air is moving laterally over a distance of 10,000 km in 10 days, or circulating once around a pressure center such as the South Atlantic Anticyclone. Thus, on time scales that exceed 10 days, a one-dimensional local model may no longer be any reasonable and must be extended to include lateral exchange with air from the continents, other latitudes, and so on.

The second way of looking at Figure 3.8 is as an average global model. The solid ground has a very small heat capacity (such as $1600\,\mathrm{J\,kg^{-1}\,K^{-1}}$ in the Gobi desert, with a penetration depth of a few meters, Albrecht, 1940) and releases most of its absorbed heat to the air within hours. On the climatic scale that we are interested in here, the solid surface appears just as a part of the atmospheric box, adding an absorption rate of sunlight proportional to the land–ocean ratio of the global surface. Then, the oceanic box plus the atmospheric box together represent the entire globe as the dynamical component of the photon-mill model.

3.6
Humidity and Latent Heat

Estimates for the climatological annual evaporation from the ocean surface vary by about 20% between different investigators and amount to approximately $E_S = 1200$ mm yr^{-1} on the global average (Baumgartner and Reichel, 1975; Peixoto and Oort, 1992). Over the subtropical South Atlantic, the values range from 1400 to 1650 mm yr^{-1}. The latent heat of evaporation from seawater depends only weakly on temperature and salinity and takes values of $L = 2500\,\mathrm{kJ\,kg^{-1}}$ at $0\,°\mathrm{C}$ or $2450\,\mathrm{kJ\,kg^{-1}}$ at $20\,°\mathrm{C}$ (Feistel et al., 2010b; IOC, SCOR, and IAPSO, 2010). The resulting mean export rate of latent heat per ocean surface area, $\rho L E_S$, is slightly less than $100\,\mathrm{W\,m^{-2}}$. Compared to this number, the sensible heat exchange between the ocean and the atmosphere is of only minor importance (Albrecht, 1940). Recall that the solar irradiation at the top of the atmosphere is 340 W m^{-2} (Section 3.1), and that 50% or 170 W m^{-2} is absorbed or reflected by the atmosphere, as shown in Figure 3.7. The ocean's albedo depends strongly on the surface roughness, the whitecap density, and the angle of radiation incidence (Albrecht, 1940; Jin et al., 2004; Emery, Talley, and Pickard, 2006). If we assume a sea-surface albedo of roughly 10% on average (Albrecht, 1940), we find that about 150 W m^{-2} of sunlight is absorbed by the ocean. Except for the small amount consumed by phytoplankton, the same power must in turn be exported from the ocean. From the latent heat of 100 W m^{-2}, above, we infer that roughly 2/3 of the energy is transferred to the atmosphere in the form of evaporation, and the remaining 1/3 by thermal radiation. The total long-wave radiation of the ocean significantly exceeds the latter number of 50 W m^{-2}, see Figure 3.7. The total intensity is available from the Stefan–Boltzmann law, such as $\sigma_{SB} T_W^4 \approx$ 400 W m^{-2} at 17 °C; the related difference of 350 W m^{-2} is then balanced by the downward radiation caused by the greenhouse effect, mainly due to atmospheric moisture, similar to the radiation transfer model shown in Figure 2.7 in Section 2.5. Schematic ocean heat balances reported by other authors may differ from our estimates. For instance, Emery, Talley, and Pickard (2006) suggest fractions of latent heat, net radiation, and sensible heat for the export from the sea which amount to 21%, 19%, and 7%, respectively, of the short-wave radiation energy received from the Sun. The deviations possibly result not only from the crude global estimates we used here but also from severe practical difficulties to precisely monitor and model the air–sea fluxes (Kleeman & Power, 1995; Worley et al., 2005). As well, simple models indicate that already the radiative transfer, quite independent of actual temperature and humidity, results in an air temperature almost equal to or even slightly higher than that of the ocean surface, so that sensible air-sea heat exchange may not contribute at all to the mean energy export from the ocean (Feistel, 2011d).

Adding the global water cycle to the photon mill increases the complexity of the model significantly and turns it into some kind of steam-engine model. In the previous sections, humidity played only an indirect role as a property that influences the value of the adiabatic lapse rate of the atmosphere, or as the substance that is predominantly responsible for the greenhouse effect. Although the four-box model, Figure 3.8, remains essentially unaltered, a new third state variable – specific humidity q – is added to the dynamic model, Equations 3.70 and 3.71, and a balance equation for the transfer of latent heat between the ocean and the atmosphere must be derived from physical laws and observational facts.

There are two important and rather general empirical rules of which such a model must take care. The first is that the temperatures of the sea surface and the atmosphere at sea level generally exhibit only small differences (order of 1–2 K) which may be explained by radiative air-sea interaction (Feistel, 2011d). The second rule is that the relative humidity of the air over the ocean is about 80%, rather independent of the season or the region (Albrecht, 1940; Gill, 1982; Mitchell, 1989; Dai, 2006; Feistel, Nausch, and Wasmund, 2008a), and is even independent of global warming, in contrast to the atmospheric water content (Santer et al., 2007).

Surprisingly, those rules seem to be inconsistent with the assumption of a dominating role of the latent heat in the ocean–atmosphere energy transfer. First, evaporation is driven by the difference between the chemical potentials of water in seawater and in humid air (Feistel et al., 2010b; IOC, SCOR, and IAPSO, 2010) rather than by the temperature difference across the surface. Thus, evaporation cooling of the ocean may establish significant air–sea temperature offsets rather than leveling them dynamically. Second, constant relative humidity of 80% implies a constant difference between the chemical potentials (Feistel et al., 2010b; IOC, SCOR, and IAPSO, 2010; Feistel, 2011b). If heat was exported from the sea at a temperature-independent rate, warming and cooling processes of the ocean would take their extremum values during the summer and winter solstice, that is, with a phase lag of the linear response of 90° behind the local irradiation signal. The rule would thus not help in explaining the observed shorter relaxation times. It is therefore necessary to briefly take a closer look at the complex processes that take place at the interface between the ocean and the atmosphere.

The *skin effect*, that is, the formation of a tiny and usually cooler layer of just a few millimeters thickness on both sides of the air–sea interface, takes a key position as a bottleneck in the global climate engine, and still poses a major challenge to irreversible thermodynamics, theoretical fluid dynamics, observational oceanography and meteorology, as well as remote sensing of the SST (Grassl, 1976; Schlüssel et al., 1990; Donlon et al., 2002; Fairall et al., 2003). In 1991 and 1992, two expeditions of r/v "A. v. Humboldt" to the subtropical North Atlantic were aimed at "Atlantic Measurement of Oceanic Radiation" (AMOR) with Eberhard Hagen as the chief scientist (Hagen et al., 1994; Hagen, Zülicke, and Feistel, 1996). The cruises took place in close cooperation with Peter Schlittenhardt of the Institute for Remote Sensing Applications (IRSA), Ispra, Italy. The research vessel was equipped with a state-of-the-art radiation thermometer to study the infrared emission of the sea surface from a short distance, together with the observation of meteorological and oceanographic conditions, and in particular to investigate the thermal skin effect by regular comparison measurements with a stirred reference vessel, Figure 3.9. Among the results, rapid fluctuations of the skin temperature on the time scale of seconds as predicted by some theoretical models could not be observed.

Exchange processes of energy and water between the ocean and the atmosphere are driven by nonequilibrium conditions at the interface. The phase equilibria between seawater, humid air, and ice can be computed very accurately and consistently from the new seawater standard TEOS-10 (IOC, SCOR, and IAPSO, 2010; Feistel et al., 2008b; Feistel, 2011b) which is based on five mutually consistent and very accurate thermodynamic formulations of the International Association for the Properties of Water and Steam:

1) a Helmholtz function, $f^F(T, \varrho)$, of fluid water (IAPWS, 2009a),
2) a Gibbs function, $g^W(T, P)$, of liquid water (IAPWS, 2009b),
3) a Gibbs function, $g^{Ih}(T, P)$, of ambient hexagonal ice (IAPWS, 2009c),
4) a Gibbs function, $g^{SW}(c, T, P)$, of seawater (IAPWS, 2008), and
5) a Helmholtz function, $f^{AV}(a, T, \varrho)$, of humid air (IAPWS, 2010).

Here, ϱ is the mass density of either pure water or of humid air, respectively, c is the absolute salinity, that is, the mass fraction of dissolved salt in seawater, $a = 1 - q$ is the

Figure 3.9 Christoph Zülicke (Humboldt-Universität Berlin) left, and Rainer Feistel (Institut für Meereskunde Warnemünde) right, on board of r/v "A. v. Humboldt" in the subtropical North Atlantic, carrying out infrared radiation measurements during the expedition AMOR-91.

mass fraction of dry air in humid air, and q is the specific humidity. The Gibbs function (2) of liquid (rather than fluid) water is derived from the Helmholtz function (1) and is redundant; it was added only for convenience and numerical speed requirements in oceanography.

The chemical potential of water in seawater is given by the formula

$$\mu_W^{SW} = g^{SW} - c \left(\frac{\partial g^{SW}}{\partial c} \right)_{T,P} \tag{3.85}$$

and correspondingly, the chemical potential of water in humid air is available from the expression

$$\mu_W^{AV} = f^{AV} + \varrho \left(\frac{\partial f^{AV}}{\partial \varrho} \right)_{a,T} - a \left(\frac{\partial f^{AV}}{\partial a} \right)_{T,\varrho} \tag{3.86}$$

The evaporation rate from the sea surface, E, expressed as volume of liquid water per time and area, is proportional to the Onsager force, that is, to the chemical potential difference (Glansdorff and Prigogine, 1971; De Groot and Mazur, 1984)

$$\varrho L E = \varepsilon \left(\frac{\mu_W^{SW}}{T_W} - \frac{\mu_W^{AV}}{T_A} \right) \tag{3.87}$$

Here, ϱ is the density of liquid water, $\varrho L E$ is the export rate of latent heat, and

$$L = T \left\{ s^{AV} - a \left(\frac{\partial s^{AV}}{\partial a} \right)_{T,P} - s^{SW} + c \left(\frac{\partial s^{SW}}{\partial c} \right)_{T,P} \right\} > 0 \tag{3.88}$$

is the specific isobaric evaporation enthalpy, that is, the latent heat of seawater in contact with humid air (Feistel et al., 2010a, 2010b). The thermodynamic relations for the specific entropies of seawater and humid air are $s^{SW} = -\left(\partial g^{SW}/\partial T\right)_{c,P}$ and $s^{AV} = -\left(\partial f^{AV}/\partial T\right)_{a,\varrho}$, respectively. Little is known about the transport coefficient ε, but below we shall derive an estimate for its effective climatological value. The latent heat, Equation 3.88, is computed from the difference between the isobaric specific heat capacity of a composite system of seawater and humid air in mutual equilibrium (sea air), c_P^{SA}, and the sum of the partial heat capacities of the two phases, c_P^{SW} and c_P^{AV} (Feistel et al., 2010a, 2010b).

The practical and theoretical problem with Equation 3.87 is that the molecular fluxes of heat and diffusion are slow compared to the actual evaporation process. For illustration, we consider a composite equilibrium system that consists of certain masses of seawater and humid air, separated by a horizontal interface. To establish nonequilibrium initial conditions, we imagine a slight homogeneous increase in the temperature while keeping the pressure and the masses of the constituents fixed, in particular the specific humidity of the air. What will happen when we watch the relaxation of this system to its equilibrium state under isobaric conditions? At the original state with the properties $T_W = T_A$ and $\mu_W^{SW} = \mu_W^{AV}$, the isobaric change in the evaporation rate with increasing temperature follows from Equations 3.85–3.88, as,

$$\varrho L T \left(\frac{\partial E_S}{\partial T}\right)_P = \varepsilon \frac{\partial}{\partial T}\left(\mu_W^{SW} - \mu_W^{AV}\right) = \varepsilon \frac{L}{T} \qquad (3.89)$$

where L/T is the evaporation entropy, Equation 3.88. As a result, water will quickly evaporate, the related loss of latent heat will cool the skin layer of the liquid, and sensible heat flow back to the liquid will also cool the adjacent skin layer of humid air such that the evaporation rate soon returns to (almost) zero. After this fast process at a timescale of seconds or less, a much slower process starts which conducts heat down the temperature gradient in the water column, and diffuses vapor down the humidity gradient in the air. If air and water are in turbulent motion, those transport processes can be accelerated by many orders of magnitude compared to the molecular transport. The resulting skin effect, and along with it the mean evaporation rate at the interface, is therefore determined by the balance of the rates of heat supply and vapor removal, controlled by turbulent exchange processes with the bulk properties of the two fluids at greater distances from the surface. In turn, the turbulence intensity depends mainly on the local wind speed, on ocean waves, and surface currents. Thus, for the practical application of the formula (3.87) to the mixed boundary layers of the air–sea interface, the transfer coefficient ε is usually parameterized in terms of the local wind speed in order to approximately evaluate the skin effect, that is, to include the influence of the complex turbulent motion (Weare, Strub, and Samuel, 1981; Baosen, 1989; Wells and King-Hele, 1990). For this reason, field measurements at sea are indispensable. Laboratory measurements with seawater were recently made by Panin and Brezgunov (2007).

We can easily estimate an effective climatological value of the coefficient ε from observational data of the evaporation rate, E, and the relative humidity, ψ, of the air above the ocean. For this purpose, we express the relative humidity in terms of the chemical potentials, Equations 3.85, 3.86. There are different definitions for ψ available from the literature; here we use the ratio of the partial vapor pressure of a given air sample to that of saturated air (WMO, 2008; IAPWS, 2010; Feistel, 2011b)

$$\psi = \frac{e}{e^{\text{sat}}} \qquad (3.90)$$

where the partial vapor pressure, $e = xP$, is defined in terms of the molar fraction of water vapor, x, and the total air pressure, P, that is,

$$\psi = \frac{x}{x^{\text{sat}}} \qquad (3.91)$$

In the ideal-gas approximation, the chemical potential of water in humid air depends on the mass fraction of dry air, a, only via the mole fraction of water vapor, x,

$$x = \frac{1-a}{1-a(1-M_\text{W}/M_\text{A})} \qquad (3.92)$$

and takes the form (Feistel et al., 2010b)

$$\mu_\text{W}^{\text{AV, id}}(a, T, P) = \mu_0 + \mu_1 T + \int_{T_0}^{T}\left(1-\frac{T}{T'}\right) \cdot c_p^\text{V}(T')\mathrm{d}T' + \frac{RT}{M_\text{W}} \ln(xP) \qquad (3.93)$$

Here, μ_0, μ_1, and T_0 are arbitrary constants subject to reference state conditions, c_p^V is the heat capacity of water vapor, R is the molar gas constant, and M_W, M_A are the molar masses of water and air, respectively. Note that P is the total pressure including the dry-air part. At saturation, the chemical potential of vapor equals that of liquid water (or seawater with zero salinity). This defines the value of x^{sat}:

$$\mu_\text{W}^{\text{SW}}(0, T, P) = \mu_0 + \mu_1 T + \int_{T_0}^{T}\left(1-\frac{T}{T'}\right) \cdot c_p^\text{V}(T')\mathrm{d}T' + \frac{RT}{M_\text{W}} \ln(x^{\text{sat}} P) \qquad (3.94)$$

If we neglect here the small vapor-pressure lowering effect of sea salt which is proportional to salinity times the osmotic coefficient of seawater, we get for the sea–air chemical potential difference the relation (Feistel et al., 2010b; IOC, SCOR, and IAPSO, 2010)

$$\mu_\text{W}^{\text{SW}} - \mu_\text{W}^{\text{AV}} \approx \mu_\text{W}^{\text{SW}}(0, T, P) - \mu_\text{W}^{\text{AV,id}}(a, T, P) = -\frac{RT}{M_\text{W}} \ln \psi \qquad (3.95)$$

Therefore, the evaporation rate E_S, Equation 3.87, is related to the relative humidity by the approximate equation

$$\varrho L E_\text{S} = -\varepsilon \frac{R}{M_\text{W}} \ln \psi \qquad (3.96)$$

From the mean global values, $\varrho L E_\text{S} \approx 100\,\text{W m}^{-2}$ and $\psi \approx 0.8$, we obtain for the effective sea–air transfer coefficient of latent heat the simple value of $\varepsilon \approx 1\,\text{kg K m}^{-2}\,\text{s}^{-1}$.

Subsaturated air in contact with liquid water is a nonequilibrium system. Evaporation of liquid water into subsaturated air is an irreversible process. We can estimate the entropy production related to this mass transfer. The thermodynamic force is given by Equation 3.87,

$$X = \frac{\mu_W^{SW}}{T_W} - \frac{\mu_W^{AV}}{T_A} \tag{3.97}$$

The evaporation mass flux at the surface is $J = \varrho E_S = (\varepsilon/L)X$. For the areic entropy production we infer the value

$$\frac{P}{A} = JX = \frac{L}{\varepsilon}(\varrho E_S)^2 \approx 0.004 \text{ W m}^{-2} \text{ K}^{-1} \tag{3.98}$$

Compared to the total entropy production of the photon mill, oceanic evaporation contributes only a minor fraction of 0.4%. The empirical rule of constant relative humidity implies a rule of constant entropy production that accompanies the latent heat export of the ocean.

While very accurate values for the chemical potentials of water in its different phases and mixtures with air or sea salt are easily available from the TEOS-10 Sea-Ice-Air source-code library (Wright et al., 2010), we provide here simplified formulas to support convenient analytical estimates. At surface pressure, assuming constant heat capacity, we get from Equation 3.93 for ideal-gas vapor with the molar fraction x in humid air the formula

$$\mu_W^V(a, T) \approx \mu_{V,0} + \mu_{V,1}T - c_P^V T \ln(T/K) + \frac{RT}{M_W}\ln(x) \tag{3.99}$$

where the numerical values are $c_P^V \approx 1900 \text{ J kg}^{-1} \text{ K}^{-1}$, $R \approx 8.3 \text{ J mol}^{-1} \text{ K}^{-1}$, and $M_W \approx 0.018 \text{ kg mol}^{-1}$. For pure liquid water with assumingly constant heat capacity, we similarly get

$$\mu_W^L(T) \approx \mu_{L,0} + \mu_{L,1}T - c_P^L T \ln(T/K) \tag{3.100}$$

where the heat capacity is $c_P^L \approx 4200 \text{ J kg}^{-1} \text{ K}^{-1}$. At the triple point, $T = T_t = 273.16$ K, the two chemical potentials must coincide for pure vapor at the partial pressure equal to the triple-point pressure, that is, $x = x_t = P_t/P = 611.657/101325$, and the temperature derivative of the difference, Equation 3.89, must provide the latent heat, $L \approx 2500 \text{ kJ kg}^{-1}$. These conditions lead to relations between the adjustable coefficients, in the form

$$\mu_{L,0} - \mu_{V,0} = -L - (c_P^L - c_P^V)T_t \tag{3.101}$$

$$\mu_{L,1} - \mu_{V,1} = \frac{L}{T_t} + (c_P^L - c_P^V)(\ln(T_t/K) + 1) + \frac{R}{M_W}\ln(x_t) \tag{3.102}$$

For the relative humidity, Equation 3.95,

$$\psi(q, T) = x(q)/x^{sat}(T) \tag{3.103}$$

as a function of the specific humidity, $q = 1-a$, Equation 3.92, we get the results

$$x(q) = \frac{q}{1-(1-q)(1-M_W/M_A)} \tag{3.104}$$

and

$$x^{\text{sat}}(T) = \exp\left\{C_0 + C_1 \ln(T/K) + \frac{C_2}{T}\right\} \tag{3.105}$$

Here, the constants are defined as

$$C_0 = \ln x_t - C_1 \ln(T_t/K) - \frac{C_2}{T_t} \approx 48 \tag{3.106}$$

$$C_1 = -(c_P^L - c_P^V)\frac{M_W}{R} \approx -5 \tag{3.107}$$

and

$$C_2 = C_1 T_t - L\frac{M_W}{R} \approx -6800 \text{ K} \tag{3.108}$$

Equation 3.103 with the numerical coefficients (3.106)–(3.108) is the final convenient formula for relative humidity we were looking for; it is approximately valid for surface pressure, that is, $e^{\text{sat}}(T) = x^{\text{sat}}(T) \times 101325$ Pa is the related expression for the saturation vapor pressure. Compared to the common Clausius–Clapeyron formula, Equation 3.105 includes additional correction terms for the heat capacities of the two fluids. More accurate empirical correlation equations for the relative humidity and conceptual questions are discussed by Sonntag (1990), Lovell-Smith and Pearson (2006), Feistel (2011b), Feistel et al. (2010b), and IAPWS (2010). We can now apply Equation 3.103 to the atmosphere.

Only 0.001% of the terrestrial water is contained in the atmosphere, which equals a total of 13,000 km^3 (Baumgartner and Reichel, 1975). Relative to the global surface area of $A = 0.51 \times 10^{15}$ m^2, that number is equivalent to 25 mm of precipitation, or to an areic mass of $m/A = 25$ kg m^{-2}. The sea-level air pressure is $P_0 = 101,325$ Pa, the gravity is $g = 9.81$ m s^{-2}, hence the mass of the atmospheric air column is $M/A = P_0/g \approx 10,000$ kg m^{-2}. The average specific humidity is then $q = m/M = 0.25\%$. The typical relative humidity of $\psi = 80\%$ observed at the sea surface corresponds to a specific humidity of $q = 0.3\%$ at 0 °C and of $q = 1.2\%$ at 20 °C, as computed from Equation 3.103. Both values are higher than the global average; clearly, sea air is more humid than continental air.

The globally observed recent mean atmospheric warming rate is 27 mK yr^{-1} over land and 13 mK yr^{-1} over the ocean (Trenberth et al., 2007). The total atmospheric moisture content over the oceans has increased by 0.41 kg m^{-2} per decade since 1988 (Santer et al., 2007). This finding is fairly consistent with the other observations. If at the sea-surface the air temperature rises by 0.13 K per decade at constant relative humidity of $\psi = 80\%$, then the related atmospheric moisture content, $m/A = q \times 10,000$ kg m^{-2}, will approximately increase from 30.1 to 30.4 kg m^{-2} at 0 °C, and from 115.5 to 116.5 kg m^{-2} at 20 °C, provided that the specific humidity is homogeneously distributed in the turbulent atmosphere model. Formally, at an air temperature of 5.8 °C the observed temperature difference of 0.13 K corresponds to 0.41 kg m^{-2} of additional water observed in the air column, Equation 3.103. In contrast, between 1900 and 2000 the global average air temperature over the ocean was 16.1 °C (NCDC, 2010).

This atmospheric moisture increase may or may not be causally related to a stronger or faster cycle of evaporation and precipitation. The mean evaporation rate over land is $E_L \approx 500$ mm yr^{-1}, the global average is $E_E \approx 1000$ mm yr^{-1} (Baumgartner and Reichel, 1975). The related residence time of water in the atmosphere is 25 mm/E_E = 1 yr/40 = 9 days. Durack and Wijffels (2010) found that the surface salinity of the global ocean has generally been increasing in arid (evaporation-dominated) regions and freshening in humid (precipitation-dominated) regions, with a spatial distribution similar to that of the surface salinity anomalies relative to the spatial mean over the globe. This has led them to suggest that the global hydrological cycle has accelerated over the past few decades, and to confirm earlier studies which reported global salinity changes and showed that subtropical surface waters are generally becoming saltier, and high-latitude waters are freshening (Antonov, Levitus, and Boyer, 2002; Curry, Dickson, and Yashayev, 2003; Boyer et al., 2005; Stott, Sutton, and Smith, 2008; Min et al., 2011).

Intensified evaporation is transferring additional latent heat from the ocean to the atmosphere; this way it enhances the warming of the air at the expense of increased cooling of the ocean water. In fact it is debated whether any ocean warming trend is already exceeding the uncertainty of the scattered data (Trenberth, 2010). Between 1993 and 2008, a statistically significant increase in the heat content of 0.64 ± 0.29 W m^{-2} was observed for the upper 700 m of the global ocean water by Lyman et al. (2010). This result is not confirmed by other authors who found even negative values between -0.01 ± 0.2 W m^{-2} and -0.16 ± 0.2 W m^{-2} for the years 2003–2008 (Knox and Douglass, 2010). Ocean warming in combination with a slightly strengthened Gulf Stream (Willis, 2010) is obviously the most plausible explanation for the rapid sea-ice melt seen in the Arctic Ocean (Piechura and Walczowski, 2009; Kwok et al., 2009; Kaufman et al., 2009; Spielhagen et al., 2011). In contrast, for the mass balance of the continental Antarctic glacier more complex processes are responsible, in particular, increasing precipitation as a result of the enhanced hydrological cycle may result in glacier growth despite of global warming (Bindschadler, 2006; Steig et al., 2008; Pritchard et al., 2009; Huybrechts, 2009; Liu and Curry, 2010; Joughin, Smith, and Holland, 2010; Hansen, 2010; Laepple, Werner, and Lohmann, 2011).

In the sense of the principle of Braun and Le Chatelier, large sheets of floating sea-ice possess an interesting thermodynamic protection mechanism against melting by the warmer ocean water below, similar to the skin effect at the sea surface. The freezing point of seawater is lowered by salinity, as approximately described by Raoult's law in the form (Feistel et al., 2008b)

$$\frac{\Delta T}{T} \approx -0.22 \times c \qquad (3.109)$$

At an oceanic salinity of $c \approx 35\,\mathrm{g\,kg^{-1}}$, the resulting freezing point is roughly at $\Delta T \approx -2°C$. When an ice cover melts from below at a temperature slightly above the freezing point of seawater but still below 0 °C, pure liquid water is released from the ice which immediately elevates the melting point according to Equation 3.109. Since pure water has a lower density than seawater, a thin fresh or brackish water layer may form and persist as a crossover layer in between the ice sheet and the seawater below, and this way protect the ice from further melting until that layer is destroyed by turbulent mixing. In contrast, the so-called Arctic amplification is a feedback process by which the diminishing sea ice is highly correlated with a near-surface air temperature rise that is almost twice as large as the global average in recent decades (Screen and Simmonds, 2010). Increases in atmospheric water vapor content, partly in response to reduced sea-ice cover, may have enhanced warming in the lower part of the atmosphere during summer and early autumn. That strong feedback is likely increasing the chances of further rapid warming and sea-ice loss.

Returning to the two climatic rules mentioned in the beginning of this section, we conclude that the temperature difference between air and water at the sea surface is amplified by evaporation but reduced by sensible heat flow within the skin layer, and by intense thermal radiation exchange similar to Figure 2.7 in Section 2.5 between the liquid skin layer and the atmospheric boundary layer (Feistel, 2011d). The effective thickness of the boundary layer is given by its specific humidity and the related absorption length of infrared light. The typically observed temperature difference in the order of 1–2 °C (Albrecht, 1940) is, therefore, a measure for the quantitative relation between latent and radiative heat transfer at the sea surface.

The second empirical rule which states a constant relative humidity of roughly 80% lacks a simple explanation. Solar heating of the ocean will also warm up the air by thermal radiation and thus will reduce relative humidity, but at the same time heating will increase specific humidity by evaporation. For keeping the humidity below saturation, mainly two air-dryer systems are at work – precipitation of condensed water or ice from clouds, and heating up dry air in the sunshine on land. The timescale for air exchange between oceans and continents by global atmospheric circulation is in the order of 10 days, a very fast process on climatic scales. It remains an interesting question why those different processes fluctuate about a constant, quasistationary value of 80% even under temporally and spatially strongly varying conditions (Mitchell, 1989).

3.7
Greenhouse Effect

Despite the fact that the atmospheric humidity essentially controls the evaporation rate from the sea surface, and in turn significantly influences the temperature of the oceans, water in the atmosphere additionally serves as inlet and outlet valves of the steam-engine Earth. The two forms, condensed water as liquid or ice, or gaseous water as atmospheric humidity, substantially affect in rather different ways the terrestrial energy balance. Water vapor is almost invisible and has only a little influence on the solar irradiation in the optical spectrum. In the infrared range, it is the strongest greenhouse gas in the atmosphere (Abbott and Fowle, 1908; Emden, 1913); it is temporally and spatially more variable than CO_2 but typically exhibits gradients on large scales, such as between the tropical and the subtropical climate belts. Water vapor mainly acts as an "outlet valve" that is regionally controlled. Either in the form of liquid droplets or solid crystals, condensed water in clouds strongly scatters visible light, increases the atmospheric albedo, and this way functions as an "inlet valve" for the terrestrial energy supply. In the infrared spectrum, the condensed water of clouds is opaque and almost perfectly absorbing; clouds therefore act also as "outlet valves." The spatial patterns and mutual correlations between those different "valves" are crucial for their interplay since sunshine is confined to the daytime hemisphere while the infrared cooling is permanent and omnidirectional. Dynamical processes of clouds are highly variable; they run on timescales in the order of hours and have spatial extensions between 10 m and 10 km; only their statistical properties are climatically relevant. Clouds effectively control the climate, and are in turn controlled themselves by complex and transient dynamical processes, such as the local weather, that represent various forms of nonlinear response to the climatic conditions (Feingold et al., 2010). The cloud–climate system is self-organized; it is the most important machinery of the steam-engine Earth, and it is its least understood feedback process (Ebeling and Feistel, 1994; Clement, Burgman, and Norris, 2009).

Due to Planck's law, the thermal radiation of the ocean occurs predominantly at wavelengths greater than 4 µm and has its maximum at about 20 µm which corresponds to an effective temperature of 287 K (Schmetz and Raschke, 1990). In this spectral band, the absorption length of infrared light in liquid water is between 0.1 and 0.01 mm. The absorption of water vapor has an "atmospheric window" between 8 and 14 µm wavelength in which remote-sensing satellites can "see" the temperature of the surface as long as it is not covered by a veil of haze, dust, or clouds. Water vapor in the atmosphere accounts for 50–60% of the terrestrial greenhouse effect, while CO_2 has infrared absorption bands at 14, 16, and 20 µm and is responsible for only one quarter (Trenberth et al., 2007). Over the oceans, where the humidity is higher, the share of vapor is still larger and the effect of CO_2 is even less relevant. At sea level in the tropics, air is hot and saturated, daytime heating is strong, and any night-time cooling is hardly noticeable. Where infrared transmission is practically already blocked by water vapor, any additional CO_2 is irrelevant.

The famous Keeling curve (Tans, 2010) tells us that the atmospheric CO_2 increased from 315 to 385 ppmv (part per million of the volume), or by 20%, over the last 50 years. Thus, the terrestrial CO_2-induced part of the total greenhouse effect increased by about 5% in those 50 years, or 1% per decade. Compared to the uncertainty of the warming caused by the "steam blanket" the Earth is wearing, this is a small but systematic influence; it may easily be exceeded or compensated by dynamical changes in the atmospheric cloud or humidity distribution. The global reserves of fossil fuel are limited; this implies also a natural upper limit for the anthropogenic CO_2 release. Political measures can slow down this process (if at all) but sooner or later coal, oil, and gas will be exploited until the deposits are exhausted. The final CO_2

level will be about the same, with or without political interference, as well as its long-term effect on the climate.

Most of the evidence for the climate impact of CO_2 comes from numerical global circulation models that compare observational data with hindcasts, with or without the CO_2 rise implemented. But those models have significant uncertainties, in particular as long as they cannot properly resolve the dynamics of atmospheric vapor or clouds (Schiermeier, 2010). No such climate model was ever able yet to predict in advance phenomena such as the exceptional heat wave, drought, and bush fires in Russia in summer 2010, or the probably related monsoon displacement that caused the vast Indus river flooding in Pakistan at the same time, even though such processes occur on large scales of thousands of kilometers and on time scales of several weeks. The unexpectedly rapid melting of the Arctic sea ice and the surprisingly slow melting, if at all, of Antarctica are other examples for the models' restricted and immature prediction capabilities.

Various paleorecords indicate temporal correlation between rising CO_2 concentrations and global warming (Tripati, Roberts, and Eagle, 2009; Bijl et al., 2010) but leave open the question for the causal relation between the two signals, see Figure 3.10. In several cases, such as the thermal maximum 56 Myr ago, it turned

Temperature and CO_2 Series from Vostok Ice Core

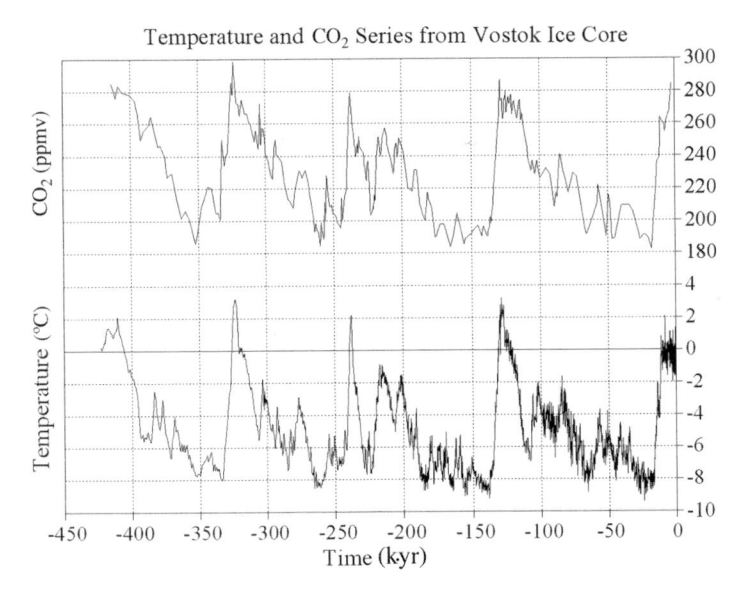

Figure 3.10 Comparison between partial pressure of atmospheric CO_2 and relative air temperature over the past 420,000 years from Vostok ice core data (Jouzel et al.,1987, 1993, 1996; Petit et al.,1999, 2000). While the statistical correlation between the two curves is obvious, the causal relation between the two signals is unclear. There is no significant time lag of the temperature signal behind the CO_2 level, on the contrary, perhaps two of the three temperature maxima (-128.5, -238.0, -322.6 kyr) may have occurred 100–200 years earlier than those of carbon dioxide (-128.4, -237.8, -323.5 kyr). While the seesaw pattern in the temperature series can be explained by the oceanic diode effect (Section 3.8), there is no similar mechanism known for the steep rise in CO_2, except a preceding warming of the ocean, or human industrialization (from 315 to 385 ppmv in 50 years).

out that the warming preceded the CO_2 rise (Secord et al., 2010; Soares, 2010). Such a delay excludes CO_2 as the possible cause of the warming and rather suggests that it may be the consequence, possibly amplified later by positive feedback between warming and CO_2 level. Rather than its cause, the dramatic decrease in atmospheric CO_2 from more than 1000 ppm to less than 500 ppm between 35 and 25 Myr ago might be the result of the Antarctic glaciation and the cooling Southern Ocean, see Table 3.3. The slight precedence of temperature changes relative to CO_2 changes indicates that the main effect is the CO_2 increase in the atmosphere which is a result of rising temperature (Soares, 2010). The existence of a glacial deep-ocean CO_2 reservoir is currently debated (Hain, Sigman, and Haug, 2011). About 15 Myr ago, when global temperatures where higher than today by 3–6 °C, CO_2 levels apparently were similar to recent values (Tripati, Roberts, and Eagle, 2009).

It is difficult to find immediate observational evidence for the CO_2 rise as the cause of the global warming and its side effects. The regions where the CO_2 effect should be strongest compared to that of atmospheric water are certainly the subtropical deserts such as the Sahara or the Gobi. Under cloud-free and dry-air conditions, the direct thermal radiation into the outer space should be higher there than anywhere else on our planet. This is a very plausible explanation for the extreme day–night temperature differences experienced in the deserts. From rising CO_2 levels in the air, we may infer that the infrared emission will be reduced, and consequently that the night temperatures will increase and the day–night differences will gradually get smaller, due to the enhanced downwelling radiation from the atmospheric CO_2. Fifty years of observation are available from the Australian outback (ABM, 2010). In the dry season, in particular in August in the Northern Territory, the minimum temperatures show no relevant trend. In Figure 3.11 we see that the day–night temperature differences increased systematically over the past five decades. These trends are just opposite of what a simple greenhouse model is suggesting; their correct explanations require a more detailed analysis of the local meteorological conditions and processes. Although this example does not disprove the causal CO_2 effect on global warming, it nevertheless clearly shows that the climate problem must not be reduced to a simple linear response behavior of all the meteorological parameters to the rising atmospheric CO_2 concentration. From this perspective, popular suggestions to counteract the global warming by large-scale technological measures of "geo-engineering" appear highly risky for their unwanted and unpredictable potential consequences (Hegerl and Solomon, 2009). Lake Aral, the Colorado, and the Jordan river are prominent examples for collateral damage caused by immature human intervention activity, fortunately on a local rather than the global scale.

A complicated and sometimes underestimated climate effect is that of clouds. They are responsible for 60% of the terrestrial albedo (Schmetz and Raschke, 1990). In the literature on the climatic role of clouds, their net effect is often described as rather neutral; their warming and cooling effects are statistically almost balanced. With respect to their altitude and structure, some clouds are classified as slightly cooling and others as slightly warming (Schmetz and Raschke, 1990). But such a classification addresses only a minor aspect of the role clouds play on the climate stage. Readers who live at higher latitudes will agree from their experience that a dense

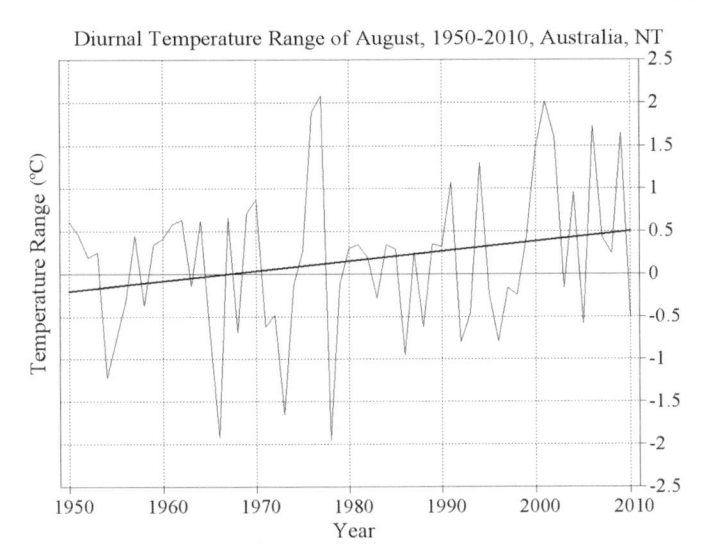

Diurnal Temperature Range of August, 1950-2010, Australia, NT

Figure 3.11 Positive climatological trend about $+1.2\,°C$ per century of mean daily min–max temperature difference anomalies in the dry season (August) in the Northern Territory of Australia, published by the Australian Bureau of Meteorology (ABM, 2010). An opposite trend would be expected as a direct result of the increased CO_2 greenhouse effects.

cloud cover is warming in winter and cooling in summer, no matter what particular cloud type is involved, and that those thermal effects of the "steam blanket" are quite substantial. Cloud cover always act as an outlet valve, but only at daytime as an inlet valve. Hence, rather than the mean frequency or the type of clouds, the temporal and spatial correlation between cloud cover and solar irradiation is highly relevant, or the phase shift of cloudiness in the diurnal and seasonal cycle. Cloud formation is often influenced by solar heating and exhibits a pronounced daily or seasonal signal, for instance in the form of afternoon showers or rainy seasons. Hence, there is a nonlinear feedback that controls the statistical phase lag between cloud cover and local irradiation received at the ground. Long-term shifts of that phase, for example, in the form of meridional displacements of climatic front zones, may significantly contribute to global warming or cooling. Large-scale displacements of clouds, droughts, and heat waves on the one hand and torrential rains and floods on the other, such as in Russia and in Pakistan in summer 2010, are natural phenomena in the history of climate; if they appear more frequently and more systematically, they may certainly influence the global energy balance via the cloud valves. In other words, the currently neutral role of clouds may possibly change into one of a self-organized key player that is much more variable and less predictable than the steady global increase in CO_2 (Kerr, 2009; Clement, Burgman, and Norris, 2009; Dessler, 2010). Water in the form of vapor and clouds may account for 75% of the total greenhouse effect (Lacis et al., 2010).

Another important global feedback loop is the atmospheric water-vapor instability. Under the assumption of constant relative humidity, the moisture content of the air increases with rising temperature according to Equation 3.103. Due to the greenhouse effect of the additional vapor, the temperature will rise further, and so on until all oceans are desiccated and the water forms a Venus-like steam-ball Earth. Vice versa, if the initial fluctuation was a cold one, more water would have been precipitated from the air, thus reducing the greenhouse warming, and cooling down the surface until all the water is eventually condensed and forms a frozen, Mars-like snowball Earth. This instability is inevitable for a photon-mill model with just two order parameters – temperature and specific humidity – if relative humidity and other relevant parameters are fixed. The real ocean–atmosphere system has more degrees of freedom and the mean state of the atmosphere is rather insensitive to temperature fluctuations, that is, it is in an asymptotically stable attractor state. Attractors that possess positive Lyapunov coefficients which amplify perturbations in certain regions of the phase space are usually chaotic (Gaspard and Nicolis, 1983), see Section 4.2 for more details. Thus, the water-vapor instability may contribute to the irregular dynamical details of weather and climate. It is an open question what processes are actually responsible for the observed global asymptotic stability of the terrestrial climate attractor. These processes reduce the effect of the explosive vapor instability to some gradually enhanced global warming of $2\,\mathrm{W\,m^{-2}\,K^{-1}}$, consistent with the assumption of a constant relative humidity discussed previously (Minschwaner and Dessler, 2003; Dessler, Zhang, and Yang, 2008; Dessler and Sherwood, 2009). On annual and even on monthly scales, the rise of water vapor in the atmosphere is highly correlated with temperature changes, in contrast to CO_2 (Soares, 2010). The water-vapor instability is a striking argument that dynamical greenhouse models for just one or three order parameters, such as those discussed in the previous sections and in particular the Arrhenius radiation model, are oversimplified and rather unrealistic global climate models. A chaotic state of so-called SOC (Bak, Tang, and Wiesenfeld, 1987) may be a more promising physical working hypothesis for the stability of the terrestrial climate attractor, as indicated by characteristic fluctuation spectra (Yano, Fraedrich, and Blender, 2001). As an example for additional relevant degrees of freedom, although water vapor in the stratosphere increased between 1980 and 2000, it decreased by about 10% after the year 2000, with a related significant impact on the greenhouse warming (Solomon et al., 2010). Similarly, the trend of increasing global evapotranspiration on land ceased in 1998 (Jung et al., 2010).

3.8
Spatial Structure of the Planet

After its formation, the fluid components of the planet Earth arranged to density-stratified sphere under the action of its own gravity (Graham, 2010). The different substances are radially separated by so-called discontinuities where certain properties such as the density, the speed of sound, or the possible types of sound waves change abruptly. The inner core is enclosed by an outer core, above which a lower and an upper mantle are located. The hot inner parts with molten rocks are separated from the colder hydrosphere and atmosphere by a thin solid crust. Within each of those layers, the geophysical fluids produce various dissipative structures, driven by two different energy sources – nuclear fusion of hydrogen to helium in the Bethe–Weizsäcker cycle inside the Sun, heating Earth from outside, and nuclear fission, in particular of uranium (U-238) to lead (Pb-206) in several steps, heating from inside Earth (O'Nions, Hamilton, and Evensen, 1984). Because of the half-life of 4.468 Gyr of U-238, one half of the original amount of nuclear fuel is still left over from the formation of Earth. The average solar flux at the atmosphere's top is 341 $W\,m^{-2}$, Section 3.1, while the geothermal heat flow from below the surface is 0.087 $W\,m^{-2}$ on the global average (Pollack, Hurter, and Johnson, 1993). From a heat conduction coefficient of solid rock in the order of 3 $W\,m^{-1}\,K^{-1}$ and a typical temperature of liquid lava in the order of 1200 K, we can estimate a global mean crustal thickness of about 30 km. Local heat flux is much higher where the solid crust is thin, and lower on the thicker continental plates, such as 0.05 $W\,m^{-2}$ in the Antarctic (Shapiro and Ritzwoller, 2004; Maule et al., 2005). With respect to a surface temperature of about 300 K, the mean global geothermal entropy export can be estimated as $\approx 3 \times 10^{-4}$ $W\,m^{-2}\,K$. It contributes a share of only 0.03% to the total terrestrial export and corresponds to perhaps 20% of the biological entropy production but is still four times higher than the dissipation caused by the human society. The question of where in Earth's interior most of the entropy production occurs is still subject to scientific debates (Stadler et al., 2010). In the Antarctic, the geothermal heat flow may melt the glacier at greater depth and form subglacial lakes. (Siegert et al., 2001; Schiermeier, 2011). A particularly exotic case is Lake Ellsworth where the vertical temperature profile is controlled by the interplay between the density anomaly of liquid water and the pressure dependence of the melting temperature (Thoma et al., 2011). For comparison with the terrestrial value, an average global heat flux of 0.018 $W\,m^{-2}$ was calculated for the Moon from measurements made by the missions Apollo 15 and 17 (Langseth, Keihm, and Peters, 1976). The lunar solid crust is estimated to be about 50 km thick. Analysis of lunar quakes suggests the presence of a solid inner and liquid outer core (Weber et al., 2011).

The basic dissipative structure-building mechanism is thermal convection, similar to the classical Benard cells, but significantly modified by the particular values of heat supply, gravity, thermal expansion, viscosity, the kind of boundary conditions, and the Coriolis force. Due to the different densities of the fluid substances, convection occurs in radially separated layers which interact at their mutual interfaces. Detailed studies of hydrodynamic instabilities are available from various review articles and textbooks (Ruelle and Takens, 1971; Swinney and Gollub, 1981; Ebeling and Klimontovich, 1984; Pedlosky, 1987; Bestehorn, 2006).

The atmospheric layer above the oceans and the continents is referred to as the troposphere and, as a result of its optical transparency, is heated from below with a strong meridional gradient from the equator to the poles. Moist air is uplifted in the intertropical convergence zone (ITCZ), precipitating much of the water, and is sinking back down as dry air in the very stationary subtropical anticyclones at about 30° latitude, see the schematic in Figure 3.12.

Deflected by the Coriolis force, the trade winds at the surface blow one part of the air back to the equator, thus closing what is called the Hadley circulation. The other

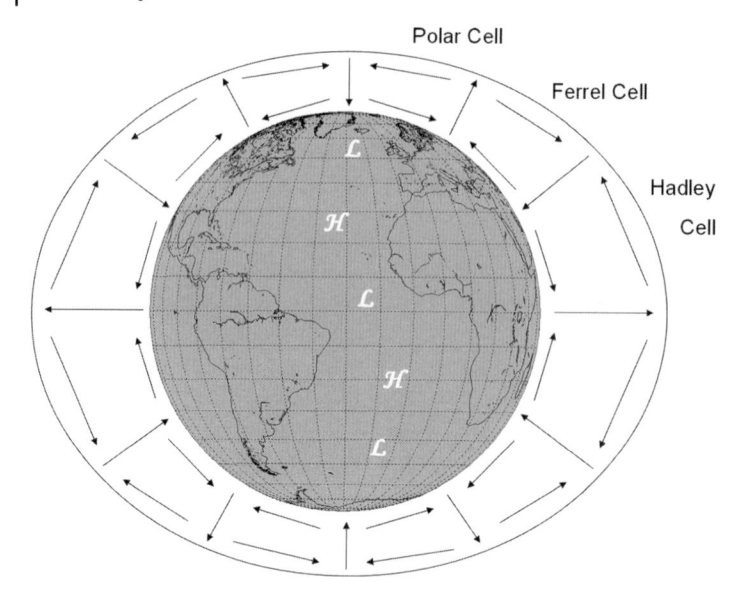

Figure 3.12 Global cellular system of air convection in the troposphere. Humid air is ascending in the low-pressure belts ($\mathcal{L}$, cyclones) at the equator and in the west-wind regions, such as the Icelandic Low. Dry air is descending in the high-pressure zones ($\mathcal{H}$, anticyclones), such as the Azores High and the South-Atlantic Anticyclone. In reality, the height of the troposphere is only 1/1000 of Earth's diameter, and is about three times higher at the equator than at the poles.

part of the descending air is moving poleward to the west-wind belt and is uplifted within its unsteady cyclones to close the Ferrel circulation. Finally, sinking cold air at the poles is also moving to the west-wind zone at about 60° latitude and forms the polar circulation (Bindschadler, 2006). Along the front zones between these circulation cells, transient jet streams emerge and decay systematically but irregularly (Raethjen, 1961).

Clouds and humid climate dominate where moist air is uplifted, while sinking dry air is related to arid climate at the ground. This rather regular and stable structure is regionally modified by mountain ranges and the global sea–land distribution; it varies systematically by global warming, and quasiperiodically by various superimposed, assumingly self-organized oscillations such as the quasibiennial oscillation (QBO, Baldwin et al., 2001), the North Atlantic oscillation (NAO, Hurrell, 1995), the El-Niño-Southern oscillation (ENSO, Diaz and Markgraf, 2000; McPhaden, Zebiak, and Glantz, 2006), or the Benguela Niño events with a period of 11–14 years (Shannon et al., 1986; Hagen et al., 2001;Hagen, Agenbag, and Feistel, 2005), and so on, at least up to the long periods of the Dansgaard–Oeschger events (Schulz, 2002; Kaplan et al., 2010) and the Heinrich events (Heinrich, 1988) with periods of about 1.5 kyr and 10 kyr, respectively, as well as the famous ice-age cycle in the order of 100 kyr (Hewitt, 2000; Zachos et al., 2001; Beal et al., 2010), see also Figure 3.10. Hardly any of those cycles is thoroughly understood yet (Burrows, 1992). The current amplification of El-Niño amplitudes (Lee and McPhaden, 2010) is considered responsible for various weather extremes observed globally. The weak vertical stability of the troposphere is one reason for its sensitivity with respect to various kinds of perturbations, for the emergence of instabilities and of spatial and temporal structures with relatively long relaxation times, such as El-Niño oscillations, hurricanes, tornados, or dust devils (Figure 3.16).

As an example for a self-organized dynamical feedback system, we may consider the hypothesis of the Benguela–St. Helena instability (Hagen *et al.*, 2001; Hagen, Agenbag, and Feistel, 2005; Feistel, Hagen, and Grant, 2003). A stable, extended high-pressure cell, the South Atlantic anticyclone (SAA), is permanently located over the subtropical South Antlantic, see Figures 3.12 and 3.13. In combination with a low-pressure system over the Kalahari desert, the SAA controls the intensity of the south-eastern trade wind (SET) parallel to the Atlantic coasts of Angola, Namibia, and South Africa and drives the northeastward Benguela current (Schell, 1968). Due to the Coriolis force, surface water is pushed offshore, the coastal sealevel is lowered, and cold water is pulled up from deeper layers to the surface and forms the Benguela upwelling system (Hagen *et al.*, 1981, 2001; Hagen, 2009), as shown in Figure 3.14.

Cold water in the oceanic surface layer results in a cooling of the atmospheric layer above. Surface air pressure is a measure for the total mass of the air column above a certain location. As the air gets colder, the density increases, and the air at higher altitudes descends to slightly lower levels. The resulting "hole" is filled with additional air from the surrounding, which in turn increases the total mass of the air column and thus the local surface air pressure. This causal chain results in higher air pressure over cold water and lower air pressure over warm water. Increasing pressure of the SAA intensifies the Benguela upwelling and the injection of cold deepwater into the surface layer, which in turn increases the air pressure. The positive

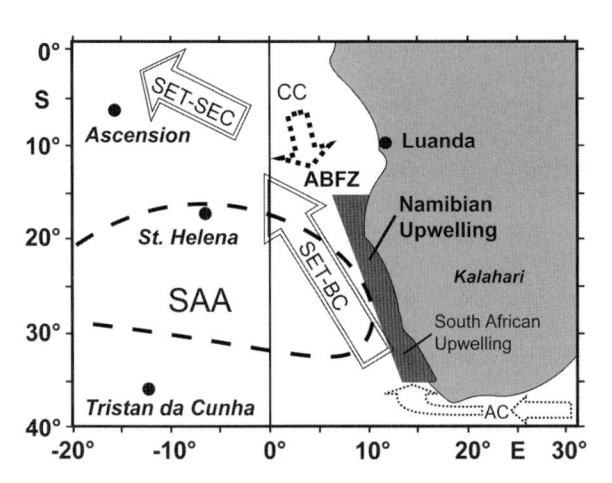

Figure 3.13 Schematic of the Benguela-St. Helena dynamical system in the South Atlantic (see text). SAA, South Atlantic Anticyclone; SET-SEC, South-East Trade wind – South Equatorial Current; SET-BC, South-East Trade wind – (cold) Benguela Current; CC, (warm) counter current; ABFZ, Angola-Benguela Frontal Zone; AC, (warm) Agulhas Current. (adapted from Feistel, Hagen, and Grant, 2003).

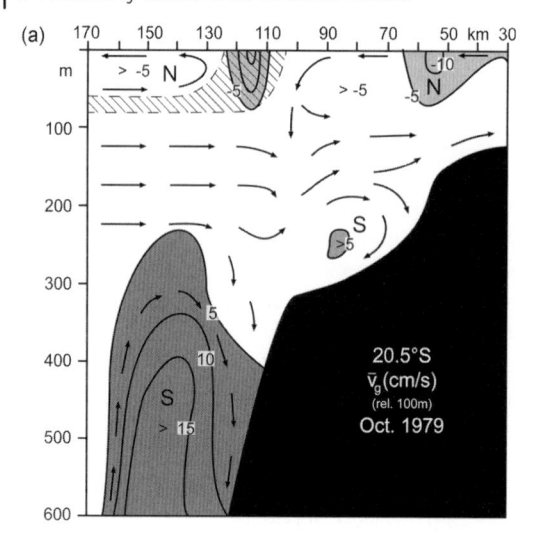

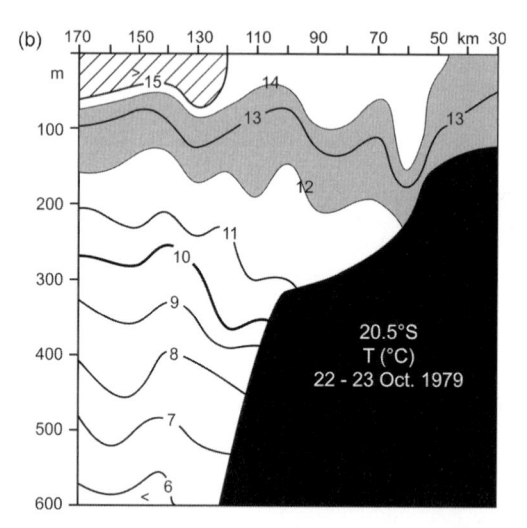

Figure 3.14 Transect perpendicular to the Benguela Current system, measured by r/v "A. v. Humboldt" off the Namibian coast 20.5° S in October 1979. The continental shelf is shown in black. (a) Arrows indicate offshore transport and upwelling circulation above 100–200 m depth, accompanied by onshore transport and downwelling below. The cores of the northward (N, negative) and southward (S, positive) currents are shown in gray. Isolines show edqual geostrophic current velocities in cm/s relative to 100 m depth. (b) Oceanic isotherms are displayed with temperatures given in °C. Cold upwelled water is shown in gray. Warmer surface water is hatched. On the shelf, isotherms are bent upward where cold deepwater rises to the surface, accompanied by the cold northward Benguela Current. Below, isotherms turn downward where the oxygen-depleted coastal countercurrent is running southward. Deepwater enriched with nutrients drives intense growth of plankton at the surface, and in turn of fish populations. In Benguela-Niño years, a thick warm surface layer prevents deepwater to rise to the surface, thus leading to a collapse of fishery. Similar conditions occur at the Humboldt Current along the Pacific coast of South America with alternating El-Niño (warm surface water) and La-Niña (cold surface water) events. Figures adapted from Hagen et al. (1981), courtesy of Eberhard Hagen.

Annual Mean Climatology, St. Helena 1893–2009

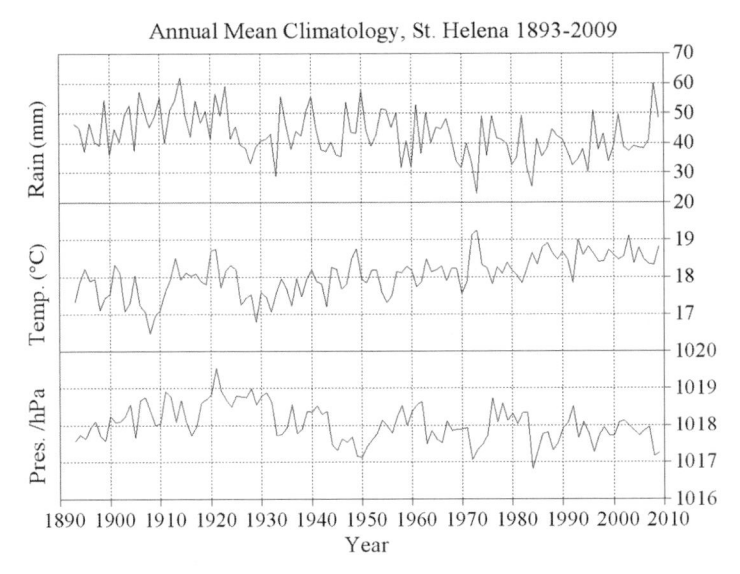

Figure 3.15 Annual mean values of sea-level pressure, station temperature and monthly precipitation on St. Helena Island, South Atlantic, 1893–2009 (Feistel, Hagen, and Grant, 2003). High air pressure is the cause for stronger trade winds and Benguela upwelling, while low temperature and high rainfall are the results. Significant correlation with the 14-year cycle of precipitation (Mathieson, 1990) is found in the pressure and temperature records. Their joint signal was extracted in the form of the monthly St. Helena Island Climate Index (HIX). Note the significant fluctuations on the decadal time scale, the correlation between warming and pressure lowering since 1930, and the recent apparent trend reversal of precipitation. The pressure value of January 2011 was the lowest on record.

feedback loop results in an instability of the force balance between the oceanic offshore transport driven by atmospheric trade winds and coastal sea-level lowering and cold upwelling. The related positive Lyapunov coefficient of the Benguela dynamical system implies chaotic fluctuations (Gaspard and Nicolis, 1983), see also Section 4.2. Such fluctuations are well reflected in the observed air pressure on St. Helena Island, as a cause of the upwelling, and in the quasi-14-year cycle of the rainy season on the island (Mathieson, 1990), as a result of the cold upwelling, as shown in Figure 3.15 (Feistel, Hagen, and Grant, 2003). Hagen et al. (2001) identified warm "Benguela Niño" events with warm coastal surface water and reduced upwelling for the years 1909, 1923, 1934, 1937/38, 1949, 1963, 1974, 1984, 1993, 1996/1997, and 1999. In contrast, exceptionally strong, cold Benguela upwelling was reported for the years 1906/1907, 1929, 1946/1947, 1955, 1969/1970, 1982, 1985, 1990, and 1992. Although the local ocean–atmosphere interaction is considered to be basically responsible for the observed behavior, external atmospheric and oceanic "teleconnections" with other global signals such as El Niño in the Pacific and the monsoon over the Indian Ocean also have significant influence, as suggested by results of statistical correlation analyses (Mo and White, 1985; Rathmann, 2008).

The strength of the SAA also controls the latitude of the oceanic subtropical front, which separates the subtropical Atlantic from the Antarctic Circumpolar Current. If this northern edge of the west-wind belt is shifting southward, e.g. as a result of global warming, the Agulhas current may carry more warm saline Indian-Ocean water into the Atlantic around the southern tip of Africa. In turn, this water supply is assumed to enhance the Atlantic MOC, and thus the global oceanic heat transport (Cheng et al., 2009; Beal et al., 2010). This example again demonstrates how the self-organisation of the global climate emerges from the complex interaction of various, very specific regional feedback processes such as the St.Helena-Benguela-Agulhas system.

Due to thermal convection, the stability of the tropospheric density stratification is almost neutral and temperature is decreasing with altitude, similar to the isentropic model of Section 3.4. The usual measure for the vertical stability is the square of the so-called Brunt–Väisälä frequency, N^2, which describes the oscillation of an adiabatically displaced air parcel under the influence of the restoring buoyancy force. If that parcel is originally at rest at a pressure level P, it has the same density ϱ as the surrounding air. When it is vertically displaced by some ΔP, its density change can be computed from the equation of state for humid air, $\varrho(a, s, P)$, at constant entropy, s, and specific humidity, $q = 1 - a$, in the form

$$\Delta \varrho = \left(\frac{\partial \varrho}{\partial P}\right)_{a,s} \Delta P = \kappa_S \varrho \Delta P = \frac{\Delta P}{c^2} \tag{3.110}$$

Here, κ_S is the adiabatic compressibility of air, and c the sound speed. The surrounding air at the pressure level $P + \Delta P$ has the density $\varrho + \Delta \varrho_E$

$$\Delta \varrho_E = \frac{d\varrho_E}{dP} \Delta P \tag{3.111}$$

where $d\varrho_E/dP$ is the environmental vertical density gradient in terms of pressure as the vertical coordinate. According to Archimedes' law, the parcel experiences a buoyancy force

$$\Delta f = g_E(\Delta \varrho_E - \Delta \varrho) \tag{3.112}$$

per volume, where g_E is the local gravity. Newton's law for the vertical motion

$$\varrho \frac{d^2}{dt^2} \Delta z = \Delta f \tag{3.113}$$

determines the oscillation frequency, N, termed the Brunt–Väisälä frequency, by the formula

$$N^2 = -\frac{\Delta f}{\varrho \Delta z} = g_E \frac{\Delta f}{\Delta P} = g_E^2 \left(\frac{d\varrho_E}{dP} - \frac{1}{c^2}\right) \tag{3.114}$$

Here, the hydrostatic approximation, $\Delta P = -\varrho g_E \Delta z$, was used. From the viewpoint of bifurcation theory, in the case of stability the two eigenvalues of the linearized dynamical problem (3.113) are located on the imaginary axis of the complex plane, and move to the origin in the case of neutral stability. For unstable conditions, they are located on the real axis with opposite signs. Thus, the vertical instability bifurcation is a kinetic phase transition from a neutral focus to a saddle point. In the stable situation, fluid parcels oscillate vertically with the frequency N after excitation, in the unstable situation, overturning convection starts and results in an altered vertical stratification.

Decreasing density with altitude, $d\varrho_E/dP > 0$, is a necessary condition for hydrostatic stability, but is not sufficient for vertical stability with respect to adiabatic excursions. In the isentropic atmosphere of Section 3.4, N^2 vanishes, that is, the environmental density gradient takes the critical value

$$\frac{d\varrho_E}{dP}\bigg|_{crit} \equiv \frac{1}{c^2} \qquad (3.115)$$

and corresponds to neutral stability of the air column, that is, thermally neutral stratification. Remember that under gravity the well-mixed isentropic atmosphere is in a nonequilibrium state because of the vertical temperature gradient, Equation 3.62. Because of the latter, this neutrally stable profile may serve as an alternative criterion for the environmental temperature gradient, such that stratification stability requires that

$$\frac{dT}{dz} > \frac{dT}{dz}\bigg|_{crit} \equiv -\Gamma \qquad (3.116)$$

if z points upward. This form of the stability condition is common in meteorology (Emanuel, 1994; Jacobson, 2005). Equations provide the most important quantitative criteria to find out whether a given stratified fluid (ocean, atmosphere) with known vertical temperature and density profiles either is in a stable state or tends to start convective motion (Figure 3.16).

In addition to the solar "floor heating" by irradiation passing the air and warming up the land and oceans beneath, the atmosphere also has a "room heating," in particular due to absorption of solar ultraviolet radiation by ozone. The latter is heating from above; the temperature is increasing with greater height, similar to the warm surface layer of turbid lakes in the summer sunshine, and results in a rather stable vertical stratification, Equations 3.114 and 3.116. That upper atmospheric layer, referred to as the stratosphere, is eroded from below by tropospheric convection in the vertical crossover zone ("tropopause"), similar to the winter convection in lakes or oceans with opposite direction, where cold surface water sinks down and erodes from above the transition layer ("pycnocline") between the mixed surface layer and the denser deep water (Reissmann et al., 2009).

An air parcel is lifted up by buoyancy, which results from the lower density of the parcel in comparison to the surrounding air. While ascending, the parcel expands adiabatically and cools down with an approximately constant lapse rate. The cooling is reduced and the buoyancy is enhanced by the release of latent heat as soon as vapor starts condensing (see Figure 3.17 for the effect of humidity on the entropy of humid air). In the stratosphere, due to the radiative heating from above, the density of air decreases with height much faster than by adiabatic expansion, and tropospheric parcels will stop rising as soon as they get too "heavy" for their vicinity, and the buoyancy vanishes. That altitude marks the tropopause, an interface between the two spheres which possess rather different structures and dynamics. At the equator, where at elevated air temperatures the troposphere is fueled with more latent heat due to the higher specific humidity, the tropopause is found at about 20 km height, while at the poles the troposphere reaches only up to 7 km, as indicated schematically in Figure 3.12.

Figure 3.16 Waterspout, locally termed "wind hose," observed over the Baltic Sea on 18 August 2010 by r/v "Maria S. Merian" (photo courtesy to Gregor Rehder, IOW).

Violent thunderstorms may penetrate even deeper into the potential-energy barrier of the stratospheric "lid." From an airplane, this can often be observed in the form of a typical anvil-head shape of the cloud top, where the vertical air motion turns horizontal. Due to warming of the troposphere and reduction in the ozone layer, the height of the tropopause has increased by several hundred meters since 1979 (Santer et al., 2003).

Air temperature has a pronounced minimum in the tropopause, typically between $-40\,°C$ and $-80\,°C$, see Figure 3.18, at which even saturated air contains very little vapor. Thus, the tropopause is very watertight and seals the troposphere which contains about 99% of the atmospheric water. At the top of the atmosphere, water molecules may dissociate under the influence of energetic solar and cosmic radiation. The liberated hydrogen atom has a small mass, m_H, and is quickly accelerated to thermal velocities v with the mean value given by the Maxwell–Boltzmann distribution, $m_H v^2 = 3kT$. Particles with a speed higher than the terrestrial escape

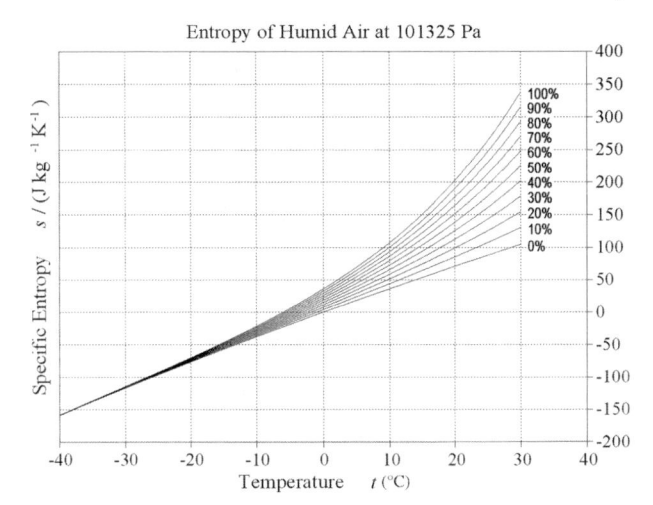

Figure 3.17 Specific entropy of humid air at atmospheric pressure as a function of the temperature for different relative humidities between dry air, $\psi = 0\%$, and saturation, $\psi = 100\%$, computed from the TEOS-10 Sea-Ice-Air library (IAPWS, 2010; Feistel et al., 2010b; Wright et al., 2010). Isentropic uplift corresponds to a horizontal line; condensation and release of latent heat occurs when the saturation curve (100%) is crossed.

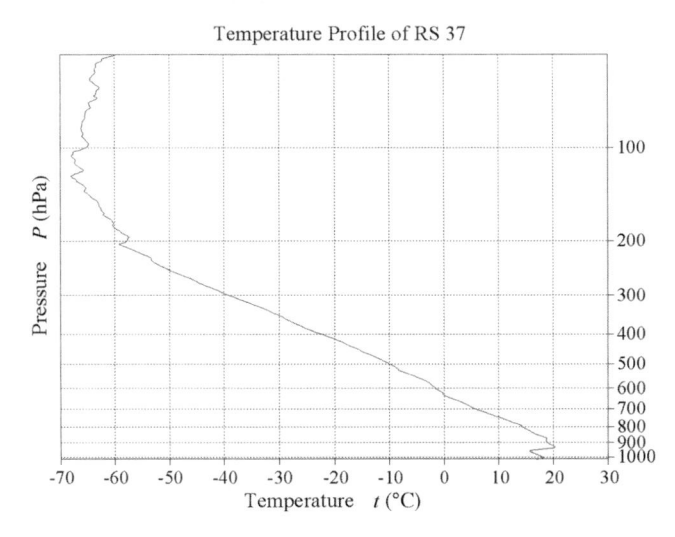

Figure 3.18 Air temperature profile up to an altitude of 20 km, measured by radiosonde #37 launched during the AMOR-92 expedition of r/v "A. v. Humboldt" on the subtropical North Atlantic off Morocco at 31°N, 10.17°W, on 21 September 1992, 12:28 UTC (Hagen, Zülicke, and Feistel, 1996). The temperature minimum indicates the tropopause, the transition layer between troposphere below and stratosphere above.

velocity of $v_E \approx \sqrt{2g_E R_E} \approx 11200$ m s^{-1} may completely disappear into the outer space (Landau and Lifschitz, 1966). In the case of water molecules, this process results in the loss of hydrogen and an excess of free oxygen atoms of inorganic origin which may form ozone particles. Thus, the tropopause acts as an "outlet valve" that protects the planet Earth against water loss by means of this form of permanent "evaporation from the atmosphere" (which, to avoid misunderstandings, is actually similar to but not the same as an evaporation process in the thermodynamic sense of the word). The current escape rate of hydrogen atoms is estimated as 3×10^{12} m^{-2} s^{-1}, or 80,000 tons per year. At this rate, it would take 2000 Gyr to lose all terrestrial water, 500 times the age of the Earth (Ebeling and Feistel, 1994). The mean thermal velocity of deuterium is only 71% of that of normal hydrogen because it is twice as heavy. Consequently, gravitational escape results in an effective isotopic fractionation of hydrogen and other elements in the atmosphere (Lammer and Bauer, 2003). Measurements of the Pioneer Venus probe in 1978 as well as ground-based spectral observations revealed that the D–H ratio of the Venus is higher than the terrestrial by two orders of magnitude (Ksanfomaliti, 1985; Donahue and Hodges, 1992), indicating high gravitational losses of the Venusian atmosphere in the past.

Current height and temperature of the tropopause constitute results of a dynamical self-organization process under the given boundary conditions. In combination with the water-vapor instability of the troposphere, Section 3.8, one can speculate about the existence of subcritical conditions under which the tropopause valve may switch open by some suitable feedback process. For instance, if the initial water content (i.e., specific rather than relative humidity) of the atmosphere is sufficiently low (such as currently at the poles), the related tropopause would be located at a reduced altitude where the air has an increased temperature (as controlled by lapse rate times altitude) and a higher specific humidity at saturation (due to the higher air temperature). That lets more water escape from the troposphere, which in turn reduces the greenhouse warming, cools the surface, and eventually reduces the atmospheric moisture (and thus the tropopause altitude) even further. This process may approach a steady state of a dry "snowball Earth" that is in radiation equilibrium with the solar irradiation, at about 255 K, Section 3.2. The opposite feedback effect would tend to a moist hot troposphere, a high cold tropopause, and only minor gravitational escape of hydrogen. While such a scenario of a low, mild, and humid tropopause may or may not be physically possible on Earth in detail, it is a matter of fact that our cosmic neighbors, Moon, Venus, and Mars, lost most of their water in the past through some open atmospheric outlet valve, in contrast to our planet (Catling and Zahnle, 2010). Gravitational loss is faster for lower escape velocities, $v \approx \sqrt{2\gamma M/R}$, for celestial objects with smaller masses, M, larger radii, R, and lower temperature at the top of atmosphere (Landau and Lifschitz, 1966). We simply do not know how narrow the window of conditions might be under which a water-tight tropopause is a stable attractor of a self-organized planetary atmosphere.

In the troposphere, dissipative structures do not only emerge on planetary spatial scales such as the climatological circulation indicated in Figure 3.12. Under suitable circumstances, transient but spectacular small-scale structures can form, such as dust devils in the African savannah, tornados along the US "tornado alley," waterspouts over the Baltic Sea (Figure 3.16), or hurricanes and typhoons over the tropical oceans. Less well known are powerful winterly polar lows (Zahn and v. Storch, 2010) or typical regional phenomena such as Baltic cyclones (Tiesel, 2008). All those examples have in common that their violence and longevity results from the Coriolis force and the broken symmetry between converging and diverging convective flow; they are not observed at the equator and decay when they approach it. An instructive universal and relatively simple thermodynamic model for cyclones, hurricanes, and tornados was proposed by Makarieva and Gorshkov (2009). We refrain here from going into those details and focus on an important and more general aspect, namely the broken symmetry between convergent and divergent horizontal currents, for example, between atmospheric lows and highs.

In the presence of horizontal surface pressure gradients, resting air experiences a restoring force in the direction of the lower pressure, Equation 3.117. As soon as the air parcel starts moving, the Coriolis force deflects the motion to the right on the northern and to the left on the southern hemisphere. As a result, the air flow is adjusting to a stationary "geostrophic" wind pattern along the isobars where the Coriolis force compensates the pressure gradient and prevents the air from leveling the pressure-field inhomogeneity. When air is circulating around a local minimum or maximum of pressure, the related angular velocity vector points downward for clockwise rotation and upward otherwise. Of that vector, for a low-pressure system, the component parallel to Earth's axis has the same direction as the terrestrial angular velocity vector, and the opposite direction for high-pressure systems, no matter on which hemisphere. Looking at Earth from outside along its axis, the air of cyclones rotates faster than the globe and that of anticyclones slower. In other words, the lower the surface pressure in the center is, the stronger is the Coriolis effect. Intensifying depressions contract and accelerate the rotation of a self-organized chimney, while even strong highs tend to expand laterally and develop only moderate winds. All the tornados ("turning winds") mentioned before are low-pressure swirls, nothing similar is known for sinking air spirals. Except for the equator, the terrestrial circulation, Figure 3.12, has low-pressure belts only in the violent and strongly fluctuating west-wind zones, in contrast to the regular subtropical trade winds. Note that the different rotation characteristic of lows and highs is associated with their convergent or divergent horizontal flows rather than with the ascending or descending motion, respectively, of the fluid in the center.

Mathematically, the same conclusion regarding the violence of cyclones can be drawn more rigorously but perhaps less intuitively. The Euler equation for the motion of a fluid reads for an observer in a rotating coordinate frame with constant planetary angular velocity, Ω (Pedlosky, 1987; Bannon, 2003)

$$\frac{\partial \mathbf{v}}{\partial t} + (v\nabla)\mathbf{v} + 2\Omega \times \mathbf{v} = -\frac{1}{\varrho}\nabla p + \mathbf{g} \qquad (3.117)$$

For slow currents apart from the equator, an approximate quasistationary solution is the geostrophic flow which is given by the horizontal balance between pressure gradient and Coriolis force

$$\left(2\Omega \times \mathbf{v} + \frac{1}{\varrho}\nabla p\right) \times \mathbf{g} \approx 0 \qquad (3.118)$$

in combination with the hydrostatic relation for the vertical component

$$\left(-\frac{1}{\varrho}\nabla p + \mathbf{g}\right)\mathbf{g} \approx 0 \qquad (3.119)$$

By means of vector analysis, Equation 3.117 can be rewritten in the form

$$\frac{\partial \mathbf{v}}{\partial t} + \frac{1}{2}\nabla(\mathbf{v}^2) + (2\mathbf{\Omega} + \mathbf{w}) \times \mathbf{v} = -\frac{1}{\varrho}\nabla p + \mathbf{g} \qquad (3.120)$$

The Coriolis force is given by the third term; it may increase or decrease depending on the angle between $\mathbf{\Omega}$ and the vorticity of the flow, $\mathbf{w} = \nabla \times \mathbf{v}$. Taking the divergence, we obtain

$$\frac{\partial(\nabla \mathbf{v})}{\partial t} + \frac{1}{2}\Delta(\mathbf{v}^2) + \mathbf{v}(\nabla \times \mathbf{w}) - (2\mathbf{\Omega} + \mathbf{w})\mathbf{w} = -\frac{\Delta p}{\varrho} + \frac{\nabla p \nabla \varrho}{\varrho^2} \qquad (3.121)$$

The terms $(\mathbf{\Omega}\,\mathbf{w})$ and Δp have the same sign if they are the dominating terms in this equation, as in the case of geostrophy, Equation 3.118. Then, a pronounced low with a pressure minimum, $\Delta p > 0$, is related to a vorticity vector, $\mathbf{w}$, in direction of the planetary angular velocity, $\mathbf{\Omega}\,\mathbf{w} > 0$, and enhances the total Coriolis force, $(2\mathbf{\Omega} + \mathbf{w}) \times \mathbf{v}$. Vice versa, high-pressure systems with $\Delta p < 0$ are slowing the geostrophic circulation, in particular at high latitudes where $\mathbf{\Omega}$ and $\mathbf{w}$ are almost antiparallel and may to some extent compensate each other. In the limit of extremely fast rotation, $(2\mathbf{\Omega} + \mathbf{w})\mathbf{w} \approx \mathbf{w}^2$, the pressure field must exhibit a steep minimum, $\Delta p > 0$. The related curl is self-sustained, independent of the planetary vorticity, of the latitude, and of the orientation of the rotation axis. As an example, the condition $2|\mathbf{\Omega}| \ll |\mathbf{w}|$ is almost satisfied for a hurricane with wind speeds of $v = 100$ km h^{-1} at a distance of $r = 100$ km from its "eye" since $|\mathbf{w}| = 2v/r = 2$ h^{-1} exceeds the value of $2|\mathbf{\Omega}| = 4\pi/(24\,\text{h})$ by a factor of approximately 4.

Convection cells confined by rotating boundaries (such as the bottom topography) develop an amazing wealth of complex flow patterns. This is also the case in the hydrosphere located beneath the atmosphere. Ocean surface currents are basically driven by friction coupling to the global wind system but are strongly modified by obstacles such as islands, continents, or ice cover. Of particular importance for the climate system is the heat storage in the surface layer, the vertical exchange, and the meridional heat transport by the Gulf Stream and other currents.

The ocean is heated from above by solar irradiation and cooled from above by net infrared radiation and latent heat export to the atmosphere, Section 3.6. The salt dissolved in oceanic seawater shifts the temperature of maximum density of water below its freezing temperature (Feistel and Wagner, 2005; IOC, SCOR, and IAPSO, 2010). Seawater with a salinity of 24 g kg^{-1} or higher, such as ocean water with typically 35 g kg^{-1}, does not pass a density maximum upon cooling down in winter, in contrast to lakes or the brackish surface layer of the Baltic Sea (Feistel, Nausch, and Wasmund, 2008a). Similar to the atmosphere, the vertical stability of the stratification can be computed from Equation 3.114. The sign of the lapse rate, Equation 3.116, depends on the thermal expansion coefficient rather than on that of the adiabatic compressibility, Equation 3.115. In general, the two quantities cannot be considered as equivalent thermodynamic properties of a fluid, as falsly suggested in some textbooks (McDougall and Feistel, 2003). At the density anomaly of liquid water at lower temperatures, the lapse rate changes its sign and the criterion (3.116) turns inappropriate for application in limnology or oceanography.

The density of seawater depends on temperature and pressure but also significantly on its salinity. The most relevant physical mechanisms that result in oceanic salinity changes are evaporation and precipitation of water at the surface, freezing and melting of sea ice, as well as mixing of different water masses, in particular with riverine freshwater. A seawater column can lose its vertical stability as a result of cooling of the surface layer, or increasing its salinity by evaporation (when water is transformed to vapor and seawater of higher salinity remains) or freezing sea-ice (when water is transformed to ice floes and brine with higher salinity is left over), or a suitable combination of both. There is also the possibility that upwelling cold seawater passes the freezing point while it is adiabatically expanding, since the adiabatic lapse rate of seawater exceeds the pressure coefficient of the freezing temperature (Feistel and Wagner, 2005; IOC, SCOR, and IAPSO, 2010). In that case, the new-formed ice will separate from the denser brine and will buoy upwards to the surface. This process is the counterpart to condensation and precipitation of water during adiabatic uplift in the atmosphere, but in the ocean such a phase separation apart from the surface is an exceptional phenomenon that may, for example, occasionally occur at the Antarctic shelf (van Ypersele, 1993).

As a result of the seasonal course of solar heating and the interaction with the atmosphere, the properties of the ocean surface are subject to a superposition of periodic, random, and systematic changes. If the surface density is lowered, the wind stirring creates a mixed layer that is separated from the deep water by a sharp interface, termed the pycnocline. Vertical transport through this potential-energy barrier is almost inhibited and the deep-water properties are sealed from the surface. If the surface density is increased, there will be a particular location where by random fluctuations the deep-water density is exceeded first. At that point, vertical stability disappears and deep convection sets in, replacing the deep ocean with denser surface water. Among the distinct regions of deep-water formation are the Greenland Sea and the Labrador Sea for the North Atlantic, the Gulf of Lions for the Mediterranean, or the Gulf of Aqaba for the Red Sea (Marshall and Schott, 1999; Manasrah *et al.*, 2004; Huang, 2010). Similar to cyclones in the atmosphere, the sinking water forms tight rotating "chimneys" of typically 10–100 km diameter that temporarily exhibit phases of violent mixing (Morawitz *et al.*, 1996; Wadhams *et al.*, 2002).

Similar to an electronic diode with one preferred current direction, the nonlinear "diode effect" of opening and closing the gate for exchange of matter and energy between the deep water and the atmosphere results in a small effective heat capacity of the ocean in warming phases (isolated low-density, warm surface layer) and a very large one during cooling (downward convection of cooler, high-density water), similar to the seesaw pattern with steep rise and sluggish decrease observed in paleoclimatic records of the temperature (Wolff *et al.*, 2010; Beal *et al.*, 2011), see Figure 3.10. Extended periods of warming, of intensified precipitation or ice melt may cause a lasting separation of the abyss from the surface layer, not only with respect to heat exchange but also regarding dissolved gases such as carbon dioxide or oxygen. Organic substance sinking down from the surface is remineralized and oxygen is depleted until the water turns anoxic and becomes enriched with hydrogen sulfide. Stagnation phases of lacking deep-water ventilation in the past left pronounced traces in the form of a significantly different sediment chemistry and interrupted fossil records. The possible role which extended anoxic deep-water conditions may have played in the various global extinction scenarios in the history of biological evolution is a matter of scientific debates (Gould, 1993; Wignall and Twitchett, 1996; Dahl et al., 2010), see also Table 3.1 in Section 3.9. In today's ocean, anoxic deep water is found in the highly productive coastal upwelling regions such as off Namibia or California (Mohrholz et al., 2007; Chan et al., 2008; Diaz and Rosenberg, 2008; Stramma et al., 2008; Lass and Mohrholz, 2008), and in seas with riverine freshwater excess such as the Black Sea or the Baltic Sea (Ryan and Pitman, 2000; Feistel, Nausch, and Wasmund, 2008).

The deep-convection cells in the North Atlantic are probably the most relevant ones for the global climate. Strength and frequency of this meridional overturning circulation (MOC) are assumed to essentially influence the transport rate of Europe's warm-water heating, the Gulf Stream, and to drive the worldwide deep circulation, referred to as the great ocean conveyor belt (Broecker, 1987; Schiermeier, 2008; Cockell, 2008; Lozier, 2010). There is a hypothesis supported by model calculations that the sink in the Greenland Sea may be blocked by global warming and the Arctic ice melt. A first data analysis suggested that significant slowing may have already occurred (Bryden, Longworth, and Cunningham, 2005) but later checks revealed that this was a premature conclusion drawn from scarce occasional observations of a strongly fluctuating process (Cunningham et al., 2007). The variability of the MOC must always be seen in the context of various other anomalies of the North Atlantic on the decadal scale (Sherwood et al., 2011; Holliday et al., 2011); the reasons and consequences of most of them are only poorly known yet.

A potential bistability of the MOC may have played a key role for the onset and termination of ice ages (Stommel, 1961; Rahmstorf, 2000; Okazaki et al., 2010; Negre et al., 2010; Huang, 2010). The water of the conveyor belt is traveling a few thousand years until it silently rises to the surface in the North Pacific, without forming spectacular violent convection chimneys. Along the way, the water is enriched with inorganic solutes of biogenic origin such as silicate (Millero et al., 2010). As a result, the sea salt of North Pacific Deep Water contains the highest silicate fraction of all ocean waters (Tsunogai et al., 1979; McDougall, Jackett, and Millero, 2009; IOC, SCOR, and IAPSO, 2010; Millero, 2010), which indicates that this global thermohaline convection cell must have been stably working for at least some millennia. Silicate excess and its measurable influence on the ocean circulation were systematically ignored by the oceanographic standards prior to the recent International Thermodynamic Equation of Seawater, TEOS-10 (IOC, SCOR, and IAPSO, 2010; Millero, 2010).

In addition to oceanic convection phenomena of global extension, there exist various self-organized structures at smaller scales. As an example, Figure 3.19 shows a "Dampfloch," also known as steam vent or spring hole. Such holes with diameters of perhaps 30–50 cm and surrounded by about six radial cracks are observed in more

Table 3.1 Selected events in the early history of Earth, the Precambrian era[a].

Age (Myr)	Event	Reference
4568.2	First solid in the solar system	Bouvier and Wadhwa (2010)
4520–4420	Giant impact forms the Moon	Bottke et al. (2010)
4510 ± 10	Giant impact forms the Moon	Touboul et al. (2007)
4500 ± 50	First terrestrial basalt	Jackson et al. (2010)
< 3850	Last universal common ancestor	David and Alm (2011)
3800	Ending of massive lunar impacts	Head et al. (2010)
3800	Isua banded-iron formation	Dymek and Klein (1988)
3700	Sulfate-reducing bacteria	Baumgartner et al. (2006)
3500	Seawater temperature of 70 °C	Robert and Chaussidon (2006)
3500	Active oceanic biology at 26–35 °C	Blake, Chang, and Lepland (2010)
3500	Apex Chert microfossils	De Gregorio et al. (2009)
3500	Oldest evidence of the magnetic field	Biggin et al. (2011)
3400	Anoxygenic photosynthesis?	Beukes (2004)
3420	Buck Reef Chert formed below 40 °C	Hren, Tice, and Chamberlain (2009)
3200	Organic-walled microfossils	Javaux, Marshall, and Bekker (2010)
3200	Onverwacht formation	Engel et al. (1968)
2700	Archaean genetic expansion	David and Alm (2011)
>2670	Eukaryotes diverge from archaea	David & Alm (2011)
>2500	Cyanobacteria emerge	David & Alm (2011)
2500	Early oxidative weathering	Reinhard et al. (2009)
2450	Cyanobacteria, eukaryotes	Dutkiewicz et al. (2006)
2400	Great oxidation event	Lyons and Reinhard (2009)
2200	Atmosphere holds 0.2% oxygen	Hazen (2010)
2150	Fossil evidence for cyanobacteria	Rasmussen et al. (2008)
2150	Vigorous oxidative weathering	Scott et al. (2008)
2100	First colonial organisms	El Albani et al. (2010)
2000	First eukaryotes	Zimmer (2009b)
2000	Preserved water in Kaapvaal craton	Lippmann-Pipke et al. (2011)
1900	Anoxic deep oceans	Lyons and Reinhard (2009)
1800	Expansion of sulfidic conditions	Scott et al. (2008)
1730 ± 50	Fossil evidence for eukaryotes	Rasmussen et al. (2008)
1500	Green algae diverge from red algae	Yoon et al. (2004)
>1500	*Akinetes* diverge from cyanobacteria	David and Alm (2011)
>1200	Red algae microfossils	David & Alm (2011)
1200	First sexual reproduction of red algae	Butterfield (2000)
1180	Sulphide-oxidising marine bacteria	Parnell et al. (2010)
1000	First non-marine eukaryotes	Strother et al. (2011)
910 ± 150	first fungi	Lücking et al. (2009)
850	Photosynthesis on land	Knauth and Kennedy (2009)
800	Rodinia break-up	Hazen (2010)
800	Seawater temperature of 20 °C	Robert and Chaussidon (2006)
750–635	Marine phosphorus maximum	Planavsky et al. (2010)
700–635	Cryogenian: "snowball Earth"	Cockell (2008)
663–551	Second great oxidation event	Scott et al. (2008)
635	First metazoans	Love et al. (2009)

a) The final 10% of the time span of this table are covered in Table 3.2 in higher resolution.

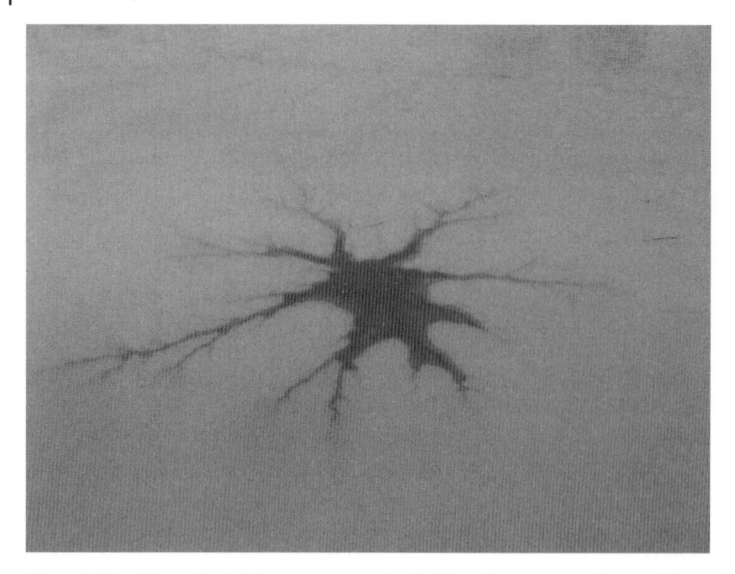

Figure 3.19 "Dampfloch" (steam vent, spring hole) in the ice cover of a pond formed by convection from below. Photo taken in January 2003 near Munich-Erding, courtesy of Martin Heß, LMU Munich, Germany.

or less regular patterns in the ice cover of shallow lakes or ponds, but only under exceptional meteorological circumstances such as those widespread in Germany in January 2009. The holes were first described by Götzinger (1909) and remained an exotic mystery until a physical explanation was given by Woodcock (1965). When a pond is covered by snow-free, optically transparent ice, solar irradiation may penetrate it and warm up the water column beneath the ice until convective overturning sets in, compare the Benard cells shown in Chapter 1. Similar but less well-understood phenomena on a larger scale are polynyas, a term borrowed from the Russian language, where open water areas form in the polar ice cover with high relevance for, for example, the formation of dense Antarctic Bottom Water (Holland, 2000; Williams and Bindoff, 2003; Williams, Carmack, and Ingram, 2007). In the case of polynyas, the upwelling of warm water is mostly caused by the wind or ocean currents in combination with topographic obstacles such as seamounts.

Underneath the atmosphere and the hydrosphere, Earth's solid crust shows a remarkable bimodal structure, Figure 3.20. The surface is divided into two different plateaus, the top of the continents at sea level and the ocean floor at about 4000 m depth, as was already noticed by Grove Karl Gilbert in 1893 (Nield, 2007). Surprisingly exactly, the top of the hydrosphere meets the steepest peak of the lithosphere's hypsographic distribution; coincidence just by chance is a poor explanation. A quick first explanation for this phenomenon exploits the fact that weathering and erosion of rocks is much faster by wind and rain than under water, such that over millions of years the top plane of the continents adjusted dynamically to the sea level. Darwin (1845) discovered a similar self-organized adjustment process to explain the observed height of coral reefs and tropical atolls. Hypothetically, one may imagine that over the

Terrestrial Hypsographic Distribution

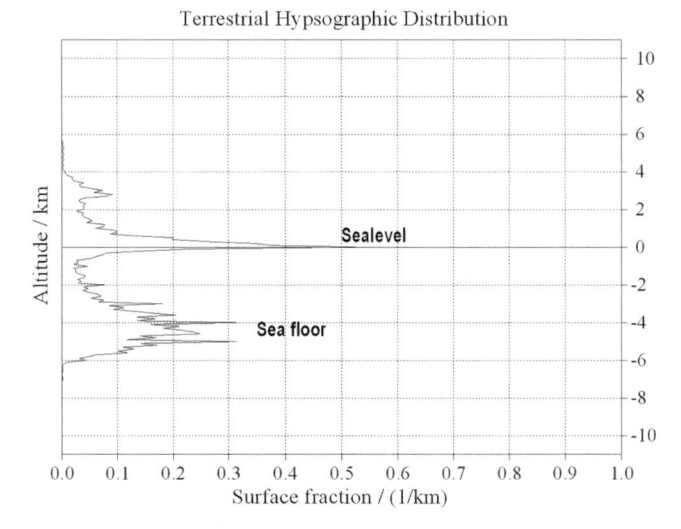

Figure 3.20 Hypsographic diagram of the terrestrial surface, $w(z)$, as a function of the altitude, z. Computed here from the digital topography ETOPO5, the density function $w(z)$ is defined such that $dA = A_E w(z)\, dz$ is the crustal area located within the altitude interval between z and $z + dz$. The two main peaks of the bimodal distribution belong to the top of the continental plates and the spreading sea floor (Sharpton and Head III, 1985).

millennia, the leveling process on land filled the ocean with eroded material and caused the sea level to rise, until this relatively fast process of conservative redistribution of water and solids has established a stable attractor state at which most of the solid surface is covered by a shallow water layer and largely protected from atmospheric denudation. But, perhaps there are also other or additional relevant mechanisms. While Earth has a bimodal hypsography and extended liquid oceans on its surface, our sister planet Venus with almost the same size has a unimodal hypsography, see Figure 3.21, and lost almost all its water (Sharpton and Head III, 1985; Ksanfomaliti, 1985). Again, just by coincidence? We shall return to this conundrum in the following section.

The dominating material of the continents is granite, while that of the sea floor is basalt. Without the different densities of the two materials, there would hardly be any reason why the one is permanently protruding from the other by 4 km, while both are floating on molten rock. The density difference between basalt and granite is about 10%, very similar to the difference between ice and water. Similar to floating tabular icebergs, see Figure 3.22, some 90% of the continental volume must exist in solid form at depths below the thin oceanic crust to provide the required buoyancy. The resulting value of 40 km thickness by which continents probably extend into the magma is just a rough estimate for the expected order of magnitude; acoustic wave tomography revealed that cratons, the ancient solid cores of the continental blocks, extend beyond 200 km depth (Hager and Gurnis, 1987; Huismans and Beaumont, 2011). About 3 Gyr old diamonds are found at depths of more than 150 km (Richardson et al., 1984; Torsvik et al., 2010), showing that solid substances survived

142 | *3 Evolution of Earth and the Terrestrial Climate*

Hypsographic Distribution of Venus

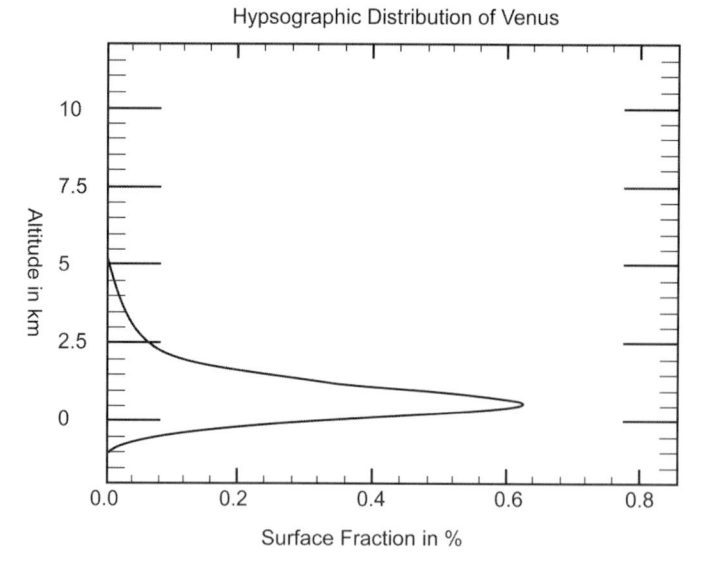

Figure 3.21 Hypsographic diagram of the Venusian surface, relative to a planetary radius of 6051 km (adapted from Sharpton and Head III, 1985).

Figure 3.22 The density difference between granite and basalt is similar to that between ice and seawater. Like suitably magnified tabular icebergs, continental plates floating on liquid magma rise to 4000 m while about 90% of the solid block is hidden beneath the molten surface to provide the necessary buoyancy. If we imagine the sea shown to be the liquid mantle (with the density of basalt), and the iceberg to be a continent (with the density of granite), its flanks were almost completely immersed in another, lighter fluid (with density of water) such as a fog layer that very precisely leaves just the iceberg's flat top face uncovered, see Figure 3.20. (photo: Antarctic Peninsula, December 2002).

at that depth over a long period of crustal evolution. Down to 250 km depth, the presence of water influences the friction and the melting points of minerals at the lower boundary of the solid crust (Green *et al.*, 2010; Green II, Chen, and Brudzinski, 2010; Reynard, Nakajima, and Kawakatsu, 2010).

The theory of self-organization of continents and oceans is in particular attributed to Alfred Wegener (1915). It took until 1968 before Wegener's continental drift hypothesis became widely accepted when the expedition of r/v "Glomar Challenger" discovered irrefutable evidence for sea-floor spreading in the Atlantic. But even a decade later, some geological textbooks still debated weaknesses of plate tectonics in comparison to more traditional explanations (Hohl, 1977). Floating on a viscous fluid mantle, continents are drifting with velocities up to 0.1 m yr^{-1} very much in the form of rigid bodies. In contrast to Wegener who suspected unknown external "cosmic" or "pole-flight" forces to cause the motion of the continents, it is generally accepted today that thermal convection of deeper layers is the reason (Closs, Giese, and Jacobshagen, 1984), even though knowledge on dynamical details is still rather limited. One may quickly think of mantle convection as a structure similar to the atmospheric circulation, Figure 3.12 (Ebeling and Feistel, 1982; O'Nions, Hamilton, and Evensen, 1984), but observation is inconsistent with such simple and regular patterns (Nowacki, Wookey, and Kendall, 2010), in addition to the substantial differences between magma and air in their viscosities, densities, or applied temperature gradients, and the related spatial and temporal scales. The very different thicknesses of the continental and the submarine crust imply locally different geothermal heat flows driving the mantle convection.

While continental rock material is relatively robust with respect to compression, it disintegrates more easily under the influence of shear or strain forces. Under this supposition, the horizontal magmatic flow beneath a continent should be either convergent or rather homogeneous, and take a similar direction as the moving plate. The displacement of the much thinner oceanic crust is probably even more similar to the viscous flow underneath. Thus, if we look at a map of continental drift and sea-floor spreading, Figure 3.23, we more or less directly watch the convection pattern driving the crustal motion. If we ignore various details and structures related to the complex shapes of the continental boundaries and the various larger and smaller plates and fractions, there remain just two global flow structures, a meridional one northward from Antarctica, and a zonal one, the widening Atlantic and the shrinking Pacific. The polar mode is visible in the successive separation of several continents from Antarctica after the Jurassic, and in particular the rapid drift of Africa, India, and Australia to the north and the related uplift of the Alps and the Himalayas. The east-west near-surface flow appears to be mainly bimodal, driving the Americas away from Europe and Africa as well as uplifting the Rocky Mountains and the Andes, while simultaneously the spreading Pacific seafloor is subducted. Somewhat idealized, the upper mantle convection cells seem to look like an orange with four deformed slices, as shown in Figure 3.23.

Deep below the mantle, in the Earth's inner core with a 1300-km radius and an estimated temperature of 6000 K (Tateno *et al.*, 2010), model computations suggest that thermal convection forms just two spatial modes, a meridional and a zonal one with typical convection velocities in the order of 10^{-10} m s^{-1}, or 3 mm yr^{-1} (Buffett,

Figure 3.23 Sea-floor spreading as suggested by the ETOPO5 bathygraphy. Arrows indicate the hypothetical convective motion of the fluid mantle beneath the solid crust. Material is apparently upwelling at the mid-ocean ridges and downwelling in the subduction zones along the continental shores such as the Pacific "ring of fire" (England and Katz, 2010).

2009; Alboussiere, Deguen, and Melzani, 2010). In contrast, a significantly faster flow of 10^{-3} m s^{-1} in the core is estimated from the wandering magnetic poles (Carrigan and Gubbins, 1984), which suggests that rather different physical processes are responsible for the spatial orientation of the convection cells and for the magnetic field (Buffett, 2010; Anderson et al., 2010; Gubbins et al., 2011).

Phase transitions of the solid and liquid minerals at different temperatures and pressures play a key role in the dynamics of the distinct spheres (Hirose, 2010; Nomura et al., 2011), that is, temperature–pressure conditions in several layers are close to the melting point of the particular mineral. Do we observe some surprising random coincidence here? While the pressure at a certain depth depends on the mass of the outer shells and cannot be altered significantly by thermodynamic processes, the local temperature is controlled by the dynamic balance between heat production, transport, and export. If we imagine a solid-state planet, its thermal conduction is much too slow to export the heat produced internally by radioactive decay. For example, 3 W m^{-1} K^{-1} may be a typical value for the heat conduction coefficient of solid rock. To conduct the mean geothermal heat flow of 0.09 W m^{-2} over a distance of 6000 km, a fictitious temperature difference of 200,000 K would be required. Hence, the interior will heat up until melting occurs and convection starts, similar to the emergence of 380 subglacial Antarctic lakes discovered so far (Wright and Siegert, 2010). On the other hand, heat transport by low-viscosity convection (such as by air or water above the crust) is much faster than the internal heat production; rapid outward heat transport cools down the layers below the convective shell (such as the solid crust below ocean and atmosphere). As a result, Earth's core and mantle are probably in states very close to the melting points of the particular minerals in the different spheres, in a quasistationary state at which the mean heat transport is balancing the heat production rate, near the kinetic phase transition between conduction and convection. If the solid rock gets too hot, the mineral will melt and faster convection will start until either the viscous heat transport can adjusted to the available heat flow, or convection stops and the rock "freezes" again. The mechanism that keeps a system close to a bifurcation point is known as SOC (Bak, Tang, and Wiesenfeld, 1987; Sprott, Bolliger, and Mladenoff, 2002). Near-critical conditions are also found, for instance, for sand piles and traveling dunes (Ball, 2009), for the salinity stratification of the Baltic Sea (Feistel and Feistel, 2006), for biological mutation rates near the "error catastrophe" (Eigen, 1971), or stop-and-go traffic on a highway (operations research, Saaty, 1966). Whether the dynamical regime of such a state is stationary, oscillating or chaotic depends on the details of the boundary conditions and material properties. The general temporal "fingerprint" of a self-organized critical state is its characteristic fluctuation spectrum.

On the horizontal interface between two convection layers that consist of fluids of different densities, lateral temperature gradients are generated by warmer upwelling from below and cooler downwelling from above (provided that the thermal expansion coefficients of those fluids are positive) (Gubbins et al., 2011). Seismic data suggest that the core–mantle interface has a bimodal structure with elevations beneath Africa and under the Pacific (Hirose, 2010). Such gradients may result in a spatial mode coupling between the convection cells in adjacent layers and in a similarity between the cell symmetries of the inner and outer spheres. It is physically plausible as well that the modality of stable convection cells may increase with the circumference of the particular spheres. Thus, we can imagine that signals generated by the temporal and spatial scales of the inner core propagate outward and become visible in the motion of the continents and the spreading oceans, and vice versa, that the positions of the floating crustal plates by thermal interaction influence the dynamics down to the inner core. Most problems and observations are still awaiting solutions and explanations. Is, for instance, the slightly faster rotating inner core (Kerr, 2005) a result of tidal friction of the crust?

It is quite obvious that the distribution of the continents strongly influences the global system of ocean currents, and in turn, by transport and release of heat, the positions and forces of the action centers of the atmospheric circulation. But as well in opposite direction, atmospheric cooling of the crust is important for the vigor and the thermodynamic efficiency of the mantle convection, in contrast to, for example, our neighbor planet Venus, where strong greenhouse insulation is suppressing the heat flow and maintaining high crustal temperatures. On our planet, we observe an amazing dynamical coupling of self-organized processes covering wide ranges of scales in time and space (King et al., 2009).

It was at 11:30 a.m. on the 20 February 1835 in a coastal forest near the town of Concepción, Chile, when Charles Darwin suddenly awoke from a nap. The ground was shaking heavily; the ocean looked like it was boiling. He could easily stand upright, but he described his feelings as similar to skating on weak ice, as if the solid Earth under his feet was a moving thin crust on a liquid. Even the oldest local people were not aware of such a strong earthquake ever before. Darwin (1845) provided a detailed description of his observations. Areas covered with mussels, formerly under shallow water, now appeared permanently exposed to the air. He did not know yet the modern term

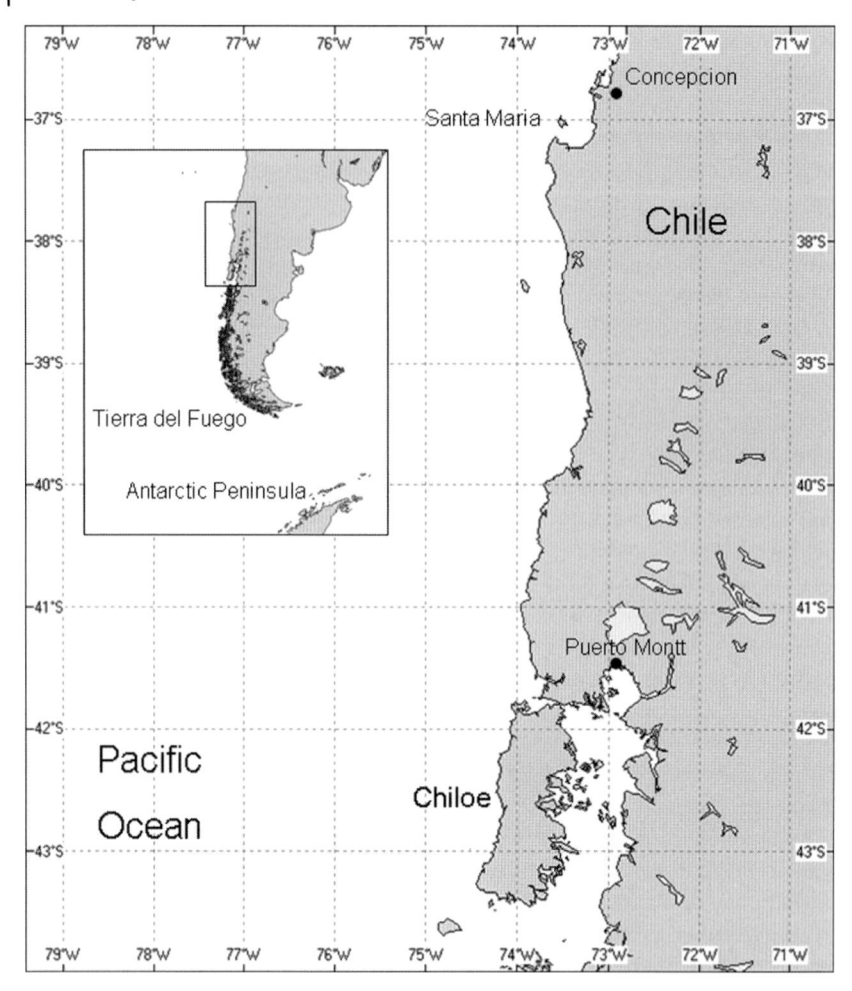

Figure 3.24 Chile's shoreline where severe earthquakes happened in 1835 and 2010. One degree latitude corresponds to 60 nautical miles, or 111 km.

"tsunami," borrowed from Japanese only after 1945, but noted that with some delay, the ocean retreated and returned twice in the form of smooth waves, washing ashore several ships. Violent spring tides formed up to 23 feet high. The quake affected an area of 720 × 400 miles, and several volcanoes turned active immediately. Land was generally uplifted by 2–3 feet, Santa Maria Island by more than 10 feet. Juan Fernandez Island, 360 miles to the north, and Chiloe Island, 340 miles southward, were also strongly hit, Figure 3.24. Later, when he crossed the Andes Mountains, Darwin found layers with fossil marine shells from the Cretaceous up to an altitude of 14,000 feet and concluded that, over a long period of time, the land must had been elevated by consecutive events similar to the one he had witnessed. If we take the end of the Cretaceous as the beginning of the Andean uplift (Thomson et al., 2010; Hoorn et al., 2010), see Table 3.2, we can infer an effective average elevation speed of 6 mm per century.

Table 3.2 Selected events in the later history of Earth, the Phanerozoic eon[a].

Age (Myr)	Event	Reference
635	First metazoans	Love et al. (2009)
635–542	First microscopic animals	Narbonne (2010)
>600	Red algal photosynthetic plastids	McFadden and van Dooren (2004)
600	Ediacaran Doushantuo animals	Bristow et al. (2009)
>551	Anoxic/oxidation oscillations	McFadden et al. (2008)
551	Oxidized ocean	McFadden et al. (2008)
550	Increase in oceanic oxygen	Kump (2010)
550	Cambrian "explosion"	Gould (1993)
540	First skeletons	Cockell (2008)
540	Color vision photopigments	Jacobs (2009)
532	Anoxic oceanic sediments	Jiang et al. (2009)
530	Burgess Shale fauna	Van Roy et al. (2010)
499	Large-scale anoxic/sulfidic conditions	Gill et al. (2011)
476	First land plants	Kenrick and Crane (2000)
472 ± 1	First land plants	Rubinstein et al. (2010)
445	Hirnantian glaciation	Finnegan et al. (2011)
438	Silurian mass extinction	Gould (1993)
434	First insects	Gaunt and Miles (2002)
420	First charcoal from burning plants	Scott and Glasspool (2006)
417 ± 45	Symbiosis of fungi with plant roots	Simon et al. (1993)
407	First winged insects	Engel and Grimaldi (2004)
400	"Explosion" of land plants	Bateman et al. (1998)
400	Dramatic rise of oxygen levels	Milton (2010), Dahl et al. (2010)
400	First tetrapods walked on land	Niedźwiedzki et al. (2010)
367	Devonian mass extinction	Gould (1993)
359	Anoxic Hangenberg event	Sallan et al. (2011)
300	Atmospheric oxygen maximum >35%	Berner (1999)
280	Antarctica located at the South Pole	Veevers (2000)
251–245	Anoxic, sulfidic ocean	Meyer et al. (2011)
250	Permean-Triassic mass extinction	Grasby, Sanei, and Beauchamp (2011)
245	Permean-Triassic mass extinction	Gould (1993)
230	First dinosaurs	Martinez et al. (2011)
214 ± 1	Manicouagan impact	Hodych and Dunning (1992)
208	Triassic-Jurassic mass extinction	Gould (1993)
165	North America separated from Africa	Sclater and Tapscott (1984)
160	First feathers	Zhang et al. (2008)
150 ± 10	*Archaeopteryx*	Wellnhofer (2008)
126 ± 1	Blossoming angiosperms	Sun et al. (2011)
125	South America separated from Africa	Sclater and Tapscott (1984)
120	First blossoms and fruits	Gould (1993)
80	Atmospheric oxygen maximum >25%	Berner (1999)
80	North America separated from Europe	Sclater and Tapscott (1984)
70	First birds *Teviornis gobiensis*	Kurochkin, Dyke, and Karhu, (2002)
65.5	Chicxulub impact	Schulte et al. (2010)
65	Cretaceous-Tertiary mass extinction	Gould (1993)

a) The final 10% of the time span of this table are covered in Table 3.3 in higher resolution.

One is strongly reminded of Darwin's report when reading recent papers on the earthquake that occurred on the 27 February 2010 at about the same location, but almost exactly 175 years later (Farías et al., 2010; Moreno, Rosenau, and Oncken, 2010). In contrast to 1835, at this time the processes before and during the event could be watched closely with sophisticated measuring techniques. The length of the terrestrial day shortened by 1.26 μs (Buis, 2010), which implies that more mass collapsed downward, probably under water, than was visibly elevated on land. A tsunami devastated Juan Fernandez Island again this time (Lorito et al., 2011). The GPS position of Concepción jumped by 2.9 m to the south-west (DGFI, 2010). With respect to the relative drift velocity of 6.8 mm yr^{-1} measured in the region before (Ruegg et al., 2009), we conclude that the plates should have been locked for at least 430 years, which is unlikely. Vice versa, 175 years of total locking result in the higher plate shear speed of 16.6 mm yr^{-1}. Drift speed observations may be biased temporarily by partial locking of the plates. Processes of periodic locking and sudden release, such as in the case of subduction earthquakes, belong to the Coulomb-friction class of self-organized oscillations, together with, for instance, violin strings (Ebeling, 1976a). The period of 160 years observed for New Zealand earthquakes is very similar to the time elapsed since Darwin's quake (Wells and Goff, 2007).

Observations fit well to model results for elastic dislocation energy stored since 1835 as a result of completely locked plates (Farías et al., 2010). The seismic moment magnitude of $M_w = 8.816$ represents the total amount of work released during the earthquake, $W = 2.1 \times 10^{22}$ N m, by the formula (Hanks and Kanamori, 1979)

$$M_w = \frac{2}{3}\lg(W/J) - 6 \tag{3.122}$$

With respect to the rupture fault width of $w = 139$ km and length of $l = 505$ km, the related mean areic accumulation rate of potential energy over a period of $t = 175$ years is $W/(w \times l \times t) = 5 \times 10^{-10}$ W m^{-2}. Compared to the mean geothermal heat flow of 0.087 W m^{-2}, we note that a thermodynamic efficiency of the heat engine "mantle convection" of just 6×10^{-7}% is apparently sufficient to move continents and successively uplift vast mountain ranges. In contrast, from the yellowish thermal radiation color of glowing lava we may estimate a typical magma temperature of about 1300 K and infer a Carnot efficiency of $1 - 300/1300 = 77$% as the theoretical upper bound for geothermal power stations. The inner core of the Earth has an estimated temperature of 6000 K, the core–mantle interface of 4000 K (Fiquet et al., 2010). The related thermal gradients provide the successive internal convection layers with free energy to overcome the viscous friction, in combination with the effective atmospheric cooling system of the crust which exports the produced entropy to the cosmic background. The underground "magma mill" has an estimated entropy export rate of 3×10^{-4} W m^{-2} K, as already mentioned in the beginning of this section, and contributes only 0.03% to the total terrestrial entropy export. The total heat flow from the core to the mantle is estimated to be about $J_Q = 5$–10 TW (Hirose, 2010). The related mean entropy export, $J_S = J_Q/4000$ K, across the lithosphere is between 2.5 and 5 μW m^{-2} K so that approximately 1% of the geothermal entropy production comes from the core and 99% from the mantle.

The Drake Passage between Tierra del Fuego and the Antarctic Peninsula, Figure 3.24, opened about 41 Myr ago (Scher and Martin, 2006). Along the submarine Drake plate, the distance between Cape Horn and the tip of the bathygraphic tongue at the South Sandwich Trench, Figure 3.23, is about 2200 km. These numbers suggest a local convection speed in the order of 50 mm yr^{-1}, a value much higher than the velocities measured at the Chilean subduction zone. The sea-floor structure between South America and Antarctica looks similar to a violent bullet puncture (Nield, 2007), see Figure 3.23. Although a rift between Antarctica and Australia had already formed 68 Myr ago after the Chicxulub impact, it was after 40–33 Myr back that the separation accelerated when the Tasmanian Seaway opened and Australia started moving northward (Müller, Gaina, and Clark, 2000; Kennett and Exon, 2004; Hill and Exon, 2004; Hassold et al., 2009).

Similar to tornadoes in the atmosphere and convection chimneys in the ocean, there are apparently also convection cells in the fluid interior with much smaller spatial scales than the global convection system. So-called hot spots, such as the Hawaiian chain of islands, indicate a locally intensified heat flow from below that does not move with the seafloor and must originate at deeper levels (Hager and Gurnis, 1987). For a resting spot in the spreading seafloor, the chain of volcanoes is parallel to the age gradient of the crustal isochrones (curves of equal age), which must correspond to the islands' weathering rate above the sea surface. Observed deviations lack final explanation but hint on the impact of specific remote processes (Whittaker et al., 2007). Quite recently, it was discovered that also rare kimberlite chimneys, where diamonds are found at great depths in the continental blocks, form a global coordinate frame of hot spots at positions that virtually ignore the plate drift (Torsvik et al., 2010). Small-scale mantle convection extending laterally over a few hundred kilometers is also suspected to produce regional periodic signals of 2–20 kyr in sedimentary sequences (Petersen et al., 2010).

3.9
Early Evolution of Earth

Chondritic meteorites are among the oldest materials in the solar system (Bland et al., 2011). The earliest piece of solid matter ever analyzed is 4.5682 Gyr old, by dating calcium–aluminium-rich inclusions in a chondritic meteorite found in Northwest Africa (Bouvier and Wadhwa, 2010). By definition, this is the age of the solar system.

It is a widely accepted theory that the solar system was formed out of the gas that was left over from a supernova explosion. Thermodynamically, a gas expansion into the vacuum is a simple but rather exotic nonequilibrium process (Landau and Lifschitz, 1966). If we remove the walls of a volume containing an ideal gas in thermal equilibrium, the particles will isotropically fly away in different directions with their instantaneous velocities, ignorant of each other. After some time, the fastest particles will form the outmost spherical wave front, followed by slower ones, and so on, such that the originally microscopic random motion of the particles is transformed into the macroscopic motion of spherical shells with equal radial velocities. If differently massive particles are involved as in the case of supernovae, the lightest particles will have the highest thermal velocities, as discussed in the context of gravitational escape in Section 3.8. As a result, the gas is fractionated into the elements and their isotopes, with hydrogen far outside and heavy atoms such as uranium at the very inner end. This "thermal" distribution of the elements, transferred from the momentum space to the configuration space, likely played a role in the structure of the initial gas cloud out of which the planets were born.

How do solids condensate out of a gas? That is a well-known physical process; it occurs whenever frost is formed in the morning, or snowflakes appear in clouds. Perhaps, the very first dust appeared in the preplanetary gas for solely thermodynamic reasons. As soon as the gas pressure is higher than the vapor pressure of the related solid or liquid condensate, the vapor turns metastable or unstable and a phase transition process is triggered. At very low densities and temperatures of the gas, the relation between the very low sublimation pressure of the solid state and the very low gas pressure is not trivially obvious. One may quickly and prematurely guess that in the almost empty interstellar space any condensed substance has a finite vapor pressure and must sooner or later evaporate into the vacuum, even at the low background temperature of the cosmos. If that was true, why then do Saturn's rings (Canup, 2010) or old distant comets (Whipple and Huebner, 1976; Weissman, 1980; Tauber and Kührt, 1987; Elliot *et al.*, 2010; Dunaeva, Antsyshkin, and Kuskov, 2010) contain a lot of ice and produce a tail when they come closer to the Sun? For the highly interesting case of water, we investigated the problem of sublimation in the limit of zero pressure and zero temperature (Feistel and Wagner, 2007). To our surprise it turned out that, theoretically, already at 20 K the vapor pressure of ice is as low as the pressure of an ideal gas which consists of a single atom in the entire universe. Obviously, any existing gas in the cosmos must be much denser than that, from which we conclude that at temperatures below, say, 50 K, water vapor cannot reasonably exist as a thermodynamically stable phase, just because our universe is not big enough to hold it at the necessarily low density. As a result, when a dilute interstellar gas cools down, its decreasing pressure will at some point exceed the even faster decreasing vapor pressure, the stability of the gas phase will turn to metastability, and condensation may start.

If a gas is thermodynamically unstable, it will spontaneously collapse and form condensed bodies, but such highly supersaturated states are rather improbable because the transition process will usually begin already at an earlier stage. If the gas is metastable, the related transition process is known as nucleation and starts from randomly formed nuclei as soon as those reach a supercritical size as a result of spontaneous density fluctuations. Subcritical droplets or crystals evaporate and disappear but supercritical ones will grow. Between the condensed objects, a selection process sets in, regarded as Ostwald ripening (Schmelzer, 1985; Mahnke and Feistel, 1985; Schmelzer and Gutzow, 1986; Schimansky-Geier, Zülicke, and Schöll, 1991), very similar to other competition processes described in Chapter 6. Due to the advancing condensation process, the pressure of the remaining gas is lowered and the related critical radius grows. As a result, the smallest nuclei turn subcritical and evaporate such that subsequently the number of nuclei decreases and their average mass increases. Between the bigger objects orbiting the Sun, the collision probability increased systematically with their cross section and further accelerated the accretion process. In a simple coagulation model, the balance equation for random inelastic binary collisions is (Greenberg et al., 1978)

$$\frac{d}{dt} N(m) = \int dm' N(m')[C(m', m-m')N(m-m') - 2C(m', m)N(m)] \quad (3.123)$$

where $N(m)dm$ is the number of objects with masses in the interval $m, \ldots, m + dm$, and $C(m, m')$ is the binary collision rate of two given objects with the masses m and m'. C is a positive and symmetric matrix. In the simplest model, C may be a product kernel, such as

$$C(m, m') = k(m \times m')^{2/3} \quad (3.124)$$

where k is a rate constant and the exponent 2/3 refers to the cross section of a volume proportional to the object's mass. Equation 3.123 implies that the total mass is conserved

$$\frac{d}{dt} \int m N(m) \, dm = 0 \quad (3.125)$$

and the total particle number is decreasing at the rate

$$\frac{d}{dt} \int dm\, N(m) = -\int dm\, dm'\, C(m, m') N(m) N(m') < 0 \quad (3.126)$$

In the initial phase, there is a vast number of tiny objects (e.g., molecules or supercritical nuclei), while in the final phase, there are few very big objects (e.g., asteroids, comets, planetoids). This process can also be modeled by percolation of random networks, see Section 4.4. In the solar system, massive impacts ended about 3800 Myr ago (Head et al., 2010; Bottke et al., 2010). With the accumulated masses of condensed objects, their gravity becomes relevant and attracts gas and other bodies. A detailed schematic of the various condensation processes involved was painted by Hohl (1981).

The planets formed in a nonlinear competition process for the available material. That process has an important and outstanding physical property in common with other evolution processes: historicity. The result of the competition depends on various specific details of the initial distribution and the later stages and boundary conditions. Fluctuations are amplified rather than attenuated by the system; information on those early fluctuations is preserved and still available, at least in principle, from the current, "final" state of the solar system. From today's shape, we are able to draw conclusions on its history. Such conclusions are impossible from, say, a gas in its equilibrium state.

To make the physical character of historicity more transparent, we consider an example. That this book is written just now and here is obviously related to the fact that mammals including *Homo sapiens* could evolve as a result of the extinction of dinosaurs at the end of the Cretaceous era, 65 Myr ago (Schulte et al., 2010). Let us for a moment follow the hypothesis of Alvarez et al. (1980) that this catastrophe was caused by an asteroid of about 10 km diameter which hit the Earth near Yucatan.

Probably, that cosmic body was also formed in the birth phase of the solar system. Halley's comet may be similar to Alvarez' comet. With a return period of 75 years, after 4568–4565 Myr, the comet had theoretically traveled $n = 60$ million times the distance to the Pluto, $d = 6 \times 10^9$ km, and back, in total $2 \times n \times d = 720 \times 10^{15}$ km. That far away, the Earth with a diameter of $2 \times R_E = 12{,}000$ km appears within a narrow opening angle of $180 \times R_E/(\pi \times n \times d) = 10^{-12}$ degrees into which the initial velocity vector must have pointed when Alvarez' comet was assembled, provided that all other circumstances remained unaltered, in order to hit the planet and kill the dinosaurs right in time for the authors to evolve, grow up, and write this book. Thus, historicity means that the present macroscopic appearance of the solar system is a result of a vast number of tiny fluctuations and collision events in the past, from the nucleation of the first droplets up to the occasional impact of heavy objects. On the other hand, at least in principle, it is possible to extract information on the detailed historical sequence of those events which by their fragile mutual interplay eventually produced the physical system as it is today. We shall return to this very fundamental aspect.

At a size larger than 150 km, the planetesimals' export of heat caused by radioactive decay became too slow and the interior started melting (Hazen, 2010). Under the influence of their own gravity, the fluid constituents of the early planets separated into layers of different densities (Solomatov, 2007; Rubie et al., 2011). The elements buried deep below the surface were almost perfectly conserved except for radioactive decay. The outer layers, in particular the atmosphere, remained in exchange with the solar wind, impact of meteorites, and so on (Halliday and Wood, 2009). Different hypotheses on the composition of the early atmosphere are discussed in the scientific literature. Here, we shall follow just one of several alternative fictitious scenarios that may prove more or less right or wrong in the future. The hypothesis appears plausible to us and is suitable as an instructive model for diverging early evolution routes of terrestrial planets. In particular we will focus on Venus and Earth under the approximation that they originally developed like twins under very similar conditions until something specific happened that separated their lifelines, suddenly but ultimately.

Let us speculate that the sister planets turned into "water worlds" after the initial phase of cooling down and the formation of a planetary solid crust (Bethel and Bergin, 2009). On Venus, the upper solid layer covered the entire planet rather homogeneously. On top of it, a global ocean condensed that may have been several kilometers deep. Without relevant obstacles in the form of separate continents, see Figure 3.21, the liquid water surface could easily compensate any lateral temperature gradients by global convection currents. As a result, the homogeneous SST of the water world did not drive any relevant deep atmospheric convection on the planetary scale. In turn, no air-drying process was available and relative humidity at the surface was established close to equilibrium, at almost 100%. The high water content caused a strong greenhouse effect and high air temperatures, together with high pressures. At high altitudes, dense clouds formed under radiative cooling and scattered or absorbed almost all of the solar irradiation within a minor penetration depth. The opaque atmosphere was heated from above rather than from below, and was stably stratified. Intense vertical convection was confined to some thin upper "mixed" layer; no significant tropopause was build up in the upper atmosphere. Hydrogen could escape from the top of the atmosphere and was replaced by evaporating water from the ocean surface. The remaining inorganic oxygen excess oxidized atmospheric gases or surface minerals, perhaps similar to processes found on Saturn's moon Rhea today (Teolis et al., 2010). Slowly but steadily, the water disappeared from the planet until the solid surface fell dry. Mantle convection was not strong enough to break up the globally closed continental crust, and was further slowed down by the high surface temperature. The result is how Venus looks today.

Soon after its initial formation, Earth suffered from a giant impact by Theia, a planet of the size of Mars (Hazen, 2010; Schönbächler et al., 2010). Parts of the two original planets were redistributed differently than before and formed a double planet, Earth and Moon. In particular, substantial parts of the formerly closed continental crust disappeared from Earth. The remaining scattered fractions floated like ice on an ocean of molten basalt and accreted to primordial continents by the mantle convection. A thin basalt crust that appeared quickly over the uncovered parts of the magma was too weak and too dense to block the continental motion. Eventually, the terrestrial hypsography received its present bimodal shape similar to Figure 3.20.

After a period of 0.7 Gyr with frequent impacts of large objects (Head et al., 2010), the early formation phase terminated. Along with the cooling down, liquid water accumulated on the surface and covered the entire globe. Earth relaxed to the same extreme greenhouse state as the Venus. While the water continuously disappeared, there came the point at which the surface of the continental blocks fell dry. As a result of solar irradiation, solid ground can heat up the atmosphere much faster and stronger than the ocean surface, reduce the relative humidity that way, turn the atmosphere dry and optically transparent, and drive monsoon-like convection by the air–sea temperature gradient. The ocean was suddenly confined to the troughs between the continents; it could no longer freely circulate on the global scale and compensation of meridional surface temperature gradients by ocean currents was hampered. A global atmospheric deep convection system switched on, similar to Figure 3.12, the greenhouse effect caused by clouds and vapor became greatly reduced, and the mean surface temperature decreased rapidly due to the water-vapor feedback effect. The related tropopause "lid" was built up and blocked any further gravitational escape of water, Figure 3.18. The sea-surface level was firmly adjusted to the upper peak of the hypsographic distribution, as shown in Figure 3.20. The effective atmosphere–ocean cooling system increased the geothermal heat flow and the mantle convection intensity. Continental drift and sea-floor spreading were established in a form similar to the present one, perhaps quasiperiodically (Nield, 2007). In the Archaean period, the solar luminosity was lower than today by 25–30%, and there is no geological evidence for high concentrations of carbon dioxide, CO_2, or methane, CH_4, at that time (Rosing et al., 2010). The evolution on Earth took a new route and opened up the gate to the origin of life. That in turn drove Earth even farther away from the fate of its twin planet Venus.

In Table 3.1 a number of events are presented that are available from the scientific literature, covering the Precambrian, the extended earliest period of the terrestrial history. It is obvious that this list is incomplete; the selected events represent a data set with significantly undersampled coverage in space and time. Evidently, conclusions drawn from that table can easily be premature and may anytime be falsified by new discoveries. As the many very recent dates of the references in the table indicate, current research is intense and rapidly progressing in that field. Nevertheless, we will continue the narrative on the early evolution based on those events, being well aware of the necessarily speculative nature of such fictitious scenarios.

To our present knowledge, it took about 1 Gyr from the giant impact to the widespread appearance of early life forms. For the origin of life, we strongly prefer the hypothesis based on submarine hydrothermal vents, similar perhaps to the conditions reported for the Apex Chert formation (De Gregorio et al., 2009). Some aspects of the earliest self-organization are more closely analyzed in Chapters 8 and 9. Hydrothermal vents offered a permanent supply of free energy in the form of thermal and chemical gradients, similar to such vents today (Bates et al., 2010). Potential "food" in the form of randomly synthesized energy-rich molecules was omnipresent. The hot and coarse rocky crust provided active catalytic surfaces and protected refuges. Those "productive" regions may have extended spatially over many thousands of square kilometers, and may have nonintermittently persisted over millions of years with their fairly constant thermochemical environments, representing an ideal natural laboratory for testing vast numbers of occasional molecular assemblies. Those fragile structures were well protected from the various violent processes that dominated the archaic world at the planetary surface. All those facts make the environment of vents a plausible stage for a slow and vulnerable initial process of molecular evolution. Provided that it really happened that way, we note a second consequence of the exceptional formation process of the Moon that possibly was crucial for the emergence of life on Earth. After the preservation of planetary ocean water and the resulting favorable temperatures under the influence of a suitably balanced land–sea surface ratio, sea-floor spreading caused by the incomplete remains of outer continental shell provided the necessary fuel for the first evolution phase. The giant impact and its hypothetical, far-reaching consequences represent an extreme example of historicity in evolution. It is part of the chaos theory in the form of Moser's (1978) consideration on the stability of the solar system.

First self-reproducing microreactors probably fed on the stock of accumulated chemical energy in the ocean and the continuous but limited influx of chemicals produced by natural sources such as hydrothermal vents. But it was only a question of time until the increasing reproduction speed of the primordial cells exceeded the available supply, and strong competition for energy efficiency must have taken place. Those which were randomly washed to the sea surface and occasionally could take chemical advantage of the energy of sunlight finally managed the global conversion to the renewable energy source by photosynthesis. All that happened within 1 Gyr after the giant impact, at 3.5 Gyr ago. After another Gyr, oxygen-producing autotrophic microbes apparently won the race, perhaps because of the high efficiency of that system, perhaps because oxygen poisoned the competitors and parasites. Widespread indication for free oxygen was available at 2.4 Gyr ago, termed the great oxidation event (GOE).

The atmospheric oxygen level rose quickly and continuously over the next several 0.1 Gyr which indicates that autotrophic organisms strongly dominated over oxygen consumers. The accelerated evolution of complex organisms was driven by the freely available energy source of oxygen respiration and cumulated in the appearance of eukaryotes 2 Gyr ago. Their efficiency, flexibility, and variability apparently reorganized the archaic ecosystem fundamentally. Probably for similar reasons as today, extended anoxic conditions developed in the oceans over the next 0.2 Gyr, where the oxygen was replaced by hydrogen sulfide (H_2S). That must have been an effective cell poison; it seems as if this self-organized catastrophe paralyzed the terrestrial marine ecosystem for the rather long period of 1 Gyr, the "boring billion" of years. Our extreme olfactory sensitivity for traces of H_2S may have its very biochemical roots in those times when that gas was a permanent deadly danger. Eventually, the recovery began with first autotrophic organisms that could escape the ocean and gradually conquered the land. Their success possibly enabled a second great oxidation event between 0.8 and 0.5 Gyr ago; the early marine evolution history repeated itself in fast motion on the solid ground. Again, phases with oxygen excess apparently accelerated the evolution while the presence of hydrogen sulfide hampered the progress (Dahl et al., 2010). Multicellular organisms and animals appeared on the scene and resulted in the Cambrian explosive expansion of marine species.

The phase between 0.7 and 0.635 Gyr ago is termed the Cryogenian (Cockell, 2008); geological evidence is available from various places that an extended global glaciation took place. It is sometimes speculated that the "snowball Earth" was triggered by atmospheric CO_2 decrease as a result of the second GOE. Ice ages appeared several times in the earlier and later history; it is not even definitely clear whether they were caused by either internal or external forcing, or both. Seen from a more general viewpoint, on a scale between 6000 and 3 K, it is obvious that the terrestrial effective atmospheric cooling established the surface temperature at a value very close to the triple point of water. Small fluctuations of the system, either by variation of the external forcing, by certain feedback loops in the ocean–atmosphere-radiation dynamics, or by the form and location of the continents may be responsible for formation, growth, or retreat of ice sheets. Those processes are sensitive and at the same time very dramatic climate indicators.

In this context, it is a very interesting observation made by the Cassini–Huygens space probe that Saturn's largest moon, Titan, has a surface temperature close to the triple point of methane (90.7 K) and possesses a troposphere and a tropopause (Lunine and Lorenz, 2009). The close similarity between Earth and Titan suggests the hypothesis that the vicinity of the triple point may be a stable, possibly chaotic, attractor for the surface temperature on planets covered with both land and extended liquid "mares," perhaps in a dynamical regime of SOC. Surprisingly, even at the rather low temperature of Titan, simple organic compounds may form spontaneously (Horst et al., 2010). Recalling the analogy of Figure 3.22, on the Jovian moon Europa, the liquid ocean under an icy crust may play a similar role as the granite crust covering liquid magma on Earth-like planets (Marion et al., 2005).

Phase transitions of the ambient water are key features of the discontinuous response of the climate system to small temperature variations (Feistel and Hagen,

1993). Formally, if we assume that the time evolution of the local surface temperature (or any other set of local properties) is subject to some nonlinear functional master equation of the general form

$$\frac{\partial}{\partial t} T(\mathbf{r}, t) = \Psi\{T(\mathbf{r}', t') | \mathbf{r}, t\} \tag{3.127}$$

we may consider a series expansion with respect to small temperature variations, $T = T^0 + \delta T$. The related first expansion term yields the linear functional master equation

$$\frac{\partial}{\partial t} \delta T(\mathbf{r}, t) = \int d\mathbf{r}' dt' \, C(\mathbf{r}, t | \mathbf{r}', t') \, \delta T(\mathbf{r}', t') \tag{3.128}$$

The related integral kernel is the "teleconnection matrix" that correlates the properties at distant locations $\mathbf{r}$ and $\mathbf{r}'$, at different times, t and t', and is defined by the Fréchet derivative

$$C(\mathbf{r}, t | \mathbf{r}', t') = \left. \frac{\delta \Psi(\mathbf{r}, t)}{\delta T(\mathbf{r}', t')} \right|_{\text{climate}} \tag{3.129}$$

taken at the particular climatic state. The problem with such an approach is that the functional derivative exhibits singularities at kinetic and thermodynamic phase transitions of the climate system, in particular in states of SOC. Then the result is that even for small δT on the right-hand side of Equation 3.128, the left-hand side does not need to remain small. This fact makes the truncated linear Equation 3.128 an inappropriate global mathematical climate model for various relevant situations. From that perspective, it is a surprising observation in the scientific literature on the climate system that practically all authors emphasize the nonlinearity of the climate dynamics while at the same time there is an extended set of articles which, from almost any available data series, construct certain "teleconnection patterns" of the form of Equation 3.128 by linear statistical methods such as empirical orthogonal functions (EOF) or principal oscillation patterns (POP). Linear correlation and response analyses, in particular with respect to the El-Niño oscillations, have even conquered the research field on historicity in the human social evolution (Hsü, 2000; Caviedes, 2001; Jones et al., 2001; Behringer, 2007; Büntgen et al., 2011). Certainly, various climatic phenomena can be described in the form of a linear response to small changes at other times or at other locations similar to Equation 3.128, or with respect to alternative observational properties. In other cases, such as the formation or disappearance of an oceanic ice cover as a result of small temperature changes, or the onset of convection, the related singularity of Equation 3.129 requires a nonlinear description. Hence, it depends on the particular, concrete circumstances under which either linear or nonlinear climate models are the tools of choice.

Much more geological, biological, and eventually even historical information has been gathered for the most recent period of 0.5 Gyr after the so-called Cambrian "explosion" from which on a rapidly increasing wealth of fossil evidence could be discovered at numerous locations all over the world. In Tables 3.2 and 3.3, just a few

Table 3.3 Selected events in the latest history of Earth, the Cenozoic era[a].

Age (Myr)	Event	Reference
65	Cretaceous-Tertiary mass extinction	Gould (1993)
65	Uplift of the Andes began	Hoorn et al. (2010)
56.3	Paleocene-Eocene thermal maximum, rapid tropical plant diversification	Jaramillo et al. (2010)
55.8	Global thermal maximum	Secord et al. (2010)
55	Global thermal maximum	Pagani et al. (2006)
51 ± 1	First mallow trees in Indian rainforest	Rust et al. (2010)
50	First daisies and sunflowers in Patagonia	Stuessy (2010)
50	Uplift of the Himalayas began	Zhisheng et al. (2001)
41	Drake Passage opened	Scher and Martin (2006)
40	Uplift of the Alps began	Laubscher (1984)
40	Upheaval of Ethiopia began	Hancock, Pankhurst, and Willetts (1983)
40	Middle Eocene climatic optimum	Bijl et al. (2010)
40	Maximum body size of mammals	Smith et al. (2010)
36	Atlantic largely had present shape	Sclater and Tapscott (1984)
36	Giant penguins	Clarke et al. (2010)
36	Sea-level oscillations by ~40 m	Peters et al. (2010)
36–35	Substantial continental ice sheets	Miller, Fairbanks, and Mountain (1987)
35	Subglacial Lake Vostok formed	Schiermeier (2011)
34	Antarctic glaciation began	Ehrmann (2000), Bo et al. (2009)
34–30	Rapid decrease in atmospheric CO_2 from ≈1500 to ≈500 ppmv	Pagani et al. (2005), Urban et al. (2010)
33.5	Tasmanian Seaway opened, Antarctic Circumpolar Current formed	Hassold et al. (2009), Hill and Exon (2004), Kennett and Exon (2004)
33	Tasmanian Seaway opened	Müller, Gaina, and Clark (2000)
26	West-Antarctic glaciation began	Barker, Diekmann, and Escutia (2007)
26–12	North-Atlantic deepwater formation	Miller and Fairbanks (1983)
23	CO_2 decrease from 600 to 340 ppmv	Kürschner, Kvaček, and Dilcher (2008)
20–15.5	CO_2 increase to 400–500 ppmv	Kürschner, Kvaček, and Dilcher (2008)
20	Expansion of grass lands and ruminants	Van Soest (1994)
18–16.6	Miocene climatic optimum	Böhme (2003)
15.5–14	CO_2 decrease to 280 ppmv	Kürschner, Kvaček, and Dilcher (2008)
14–13.5	Abrupt European cooling by >7 °C	Böhme (2003)
13.8–10.4	Four cold South-Atlantic deepwater events	Westerhold, Bickert, and Röhl (2005)
9.6	Enhanced Orange-River erosion	Robert, Diester-Haass, and Paturel (2009)
8.9	Development of Benguela upwelling	Robert, Diester-Haass, and Paturel (2009)
5.6 ± 0.3	Significant sea-level lowering	Robert, Diester-Haass, and Paturel (2009)
5.6	Mediterranean fell dry	Meijer and Krijgsman (2005)

a) The final 10% of the time span of this table are covered in Table 3.4 in higher resolution.

key steps are reported for illustration of the enormous self-acceleration of the biological evolution after the Cambrian explosion. Rapid diversification of plants and animals was particularly pronounced in the Tertiary, after the extinction of the dinosaurs, in contrast to the rather regular, apparently easy-going pace of the continental drift over millions of years of terrestrial history before.

The spectacular "Tertiary explosion" of plants and animals was virtually terminated after the violent opening of the Drake Passage. Polar heating by a warm ocean current in the form of a putative South-Atlantic counterpart of today's Gulf Stream was discontinued. The subsequent Antarctic glaciation (Ehrmann, 2000; Barker, Diekmann, and Escutia, 2007; Bo et al., 2009; Dallai and Burgess, 2011) developed over an extended period of time and occurred in conjunction with the opening of Tasmanian Seaway and the resulting Antarctic Circumpolar Current (Veevers, 2000; Hill and Exon, 2004; Kennett and Exon, 2004; Hassold et al., 2009). The related climate change and especially the global lowering of sea level resulted in the conversion of vast regions of shallow coastal oceans into continental lowlands. Due to the gentle cross-shore slope of the terrestrial hypsographic distribution, Earth is characterized by an exceptional sensitivity of the land–ocean surface area ratio with respect to small changes of the sealevel (Sharpton and Head III, 1985), see Figure 3.20. After the opening of the Drake Passage, a long sequence of events affected substantially the climatic conditions in and around Africa, from the uplift of the Alps and the Ethiopian highlands to the Mediterranean desiccation and the connection of the Red Sea to the Indian Ocean. Similarly, the closing of the Isthmus of Panama was followed by the glaciation of Greenland and the onset of ice-age oscillations, Table 3.4. It may well be that the violent break-through of the Drake Passage, beyond its drastic effects on the global oceanic circulation and hence on the climate, also altered the regional mantle convection pattern and the drift of the African continent, see Figure 3.23. In turn, the closing of the equatorial gateway between the Pacific and the Atlantic must have had significant impact on the basin-scale current systems and El-Niño-like phenomena in the two oceans, with severe consequences for the global climate.

The beginning of the social evolution may be related to the development of a special human lifestyle in the African savannah that required the loss of fur (Reichholf, 2004) and the evolution of skin pigments, 1.2 Myr ago (Rogers, Iltis, and Wooding, 2004; Jablonski, 2006), as a joint visible characteristic of all the recent human races. Apparently, dramatic climatic changes in Africa, such as those resulting stringently from dissication ("Messinian Salinity Crisis") and subsequent flooding ("Zanclean Flood") of the Mediterranean, from the formation of the oceanic Red Sea between Ethiopia and Arabia, as well as from the later ice-age cycles, should have posed an immense selective pressure on the early human populations. A striking impression from Table 3.4 is the historicity of the evolution of humans; we may owe our present existence to an exceptional series of very specific geological and climatic circumstances (deMenocal, 2011). If we combine this speculation with the tiny chance that our planet formed the way it did when it collided with the Moon, and that the millions of years of dinosaur dominance ended timely by a suitable cosmic impact, we may come to the conclusion that intelligent extraterrestrial life similar to ours is rather unlikely.

For illustration, the incredibly long time span of about 4.5 Gyr of terrestrial evolution is often mapped to better comprehensible intervals of 24 hours or 4.5 years (Sagan, 1978; Ebeling and Feistel, 1982; Gould, 1993). As an alternative intuitive picture, we may take advantage of the fact that the north-eastern German country, Mecklenburg-Vorpommern, where the authors lived and worked for many years, has

Table 3.4 Selected events in the latest history of Earth from Pliocene to the first historical event that was also recorded by means of human language (age given before present).

Age (Myr)	Event	Reference
5.6	Mediterranean fell dry	Meijer and Krijgsman (2005)
5.33	Zanclean flood of the Mediterranean	Garcia-Castellanos et al. (2009)
5.5–5	Antarctic Circumpolar Current slacked	Hassold et al. (2009)
5	Red Sea turned oceanic	Horowitz (2001)
4.4	*Ardipithecus* from Awash, Ethiopia	Gibbons (2009)
3.6	Bipedality of *Australopithecus*	Haile-Selassie et al. (2010)
3.5	Isthmus of Panama closed	Campbell et al. (2010), Hoorn et al. (2010)
>3.39	Stone-tool scraps in Dikika, Ethiopia	McPherron et al. (2010)
3.2	Skeleton of "Lucy" from Afar, Ethiopia	Johanson and Edey (1990)
3	Greenland ice cap formed	Lunt et al. (2008)
2.6	First stone tools from Afar, Ethiopia	Semaw et al. (2003)
2.5 ± 0.1	Quaternary ice-age oscillations began	Hewitt (2000), Horowitz (2001)
2	Genetic bottleneck of *Homo erectus*	Hawks et al. (2000)
2	Human spoken language	Janson (2006)
1.8	Human spoken language	Berger (2008)
1.75	*Homo erectus* from Dmasini, Georgia	Vekua et al. (2002)
1.5	Hand-axe technology in Africa	Scott and Gibert (2009)
1.2	Human gene for skin pigmentation	Rogers, Iltis, and Wooding (2004)
0.64	Yellowstone super-vulcano erupted	Smith et al. (2009)
0.6	*Homo heidelbergensis* in Europe	Sirocko (2010)
0.44–0.27	Neanderthals separated	Reich et al. (2010)
0.42	Oldest preserved Antarctic glacier ice	Petit et al. (1999)
0.4–0.3	European humans habitually used fire	Roebroeks and Villa (2011)
0.28	*Denisova* humans in Altai, Siberia	Reich et al. (2010)
0.2–0.13	Second ice age: Riss/Illinoian	Figure 3.10
0.17	Humans began wearing clothes	Toups et al. (2011)
0.16	*Homo sapiens* from Awash, Ethiopia	White et al. (2003b)
0.16	Heat treatment of stone tools, in Africa	Brown et al. (2009)
0.15	Women's common genetic root: "Eve"	Wells (2002)
0.13	Sahara: a land of lakes and rivers	Balter (2011)
0.125	Stone tools at the Persian Gulf	Armitage et al. (2011)
0.11–0.012	First ice age: Würm/Wisconsin	Denton et al. (2010), Figure 3.10
0.059	Men's common genetic root: "Adam"	Wells (2002)
0.045–0.042	Human settlement of Australia	O'Connell and Allen (2004)
0.032	Neanderthals in Byzovaya, Siberia ?	Slimak et al. (2011)
0.04	Extinction of Neanderthals	Sirocko (2010)
0.015	Human settlement of America	Hubbe, Neves, and Harvati (2010)
0.011–0.008	Flooded forests in the Baltic Sea	Tauber (2007, 2011)
0.0082	Abrupt Greenland cooling by –6 °C	Thomas et al. (2007), Cheng et al. (2009)
0.0076	"Noah's Flood" of the Black Sea	Ryan and Pitman (2000)
0.006	Human written language	Janson (2006)

Figure 3.25 In north-eastern Germany, Mecklenburg-Vorpommern has about 4500 km of century-old tree-bordered alley roads which may serve to illustrate a journey through 4500 million years of terrestrial evolution, including a "Phanerozoic Park."

a total of 4500 km of beautiful alley roads (LSBV, 2010), as shown in Figure 3.25. They make conveniently accessible the landscape's various lakes, forests, and Baltic coast lines that were mainly shaped during and after the last ice age. Similarly, one may also imagine a car ride over a distance of 4500 km such as from Miami, Florida, to Vancouver, British Columbia.

On the long way to the finish line, after 1000 km of driving the car through a hostile desert without respirable air the first marine biology becomes visible. After another 1000 km, the great oxidation event happens at almost half the total distance. It takes yet another tedious day trip of 1000 km to pass the boring billion until the first fungi conquer the land. After crossing within an hour the Cryogenian snow range of 70 km length, a more appealing landscape opens before our eyes with the second oxidation event, and just 100 km later, we arrive at the Cambrian zoo with a wealth of living marine creatures. The remaining 550 km are experienced as an exciting safari tour through the "Phanerozoic Park," including occasional "Death Valleys." Behind the 400 km mark, after having traveled already 90% of the distance, the land is suddenly covered with green plants where animals and flying insects are living. Reptiles grow larger and larger; dinosaurs show up 230 km before the final destination. Still 150 km to go, and birds appear in the sky, soon followed by a garden with flowers and fruits. Out of a sudden, 65 km away from the goal, the dinosaur paradise ends in the short but disastrous Chicxulub catastrophe and changes into a different and very flexible Cenozoic world of new animals and plants. On the last 2.4 km, snow accumulations cover the road every 40–100 m. In between, walking on two legs, naked humans appear on the scene; they quickly invent some clothes to wear and learn to light the fire.

Figure 3.26 Inclined tree trunks of a submerged forest at an ancient river bed, dated about 10,000 years back, discovered in the Baltic Sea at about 21 m water depth (Tauber, 2011). Photo taken in the Kadet Channel in 2005, courtesy of Franz Tauber, IOW.

Noah's flood is perhaps the first event in natural history which was conveyed by means of symbolic rather than structural information (Chapter 8), namely first orally from generation to generation, and later in written documents such as the Bible (Ryan and Pitman, 2000; Haarmann, 2003; Schoppe and Schoppe, 2004). Beyond the Black Sea where no evidence was found yet, also other candidates exist where settlements and forests were flooded at about that time, 8–11 m back on our scale, such as in the Baltic Sea by the Litorina transgression (Westman et al., 1999; Küster, 2002; Tauber, 2007, 2011; Lübke, Schmölck, and Tauber, 2011), see Figure 3.26, possibly related to the 8.2 kyr event (Thomas et al., 2007; Dominguez-Villar et al., 2009), see Table 3.4. At this time, agriculture spread from Anatolia and written language developed (Gray and Atkinson, 2003; Schoppe and Schoppe, 2004; Gadjimuradov and Schmoeckel, 2005), see also Chapter 8. Two meters before the target line humans mark the year zero of the calendar. Only over the final dozen of centimeters of the entrance sill, they carry along with them the scientific concepts of Darwinian evolution, thermodynamics, quantum mechanics, and theory of relativity in their minds and books. Unfortunately we lack of prediction capability to tell how that road might continue over the next meters and kilometers into the future.

4
Nonlinear Dynamics, Instabilities, and Fluctuations

Der Vorgang ist ganz leicht erklärlich.
Der Natur riss einfach die Geduld.
(The case is easily explained:
Nature's at the end of tether.)
Erich Kästner: Drei Männer im Schnee

4.1
State Space, Dynamic Systems, and Graphs

The concepts of nonlinear dynamics, instabilities, and fluctuations played an important role in the theory of self-organization and evolution. After the pioneering work of Poincaré and Boltzmann in the nineteenth century, we mention in particular the role of the schools of Andronov and Prigogine in the twentieth century. On the basis of the introductory discussion in Section 1.3, we will first introduce several basic concepts of the deterministic and stochastic dynamics. More details and many applications will be discussed later. Let us first – in a more rigorous way than in Section 1.3 – introduce the concepts of state space, dynamical models, and trajectories.

The geometrical interpretation of the trajectories of a dynamical system as orbits in an appropriate state space, commonly termed the phase space, appears to be one of the most important instruments for the investigation of dynamical processes. Generalizing the concepts of classical mechanics, Poincare, Lyapunov, Barkhausen, Duffing, van der Pol, Andronov, Witt, Chaikin, Birkhoff, Hopf, and others developed the basis for the modern theory of dynamical systems. As pointed out in Section 1.3, not only theoretical mechanics but also theoretical biosciences contributed to this method. We mention in particular the pioneering work by Lotka, Volterra, Fisher, Wright, Haldane, Rashevsky, and others.

Here it is not intended to develop in detail the general theory of dynamical systems, which is meanwhile quite independent of physics and biosciences and has found applications in most branches of science. In particular, the theory of dynamical systems is also an essential part of synergetics founded by Haken, based on laser physics (Haken, 1970, 1978, 1981, 1988) and the theory of self-organization

Physics of Self-Organization and Evolution, First Edition. Rainer Feistel and Werner Ebeling.
© 2011 Wiley-VCH Verlag GmbH & Co. KGaA. Published 2011 by Wiley-VCH Verlag GmbH & Co. KGaA.

developed by Prigogine, Glansdorff, and Nicolis, backing on the concepts of irreversible thermodynamics (Glansdorff and Prigogine, 1971; Nicolis and Prigogine, 1977).

Rather than going into details we will take here a more general system-theoretical view following (Neimark, 1972; Butenin, Neimark, and Fufayev, 1976; Neimark and Landa, 1987; Ebeling and Sokolov, 2005). Let us consider a dynamical system, assuming that its state at a time t is given by a set of time-dependent parameters (coordinates). We consider this set as a vector, $\mathbf{x}$, and write $[\mathbf{x}(t_1), \mathbf{x}(t_2), \ldots, \mathbf{x}(t_n)]$ for a given sequence the coordinates take in the course of time. The set of state vectors $\mathbf{x}(t)$ spans a vector space, $\mathbf{X}$, which will be termed the state space or phase space of the system. Let us assume now that the state at a time t, given by $\mathbf{x}(t)$, and the state at a later time $t + \delta t$, given by $\mathbf{x}(t + \delta t)$, are connected in a causal way. In particular, we assume that the state $x(t + \delta t)$ is given as a function or a more general mapping of the state $\mathbf{x}(t)$ and certain additional parameters which we denote by the vector $\mathbf{u}(t) = [u_1, u_2, \ldots, u_m]$. The set of all possible parameter values $\mathbf{u}$ forms the control space of the system. The parameters $\mathbf{u}$ take into account the influence of the environment and possibly also actions of external control. We assume that the connection between $\mathbf{x}$ and $\mathbf{x}(t + \delta t)$ is given by a dynamical map $\mathbf{T}$,

$$\mathbf{x}(t + \delta t) = \mathbf{T}(\mathbf{x}(t), \mathbf{u}, \delta t) \tag{4.1}$$

The set $\{\mathbf{X}, \mathbf{T}\}$ is referred to as the dynamical model of the system (Neimark, 1972; Neimark and Landa, 1987; Ebeling, Engel, and Feistel, 1990). The choice of the phase space is not unique and is in general a rather difficult problem. Even if we have a set of observations and measurements, there are still many possibilities to make a choice for the phase space. For example, one may think of the movements of the planets, which were first modeled by so-called epicycles relative to the Earth and only much later by Newton's laws with respect to Copernicus' sun-centered reference frame. The time increments δt may be continuous or discrete. In the case of continuous time, the states form a continuous orbit. The representation of processes by orbits (trajectories) in a phase space requires a continuous model of the process. Sometimes the trajectories are given only by a smoothed sequence of (discrete) observations. In the case that the observations are made at fixed time steps, δt, the trajectory is a sequence of events in the form

$$\mathbf{x}(t_1), \mathbf{x}(t_2), \ldots, \mathbf{x}(t_n)$$

Such time series may be given by recording the outputs of some measuring instruments in certain time intervals, but also by a stroboscopic observation of continuous trajectories or by periodic reports. There are intrinsically discrete processes such as annual reports on the industrial production of a country. Examples of models with discrete time steps will be considered later on. In the rest of this chapter, we restrict ourselves to the mathematically more convenient case of continuous time. As already mentioned, the existence of a dynamical map $\mathbf{T}$, which defines the state at the time $(t + \delta t)$, that is, $\mathbf{x}(t + \delta t)$, by means of the state at an earlier time t, that is, $\mathbf{x}(t)$, expresses the causality of the dynamical process under consideration. In the real world, we observe two different cases of causal relations between $\mathbf{x}(t)$ and $\mathbf{x}(t + \delta t)$. If the relation between the two quantities is unambiguous, that is, if the future state is given by the map **T** in a unique way by the initial one, we speak about deterministic models. Otherwise, if there are several possibilities for the future state $\mathbf{x}(t+\delta t)$ depending on chance, we speak about stochastic models. A typical example for deterministic-type models is given by Onsager's relaxation equations, which we derived in Chapter 2. They are a set of first-order differential equations and read

$$\frac{dx_i}{dt} = -\sum_j \lambda_{ij} x_j \tag{4.2}$$

In this equation, there appears a set of relaxation variables, x_i, and a matrix, λ_{ij}, which defines their mutual relations. This structure suggests an alternative approach to describe the state of a system in a graphical way that is based on set and graph theory and operates with the relations between elements and sets as the primary aspect of a structure (Harary, Norman, and Cartwright, 1965; Laue, 1970; Hartmann et al., 2009). In order to characterize a structure, we introduce relations, operations, and graphs. This idea of structure reflects the most abstract properties of a system. Due to their abstract nature, all the ideas and results can be transferred to other systems and comparisons are possible. In general, we can distinguish between spatial, temporal, causal, and functional structures.

To illustrate the approach, we use graphs that represent the elements and their connections by geometrical symbols, nodes, and edges (Harary, Norman, and Cartwright, 1965; Laue, 1970; Casti, 1979; Ebeling and Feistel, 1982). As examples we may take the relations expressed by the matrix λ_{ij} in the relaxation equations (4.2), which may be represented by a graph. Graphs are also widely used for representing ecological or economic networks, representing the flow of materials and outcomes. In most natural and socioeconomic systems, the relations between the objects and agents can be represented in a network form. In some cases, information flows, for example, price signals between market participants, are exchanged. The structure of the network then describes a situation with local interaction (not every agent is informed about all the other agents but the agents do not act independently) (Kirman, 2003). Socioeconomic networks can also describe a variety of different actions of agents that influence other agents. The diffusion of technologies over firms can be described as a network of actions from the formation of a company (with a certain technology) over knowledge transfer between companies (in the form of imitation or merging) to the termination of companies due to technological competition. We will come back to such a network interpretation in Section 4.3.

Maybe it is worthwhile to underline that it is not possible to give a complete structural description of living objects by binary relations to which graphs belong as representatives. Graphs are only very useful tools for analyzing those systems. Of course we cannot describe complex objects merely by graphs because their binary relations are ambiguous, and ternary and higher order relations may be relevant. Nevertheless, graphs are very helpful for the representation of complex structures. Before the elements of the theory of graphs are explained, it is necessary to introduce the two notions, *set* and *relation*.

What we understand by a *set* is nearly the same as in common language. A set is always abstract and is specified by its elements. Let a set M be given, which consists of the elements $a_1, a_2, a_3, \ldots$, and is symbolically written in the form

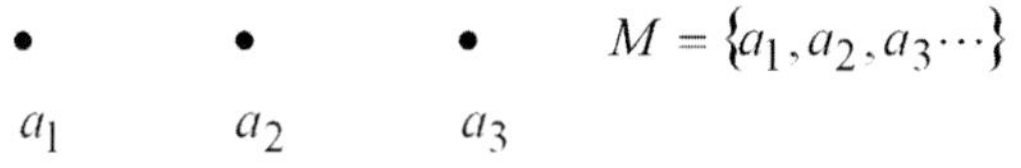

•   •   • $M = \{a_1, a_2, a_3 \cdots\}$

a_1 a_2 a_3

Figure 4.1 A set of three objects that are represented symbolically by three points in an abstract space.

$M = \{a_1, a_2, a_3, \ldots\}$. The elements a_i, $i = 1, 2, 3, \ldots$ describe any things or ideas that build the set M. The number of elements determines whether we have a finite or an infinite set. Our investigations in this section are primarily related to finite sets. If we loosely speak about sets in the following, finite sets will be meant. Graphically, the elements of a set can be described by vertices located in an n-dimensional space. This way we assume that a one-to-one relation exists between the vertices and the elements of the set. An example is given in Figure 4.1.

The elements of a set can be grouped into pairs. Let us consider, for example, a set $M = \{a_1, a_2, \ldots, a_5\}$ with the pairs $[a_1, a_3]$ and $[a_2, a_5]$. The elements a_1 and a_2 of M belong to the first element of the set of pairs associated with M, and a_3 and a_5 to the second one. The inverse pair $[a_5, a_3]$ in general describes a different relation between the elements a_5 and a_3 than $[a_3, a_5]$, that is, the elements a_i and a_j of an ordered pair $[a_i, a_j]$ are not exchangeable with each other without affecting the kind of mutual relation of the elements a_i and a_j.

Let a, b be two elements of a set M. Let R be a subset of ordered pairs of M, such as $[a, b]$. Then R is termed a *binary relation* on the set M. Several useful and rather general properties can be defined that a relation R may or may not possess (Görke, 1970; Ebeling and Feistel, 1982). We use the symbols $\in$ for "is an element of" and $\notin$ for "is not an element of."

A relation R is *symmetric* if $[a, b] \in R$ implies that also $[b, a] \in R$. Example: If a person a is married to a person b, then b is also married to a, that is, the binary relation "married" is symmetric. If a person a loves a person b, this does not necessarily imply that b loves a too. The relation "loves" is in general not symmetric.

A relation R is *asymmetric* if $[a, b] \in R$ implies that $[b, a] \notin R$. Example: If a person a is the daughter of a person b, then b cannot be the daughter of a. The relation "daughter of" is asymmetric.

A relation R is *antisymmetric* if $[a, b] \in R$ and $[b, a] \in R$ implies that $a = b$. Example: If a body a is not warmer than a body b, and b is not warmer than a, then a has the same temperature as b. The relation "not warmer than" is antisymmetric.

A relation R is *transitive* if $[a, b] \in R$ and $[b, c] \notin R$ implies that $[a, c] \in R$. Example: If a person a is an ancestor of a person b, and b is an ancestor of a person c, then a is an ancestor of c. The relation "ancestor of" is transitive. If a person a is an enemy of a person b, and b in turn is an enemy of a person c, this does not necessarily mean that a is an enemy of c. Often quite the contrary is true, as the saying "my enemy's enemy is my friend" is suggesting. Usually, the relation "enemy of" is not transitive.

A relation R is *reflexive* if $[a, a] \in R$ for each $a \in M$. Example: A material a may substitute a material b for a certain purpose. Trivially, any material can substitute itself, so "substitute of" is reflexive. A person a may be able to cut the hair of a person b. Most persons are unable to cut their own hair; hence, "cutting hair" is in general not reflexive.

A relation R is *irreflexive* if $[a, a] \notin R$ for each $a \in M$. Example: No person can be its own parent, so the relation "parent of" is irreflexive.

A relation R is *linear* if at least one of $[a, b] \in R$, $[b, a] \in R$, or $a = b$ holds for each pair $a, b \in M$. Example: For any events a, b in time, either a is earlier than b, or b is earlier than a, or $a = b$. The relation "earlier than" is linear. In the theory of relativity, this linearity gets lost because the temporal sequence of certain events may depend on the observer's reference frame.

- A relation that is linear, transitive, and irreflexive is termed an *irreflexive order relation*.
- A relation that is asymmetric, transitive, and irreflexive is termed an *irreflexive semiorder*.
- A relation that is linear, transitive, antisymmetric, and reflexive is termed a *reflexive order relation*.
- A relation that is transitive, antisymmetric, and reflexive is termed a *reflexive semiorder*.
- A relation that is transitive, symmetric, and reflexive is termed an *equivalence relation*.

Relations such as those defined above play a fundamental role for, example, the mathematical formulation of causality and irreversibility of physical processes. Causality and irreversibility are asymmetric relations (Lieb and Yngvason, 1999; Thess, 2007). For a macroscopic system described by 10^{23} momentum and position variables $(\mathbf{p}, \mathbf{q})$, the existence of an order relation between subsequent mechanical states, $(\mathbf{p}, \mathbf{q}) \to (\mathbf{p}', \mathbf{q}')$, is not a trivial problem and was subject to fundamental physical debates by Poincaré, Einstein, Zermelo, Prigogine, and others. For reversible processes, an equivalence relation is formed by the pairs of physical states that are connected with each other by adiabatic transitions (Caratheodory, 1909; Margenau and Murphy, 1943; Strehlow and Seidel, 2007). In the framework of group theory (Landau and Lifschitz, 1967b; Kurosch, 1970, 1972), another equivalence relation can be defined by all sets of generalized coordinates $(\mathbf{p}, \mathbf{q})$ of a given system that result from the use of different reference frames by an observer. For example, the inconsistency between those equivalence relations for Newton's mechanics (the group of Galilean transformations), on the one hand, and Maxwell's electrodynamics (the group of Lorentz transformations), on the other hand, had once inspired Einstein to develop his theory of relativity.

The question whether evolution is a process that systematically leads to more order or higher complexity is similarly related to the problem of whether an order relation, or at least a semiorder, can formally be associated with the physical states passed subsequently by the evolving system (Ebeling and Feistel, 1982). Starting from the fact that Darwinian selection, that is, the replacement of a particular species by a fitter successor, is an asymmetric relation between the former and the later state, an axiomatic, Boolean reaction system can be defined axiomatically and investigated under this specific viewpoint (Feistel, 1979) (see also Chapters 5–7). First results indicate that the existence of such order relations, that is, of universal evolution principles, is in general not granted by dynamical laws in the form of chemical

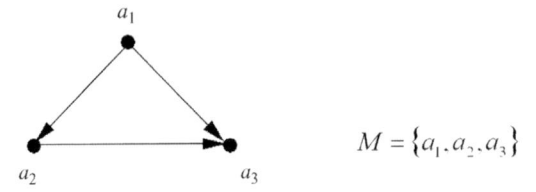

Figure 4.2 Directed graph consisting of three arrows (directed edges) pointing at three objects (nodes).

reaction kinetics, in particular because transitivity of the binary relation "is fitter than" implies additional restrictive requirements.

Binary relations and their properties can conveniently be represented by graphs. As an example, we consider a set of objects with a certain structure between them. The structure can be specified either in the form of relations, those could be predator-prey relations, trade relations, or information flows. Many relations in nature and society are directed. Let the set M consists of three objects a_1, a_2, and a_3. There are given the following relationships: a_1 has strong relations to a_2 and a_3; a_3 has strong relations to a_2. The ordered pairs for this relation follow to be $[a_1, a_3]$, $[a_1, a_2]$, and $[a_3, a_2]$. We can represent this relationship geometrically in the form of a directed graph as shown in Figure 4.2.

A graph consists of the elements of the set and the relations between them. In such a way, a graph can visually represent specified relationships between elements of a set. In our example, Equation 4.2, the relations are expressed in terms of the matrix λ_{ij}. In general, a graph is a structure model that reflects the structure of the investigated object. In order to get this structure, we have to consider the relations between the corresponding partial objects.

As a system we usually consider a model that describes the original object or process in a simplified way. A system is in general defined by the set of its elements and the set of relations between its elements. A dynamical system includes a map depending on the time and provides an instruction for predicting future states. Obviously, different degrees of fine-tuning, complexity, and accuracy of the model with respect to the original, real template are possible. A model is termed homomorphous if a one-to-one mapping between model and original is possible. The limiting case of maximum adjustment of the model to the original with respect to the structure is regarded as isomorphous.

We will see later that graphs are very useful tools for the characterization of the structure of dynamical systems. Detailed examples are discussed in Section 4.4 as well as in Section 6.5 on the competition in networks of interacting species. In Section 4.2, we restrict ourselves to the simplest case of deterministic models.

4.2
Deterministic Dynamic Systems

From the point of view of physics, the simplest example of a deterministic dynamics is given by the friction-dominated motion of a particle (neglecting its inertia) with the coordinate x in a potential field U(x) that has the equation of motion (ϱ – friction coefficient):

$$\frac{dx(t)}{dt} = -\frac{1}{\varrho}\frac{dU(x)}{dx} \tag{4.3}$$

In this case, the orbits are given by the gradient lines (the lines of steepest descent of the potential). The orbits are attracted by the minima of the potential. In general, the minima correspond to points that will therefore be referred to as point attractors. Iterated maps, operating on discrete time, are another relatively simple example of deterministic systems (Collet and Eckmann, 1980; Schuster, 1984).

A less simple example for dynamical systems is given by mechanical systems. Most mechanical systems are described by a *Hamilton dynamics*, which is defined by a scalar function H on a phase space of f coordinates,

$$\mathbf{q} = (q_1, q_2, \ldots, q_f)$$

and f momenta

$$\mathbf{p} = (p_1, p_2, \ldots, p_f)$$

The so-called Hamiltonian, H, defines the dynamics by the canonical equations,

$$\frac{dq_i}{dt} = \frac{\partial H}{\partial p_i} \quad \frac{dp_i}{dt} = -\frac{\partial H}{\partial q_i} \tag{4.4}$$

where $i = 1, 2, \ldots, f$. By integration of the Hamiltonian equations at a given initial state, $\mathbf{q}(t)$ and $\mathbf{p}(t)$, we may calculate the state at $t + \delta t$ in a unique way. One of the most recent findings for Hamiltonian systems is that despite of the deterministic connection between initial and future states, the predictability of future states is quite limited (Hoover, 2001; Landa, 2001).

The dynamics of Hamilton type is a rather special one. In many cases – we may think here, for example, of chemical and ecological processes – the dynamics is given by more general types of autonomous differential equations that cannot be derived from solely one function (such as, e.g., a potential or a Hamilton function). For example, the time evolution of the state vector $\mathbf{x}(t)$ may be given by the system of autonomous differential equations as

$$\frac{dx_i(t)}{dt} = F_i(x_1, x_2, \ldots x_f, \mathbf{u}) \tag{4.5}$$

where $i = 1, 2, \ldots, n$, and may be written in a more compact vector form as

$$\frac{d\mathbf{x}}{dt} = \mathbf{F}(\mathbf{x}, \mathbf{u}) \tag{4.5a}$$

However, even a dynamics defined by such a quite general differential equation is still a rather special form of the map (2.69). One may obtain a differential equation if the map **T** for small δt has the Taylor expansion,

$$\mathbf{x}(t + \delta t) = \mathbf{x}(t) + \mathbf{F}(\mathbf{x}, \mathbf{u})\delta t + O(\delta t^2) \tag{4.6}$$

The class of systems described by a system of differential equations is rather large. We have to remember that all differential equations of higher order as well as nonautonomous differential equations may be reduced to systems of the type (4.5). Even partial differential equations may be reduced to ordinary differential equations, at least in some approximation. The standard method that allows the reduction of an (infinitely dimensional) partial differential equation to a finite-dimensional system of ordinary differential equations is the procedure proposed by the Russian mathematician Galerkin. For the reasons given, we consider systems of the type (4.5) as a sufficiently general basis for our further considerations of dynamical systems.

Further, we will assume here that the conditions for the validity of Cauchy's theorem are fulfilled that guarantees existence and uniqueness of the solutions of Equation 4.5. Then, for given initial conditions, $\mathbf{x}(t_0) = \mathbf{x}_0$, there exists exactly one solution

$$\mathbf{x}(t) = \mathbf{x}(t; \mathbf{u}, \mathbf{x}_0)$$

Since $\mathbf{x}(t)$ is time-dependent, we obtain a trajectory in the state space. The theorem about uniqueness has an important topological consequence: Any crossing of trajectories is forbidden since otherwise at the crossing point the system would possess two different solutions, which is actually excluded by Cauchy's theorem. In the special case of two dimensions, every closed trajectory separates the motion in the compact region inside from that outside, depending on the position of the starting point.

The manifold of motions described by Equation 4.4 defines a field of trajectories. In general, the functions $\mathbf{F}$ are nonlinear and may be quite complicated. Thus, the solution cannot be given analytically. However, several general conclusions on the properties of the dynamical system may be obtained even without knowledge of the explicit solution. This is possible on the basis of the so-called qualitative theory of dynamical systems (Andronow, Chaikin, and Witt, 1965; Neimark and Landa, 1987). One may answer questions such as:

- What are the stationary states of the dynamical system, are they stable?
- Do periodical motions exist, are they stable, are there chaotic motions?
- Is it possible that the system reaches certain boundaries in the state space?
- What is the dependence of the motion on the control parameters?

Answers to these questions may be achieved by a purely qualitative analysis of the geometry of the field of trajectories (Anishchenko et al., 2002).

Of special interest are stationary points, where the system is at rest, and especially stable stationary points that may be the targets of the motion at long time. We term these states point attractors. There may exist other attracting manifolds of dimension $d < n$. Attractors are closed bounded sets that are attracting and are invariant with respect to the dynamics (Anishchenko et al., 2002). In other words, if a point belongs at a time t to an attractor, it will stay on it for any time. Attractors have an attracting basin that consists of all trajectories that asymptotically approach the attractor in the long-term limit. To the class of attractors there belong stable stationary points as discussed above, stable limit cycles, stable tori, and so-called strange or chaotic attractors. What kinds of attractors are possible in a dynamical system depends on the dimension n of the state space and on the functions **F**. The investigation of the attractors of a dynamical system and of their properties is the main aim of the qualitative theory (Anishchenko, 1995; Anishchenko et al., 2002).

Also of special interest is the localization of separatrices and saddle points. Saddles are intersection points of separatrices. They, on the other hand, divide the state space into the regions of attraction of different stable manifolds. To acquire knowledge about the dynamical system, the information available on saddles and separatices is therefore crucial. Saddles and separatrices can be reached only if the initial position is located at a saddle or at the separatrix. The situation may change when stochastic influences are included. In the frame of stochastic theory, deterministically unstable manifolds may be reached and may be crossed due to fluctuations. Therefore, stochastic perturbations may lead to transitions between several regions of (deterministic) attraction.

A very useful method of qualitative dynamics is the investigation of the vector field $\mathbf{F}(\mathbf{x}, \mathbf{u})$. The system of differential equation(4.5) defines a flux in the state space. The trajectories are tangential to the vector field **F**. Investigating the divergence of this field, we find three principal cases for the local behavior of the vector field:

1) Locally expanding dynamics with the property

$$\nabla \mathbf{F}(\mathbf{x}, \mathbf{u}) > 0 \tag{4.7}$$

locally conservative dynamics with

$$\nabla \mathbf{F}(\mathbf{x}, \mathbf{u}) = 0, \quad \text{and} \tag{4.8}$$

2) locally contracting dynamics with

$$\nabla \mathbf{F}(\mathbf{x}, \mathbf{u}) < 0 \tag{4.9}$$

> We underline that this is a local property, the character may change in other regions of the space, and in particular, it may change along trajectories.

The sign of $\nabla \cdot \mathbf{F}(\mathbf{x}, \mathbf{u})$ determines the relative change of a small volume element of the state space, which is given by

$$\Delta\Omega(\mathbf{x}; \mathbf{u}; t + \delta t) = \Delta\Omega(\mathbf{x}; \mathbf{u}; t) + \nabla \cdot \mathbf{F}(\mathbf{x}, \mathbf{u})\, \delta t \tag{4.10}$$

If the vector field is everywhere free of sources (the divergence vanishes everywhere), the dynamics is termed conservative. Reversible motions, such as the Hamiltonian dynamics, Equation 4.4, are necessarily conservative. For all members of this class, the relation

$$\nabla \cdot \mathbf{F}(\mathbf{x}, \mathbf{u}) = 0 \tag{4.11}$$

holds for all points of the state space.

A typical example of a contracting dynamics is a dynamical system with an attractor. In this case, a small volume element of the state space is always shrinking with time. For an attractor, the inequality

$$\nabla \cdot \mathbf{F}(\mathbf{x}, \mathbf{u}) < 0 \qquad (4.12)$$

is valid (Anishchenko, 1995; Anishchenko et al., 2002). Dynamical systems with this property are termed dissipative. Irreversible motions correspond to dissipative systems; their dynamics corresponds to a movement toward the attractor. The simplest example is an oscillator with linear friction,

$$\dot{q} = v \quad \dot{p} = -\gamma_0 p - kq \qquad (4.13)$$

where the momentum is $p = mv$, m is the particle mass, and $\gamma_0 = \varrho/m$ is the collision frequency connected with the friction constant ϱ. Accordingly, we get

$$\nabla \cdot \mathbf{F}(\mathbf{x}, \mathbf{u}) = -\gamma_0 \leq 0 \qquad (4.14)$$

Two other important classes of dynamical systems are the gradient systems

$$\frac{d\mathbf{x}}{dt} = -f \, \nabla V(\mathbf{x}, \mathbf{u}) \qquad (4.15)$$

and the canonical–dissipative systems, (H – Hamiltonian),

$$\frac{d\mathbf{q}}{dt} = \frac{\partial H}{\partial \mathbf{p}} \qquad (4.16)$$

$$\frac{d\mathbf{p}}{dt} = -\frac{\partial H}{\partial \mathbf{q}} + f(H)\frac{\partial H}{\partial \mathbf{p}} \qquad (4.17)$$

Here, $f(H)$ is a certain function that characterizes the dissipative properties of the system (Graham, 1973, 1981; Ebeling, 1981; Feistel and Ebeling, 1989). A simple case with

$$H = \frac{p^2}{2m} + \frac{m}{2}\omega_0^2 q^2 \quad f(H) = -\gamma_0 = \mathrm{const}$$

leads us to a damped Hamiltonian dynamics of the type of Equation 2.81. Another interesting case, closely related to Rayleigh's theory of self-sustained sound oscillations, is $f(H) = a - bH$. These dynamical systems are regarded as *canonical–dissipative systems* (Ebeling and Sokolov, 2005), they have attractors (limit cycles) located on the ellipse

$$H = H_0 = \frac{a}{b} \qquad (4.18)$$

with the exact stable periodic solutions

$$p(t) = \sqrt{\frac{2ma}{b}} \cos(\omega_0 t + \delta) \qquad (4.19)$$

$$q(t) = \sqrt{\frac{2a}{\omega_0^2}} \sin(\omega_0 t + \delta)$$

Gradient systems may have only a very special type of attractors, namely, stationary points, corresponding to the minima of the potential. The potential $V(\mathbf{x}, \mathbf{u})$ is a Lyapunov function of the motion, since it follows from Equation 2.83 that

$$\frac{dV}{dt} \leq 0 \quad (4.20)$$

Any motion is therefore accompanied by a monotonous decrease of V. This process continues up to the point where a stable point (a minimum of V) is reached. In the case of canonical–dissipative systems, the dynamics consists of a conservative part and a dissipative part. For two-dimensional dynamical systems ($n = 2$), this corresponds to a representation of the vector field $\mathbf{F}$ by a divergence-free and a rotation-free component. For the (modified) Rayleigh system of stable oscillations discussed above, the attractor is a limit cycle and $(H - H_0)^2$ is a Lyapunov function.

In the following part, we are concerned with the topics stability of motion and Lyapunov exponents. Since the states and the trajectories of dynamical systems are never exactly known and are subject to stochastic perturbations, the stability of motion with respect to small changes is of significant interest. The first mathematical investigation of dynamical stability was given by the Russian mathematician Lyapunov in 1892 and nearly at the same time by the French theoretician Poincaré. A remarkable contribution to the stability theory was given in the thirties of the twentieth century by the Russian school founded by *Mandelstam, Andronov, Witt*, and *Chaikin* in Moscow and Gorki and by another school founded by *Krylov, Bogolyubov*, and *Mitropolsky* in Kiev.

In order to give the main ideas of stability theory, let us consider two trajectories $\mathbf{x}(t; \mathbf{x}, t)$ and $\mathbf{x}(t; \mathbf{x} + \mathbf{q}, t)$, which at the initial time t differ by a small vector $\mathbf{q}$. In order to investigate the stability according to Lyapunov, we calculate the time development of the distance vector

$$\mathbf{q}(t) = \mathbf{x}(t; \mathbf{x} + \mathbf{q}, t) - \mathbf{x}(t; \mathbf{x}, t) \quad \mathbf{q}(t_0) = \mathbf{q}_0 \quad (4.21)$$

The motion is termed globally stable in the sense of Lyapunov if for all t and any $\varepsilon > 0$, there exists an $\eta(\varepsilon, t_0)$ such that from

$$|\mathbf{q}(t_0)| < \eta \quad (4.22)$$

it follows that

$$|\mathbf{q}(t)| < \eta \quad (4.23)$$

for any $t > t_0$. If such an η does not exist, the motion is regarded as unstable in the sense of Lyapunov. In the special case that

$$\lim_{t \to \infty} |\mathbf{q}(t)| = 0 \quad (4.24)$$

the motion is termed asymptotically stable.

From the equations of motion, we get linear approximation

$$\dot{q}_i(t) = F_i(\mathbf{x}+\mathbf{q}) - F_i(\mathbf{x}) = \sum_j J_{ij}(\mathbf{x}) q_j \tag{4.25}$$

where the elements of the Jacobi matrix are given by

$$J_{ij}(\mathbf{x}) = \frac{\partial F_i}{\partial x_j} \tag{4.26}$$

In the special case that we are interested in the stability of a singular point

$$\mathbf{F}(\mathbf{x}) = 0 \tag{4.27}$$

which may be considered as a special (constant) trajectory, the Jacobi matrix has constant matrix elements. In this case, Equation 4.25 is a system of linear homogeneous differential equations with constant coefficients. It is well known that then the stability is determined by the eigenvalues of the Jacobi matrix. With the standard ansatz

$$\mathbf{q}(t) \approx \exp(\lambda t) \tag{4.28}$$

we get the characteristic equation

$$\det\left(J_{ij} - \lambda \delta_{ij}\right) = 0 \tag{4.29}$$

The roots of this equation are in general complex. A necessary and sufficient condition for the asymptotic stability of the stationary point (singular point) is that all roots have negative real parts,

$$\text{Re}\,\lambda_i < 0 \tag{4.30}$$

for all $i = 1, \ldots, n$. Already one eigenvalue with a positive real part is sufficient to turn the stationary point unstable, since any small deviation in the corresponding direction will be amplified. If the eigenvalue has an imaginary part, the increase ($\text{Re}\,\lambda_i > 0$) or decrease ($\text{Re}\,\lambda_i < 0$) will be oscillatory. In the case of $\text{Re}\,\lambda_i = 0$, no conclusion about stability is possible in the framework of the linear theory (Butenin, Neimark and Fufayev, 1976). Since for gradient systems, the Jacobi matrix is symmetric,

$$J_{ij}(\mathbf{x}) = J_{ji}(\mathbf{x}) \tag{4.31}$$

the corresponding eigenvalues are always real. Correspondingly, the stationary points are always nodes or saddles. For (conservative) Hamiltonian systems, the trace of the Jacobian is zero:

$$\text{Tr}\{J\} = 0 \tag{4.32}$$

The eigenvalues come in pairs located on the imaginary axis. This means that Hamiltonian systems do not possess stable singular points, all unstable points are of saddle type. If one is interested in the stability of periodic orbits

$$\mathbf{x}(t+T) = \mathbf{x}(t)$$

we may start again from Equation 4.25. However, the Jacobi matrix is now a periodic function of time with the period T. In this way, the perturbations satisfy a system of linear differential equations with periodic coefficients. In that case, the stability is known to depend on the so-called Floquet coefficients (Berge et al., 1984). A stable closed orbit is termed a limit cycle.

The real parts of the eigenvalues of the Jacobi matrix and the Floquet exponents are special cases of a more general concept, the so-called Lyapunov exponents. In order to explain this, we go back to Equation 4.25, which symbolically reads

$$\frac{d\mathbf{q}}{dt} = \mathbf{J}(\mathbf{x}(t))\,\mathbf{q} \tag{4.33}$$

Here, $\mathbf{x}(t)$ is the trajectory which we like to investigate. This trajectory is the solution of our original equation

$$\frac{d\mathbf{x}}{dt} = \mathbf{F}\,(\mathbf{x}, \mathbf{u})$$

In general, the simultaneous analytical solution of both equations will be impossible, one has to refer to numerical calculations. The norm of the solution of Equation 4.33 will in general behave exponentially in time

$$|\mathbf{q}(t)| \approx \exp(\lambda t) \tag{4.34}$$

Therefore, we define the characteristic quantities

$$\lambda = \lim_{t \to \infty} \frac{1}{t} \ln \frac{|\mathbf{q}(t)|}{|\mathbf{q}(0)|} \tag{4.35}$$

as *Lyapunov exponents*. They characterize the long-term behavior of the deviations in linear approximation. In dependence on the initial conditions for $\mathbf{q}(t)$, the exponents λ may take a number of discrete values, λ_i, $i = 1, \ldots, n$, which are termed the *spectrum of the Lyapunov exponents*. The method described here is based on the original investigations of *Lyapunov* and on a statement proven in 1968 by *Oseledec* (Anishchenko et al., 2002).

The sum over all exponents is related to the divergence of the vector field averaged along the trajectories

$$\langle \nabla \mathbf{F} \rangle = \sum_i \lambda_i \tag{4.36}$$

Therefore, we have in the case of conservative systems

$$\sum_i \lambda_i = 0 \tag{4.37}$$

and for dissipative systems

$$\sum_i \lambda_i < 0 \tag{4.38}$$

The fundamental works of *Lorenz*, *Ruelle*, and *Takens* showed that for dimensions of the state space $n \geq 3$, there exist systems of which the maximum Lyapunov exponent is positive. The corresponding attractors were termed strange attractors (Eckmann and Ruelle, 1985; Anishchenko, 1995; Ott, 1993; Anishchenko et al., 2002).

In the following text, we assume that the Lyapunov exponents are ordered with respect to their index 1, ..., n in the form of a nondecreasing series. In symbolic notation, the following possibilities exist for the signs of the Lyapunov exponents:

1) Stable singular points: $(-, -, -, \ldots, -)$.
2) Stable limit cycles: $(0, -, -, \ldots, -)$.
3) Stable m-torus: $(0, 0, \ldots, 0, -, - \ldots, -)$, (with m vanishing exponents, i.e., m − zeros)
4) Chaotic attractor: $(+, 0, -, -, \ldots, -)$
5) Chaos of higher order: $(+, \ldots, +, 0, -, \ldots, -)$

If at least one (the largest) Lyapunov exponent is positive, we will say that the motion is chaotic. The *Hausdorff dimension* D_H of chaotic attractors is in general a noninteger number (Schuster, 1995; Ott, 1993). If at least one exponent is positive we know that $D_H > 1$ holds for the dimension. If j exponents are positive, that is, the ordered exponents satisfy the inequality

$$\lambda_1 \geq \cdots \geq \lambda_j > 0 > \lambda_{j+1} \geq \cdots \geq \lambda_n \tag{4.39}$$

then the dimension of the attractor will in general lie between j and $j + 1$ that is, the inequality $j < D_H < j+1$ holds. This is due to the following rule: If the sum of the j largest Lyapunov exponents is positive, then a small volume of the dimension j enclosing a phase point of the (chaotic) trajectory is expanding. The quantity

$$D_L = j + \frac{1}{|\lambda_{j+1}|} \sum_{i=1}^{j} \lambda_i$$

is termed the Lyapunov dimension of the attractor (Eckmann and Ruelle, 1985), where j is defined by the inequality (4.39). For stable singular points, we define $D_L = 0$; for stable limit cycles, one gets $D_L = 1$; and for stable m-tori, we find $D_L = m$ (Anishchenko et al., 2002).

For chaotic attractors of systems defined by n differential equations, the dimension D is a noninteger number with $2 < D < n$. The sum of the positive Lyapunov exponents

$$H_P = \sum_{i=1}^{j} \lambda_i \tag{4.40}$$

we term the *Pesin entropy* (Steuer et al., 2001a, 2001b). In most cases, the Pesin entropy is equal to the *Kolmogorov entropy*. The Kolmogorov entropy was originally defined in terms of information theoretical methods (see Chapter 8) and is closely related to the problem of the predictability of motions (Schuster, 1995; Kantz and Schreiber, 1997).

4.3
Stochastic Models for Continuous Variables and Predictability

The description of processes on the deterministic level is in some sense incomplete since stochastic influences are always present under real conditions (Stratonovich, 1961, 1963, 1967; van Kampen, 1992; Ebeling and Feistel, 1982). On the other hand, all real processes of self-organization and evolution belong to the class of stochastic processes since they are subject to random influences. Therefore, the study of stochastic models is mandatory for our book.

We introduce here only principal aspects of stochastic descriptions. Details and examples of stochastic descriptions are given later. Besides the general arguments given above, there are several special reasons for the importance of stochastic models:

Meso- or macroscopic variables represent the net effect of a large number of microscopic degrees of freedom that are subject to thermal fluctuations. All variables depending on the number of particles are necessarily discrete. This gives rise to a special kind of random effects, the so-called shot noise. The intrinsic quantum character of the microscopic dynamics leads also to specific stochastic effects. A mesoscopic or macroscopic system is, as a rule, embedded in a very complex surrounding leading to stochastic external forces.

Any part of our universe is filled with thermal photons forming the sea of background radiation (about 500 photons with a temperature of 2.7 K in 1 cm^3). These photons interact stochastically with any system under investigation and there is principally no way to get rid of such influences. Measurements provide results only within finite uncertainties. Usually the initial conditions of a real dynamical system are not exactly known. Positive Lyapunov exponents may subsequently amplify unknown initial deviations.

Due to these stochastic influences, the future state of a dynamical system is in general not uniquely defined in terms of its past. In other words, the dynamic map defined by Equation 4.1 is ambiguous. A given starting point $\mathbf{x}(0)$ may be the origin of several different trajectories. The choice between the different possible trajectories is a random event that is unknown until it has actually happened. In this way, the term trajectory loses its precise, unambiguous meaning and should be supplemented in terms of probability theory. What can we say about the predictability of stochastic processes? Unique predictions of stochastic trajectories are in principle impossible. All possible predictions are about probabilities and mean values. As an example of a stochastic trajectory of a scalar variable, let us look at the temperature evolution in the United States in time interval including the last century (see Figure 4.3).

Here we will develop first some general tools for the description of stochastic processes and will then come back to the problem of predictions. Several concrete stochastic models are described in detail in Chapters 5–10. Here we introduce only the basic ideas. In the framework of stochastic models, we describe the state of the system at a time t by the probability density $P(\mathbf{x}, t; \mathbf{u})$. By definition, $P(\mathbf{x}, t; \mathbf{u}) \, d\mathbf{x}$ is the probability of finding the trajectory at the time t in the interval $(\mathbf{x}, \mathbf{x} + d\mathbf{x})$. Instead of the deterministic equation for the state evolution, we now get a differential equation for the probability density $P(\mathbf{x}, t; \mathbf{u})$. In order to find this equation, we

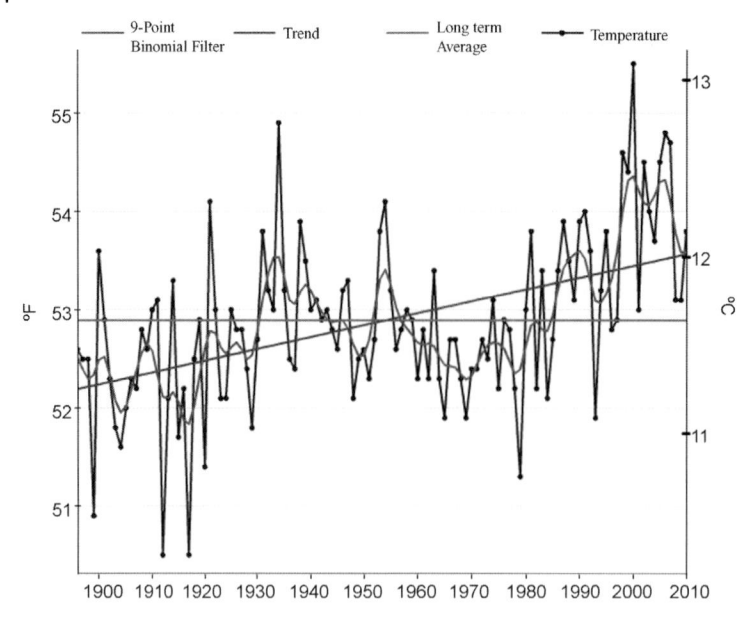

Figure 4.3 Stochastic evolution of the mean yearly air temperatures in the United States calculated from measurements of a network of about 1200 institutions (from NOAA, 2010). Beside the actual trajectory we also show a short-time average that shows some tendency of increase in the last 20 years and a long-time average that is more or less constant. We see how difficult and even problematic the prediction of a stochastic process based on a finite time window of observations might be.

introduce the idea of stochastic forces due to *Paul Langevin* developed in 1907. Instead of Equation 4.5, we now write

$$\dot{x}_i(t) = F_i(\mathbf{x}, \mathbf{u}) + \xi_i(t) \tag{4.41}$$

where $\xi_i(t)$ is the ith component of a stochastic force with zero mean value

$$\langle \xi_i(t) \rangle = 0 \tag{4.42}$$

The latter condition guarantees that the averaged trajectories satisfy the equation

$$\langle \dot{x}_i(t) \rangle = \langle F_i(\mathbf{x}, \mathbf{u}) \rangle \approx F_i(\langle \mathbf{x} \rangle, \mathbf{u}) \tag{4.43}$$

In order to derive the desired equation for the probabilities, we consider an ensemble of N representative points in the state space, each corresponding to one particular system. At $t=0$, the points may be distributed corresponding to some initial probability $P(\mathbf{x}, t = 0; \mathbf{u})$. The fraction of points in a given volume V at the time t is

$$N_V(t) = N \int_V d\mathbf{x}\, P(\mathbf{x}, t; \mathbf{u}) \tag{4.44}$$

and the time derivative is given by the surface integral

$$\frac{d}{dt} N_V(t) = -N \oint_S d\mathbf{O}\, \mathbf{G}(\mathbf{x}, t; \mathbf{u}) \tag{4.45}$$

where $\mathbf{G}$ is the probability flow vector. Making use of the Gauss theorem, we infer the local probability balance equation

$$\frac{\partial}{\partial t} P(\mathbf{x}, t; \mathbf{u}) = -\nabla \cdot \mathbf{G}(\mathbf{x}, t; \mathbf{u}) \tag{4.46}$$

In the special case that there are no stochastic forces, the flow is proportional to the deterministic field, that is,

$$G_i(\mathbf{x}, t; \mathbf{u}) = F_i(\mathbf{x}, t; \mathbf{u})\, P(\mathbf{x}, t; \mathbf{u}) \tag{4.47}$$

and Equation 4.46 takes the form of a hydrodynamic continuity equation in the state space for the representative points with the density P and the velocity $\mathbf{F}$. Including now the influence of the stochastic forces, we assume here *ad hoc* additional diffusive contribution to the probability flow that is directed downward the gradient of the probability density

$$G_i(\mathbf{x}, t; \mathbf{u}) = F_i(\mathbf{x}, t; \mathbf{u})\, P(\mathbf{x}, t; \mathbf{u}) - D \frac{\partial}{\partial x_i} P(\mathbf{x}, t; \mathbf{u}) \tag{4.48}$$

This is the simplest ansatz consistent with Equation 4.43 for the mean values. The relation of the "diffusion coefficient" D to the properties of the stochastic force is discussed later.

By inserting Equation 4.48 into Equation 4.46 we get a partial differential equation that is known as *Smoluchowski–Fokker–Planck equation*, or just *Fokker–Planck equation*. This way we have found a closed equation for the probabilities. The Smoluchowski–Fokker–Planck equation is consistent with the deterministic equation and can, at least approximately, take into account stochastic influences.

More rigorous statistical–mechanical derivations of diffusion-type equations based on the Liouville equation and the Zwanzig projection technique were first given in (Ebeling, 1965; Falkenhagen and Ebeling, 1971). On the basis of a given probability distribution $P(\mathbf{x}, t; \mathbf{u})$, we may define the mean value of any function $g(\mathbf{x})$ by the integral

$$\langle g(\mathbf{x}) \rangle = \int d\mathbf{x}\, g(\mathbf{x}) P(\mathbf{x}, t; \mathbf{u}) \tag{4.49}$$

Further, we may define standard statistical expressions such as, for example, the dispersion, and in particular the mean uncertainty (entropy) in the form

$$H = -\langle \log P(\mathbf{x}, t; \mathbf{u}) \rangle \tag{4.50}$$

The approach based on the Fokker–Planck equation is the simplest but not the only one to include probabilities in dynamical models. There exists a different approach that originally goes back to Markov who studied first discrete-time stochastic processes, nowadays termed Markov chains. Markov's model was generalized to continuous time processes by Chapman and Kolmogorov. The basic idea of Markov, Chapman, and Kolmogorov is the introduction of transition probabilities. In order to explain the idea, we consider the simplest case of a random variable having the values $\mathbf{x}_1, \mathbf{x}_2, \ldots, \mathbf{x}_n$ at the subsequent times $t_1 < t_2 < \cdots < t_n$. From the knowledge of the joint probability-distribution density (van Kampen, 1981)

$$P_n(\mathbf{x}_1, t_1; \mathbf{x}_2, t_2; \ldots, \mathbf{x}_n, t_n) = \langle \delta(\mathbf{x}_1 - \mathbf{x}_1(t_1)) \cdots \delta(\mathbf{x}_n - \mathbf{x}_n(t_n)) \rangle \quad (4.51)$$

we would obtain the full information on the dynamics of the system, where $\mathbf{x}_1, \mathbf{x}_2, \ldots, \mathbf{x}_n$ are n coordinates describing the system at the subsequent times $t_1 < t_2 < \cdots < t_n$.

The probability to find the system at the time t_1 within an infinitesimal vicinity of $\mathbf{x}_1$, and correspondingly at the states $\mathbf{x}_2, \ldots, \mathbf{x}_n$ at the later times t_i, is given by

$$P_n(\mathbf{x}_1, t_1; \mathbf{x}_2, t_2; \ldots, \mathbf{x}_n, t_n) \, d\mathbf{x}_1 d\mathbf{x}_2 \cdots d\mathbf{x}_n$$

In Equation 4.51, the brackets on the right-hand side indicate averaging over an ensemble. There, $\mathbf{x}_1(t_1), \mathbf{x}_2(t_2)\ldots$ stand for the stochastic realizations of the number belonging to the ensemble at different times. The delta-functions map that part of the realizations onto the values $\mathbf{x}_i$ to the coordinate $\mathbf{x}_1(t_1)$. The joint probability (4.51) defines an infinite hierarchy of probability densities (van Kampen, 1981). Full knowledge means that, for example, arbitrary time-dependent moments of the stochastic process may be calculated in the form

$$\langle A \rangle(t_1, t_2, t_k) = \int A \, P_k(\mathbf{x}_1, t_1; \mathbf{x}_2, t_2; \ldots, \mathbf{x}_k, t_k) \, d\mathbf{x}_1 d\mathbf{x}_2 \cdots d\mathbf{x}_k \quad (4.52)$$

Thereby, the value of k describes how many times are necessary to be included, depending on the number variables that occur as arguments inside the brackets. In most physical applications, values $k \leq 2$ occur. To give a rigorous definition of a Markov process, the conditional probability density P is introduced (Risken, 1984)

$$P(\mathbf{x}_n, t_n | \mathbf{x}_{n-1}, t_{n-1} \ldots \mathbf{x}_1, t_1) = \left\langle \delta(\mathbf{x}_n - \mathbf{x}_n(t)) |_{\mathbf{x}_{n-1} = \mathbf{x}_{n-1}(t), \ldots, \mathbf{x}_1 = \mathbf{x}_1(t)} \right\rangle \quad (4.53)$$

In contrast to (4.51), here P provides the probability to find the system at the time t_n in the neighborhood of $\mathbf{x}_n$ under the condition that at the earlier times $t_1 < t_2 < \cdots t_{n-1}$, it took exactly the sequence of states $\mathbf{x}_1, \mathbf{x}_2, \ldots, \mathbf{x}_{n-1}$. Then we find the following identity

$$P_n(\mathbf{x}_1, t_1; \mathbf{x}_2, t_2; \ldots; \mathbf{x}_n, t_n) = P(\mathbf{x}_n, t_n | \mathbf{x}_{n-1}, t_{n-1} | \cdots | \mathbf{x}_1, t_1) \times$$
$$P_{n-1}(\mathbf{x}_1, t_1; \mathbf{x}_2, t_2; \ldots; \mathbf{x}_{n-1}, t_{n-1}) \quad (4.54)$$

On the basis of the relation (4.54), a Markovian process can be defined. If the conditional probability depends only on the value $\mathbf{x}_{n-1}$ of the latest past

$$P(\mathbf{x}_n, t_n | \mathbf{x}_{n-1}, t_{n-1} \ldots \mathbf{x}_1, t_1) = P(\mathbf{x}_n, t_n | \mathbf{x}_{n-1}, t_{n-1}) \quad (4.55)$$

rather than on any earlier states, we refer to the process as Markovian (of first order).

In that case, the joint probability density can be reduced to a simple product of pairwise conditional probabilities at subsequent times and to an initial probability density at the time t_1

$$P_n(\mathbf{x}_1,t_1;\mathbf{x}_2,t_2;\ldots;\mathbf{x}_n,t_n) = P(\mathbf{x}_n,t_n|\mathbf{x}_{n-1},t_{n-1}) \cdots P(\mathbf{x}_2,t_2|\mathbf{x}_1,t_1)P_1(\mathbf{x}_1,t_1) \tag{4.56}$$

Markovian (first order), therefore, means that the described process possesses the memory of one single transition. The step from $\mathbf{x}_{n-1}$ at the time t_{n-1} to $\mathbf{x}_n$ at t_n is independent of the path by which the state $\mathbf{x}_{n-1}$ was approached. In other words, the transition probability from the present state to a future state is not influenced by any knowledge about the states in the past. $P(\mathbf{x}_2,t_2|\mathbf{x}_1,t_1)$ is often regarded as the transition probability; here we also will make use of this terminology.

Let us emphasize that the concept of Markov processes refers to a property of the model we apply for the description of reality rather than to a property of the physical system under consideration (van Kampen, 1981). If some given physical process cannot be described by a Markov relation in an appropriate state space, it may often be possible to embed the process in an extended Markovian model by introducing additional degrees of freedom. This way a non-Markovian model can be converted, by enlargement of the number of variables and expansion of the phase space, into a more general model with Markovian properties. Vice versa, if a Markovian model is given, the dynamic elimination of fast variables ("slaved modes") usually results in a reduced but "non-Markovian" system with memory. We already note here that the basic equation of statistical physics, the Liouville equation which is introduced in the next chapter, is of Markovian character.

In order to derive a kinetic equation for the time evolution of the probability density, we consider the following identity

$$P_3(\mathbf{x}_3,t_3;\mathbf{x}_1,t_1) = \int P_3(\mathbf{x}_3,t_3;\mathbf{x}_2,t_2;\mathbf{x}_1,t_1)\,d\mathbf{x}_2 \tag{4.57}$$

Assuming Markovian character of the considered process, we obtain

$$P(\mathbf{x}_3,t_3|\mathbf{x}_1,t_1)P_1(\mathbf{x}_1,t_1) = \int P(\mathbf{x}_3,t_3|\mathbf{x}_2,t_2)P(\mathbf{x}_2,t_2|\mathbf{x}_1,t_1)P_1(\mathbf{x}_1,t_1)d\mathbf{x}_2$$

For arbitrary $P_1(\mathbf{x}_1,t_1)$, therefore, the Chapman–Kolmogorov equation for the transition probabilities follows to be

$$P(\mathbf{x}_3,t_3|\mathbf{x}_1,t_1) = \int P(\mathbf{x}_3,t_3|\mathbf{x}_2,t_2)P(\mathbf{x}_2,t_2|\mathbf{x}_1,t_1)d\mathbf{x}_2 \tag{4.58}$$

This is a fundamental, closed nonlinear integral equation for the transition probability P. This equation is universally valid for any first-order Markov process. Similar equations hold true for the functions P_1 and P_2, in the form

$$P_2(\mathbf{x}_3,t_3;\mathbf{x}_1,t_1) = \int P(\mathbf{x}_3,t_3|\mathbf{x}_2,t_2)P_2(\mathbf{x}_2,t_2;\mathbf{x}_1,t_1)d\mathbf{x}_2 \tag{4.59}$$

An additional integration over $\mathbf{x}_1$ leads us to back to

$$P_1(\mathbf{x}_3, t_3) = \int P(\mathbf{x}_3, t_3 | \mathbf{x}_2, t_2) P_1(\mathbf{x}_2, t_2) d\mathbf{x}_2 \tag{4.60}$$

Note that Equation 4.60 is a Fredholm integral equation of the second kind for P_1, that P is a nonnegative integral kernel of Equation 4.60, and that P_1 is a nonnegative eigenfunction of the related integral operator, which evidently possesses a unity eigenvalue. For Markovian processes, Equations 4.58–4.60 are the starting points to find dynamical laws for the time evolution of the particular probability densities.

Now a second simplification will be made. Most stochastic problems are uniform in time, that is, they are invariant under a time shift $t \to t + \delta t$. Successive events in such "autonomous" systems depend only on time differences rather than on the absolute time. Those stochastic processes are termed stationary processes. As a consequence, the two-time density as well as the transition probability will be functions of the time difference only

$$P_2(\mathbf{x}_2, t_2; \mathbf{x}_1, t_1) = P_2(\mathbf{x}_2, t_2 - t_1; \mathbf{x}_1)$$

and

$$P(\mathbf{x}_2, t_2 | \mathbf{x}_1, t_1) = P(\mathbf{x}_2, t_2 - t_1 | \mathbf{x}_1) \tag{4.61}$$

Since both functions are equivalent in their dynamical behavior, for convenience, we further on omit the subscript and the initial condition whenever possible to avoid too many functional arguments.

In the limit that the time difference between the two subsequent events is small, $\delta t = t_2 - t_1 \to 0$, the transition probability in the kernel of the integrals (4.58) and (4.59) can be expanded into a series of small δt,

$$P(\mathbf{x}_3, \delta t | \mathbf{x}_2) = 1 - \delta t \int d\mathbf{x}\, W(\mathbf{x} | \mathbf{x}_2) \delta(\mathbf{x}_3 - \mathbf{x}_2) + W(\mathbf{x}_3 | \mathbf{x}_2)\, \delta t + O(\delta t^2) \tag{4.62}$$

Here, the existence of a function $W(\mathbf{x}'|\mathbf{x})$ is assumed that occurs as the probability for a transition $\mathbf{x} \to \mathbf{x}'$ per unit time. The first item in front of the Dirac delta function is the probability for remaining in the state $\mathbf{x}_2$ during δt. The second item stands for the transition rate from $\mathbf{x}_2$ to a different $\mathbf{x}_3$.

Inserting the expansion (4.62) into the Chapman–Kolmogorov equation (4.58) provides the dynamical equation for the probability density

$$\frac{\partial}{\partial t} P(\mathbf{x}, t) = \int d\mathbf{x}' \{ W(\mathbf{x}|\mathbf{x}') P(\mathbf{x}', t) - W(\mathbf{x}'|\mathbf{x}) P(\mathbf{x}, t) \} \tag{4.63}$$

This equation is commonly known as the *Pauli equation* or *master equation* since it plays a fundamental role in the theory of stochastic processes. The integration is performed over all possible states $\mathbf{x}'$ that are attainable from the state $\mathbf{x}$ by a single jump. It is a linear equation with respect to P and determines uniquely the evolution of the probability density. The right-hand side consists of two parts, the first stands for the gain of probability due to transitions $\mathbf{x}' \to \mathbf{x}$, whereas the second describes the loss due to reversed events.

Equation 4.63 needs still further explanation by the determination of the transition probabilities per unit time corresponding to the special physical situation. This problem is the subject of Chapters 5 and 6. In many cases, the transition probability is a quickly decreasing function of the jump distance $\Delta \mathbf{x} = \mathbf{x} - \mathbf{x}'$. By using a series expansion with respect to $\Delta \mathbf{x}$ and to the moments of the transition probability, one can transform Equation 4.63 to an infinite Taylor series. This is the so-called *Kramers–Moyal expansion*,

$$\frac{\partial}{\partial t} P(\mathbf{x}, t) = \sum_{m=1}^{\infty} \frac{(-1)^m}{m!} \sum \frac{\partial^m}{\partial x_{i_1} \cdots \partial x_{i_m}} M_{i_1 \ldots i_m} P(\mathbf{x}, t) \tag{4.64}$$

where

$$M_{i_1 \ldots i_m}(\mathbf{x}) = \int d\Delta \mathbf{x}\, d\Delta x_{i_1} \cdots d\Delta x_{i_m}\, W(\mathbf{x} + \Delta \mathbf{x} | \mathbf{x}) \tag{4.65}$$

are the moments of the transition probabilities per unit time.

According to a theorem by *Pawula*, there are two possibilities when homogeneous Markov processes are considered (Risken, 1984):

1) All coefficients of the Kramers–Moyal expansion are different from zero.
2) Only two coefficients in the expansion are different from zero.

In the first case, we have to deal with the full master equation. In the latter one, the Markovian process is termed diffusive, which is of special interest to us, and leads to the following second-order partial differential equation:

$$\frac{\partial}{\partial t} P(\mathbf{x}, t) = \sum_i \frac{\partial}{\partial x_i} [M_i(\mathbf{x}) P] + \sum_{i,j} \frac{\partial^2}{\partial x_i \partial x_j} [M_{ij}(\mathbf{x}) P] \tag{4.66}$$

This is a generalization of the Smoluchowski–Fokker–Planck equation given above. For its solution, we need of course initial conditions $P(\mathbf{x}, t = 0)$ and boundary conditions that take into account the underlying physics. Writing Equation 4.66 again in the form of a continuity equation (4.46), we find the components for the vector of the probability flow,

$$G_i(\mathbf{x}, t) = M_i(\mathbf{x}) P(\mathbf{x}, t) + \sum_j \frac{\partial}{\partial x_j} [M_{ij}(\mathbf{x}) P(\mathbf{x}, t)] \tag{4.67}$$

The rigorous mathematical theory of Equation 4.63 was developed by Kolmogorov and Feller; therefore, one often speaks of the *Kolmogorov–Feller equation*. In physics, however, this equation was used much earlier by *Einstein, Smoluchowski, Fokker,* and *Planck* for the description of diffusion processes and Brownian motion (Chandrasekhar, 1943). Because of this original physical relation, the coefficients $M_i(\mathbf{x})$ and $M_{ij}(\mathbf{x})$ are often termed drift coefficients and diffusion coefficients, respectively. The solutions of partial differential equations of the Smoluchowski–Fokker–Planck type are subject to several theorems (van Kampen, 1981). Under quite general conditions, the solution is a unique function, that is, $P(\mathbf{x}, t)$ may be uniquely be predicted from $P(\mathbf{x}, 0)$. In difference to the deterministic picture, we can predict now a probability rather than the future state itself.

Let us come now to the crucial question about the predictability of stochastic processes. In difference to deterministic dynamic processes, processes that are subject to stochastic influences cannot be predicted exactly (Kantz and Schreiber, 1997). The future of stochastic processes is uncertain to some extent and this uncertainty can only be reduced by some additional information but we cannot make predictions that are 100% certain. All evolutionary processes belong to the class of stochastic processes. Due to the relevance of predictability for all ranges of nature and society, there exist many special investigations (Ebeling, Freund, and Schweitzer, 1998).

Let us summarize our findings from the studies given above: In a stochastic description, instead of a state vector $\mathbf{x}(t)$ which is determined by ordinary differential equations, we have now a probability distribution $P(\mathbf{x}, t)$ that is determined by partial differential equations or by integral equations. Therefore, we can calculate only a probability distribution for the states in the future and not the states themselves.

In later chapters, we give many examples, so let us discuss here only one which is connected with Figure 4.3 that shows the stochastic temperature evolution for about 100 years. The evolution of the temperature is part of a problem of special public interest, namely, the evolution of our climate. We touched this problem already in Chapter 3. Essentially beginning with the IPCC Report 2007 (Report of the Intergovernmental Panel on Climate Change) and the international conferences in Stockholm 2009 and so on, the question of the future evolution of the temperature on earth is particularly a hot topic. Let us take this as an example in order to illustrate some problems connected with the prediction of evolutionary processes. This consideration is based on the facts of the climatic evolution explained in Chapter 3. As we have shown there, the evolution of our climate is an extremely slow process, if measured in units of human life times, and is further a stochastic process subject to strong fluctuations (Ebeling and Feistel, 1994, 2008). One cannot speak about a stationary state. However, looking at the last 100–10 000 years, we see some stationarity of mean values that show slow trends. As a matter of fact we observe a slow increase of the temperature by about 2 degrees since 1850. Following the predictions of experts (IPCC, 2007), this increase will continue at least another 100 years. With some probability this increase is even amplified, so that with some probability up to 2100, an increase by about 3 degrees may be predicted (Lanius, 1994a, 1994b; Ebeling and Feistel, 2008).

Let us look at the question of the temperature in this century evolution in more detail.

As an illustration for the evolution during a century, Figure 4.3 shows the mean yearly temperatures in the last century in the United State (NOAA, 2010). Evidently we have to do with a stochastic process with strong fluctuations and a small but significant trend of the mean by about plus 2 degrees during a century. A similar but much smaller increase of the mean we observe in data from the observatory in Warnemünde (Tiesel, 2008); maritime influences have a damping influence here on the general temperature increase (Hagen and Feistel, 2008; Feistel, Nausch, and Wasmund, 2008a).

Summarizing the data situation we see a significant increase since the 1980s of the last century; however, a similar increase was observed also in the years 1920–1940. This way we see that predictions of the temperature evolution based on data of just one century are extremely difficult (Werner et al., 1999). As a matter of fact, the increase over the last and for this century seems to be certain. A possible explanation of this so-called global warming may be based on the greenhouse effect (Lanius, 1994a, 1994b, 2009; Behringer, 2007).

In order to illustrate the difficulties of a prediction of temperature evolution in the present century, we estimated the probability of a definite increase of the global temperature of the earth up to the year 2100. We estimated this probability distribution for a prediction of the temperature on the basis of the data given in the IPCC Report (Ebeling and Feistel, 2008). In agreement with the IPCC Report, we assume that the maximum of the probability is about an increase by 3 K. However, as we see from the distribution, the interval between 1 and 6 K is still in the range of probable futures (Figure 4.4). This underlines how uncertain predictions are and how dangerous political statements might be that the increase can be kept below 3 K. The curve demonstrates the stochastic character of the temperature trajectory and the uncertainty of any prediction.

It is important to understand that for a stochastic process nothing can be totally excluded. In general, there exists always some probability that unexpected (and sometimes unwanted) future evolutionary pathways are realized. Let us look again at our example, the evolution of the temperature in this century. In order to illustrate that we cannot exclude even a global cooling for the long term, we investigate the

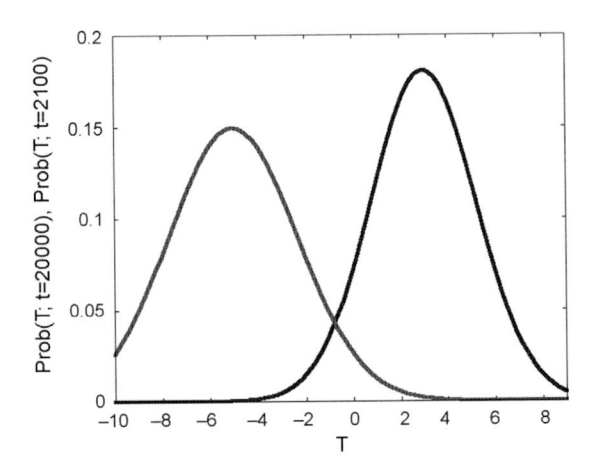

Figure 4.4 Estimate of the probability for the change of the global temperature of the earth up to the year 2100 (right curve from Ebeling and Feistel, 2008). The maximum of the probability is about 3 K according to the IPCC Report, the interval between 1 and 6 K is still in the range of probable futures and beyond that the probabilities are small. For illustration, we add an estimate for the temperatures in the year 20 000, corresponding probably to another glacial period (estimated from the temperature evolution of Antarctica, Figure 4.5).

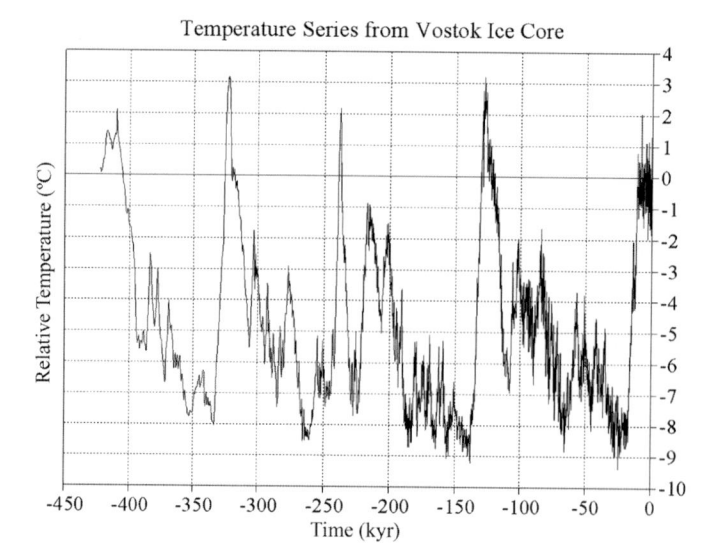

Figure 4.5 Temperature in Antarctica during the past half a million of years from ice core drilling measurements at the Russian–French station "Vostok" (Jouzel *et al.*, 1987, 1993, 1996; Barnola *et al.*, 2003; Petit *et al.*, 1999, 2000; Behringer, 2007).

temperature history on the long scale of several hundred thousands of years. Looking at Figure 4.5, we see quasiperiodic temperature cycles repeating in periods of about 100 000 years.

The data demonstrated in Figure 4.5 show the results from ice core drilling at the Russian–French research station "Vostok" at Antarctica over the last 420 000 years (Barnola *et al.*, 2003; Behringer, 2007). We observe very large fluctuations of the temperature in Antarctica on the scale of hundred thousands of years and may assume that the scale of fluctuations in our geographic regions is at least similar. This makes in fact any long-term prognosis based on only data from the last hundred years – as we tried above – highly questionable. Any prognosis based on the data from long-time observations suggests that we possibly have to expect for the next 20 000 years another cold period with mean temperatures going down by around 5 degrees and dramatic consequences for our geographic region. An estimation of the expected temperature distribution in about 20 000 years is shown in Figure 4.4.

Let us underline that the consideration given above was not based on any study of the possible causal influences on the temperature trajectory, as the greenhouse effect and the influence of modern civilization at the climate as studied, for example, in Chapter 3 as well as in several books (Lanius, 1994a, 1994b; Ebeling and Feistel, 1994; Behringer, 2007). The present consideration was entirely based on an investigation of stochastic temperature curves in the past.

We studied in this section tools for the treatment of stochastic processes and considered the evolution of the mean temperature on our Earth as an example of a stochastic process. Let us compare the new tools with the deterministic models. Obviously, the stochastic approach contains more information about the considered systems due to the inclusion of fluctuations into the description. Besides moments of the macroscopic variables, it enables us to determine correlation functions, spectra that will provide knowledge on the functional dependence of the fluctuation behavior at different times. On the other hand, the stochastic character limits our possibilities for making predictions on the future.

Summarizing we may state: Some physical phenomena can be explained only by taking into account fluctuations. The stochastic approach on a mesoscopic or macroscopic level often delivers more elegant solutions than the microscopic statistical approach. Inclusion of fluctuations of the macroscopic variables does not necessarily enlarge the number of relevant variables but changes only their character by transforming them into stochastic variables. In comparison to the deterministic models, the fundamental difference is the permeability of separatrices. This statement refers especially to nonchaotic dynamics, for instance, if dealing with one or two order parameters. With a certain probability, stochastic realizations reach (or cross) unstable points, saddle points, and separatrices which is impossible in the deterministic description. Stochastic effects make it possible to escape from the regions of attraction around stable manifolds. Physical situations in which that circumstance is relevant are, for example, nucleation processes or chemical reactions where energetically unfavorable states have to be overcome. In Chapters 5–9, we present a more detailed analysis of stochastic phenomena connected with the evolution and many examples of processes induced by noise.

4.4
Graphs – Mathematical Models of Structures and Networks

In Section 4.3, we considered only continuous variables and developed the tools for the treatment of such processes. Here we study variables that may assume only discrete values, for example, integer values, where the stochastic description is even an indispensable tool. Further, we pay more attention to the methods of graph theory. Some elements of the theory of sets and graphs were introduced already in Section 4.1. Here we are in particular interested in structures with random relations and several applications (Feistel and Sändig, 1977; Feistel, 1979; Sonntag, Feistel, and Ebeling, 1981; Ebeling and Feistel, 1982; Sonntag, 1984; Hartmann-Sonntag, Scharnhorst, and Ebeling, 2009). As an example, we may take again the system of relaxation equations described by Equation 4.2. As proposed in Section 4.1, we represent the elements of our system as vertices (nodes) of a graph. The relation between the elements i and j is shown by an edge from i to j (Figure 4.6).

The element i has a connection to the element j. The connections between all elements i and to j are described by a matrix, the adjacency matrix $\mathbf{A} = \{a_{ij}\}$, which has an element 1 if there is a connection (an edge) and an element 0 if there is no connection (no edge). A node is regarded as a source if there are only outgoing edges (see Figure 4.7).

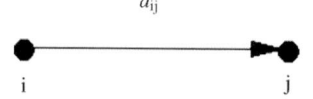

Figure 4.6 An edge that goes from node *i* to node *j* represents a binary relation between the elements *i* and *j*.

In Figure 4.7, the node *i* is a source. The sum of connections corresponds to the sum of arrows that leave the vertex *i*, this in turn corresponds to the number of nonvanishing elements of row *i* of the adjacency matrix **A**. A node is regarded as a sink if there are only incoming edges (see Figure 4.8).

The sum of these connections corresponds to the sum of the elements of the column *i* of the adjacency matrix.

Connections corresponding to mutual relations of two elements can be described by a graph picture as demonstrated in Figure 4.9.

We can calculate the number of cycles of the length *l* from the *l*th power of the adjacency matrix A^l.

Graphs can be classified qualitatively by their structure (Harary, Norman, and Cartwright, 1965; Saaty, 1966; Laue, 1970; Gantmacher, 1972; Feistel and Sändig, 1977; Feistel, 1979; Ebeling and Feistel, 1982; Jain and Krishna, 2003).

A directed graph is a *tree* if each vertex has at most one incoming edge (*divergent tree*) or at most one outgoing edge (*convergent tree*). For example, the descendants of a bacterium can be represented as a tree because each of them has exactly one parent cell.

A *path P* is a sequence of consecutive edges between two vertices. The number of edges of a path is termed its *length L(P)*. A *cycle* is a path that leads from a given vertex back to it. A cycle of the length $L = 1$ is termed a *loop*. Trees are cycle-free graphs.

A directed graph is *connected* if for any two of its vertices, i, j, there is a path from *i* to *j* or a path from *j* to *i*. A *component* of a graph is a maximum connected subgraph of a

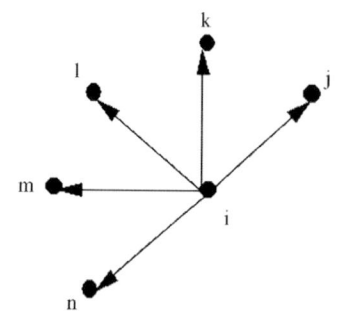

Figure 4.7 Graph of a source, that is, a node with outgoing edges only.

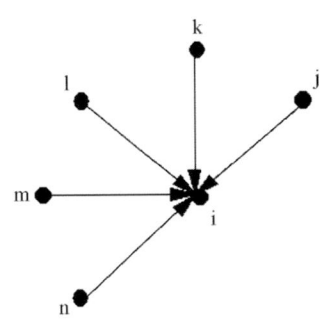

Figure 4.8 Graph of a sink, that is, a node with incoming edges only.

disconnected graph. There exist no paths between vertices belonging to different components. A component corresponds to an equivalence relation between its elements.

A directed graph is *irreducible* if for any two of its vertices, i, j, there is a path from i to j and a path from j to i. Irreducible graphs are connected. A *cluster* (or *bond*) of a graph is a maximum *irreducible* subgraph of a reducible graph. A cluster corresponds to an equivalence relation between its elements.

A directed graph is *completely reducible* if each of its components is a cluster.

A directed graph is *elementarily reducible* if each of its clusters contains exactly one vertex. An elementarily reducible graph is cycle-free. Trees are elementarily reducible graphs. The edges of a connected, elementarily reducible graph correspond to a semiorder between its vertices.

A *reduction graph* R can be associated to a given directed graph G. The vertices of R are defined as the clusters of G, and the edges of R represent existing paths between vertices of different clusters of G. A reduction graph is elementarily reducible and corresponds to a semiorder between the clusters of G.

The uniqueness of solutions of the master equation (4.63) depends essentially on the reducibility of the graph related to the transition matrix W. Only for irreducible graphs, the solution is unique and the resulting probability distribution is positive.

We introduce now several important terms characterizing a network: If K is the number of edges in the graph and S is the number of vertices, we describe

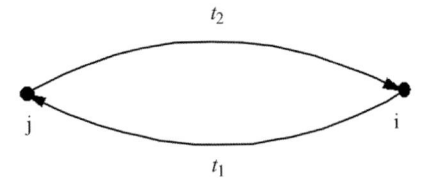

Figure 4.9 Graph of two objects with mutual connections.

the considered graph by $D(S, K)$. As a simple and in the same way important measure,

$$C = \frac{K}{S} \tag{4.68}$$

the connectivity has rendered. The maximum number N of nonparallel edges that a graph can possess, $0 \leq K \leq N$, is

1) $N = S^2$ if the graph is directed and may have loops,
2) $N = S(S-1)$ if the graph is directed without loops, and
3) $N = S(S-1)/2$ if the graph is undirected and has no loops.

The connectivity, or mean coordination number, can reasonably be classified as (Feistel, 1979):

1) $C \ll 1$, very weakly coordinated,
2) $C < 1$, weakly coordinated,
3) $C = 1$, coordinated,
4) $1 < C \ll S/2$, strongly coordinated,
5) $1 \ll C < S/2$, very strongly coordinated,
6) $C = S/2$, completely coordinated,
7) $S/2 < C < N/S$, almost compact, and
8) $C = N/S$, compact.

Most networks we are interested in here are weakly or very weakly coordinated ones. If the number of vertices is large, a "thermodynamic limit" with $S \to \infty, K \to \infty$ can be considered at constant connectivity, C. Evidently, this limit can be carried out only for graphs that are strongly coordinated, $C \ll S/2$, at most.

For some applications, we introduce several simplifications. First we exclude the relation of a node to itself (loop, reflexive relation) from our consideration, this leads to a special type of graphs. Further we consider only graphs with distinguishable vertices and indistinguishable directed edges. We exclude loops and parallel edges. We concentrate in the following on random graphs. The capacity of a network (graph) - the number of its elements- is one of its important functions. Of relevance for the structure is also the connectivity of a network. Already in 1970 Gardner and Ashby investigated the probability of the stability of large networks, in dependence on capacity and connectivity. Computer simulations and some probability statements about the structural behavior of the networks may be found in some earlier work (Sonntag, Feistel, and Ebeling, 1981; Sonntag, 1984; Hartmann-Sonntag, Scharnhorst, and Ebeling, 2009).

Isolated, connected parts of the digraph $D(S, K)$ consisting of s vertices and k arcs are called components $d(s, k)$. Let us note here that the basic problem in the structural description of a network is to identify the number and size of its components. The existence, number and size of components in a graph stands in some sense for the opportunities and obstacles to communicate, to exchange information or/and to interact.

4.4 Graphs – Mathematical Models of Structures and Networks

Let S_k^s be the number of components with s vertices and k edges. So we can write:

$$\sum_{s,k} s\, S_k^s = S; \quad \sum_{s,k} k\, S_k^s = K \tag{4.69}$$

We denote the mean values of the frequency of the component $d(s, k)$ by

$$H_k^s = \langle S_k^s \rangle = \sum_r r\, P_{k,r}^s \tag{4.70}$$

with $P_{k,r}^s$ being the probability that in a special digraph $D(S, K)$ a number of $S_k^s = r$ components of the kind $d(s, k)$ can occur. As the mean number of components of the digraph $D(S, K)$ we define

$$\eta = \sum_s \sum_k H_k^s \tag{4.71}$$

Apart from the trivial components, single vertices ($d(1, 0)$) and single edges ($d(2, 1)$), the digraph contains a "structured part", whose number of components is written as follows:

$$L = \eta - H_0^1 - H_1^2 \tag{4.72}$$

The probability distribution $P_{k,r}^s$ is not known for the finite digraphs. In the limit $S \to \infty$ at constant C, the probabilities for the emergence of different finite large components are uncorrelated, therefore we have to work with a Poisson distribution for the possibility of r components with s vertices and k arcs in the whole graph (Figure 4.10) (Sonntag, Feistel, and Ebeling, 1981):

$$\lim_{S \to \infty} P_{k,r}^s = \frac{(H_k^s)^r}{r!} \exp(-H_k^s) \quad \text{for} \quad S \gg s \tag{4.73}$$

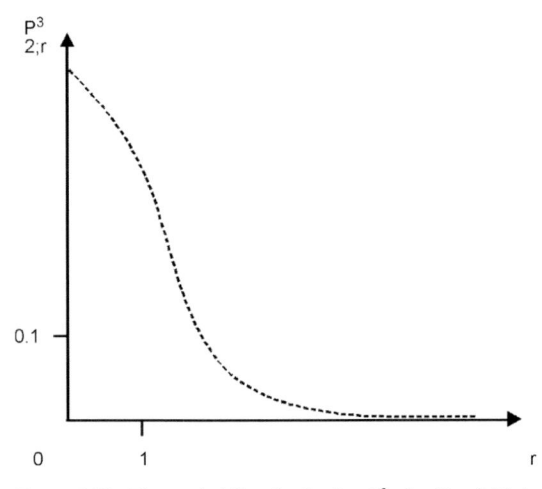

Figure 4.10 The probability distribution $P_{2,r}^3$ for $C = 0.05$ that r components $d(3, 2)$ occur.

In dependence on the connectivity C, we can for instance represent $P^1_{0;r}$, $P^2_{1;r}$, $P^3_{2;r}$, $P^3_{3;r}$ over r. So we can show the building of components in dependence on the connectivity C.

In detail, we may derive the following statements (Sonntag, Feistel, and Ebeling, 1981):

In the limit $S \to \infty$, the mean number of components η is in practice determined only by semicycle components (trees). So we can use the approximation

$$\eta = \sum_{s=1}^{S} H^s_{s-1} + 0(S^0) \tag{4.74}$$

In the limit $S \to \infty$, $C = $ const, the number of nontrivial components follows to be

$$L = \eta - S \exp(-2C) - SC \exp(-4C) \tag{4.75}$$

By a constant number of S of individuals, firms (vertices) the building of relations between these individuals, firms corresponds to an increasing number K. At constant S, the value of K is proportional to C

$$C = \frac{K}{S} \sim K; \quad S\text{-const} \tag{4.76}$$

In this way, the connectivity C increases during the development of relations. We will show, that firstly the number of small components increases and then decreases with increasing C. Large components become important. At the end all components are "absorbed" into one large component. For our applications, large and strongly connected components $d(s,k)$ are of most interest. That means a great number of relations exist $k \geq s$, so that the components are strongly connected. Strongly connected components are characterized by the appearance of cycles. For $S \to \infty$, the frequency of cycles of the length l in a graph $D(S,K)$ is

$$H^{(l)}_C = \frac{C^l}{l} \tag{4.77}$$

and for semicycles with the length l

$$H^l_C \sim \frac{(2C)^l}{2l} \tag{4.78}$$

The number of vertices of $D(S,K)$ belonging to cycles is

$$Z_C = C^3(1-C) \quad \text{for} \quad S \to \infty; \quad C < 1/2 \tag{4.79}$$

and of those belonging to semicycles

$$Z_0 = 4C^3(1-2C) \quad \text{for} \quad S \to \infty; \quad C < 1/2 \tag{4.80}$$

The number of vertices of $D(S,K)$ belonging to one tree in $D(S,K)$ referred to S is equal to the value given by Erdös and Rényi (1960) (Figure 4.11):

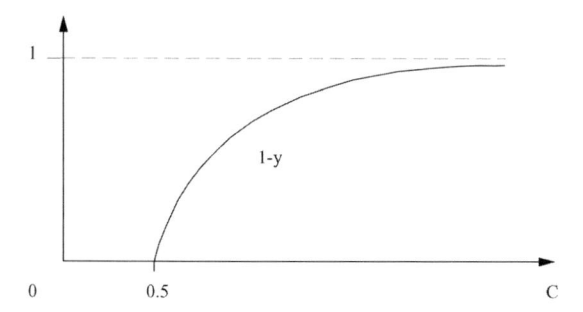

Figure 4.11 The relative number of vertices of $D(S, K)$ belonging to a tree (y) in dependence on the connectivity C (Erdös and Rényi, 1960).

$$y = \lim_{S \to \infty} \frac{R_{K,S}}{S} = \begin{cases} 1 & C \leq 1/2 \\ x(C)/2C & C > 1/2 \end{cases} \tag{4.81}$$

with

$$R_{K,S} = \sum_{s=1}^{S} s\, H_s^{s-1}$$

$$xe^{-x} = 2Ce^{-2C}$$

$$x(C) = \sum_{s=1}^{\infty} \frac{s^{s-1}}{s!} \left(2Ce^{-2C}\right)^s$$

The graphical representation shows analogies with a phase transition of the second kind – a fact that becomes particular obvious if such networks are considered by means of the percolation theory (Sonntag, 1984). For $C < 1/2$, the probability is zero that a selected vertex is part of a cycle. For $C \geq 1/2$, the probability increases up to the value one.

On the contrary, for $C < 1/2$, a vertex belongs to a tree with the probability one. For $C \geq 1/2$, that probability goes to zero. For $C > 1/2$, the graphs $D(S, K)$ consist of one large component. For $C > 1/2$ with growing C value, one tree after the other is "absorbed" by the large component. With an increasing number of edges (relations), more and more trees are linked up with one "macroscopic" component until, finally, all firms are linked with each other.

The number of components in directed graphs also corresponds to the value given by Erdös and Rényi (1960) for undirected graphs (Figure 4.12)

$$\bar{\delta} = \lim_{\frac{K}{S} \to C} \frac{\delta}{C} = \begin{cases} 1 - C, & 0 \leq C \leq 1/2 \\ \frac{1}{2C}\left(x(C) - \frac{x^2(C)}{2}\right), & C > 1/2 \end{cases} \tag{4.82}$$

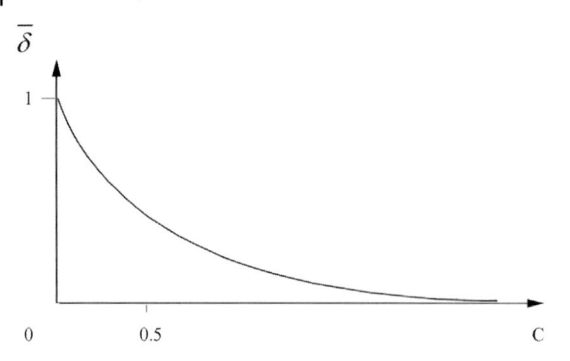

Figure 4.12 The relative number of components $\bar{\delta}$ of a graph in dependence on C.

For the examination of the developing of connections, it is interesting to inspect the fraction of vertices that belong to the structured subgraph. The number of the isolated vertices can be determined. These are vertices that do not belong to the connection net. In the same way, the number of components can be received minus the isolated vertices of the graph. This is the part of the graph that includes the economic connection network. If we understand the connections as a manifold of possibilities, we can arrive at some conclusions. Particularly, we will study the network of interaction between elements in the course of time. For the case of a constant number of vertices, the development of connections corresponds to a rising number K. On the other hand, K is proportional to C. That is why, C rises (the connectivity) by developing connections. As the calculation results show, first the number of small components increases and later decreases with the rising C. The number of the large components rises up to the moment when all are connected to one very large component. Inside this large component, the degree of connectedness of the network plays a role. A sign for this strong connectedness is the occurrence of cycles in the graph. As we have shown, the frequency of cycles of the length l is a simple function of the connectivity C. Likewise the probability can be given that a vertex belongs to a cycle. For $C<0.5$, the probability is zero, which means that the number of the vertices is irrelevant. By an increasing development of connections ($C>0.5$), this probability rises up to 1. In Chapters 5–9, there will be several applications of these findings, in particular also to food webs and catalytic networks.

4.5
Stochastic Models for Discrete Variables

Let us turn now to the problem of formulating stochastic models with discrete variables. We will show that here the methods of graph theory are very useful. First, we develop several basic concepts for the description of discrete systems, then we study applications to stochastic models of discrete systems. There exists a large class of modeling systems for Markovian processes in a discrete state space. This concerns the atomic processes behind extensive thermodynamic variables, such as particle numbers in chemical reacting systems but includes in particular also models of bioecological and socioeconomic processes. One may even say that the mathematical foundations for modeling such systems are intimately connected with the description of biological and ecological systems (Bartholomay, 1958a, 1958b, 1959) that later found widespread application to socioeconomic systems (Weidlich, 2000; Bruckner et al., 1996; Hartmann-Sonntag, Scharnhorst, and Ebeling, 2009). Let us introduce a set of objects, say species, each of them being present with N_i particles or individuals, which is referred to here as the *occupation number* of the related state. Further, let $P(N_1, \ldots, N_s; t)$ be the probability to find in the system an occupation distribution characterized by the occupation numbers $\mathbf{N} = (N_1, \ldots, N_s)$. As an example we refer to the large class of birth-and-death processes where $\mathbf{N} = (N_1, \ldots, N_s)$ are natural numbers that change during a single transition only by $+1$ or by -1. A special stochastic process that was termed "birth-and-death process" is of particular importance for our study of stochastic effects in economic processes, in particular in innovation processes. A birth process in this sense is a random appearance of a new element in a system and a death process is the disappearance of an element. Processes of this kind play a big role in biology, ecology, and sociology. Economic processes are another field where processes of this type are relevant. Economic growth is characterized by structural changes based on the introduction of new technologies in the economic world. To describe evolution problems, one has to determine the system, the elements, and their interactions. Here we consider so far abstract objects (molecules, individuals, production units, etc.). The objects are introduced as elementary units that play the role of players in the evolutionary game. But to describe it as an evolutionary process, one has to consider the microscopic level, which means we have to identify the microscopic carriers of changes.

As the basic tools for the modeling of these processes, we use so-called urn models. Already in 1907, the physicists *Paul* and *Tatjana Ehrenfest* developed a simple model for the diffusion of N molecules. Two urns A and B were isolated with respect to exchange with the surrounding. With respect to mutual exchange, there were permeable connections between A and B. Because of the isolation of the two urns, the total number of molecules in A and B remains constant. After regular time intervals, a molecule is randomly (that means with a probability of $(1/N)$) chosen and changes from its original urn to the other urn. The basic mechanism of this model is that in the next step a molecule moves from one urn to the other urn with a chance that is directly proportional to the number of molecules contained in the urn at that moment of time.

The model of the Ehrenfests represents the prototype for the investigation of decision processes between the possibilities A and B, respectively, between *yes* and *no* (Ebeling, Karmeshu, and Scharnhorst, 2001). For example, the decision may be to accept a new technology or not. In 1926, Kohlrausch and Schrödinger gave a continuous diffusion-approximation for such processes. Feller (1951) formulated a realistic variant of this model. He used a discrete Markovian process with continuous time. The time between the molecule crossings was exponentially distributed. Surveys of biological applications of birth-and-death-type processes were given by Bartholomay (1958a, 1958b, 1959) and Eigen (1971). Many applications to social processes were surveyed by Weidlich (2000).

The stochastic approach that we are using here was developed in particular in the context of general models of evolutionary processes and in particular biochemical processes (Ebeling and Feistel, 1974, 1975, 1977, 1979, 1982; Ebeling, Feistel, and Jiménez-Montaño 1977; Ebeling, Sonntag, and Schimansky-Geier, 1981; Feistel and Ebeling, 1977, 1978a; Feistel 1983a, 1983b, 1983c, 1983d, 1985; Ebeling and Sonntag, 1986). Later, the model found numerous applications in modeling scientific evolution (Ebeling and Scharnhorst, 1986; Bruckner, Ebeling, and Scharnhorst, 1990) and technological evolution (Bruckner, Ebeling, and Scharnhorst, 1990; Bruckner et al., 1996; Pyka and Scharnhorst, 2009).

For our stochastic picture, we can use the ideas on the occupation number space developed above. Contrary to the deterministic models, this stochastic description offers the advantage that at finite times new technologies (sorts) can arise or die out. Let us introduce a set of technologies numbered by $i = 1, 2, \ldots, s$. By $N_i(t)$, we denote the number of production units that use the technology i. (Note that a technology is corresponding here to an urn in Ehrenfest's problem). These numbers are termed occupation numbers. They are functions of time. The occupation numbers are either positive or zero,

$$N_i(t) = \{0, 1, 2, \ldots\} \tag{4.83}$$

Now, the state of the system at the time t can be described by the probability distribution of the occupation numbers

$$P(N_1, N_2, \ldots, N_s; t) = P(\mathbf{N}; t) \tag{4.84}$$

As elementary processes we consider processes during which only one occupation number can change and as transition processes we consider processes during which at most two occupation numbers can change (Figure 4.13),

addition: $(N_i) \rightarrow (N_i + 1)$,
removal: $(N_i) \rightarrow (N_i - 1)$, and

substitution: $\begin{pmatrix} N_i \\ N_j \end{pmatrix} \rightarrow \begin{pmatrix} N_i - 1 \\ N_j + 1 \end{pmatrix}$

As a special model subclass, we may assume, like Ehrenfest, that the total number of objects participating in the process (e.g., the number of reacting molecules or in a socioeconomic context the number of plants) is constant during the processes we consider. Manfred Eigen referred to this condition as the "constant overall-number condition" (Eigen, 1971). In particular, decision processes occur in the system as transitions which conserve the total number. If we assume that all decisions that lead to a change of the set of the occupation numbers depend mostly on the current state, we can apply the concept of Markov processes. The microscopic economic processes (substitution or innovation activities) may be considered as such a kind of elementary processes. The dynamics of the system can be described with the help of the master

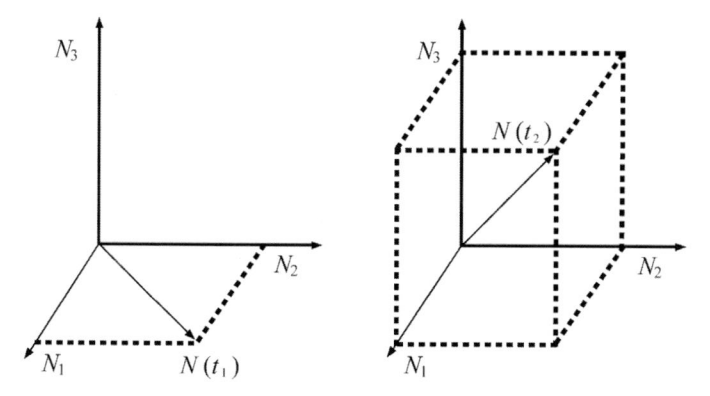

Figure 4.13 The occupation number space: The distribution of individuals over different types is represented by a vector in the positive cone. The appearance or extinction of a type can be described as approaching or leaving one of the edges of the positive cone.

equation. This equation is a balance equation between formation and decay processes:

$$\frac{\partial P(\mathbf{N};t)}{\partial t} = \sum_{\mathbf{N}'} \{W(\mathbf{N}|\mathbf{N}')P(\mathbf{N}';t) - W(\mathbf{N}'|\mathbf{N})P(\mathbf{N};t)\} \quad (4.85)$$

where the vector $\mathbf{N}$ is

$$\mathbf{N} = \{N_1, N_2, \ldots, N_s\}$$

Per unit time, the transition probabilities for the system to change from the state $\mathbf{N}'$ to the state $\mathbf{N}$ or vice versa are expressed by $W(\mathbf{N}|\mathbf{N}')$ and $W(\mathbf{N}'|\mathbf{N})$, respectively.

An important general property of discrete stochastic processes can immediately be derived from Equation 4.85. We may ask for the probability that a system remains for a certain time in a given state $\mathbf{N}$. The mean value of this distribution is known as the waiting time or residence time or escape time with respect to that particular state. To compute it, we consider the fictitious case that all neighbor states are unoccupied, $P(\mathbf{N}';t) = 0$ for any $\mathbf{N}' \neq \mathbf{N}$, such that there are no transition processes that could increase the probability $P(\mathbf{N};t)$. Initially, the considered state may definitely be occupied, $P(\mathbf{N};0) = 1$. Under those conditions, the solution of Equation 4.85 is easily found in the form

$$P(\mathbf{N},t) = P(\mathbf{N},0)\exp\left\{-t \sum_{\mathbf{N}' \neq \mathbf{N}} W(\mathbf{N}'|\mathbf{N})\right\}$$

The related residence time in the state $\mathbf{N}$ is $\tau = \left[\sum_{\mathbf{N}' \neq \mathbf{N}} W(\mathbf{N}'|\mathbf{N})\right]^{-1}$. This distribution is fundamental for an effective numerical simulation method of discrete stochastic processes (Feistel, 1977a, 1979; Feistel and Ebeling, 1977, 1978a; Gillespie, 1977; Ebeling and Feistel, 1982). This algorithm is usually termed *Gillespie algorithm* now, but in fact it was independently developed and used at the same time also by one of us.

The transition probabilities may be complicated functions (Weidlich, 2000). In most of our applications, we consider in this book a subclass where the transition probabilities are modeled as polynomials (Ebeling and Feistel, 1977, 1982; Jiménez-Montano and Ebeling, 1980; Ebeling and Sonntag, 1986; Heinrich and Sonntag, 1981). Examples for the transition probabilities of special processes are:

1) Spontaneous generation

$$W(\ldots, N_i+1, \ldots, N_j, \ldots, N_k, \ldots | \ldots, N_i, \ldots, N_j, \ldots, N_k, \ldots) = A_i^{(0)}$$
(4.86)

Self-reproduction

$$W(\ldots, N_i+1, \ldots, N_j, \ldots, N_k, \ldots | \ldots, N_i, \ldots, N_j, \ldots, N_k, \ldots) = A_{ij}^{(1)} N_j + E_i^{(1)} N_i$$
$$E_i^{(1)} = A_i^{(1)} + B_{ij}^{(1)} N_j + C_{ijk}^{(1)} N_j N_k$$
(4.87)

Decay

$$W(\ldots, N_i-1, \ldots, N_j, \ldots, N_k, \ldots | \ldots, N_i, \ldots, N_j, \ldots, N_k, \ldots) = E_i^{(2)} N_i$$
$$E_i^{(2)} = A_i^{(2)} + B_{ij}^{(2)} N_j$$
(4.88)

Conversion

$$W(\ldots, N_i+1, \ldots, N_j-1, \ldots, N_k, \ldots | \ldots, N_i, \ldots, N_j, \ldots, N_k, \ldots) = E_j^{(3)} N_j$$
$$E_j^{(3)} = A_{ij}^{(3)} + B_{ij}^{(3)} N_i + B_{ik}^{(3)} N_k + C_{ijk}^{(3)} N_i N_k$$
(4.89)

with $j \neq i$; $k \neq i$; j.

Expressions for the various coefficients can be formulated for special cases of interest. They may be functions of the total number of individuals and of the system parameters.

For catalytic networks, the expressions in relation (4.87) correspond to the process of the spontaneous self-reproduction, catalytic self-reproduction, and error reproduction. The terms in Equation 4.88 correspond to the processes of spontaneous and catalyzed decay. The terms in Equation 4.89 correspond to simple mutation processes without reproduction, mutation processes with reproduction, and ternary reproduction processes, which play a role with regard to processes with constant overall particle number.

The transition probabilities are formulated here in a general form. As a special case we can get the stochastic equations from this ansatz that correspond to the deterministic Eigen model (Eigen and Schuster, 1978) with the condition of constant overall particle number (Jiménez-Montaño and Ebeling, 1980; Ebeling, Sonntag, and

Schimansky-Geier, 1981; Heinrich and Sonntag, 1981). All this will be explained in more detail in later chapters.

Here we mention additionally that the formalism of generating functions is an elegant tool for manipulating discrete-valued master equations. With the help of the s-dimensional generating function

$$F(s_1, s_2, \ldots, s_s; t) = \sum_N s_1^{N_1} s_2^{N_2} \cdots s_s^{N_s} P(N_i, t) \quad 0 \leq s_i \leq 1 \tag{4.90}$$

we can write the master equation with the transition probabilities as follows:

$$\frac{\partial}{\partial t} F(\mathbf{s}; t) = \sum_{i \neq j} \left\{ \begin{aligned} & A_i^{(o)}(s_i - 1) F + A_i^{(1)} s_i (s_s - 1) \frac{\partial F}{\partial s_i} \\ & + A_{ij}^{(1)} s_j (s_i - 1) \frac{\partial F}{\partial s_j} \\ & + B_{ij}^{(1)} s_i s_j (s_i - 1) \frac{\partial^2 F}{\partial s_i \partial s_j} \\ & + C_{ijk}^{(1)} s_i s_j s_k (s_i - 1) \frac{\partial^3 F}{\partial s_i \partial s_j \partial s_k} \\ & + A_i^{(2)} (1 - s_i) \frac{\partial F}{\partial s_i} \\ & + B_{ij}^{(2)} s_j (1 - s_i) \frac{\partial^2 F}{\partial s_i \partial s_j} \\ & + A_{ij}^{(3)} (s_i - s_j) \frac{\partial F}{\partial s_j} \\ & + B_{ij}^{(3)} (s_i - s_j) s_i \frac{\partial^2 F}{\partial s_j \partial s_i} \\ & + \bar{B}_{ik}^{(3)} s_k (s_i - s_j) \frac{\partial^2 F}{\partial s_k \partial s_j} \\ & + C_{ijk}^{(3)} s_i s_k (s_i - s_j) \frac{\partial^3 F}{\partial s_i \partial s_j \partial s_k} \end{aligned} \right\} \tag{4.91}$$

In dependence on the values of the coefficients in these expressions, the class of processes included by this formalism is very large. So, everything depends on the detail, but this is a problem to be discussed more thoroughly later in Chapters 5–9.

A special stochastic process called "birth-and-death process" is of particular importance to our study of stochastic effects in economic processes, and in particular in innovation processes. A birth process is a random appearance of a new element in a system. A death process is the disappearance of an element. Processes of this kind play a big role in biology, ecology, and sociology and are also relevant to the field of economic processes (Pyka and Scharnhorst, 2009). Economic growth is characterized by structural changes based on the introduction of new technologies in the economic world. To describe technological evolution, one has to determine the system, the elements, and their interactions. Here we consider plants and technologies. We consider firms composed from different plants. The plants are introduced as elementary units that play the role of decision carriers according to market conditions (choosing a new technology or not). Plants also play the role of users of a particular technology. The technologies are understood as the different types present in the system. The plants are the elements or representatives of these technologies. Under this perspective, different technologies may compete for plants applying them, similar to the biological viewpoint of "selfish" genes that exploit phenotypes to reproduce. This perspective differs from the way one usually thinks about technological change, where the firm is central item. The underlying process is still a decision made by plants or firms. However, the model approach constructs an alternative perspective of it. Let us note here that this perspective is quite normal for any population-dynamic approach that deals with types (groups or species) and elements (individuals). However, in contrast to biological processes, human beings, organizations, and firms are not bound to a certain type or group they first belong to. In contrast to individuals of biological species, they have the opportunity to change the group they belong to. It is this kind of transition behavior that makes the model particularly relevant for socioeconomic applications. Let us note further that we arbitrarily use the notion of a plant or production unit as the simplest element in the system. By assuming that firms consist of several plants or production units, growth processes of firms are also covered by the model approach. Technological change is usually considered as a macroeconomic change process. However, in order to describe it as an evolutionary process, one has to consider this process at the microscopic level. This means that we have to consider the microeconomic carriers of technological changes. In the framework we present here, these are plants.

Finally, we mention again that the basic ideas for modeling those processes go back to century-old urn models developed by physicists.

4.6
Stochastic Processes on Networks

Stochastic processes on networks are of special interest for evolutionary processes (Erdös and Rényi, 1960; Feistel, 1979, 1985; Sonntag, Feistel, and Ebeling, 1981; Heinrich and Sonntag, 1981; Ebeling and Feistel, 1982; Sonntag, 1984; Ebeling and Sonntag, 1986; Ebeling *et al.*, 2006; Pyka and Scharnhorst, 2009; Ibe, 2011). Indeed most of the processes discussed in the previous sections may also be interpreted as processes on networks. Further, most of these processes may be represented in an elegant way by graphs. In this context, effects in small and sensitive networks play a special role (Hartmann-Sonntag, Scharnhorst, and Ebeling, 2009). Let us discuss what we mean by these terms:

Many complex systems display a surprising degree of tolerance to errors. The results indicate a strong correlation between robustness and network topology. In particular, scale-free networks are more robust than random networks against random node failures, but are more vulnerable when the most-connected nodes ("hubs") are targeted (Albert and Barabasi, 2002).

In small networks, any node or edge plays a specific role; here the addition or removal changes drastically the properties of the whole system. Another problem where stochastic effects play a big role is the question how a single new mutant can win the selection process, how can a new web site become a giant cluster among other old important web sites (clusters)? Is it possible to overcome the "once-forever" behavior by stochastic effects? The addition or removal of sensitive nodes or edges is a subject of investigation in big networks. With sensitive we mean elements that play a key role in the network. Some examples are discussed later. The main and most dramatic example of a sensitive network is the *primordial soup*, the origin of the evolution on Earth. In this soup, billions of spontaneously generated organic molecules and macromolecules are floating, created under the conditions of the early Earth. They are connected by catalytic relations and this forms a gigantic but significantly underoccupied network with the ability to evolve (Ebeling and Feistel, 1982). Morphogenesis of multicellular organisms can be modeled as growing networks of interacting cells, in particular neuronal networks. Models of these evolutionary processes are discussed in Chapters 5–9. Another important example is metabolic networks (Heinrich and Schuster, 1998). To give just one example, Wagner and Fell (2001) studied the clustering coefficient focusing on the energy and biosynthetic metabolism of the *Escherichia coli* bacterium. They found that in addition to the power-law degree distribution, the undirected version of this substrate graph has a small average path length and a large clustering coefficient.

Cellular networks can be subject to random errors as a result of mutations or protein misfolding, as well as harsh external conditions eliminating essential metabolites. Jeong *et al.* (2000) studied the responses of the metabolic networks of several organisms to random and preferential node removal. Removing up to 8% of the substrates, they found that the average path length did not increase when nodes were removed randomly, but it increased rapidly upon removal of the most-connected nodes, attaining a 500% increase with the removal of only 8% of the nodes. Similar results have been obtained for the protein network of yeast as well (Vogelstein, Lane, and Levine, 2000). Solé and Montoya (2001) studied the response of food webs to the removal of species (nodes). The results indicate that random species removal causes the fraction of species contained in the largest cluster to decrease linearly. However, when the most-connected (keystone) species are successively removed, the relative size of the largest cluster decays quickly. A complementary aspect is the controllability of complex networks by selected "driver nodes" (Liu, Slotine, and Barabási, 2011).

The error-and-attack tolerance of the internet and the World Wide Web was investigated by Albert, Jeong, and Barabási (2000). The internet is occasionally subject to hacker attacks targeting some of the most-connected nodes. They show that the average path length on the internet is unaffected by the random removal of as many as 60% of the nodes, while if the most-connected nodes are eliminated (attack), the average path length peaks at a very small fraction of removed nodes. Albert, Jeong, and Barabási (1999) investigated the World Wide Web and showed that the network survives as a large cluster under high rates of failure, but under attack, the system abruptly crashes down. These authors write: *The result is, that scale-free networks display a high degree of robustness against random errors, coupled with a susceptibility to attacks.*

Bianconi and Barabási (2001) showed the existence of a close link between evolving networks and an equilibrium Bose gas. Maurer and Hubermann (2000) present a dynamical model of web site growth in order to explore the effects of competition among web sites. They show that under general conditions, as the competition between sites increases, the model exhibits a sudden transition from a regime in which many sites thrive simultaneously to a "winner-takes-all market," in which a few sites grab almost all the users, whereas most other sites go nearly extinct. This prediction is in agreement with empirical data measurements on the nature of electronic markets. In the following text, we give a stochastic simulation of this model. Dorogovtsev, Mendes, and Samukhin (2003) developed a statistical mechanic approach for random networks. They underline the important fact that this differs crucially from the situation for growing networks. The latter, while growing, self-organize into scale-free structures in a wide range of parameters without condensation.

In this section, we explain how to develop a theory to describe stochastic processes in a network of interacting units. We need a stochastic picture that allows the clarification of the role of fluctuations for the survival of new species in evolution and in particular of innovations. We refer to the theory of complex networks (Pyka and Scharnhorst, 2009) and introduce the notion of sensitive networks (Hartmann-Sonntag, Scharnhorst, and Ebeling, 2009) in this context. Sensitive networks are those in which the introduction or the removal of a node/vertex dramatically changes the dynamic structure of the system. As an application we consider interaction networks of forms and technologies and describe technological innovation as a specific dynamic process. The master-equation formalism and the theory of birth-and-death processes are the mathematical instruments used here.

As we explained in the introduction to this book, "self-organization" in our understanding is the spontaneous formation of structures. Further, an "innovation", in a general system-theoretical understanding, is the appearance of a new species, of a new mode of behavior, of a new technology, of a new product, or even only of a new idea. In this understanding, an "innovation" is the driving element in biological and technological evolution.

For the applications to be given later, we distinguish between two states of a vertex, occupied and unoccupied. The states are marked as shown in Figure 4.14.

We also distinguish two states for the edges of the graph. Edges that go out from unoccupied vertices ($N_i = 0$) are omitted because they are inactive (cannot work). Active (working) edges are characterized by the graphic representation shown above. For socioeconomic networks, unoccupied vertices stand for new technologies that have not yet been discovered. They can be understood as hidden, future

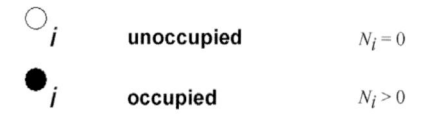

○$_i$ unoccupied $N_i = 0$

●$_i$ occupied $N_i > 0$

Figure 4.14 Occupied and unoccupied vertices.

possibilities. The model does not allow the prediction of a certain new technology but it can make statements how the system handles the appearance of new technologies in general.

Eventually, we get a graph that describes the whole system at a fixed time t. If the color of a vertex is changed, new connections (elementary processes) can come up or former connections may no longer continue. Processes can spread the elements on the unoccupied vertices or they can select them out of the set of occupied vertices. The basic structure of the network is the "maximal" graph (all vertices i are occupied, all reactions can work).

In a network with a great number of nodes s, but a small overall number of elements N (individuals, organizations, plants) ($N \ll s$), a lot of vertices are not occupied. Let us consider a graph with ($N \gg s$) and reduce the overall number of elements to a state ($N \ll s$). In general, vertices such as i and j become unoccupied by processes corresponding to the following transition rates:

$$A_i^{(2)} N_i, \; B_{ij}^{(2)} N_i N_j, \; A_{ij}^{(3)} N_j, \; B_{ij}^{(3)} N_i N_j, \; \bar{B}_{ik}^{(3)} N_k N_j, \; C_{ijk}^{(3)} N_i N_j N_k$$

These are processes of decay and of conversion. If only a certain part of the process works because of the small overall number, we get a graph with a lot of components. With a decreasing number of elements (individuals, organizations, and plants), the maximal graph with few components develops to a graph with a lot of components. The "minimal" graph we can get (maximally decomposed) consists of self-reproduction processes, spontaneous generation (simple innovations), sponsored innovation processes, and sponsored innovation processes with self-reproduction (self-reproducing process). Vice versa, vertices can be occupied by the following processes: spontaneous generation (simple innovation), $A_i^{(0)}$, error reproduction, $A_{ij}^{(1)} N_j$, mutation without reproduction (innovation without self-reproduction), $A_{ij}^{(3)} N_j$, and mutation without reproduction but with catalytic support (sponsored innovation without self-reproduction), $\bar{B}_{ik}^{(3)} N_k N_j$.

We assume that the processes through which the vertices can be occupied are very rare. This means spontaneous generations and mutations have a small probability. After a relatively short time, the components are in a local equilibrium. New components can be occupied (second process). This is a hopping process between the components. Certain components die out under selection pressure, others survive and grow.

Now we show a representation of the transition probabilities in the network picture. We consider a system with s elements (sorts, plants) that interact. This system can be described by a graph in which each element i corresponds to a vertex of the number i. We mark the transition probabilities for different processes by edges of different type in order to distinguish these probabilities as shown in Figure 4.15.

The transition from unoccupied ($N_i = 0$) to occupied ($N_i > 0$) systems is of special interest for underoccupied systems. Thus, we get a graph that describes the whole system at a fixed time t. If the color of a vertex is changed, new reactions (connections) can flame up or former reactions (connections) may no longer go on. Processes can

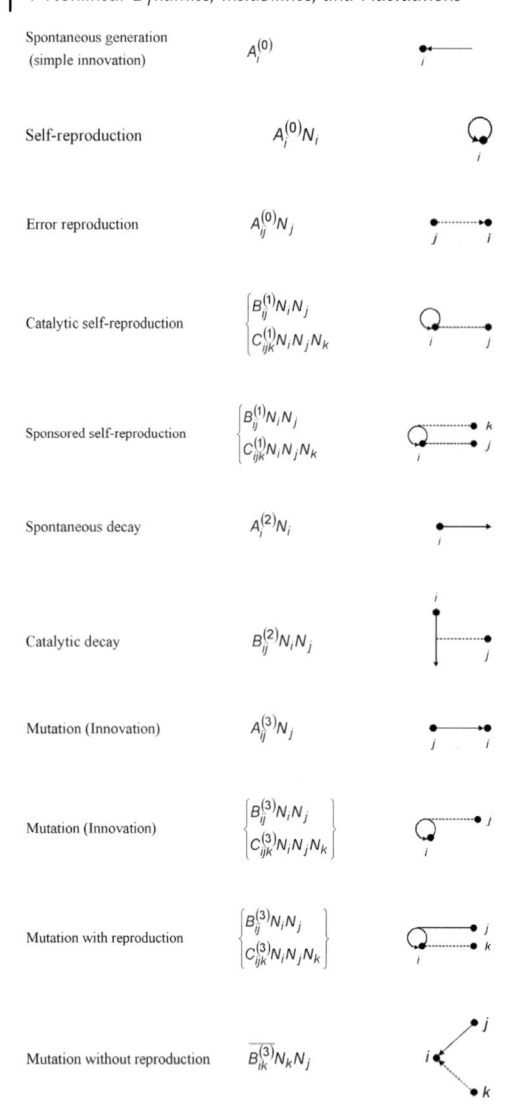

Process	Symbol
Spontaneous generation (simple innovation)	$A_i^{(0)}$
Self-reproduction	$A_i^{(0)} N_i$
Error reproduction	$A_{ij}^{(0)} N_j$
Catalytic self-reproduction	$\begin{cases} B_{ij}^{(1)} N_i N_j \\ C_{ijk}^{(1)} N_i N_j N_k \end{cases}$
Sponsored self-reproduction	$\begin{cases} B_{ij}^{(1)} N_i N_j \\ C_{ijk}^{(1)} N_i N_j N_k \end{cases}$
Spontaneous decay	$A_i^{(2)} N_i$
Catalytic decay	$B_{ij}^{(2)} N_i N_j$
Mutation (Innovation)	$A_{ij}^{(3)} N_j$
Mutation (Innovation)	$\begin{cases} B_{ij}^{(3)} N_i N_j \\ C_{ijk}^{(3)} N_i N_j N_k \end{cases}$
Mutation with reproduction	$\begin{cases} B_{ij}^{(3)} N_i N_j \\ C_{ijk}^{(3)} N_i N_j N_k \end{cases}$
Mutation without reproduction	$\overline{B_{ik}^{(3)}} N_k N_j$

Figure 4.15 Stochastic transition processes related to self-reproduction.

spread the particles (plants) on the unoccupied vertices or they can select them from the occupied vertices.

A special topic of large interest for applications is the theory of *small and sensitive networks*. In connection with our special interest in evolution processes, we discuss now this special type of networks. The term "sensitive" networks denotes, as pointed out above, networks that are sensitive to the introduction or removal of one or few nodes or edges, or in a more general context, to the occupation or leaving of a node. Specifically, sensitivity is linked to the question of whether a node (a species, a mode of behavior, an idea, a technology) is occupied by at least one individual or not. We show in later chapters that this problem is of relevance for the modeling of the origin of life and in particular also of interest for modeling innovation processes (Hartmann-Sonntag, Scharnhorst, and Ebeling, 2009). A problem of relevance in this context is the problem of error tolerance.

Hartmann-Sonntag, Scharnhorst, and Ebeling (2009) and Ebeling *et al.* (2006) presented detailed investigations on the structure and dynamical behavior of sensitive networks. They introduce a stochastic picture that allows for the clarification of the rule of fluctuations for the survival of innovations in such a nonlinear system. Random graph theory, percolation, master equation formalism, and the theory of birth and death processes were used. Using the traditional formalism of statistical mechanics, Dorogovtsev, Mendes, and Samukhin (2003) constructed a set of equilibrium statistical ensembles of random networks without correlation and found their partition function and main characteristics.

Several interesting applications were given to economic problems, in particular those connected with innovation processes. In economics, innovation is mainly understood as technological innovation describing the introduction of new technologies, products, and production processes. The differentiation between invention and innovation relates innovation to the economic exploitation of new ideas. However, it is also possible to look at innovations from a more general, evolutionary point of view (Saviotti and Mani, 1995; Bruckner *et al.*, 1996; Hartmann-Sonntag, Scharnhorst, and Ebeling, 2009). Ziman gives one example for such an approach when he writes "Go to a technological museum and look at the bicycles. Then go to a museum of archeology and look at the prehistoric stone axes. Finally, go to a natural history museum and look at fossil horses. In each case, you will see a sequence, ordered in time of changing but somewhat similar objects" (Ziman, 2000).

Figure 4.16 illustrates an evolutionary approach. The system is composed of a large set of enumerable types that are represented as nodes in a network. At a certain point in time, only a small part of these nodes are populated that is, they are active. A populated node is represented by a fat node and a nonoccupied node by an empty circle. This way our network consists of fat/occupied and empty/nonoccupied nodes with certain links. The whole picture changes in time. The pattern of interaction between them (including processes of self-influence) determines the dynamic composition of the system. It is visualized in terms of (active) links between the nodes. This active network will produce a dynamics that has a certain set of stable states. We assume that the activated part of the network is embedded in a much larger network of inactive nodes and links. The inactive nodes represent future possibilities in the evolution of the system. An innovation appears when an unoccupied node becomes occupied for the first time. With this first occupation, the set of links connecting the "new" node with already occupied nodes also becomes activated. It is readily apparent that such an event changes the whole composition of the system. Accordingly, the stable state that the system might have reached already becomes instable and the system searches for a new stable state. If we assume that the interaction between the nodes (types) is a competitive one, the stable state of a certain activated network can also include the deactivation of certain nodes. Types (nodes) that are selected out will transit to a nonoccupied inactive status.

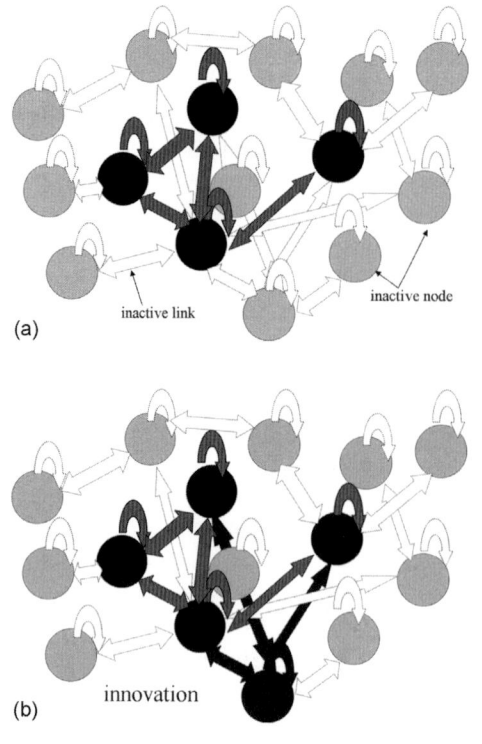

Figure 4.16 Illustration of a sensitive network. Embedded in a large network of inactive nodes and links, the activated part of the network (a) changes its composition when an innovation is emerging (b).

However, not every change is an innovation. In this chapter, we follow a system theoretical approach to innovation. In this framework, innovation is something new to the system, and most essentially, the emergence of an innovation changes the state of the system dramatically. In other words, the actual state of the system becomes unstable and a transition to a new state occurs. To define an innovation, we first have to define the state of the system. Here, we again choose a very specific approach. We represent the state of the system as a point in the high-dimensional occupation number space. In this space, a coordinate axis is attached to a certain type i of elements (with $i = 1, 2, \ldots, s$ being natural numbers). The occupation numbers are represented on this axis.

To describe technological innovation, we have to ask for a respecification of this abstract concept. For socioeconomic systems, the axes of the state space refer to different possible taxonomies. For instance, an axis i can represent a certain technology from a set s, of different technologies present in the system.

With such a taxonomy, many competition processes can be described. The carriers of this competition processes are different species or firms using different technologies and competing with their products on a market.

Formally, we can find each type i in N_i exemplars of elements in the system. The exemplars may be individuals, but also organizational and institutional units like firms and groups. N_i, the occupation numbers, are functions of time. They are positive or zero. A complete set of occupation numbers $N, N_2, \ldots, N_s$ at a fixed time characterizes the occupation state of this system. The time-dependent change of the occupation numbers is described by the movement of this point in the space. The whole motion takes place on the nonnegative cone K of the space. In this picture, we can describe the case that a type i is not present in the system at time t. That means the type i is occupied with the number zero ($N_i = 0$). We call a system an underoccupied system if we can make the assumption that the sum of the occupation numbers is essentially smaller than the total number of the possible elements (Ebeling and Sonntag, 1986). In this picture, an "innovation" is an occupation of a nonoccupied type. In deterministic systems, zero occupation can only be achieved in the limit of infinite time $t \to \infty$, if a sort died out (zero can be a stable stationary state). For finite times $t > 0$, types cannot arise, if they are not in the system at time $t = 0$, and present types cannot die out. The situation is different if we use the stochastic picture. In stochastic systems, the zero state can be reached in finite times t. A stochastic description offers the advantage that new sorts (innovations) can arise or die out at finite times. Therefore, the stochastic description is especially suited for evolutionary processes and in particular for innovation processes. Let us note here that any innovation will change the taxonomy of types in the system. Innovation has to do with uncertainty and its prediction is impossible. With the notion of an underoccupied system, we escape the problem of determining a priori the place or kind of an innovation. Instead, we equip the system with a reservoir of possible innovations. Which of these possibilities turns into a realization remains uncertain. In some respects, this is a trick to avoid the problem with a changing taxonomy. There are other possibilities to escape this problem, for example, so-called continuous models operating on a characteristic space as we discussed elsewhere (Feistel and Ebeling, 1989; Ebeling et al., 1999), see Chapter 7. However, the discrete approach has, in our respect, certain advantages, as is discussed later.

In an underoccupied system, most elements have, at a given time t, the occupation number zero. So we can change from the high-dimensional cone K to a lower-dimensional cone K^+ with only positive occupation numbers. Accordingly, the time-dependent variation of the system can be described as a switching of the state point on the edges of the cone K. If K^+ is an element of the set of all possible cones, we observe a switch from one subcone to another. As the process is discrete, it is a hopping on the edges of different positive cones. In Figure 4.13, we visualize such a process for three dimensions. At any point in time, the state of the system is represented by a certain vector $N(t)$. In a stationary state, the endpoint of this vector defines a positive cone. In our example, the vector moves in the plane spanned by N_1 and N_2. An innovation opens up a new dimension of the system. In our example, a new third type is introduced into the system. After the innovation, the vector is moving in the space defined by N_1, N_2, and N_3. In general, we can assume that the system operates in a multidimensional space where the cone can have a very complicated shape, and the vector $N(t)$ jumps between the edges of this cone. The hopping process visualizes the transition between one stable stationary state and another stable stationary state. In this sense, innovation is the outcome of a process of destabilization. In the framework we propose in this chapter, innovations are seen as stochastic instabilities. The changes occurring in the occupation number space result from interactions of the different types present in the system. These interactions can be visualized as graphs or networks where the nodes represent the types and links between them represent different forms of interactions. In the network picture, an innovation corresponds to the appearance of a new node and the activation of a link to this node. The models we present in Section 4.3 allow us to differentiate between different processes that finally introduce such a new node. The conceptualization of types as elements (nodes) of a network represents a graph theoretical approach to the dynamics of the system.

In the last years, complex systems in nature and society have been carefully investigated. Already in the 1970s, theories of self-organization were used to build a bridge between social and natural systems investigations. As part of this development, complex networks have been investigated. Recently, as a new branch in complexity theory, complex networks have been reconsidered and extensively studied (Bornholdt and Schuster, 2003). They seem to be particularly relevant for the study of innovation processes (Pyka and Scharnhorst, 2009). In the context of complexity theory, the concept of networks has not only been used as an easy-to-use metaphor. As Bornholdt and Schuster note: "Recent advances in the theory of complex networks indicate that this notion may be more than just a philosophical term. Triggered by recently available data on large real world networks (e.g., on the structure of the internet or on the molecular networks in the living cell) combined with fast computer power on the scientist's desktop, an avalanche of quantitative research on network structure and dynamics currently stimulates diverse scientific fields." (Bornholdt and Schuster, 2003). Social networks form one important area of application of complex networks theory. Of great interest is in particular the structural analysis of systems (Albert and Barabási, 2002). In the very beginning, investigations of large complex systems were done by random graphs. More and more it became possible to analyze real complex systems and large systems too. With the development of computer capacity, the amount of empirical data increases. It becomes possible to compare the theoretical results by random graph theory with that of the real data analysis. Obviously, more than pure randomness exists. Organization principles and rules of system evolution play a decisive role, leading to small-world behavior and scale-free networks. Our world is not a random world. Other evolutionary principles are of great interest. In addition, it is evident that a theory of evolving networks may give a more realistic approach to real systems. This is why, we give special consideration to evolving networks here.

From the analysis of empirical data, we learn that many real-networks have a small-world character. The small-world concept describes the fact that despite their often large size, in most networks, there is a relatively short path between any two nodes. The small-world property characterizes most complex networks. For example, the chemicals in a cell are typically separated by only three reactions, or in a more exotic case, the actors in Hollywood are on average within three costars from each other. The small-world concept corresponds to our observations; it is a structural, not an organizing, principle (Albert and Barabasi, 2002).

Not all nodes in a network have the same number of edges (the same node degree). The spread in node degrees is characterized by a distribution function, $p(k)$, which gives the probability that a randomly selected node has exactly k edges. Since in a random graph the edges are placed randomly, the majority of nodes have approximately the same degree, close to the average degree of the network. The degree distribution of a random graph is a Poisson distribution with the peak over the average degree. In real networks, the distributions of the edges are more complicated. Important results were obtained by the analysis of large real systems: the degree distribution deviates significantly from a Poisson distribution and follows general structural rules in many cases. Many large networks are scale-free, that is, their degree distribution follows a power law. In addition, even for those networks for which $p(k)$ has an exponential character, the degree distribution significantly deviates from a Poisson distribution achieved by random graph theory for such systems. Scale-free networks express a hierarchy between the nodes. Not every node is important at the same level. Accordingly, not each of the links between the types has the same importance. There are very sensitive relations or elements too. As mentioned previously, we consider networks as "sensitive" if their properties depend strongly on the introduction or removal of one or a few nodes or edges, or on changes of the occupation of nodes. We will show that for the evolutionary character, the description of the time behavior by master equations on occupation number spaces is an appropriate tool. The discrete character of occupation number description allows for an appropriate description of the introduction respectively of removal of relations, edges, and so on. We will analyze not only the steady states of our stochastic systems but also the time evolution. Albert and Barabási (2002) also refer to approaches with master and rate equations. They write that, in addition, these methods, not using a continuum assumption, appear more suitable for obtaining exact results in more challenging network models. In addition, they mentioned that the functional form of the degree distribution, $p(k)$, cannot be guessed until the microscopic details of the network evolution are fully understood. According to our point of view, the method of master equations is an excellent tool to use in the investigation of many open questions and is able to bring much light to bear on this subject. For example, by using this discrete approach, we have the chance to get statements about the kind of fluctuations. Evidently, this is one of the most important questions.

Our aim here is the calculation of the role of fluctuations by the use of the master equation approach. This way, we can make statements as to how the systems differ from linear systems which obey a Poisson distribution. In principle, by investigating the fluctuations, correlations, and spectral densities, we are able to study several microscopic events. One of the questions to solve is which fluctuation effects produce power-law distributions. We see a deep connection of these network systems to systems that produce $1/f$-noise. We remember that $1/f$-noise is a stochastic process with a specific power-law spectrum (Watts and Strogatz, 1998). A characteristic property of processes that produce $1/f$-noise are long range-correlations. We suppose that the scale-free networks and small-world behavior may have some relation to this. We remember that in small-world networks, the degree function obeys a power law; there exists a small pathway between every two of the elements.

In investigating self-organization and evolution processes in networks, our basic approach is that we understand the corresponding networks as dynamic or, more precisely, as evolutionary systems. This dynamic and evolutionary approach allows us to make statements about innovation processes, special competition effects, the sensitivity of networks, the constraints of growth, and the fitness of network systems.

"As complex we describe holistic structures consisting of many components, which are connected by many (hierarchically ordered) relations respectively operations. The complexity of a structure can be seen in the number of equal respectively distinct elements, in the number of equal respectively distinct relations and operations, as well as in the number of hierarchical levels. In the stricter sense, complexity requires that the number of elements becomes very large (practically infinite)" (Ebeling, Freund, and Schweitzer, 1998). We are interested here in the origin of complex structures, and in the development of order (information). As mentioned already, Ilya Prigogine in collaboration with his coworkers did pioneering work in the investigation of self-organizing systems (Nicolis and Progogine, 1977). Further important work on the self-organization of macromolecules and the evolution of networks in the primary soup has been done by Eigen (1971) in collaboration with Schuster (Eigen and Schuster, 1977, 1978). A somewhat different view on this was developed in the formulation of the synergetics by Haken (1978). The investigation of such systems shows that the formation of order in complex systems can be allocated to physical processes that play a role far from equilibrium. We underline that biological, just as socioeconomic, processes can be investigated with the help of the theory of self-organization because they obey the valid physical and chemical laws. However, processes that include the real life (biological and socioeconomic systems) also obey other rules and laws, which are not solely determined by physics. This is already evident from the very general character of the structures we consider here. In this chapter, we formulated the ideas of structure and dynamical systems mathematically and kept away from the concrete situations that appear in real systems.

5
Self-Reproduction, Multistability, and Information Transfer as Basic Mechanisms of Evolution

Life is that the form is maintained through the change of substance.
                    Thomas Mann: The Magic Mountain

5.1
The Role of Self-Reproduction and Multistability

There are several elementary processes that play a fundamental role in evolution, among them are the processes of self-reproduction and processes that show bistability or multistability. Further, we mention the role of oscillations. These processes are evidently a *conditio sine qua non* for all higher forms of self-organization such as the evolution of life. We do not know any form of life that does not use processes of self-reproduction and multistability and many processes in cells are connected with oscillations.

Let us start with several examples for self-reproduction processes. The first example is a simple reproduction as, for example, the birth of an identical copy of the individual of a species, and the second one is an autocatalytic chemical reaction

$$\begin{aligned} X &\to 2X \\ A + X &\to 2X \\ X &\to D \end{aligned} \qquad (5.1)$$

In the first example, we may think of the division of a cell, a bacterium, or the synthesis of a virus copy in a host cell. The chemical reaction that we present as the second example is a so-called autocatalytic reaction; the existence of the chemical species X favors its own production. An example of an autocatalytic process is the reaction of an ester (ethyl ethanoate) with a free hydrogen ion in aqueous solution,

$$CH_3COOC_2H_5 + H_2O + H^+ \to CH_3COO^- + C_2H_5OH + 2H^+ \qquad (5.2)$$

In this process, we observe an autocatalytic increase of the number of free protons in the system.

Another example from biochemistry is the self-reproduction of polynucleotides which is of fundamental importance for the evolution of living organisms.

Physics of Self-Organization and Evolution, First Edition. Rainer Feistel and Werner Ebeling.
© 2011 Wiley-VCH Verlag GmbH & Co. KGaA. Published 2011 by Wiley-VCH Verlag GmbH & Co. KGaA.

The standard way for the quantitative description of self-reproduction in many-particle (macroscopic) systems is by densities or concentrations. The dynamics of such variables is described by ordinary differential equations which we analyzed in Chapter 2. Dynamical models for processes of self-reproduction and growth were first developed in the nineteenth century for modeling processes of growth or decay of populations by *Verhulst* and *Pearl*.

Another type of elementary chemical processes that play a key role in the evolution are reactions that show bistable behavior such as

$$A + 2X \rightarrow 3X, \quad X \rightarrow D \tag{5.3}$$

This example of a bistable chemical reaction is based on second-order autocatalysis. As we show explicitly in the next paragraph, this reaction may run in two regimes, one that leads to a rather high concentration of the substance X and one that leads to a low concentration. The two regimes correspond to two alternative attractors of the dynamics. These two possibilities may be used as the basis for storing one bit of information. Another example for a multistable reaction is the formation of a polynucleotide molecule from the building blocks A, C, G, and U, where the symbols stand for the monomers, the nucleotides adenine, cytosine, guanine, and uracil. The RNA molecule is a nucleic acid that has a length of typically 100 monomers; it consists of the units A, C, G, and U in different spatial arrangements (sequences):

$$v_1 A + v_2 C + v_3 G + v_4 U \rightarrow GCAACUG\ldots UAG \tag{5.4}$$

We may consider this as a multistable reaction since from the same ingredients in a different realization of the reaction, we may get a completely different outcome such as

$$v_1 A + v_2 C + v_3 G + v_4 U \rightarrow AGCACUG\ldots UAG \tag{5.5}$$

In this realization, just by chance a nucleotide A came first and then came a G and then a C nucleotide. In spite of the same gross content of A, C, G, and U, this is a completely different molecule which may have different physical and chemical properties. The binding energies between different pairs of nucleotides are very similar such that there is no energetically preferred equilibrium sequence, and each arbitrary sequence formed of the given building blocks is a stable isomer molecule. The formation of DNA molecules is quite similar; they consist of the nucleotides A, G, C, and T, where T stands for the amino acid thymine. This multistability of the chain molecules, RNA and DNA, facilitated their functioning as storage media for symbolic information (see also Section 5.5; Chapters 8 and 9), and this way enabled the origin of life on Earth.

DNA molecules are much longer than RNA molecules and consist of double strands, as we know from the *Watson* and *Crick* model. We also know – and these ideas go back to *Delbrück*, *Gamov*, and in particular to *Schrödinger* – that the DNA is the carrier of the memory in the genetic material of the higher organisms.

We have seen that self-reproduction and multistability play a basic role in the processes of self-organization and evolution, and deserve our special interest. So far, we discussed only the principal role of elementary processes of self-reproduction and multistability. Let us now discuss in more detail some ways of their theoretical description and their specific role in self-organization processes.

5.2
Deterministic Models of Self-Reproduction and Bistability

We first consider a process in which a number density increases or decreases due to the self-reproduction or decay

$$X \rightarrow 2X$$
$$X \rightarrow D \tag{5.6}$$

with the reaction rates k' and k''. The corresponding dynamic equation is very simple:

$$\frac{d}{dt}x(t) = kx(t), \quad k = k'-k'' \tag{5.7}$$

with the simple exponential solution, see Figure 5.1,

$$x(t) = x(0) \exp(kt) \tag{5.8}$$

Let us now consider a different example, a reversible chemical reaction, which also includes a backward reaction

$$A + X \leftrightarrow 2X, \quad X \rightarrow D \tag{5.9}$$

The corresponding dynamical equation reads

$$\frac{d}{dt}X(t) = k_1 AX(t) - k_2 X(t) - k'_1 X^2(t) \tag{5.10}$$

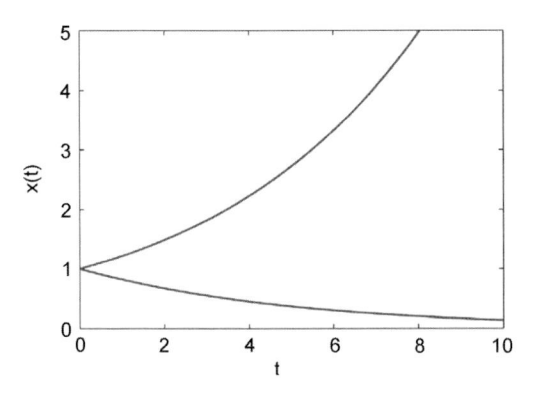

Figure 5.1 Growth (rate $k=0.2$) and decay (rate $k=-0.2$) in a simple self-reproduction process.

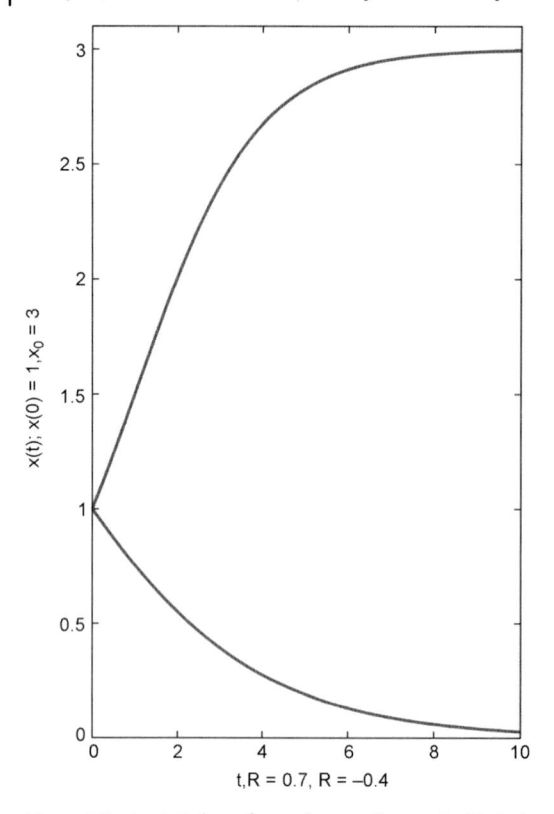

Figure 5.2 Logistic law of growth according to the Verhulst–Pearl equation for the rates $R = 0.7$ (upper curve) and $R = -0.4$ (lower curve).

Here, X and A denote the corresponding concentrations. For a constant value of A, the solution reads

$$\frac{X(t)}{X(0)} = \frac{X_0}{X(0) - [X(0) - X_0]\exp(-Rt)} \tag{5.11}$$

where the newly introduced variables are $R = k_1 A - k_2$ and $S = k'_1$, $X_0 = R/S$. An example is shown in Figure 5.2.

The attractors of the first-order autocatalytic reaction system are demonstrated in Figure 5.3 as a function of the rate R. The abscissa $x = 0$ is stable only for $R < 0$; the line $x = R/S$ is the stable attractor for $R > 0$. The point $R = 0$ marks the transition between two branches of the reaction. This point may be considered as a second-order kinetic transition (Haken, 1970). This transition bears some analogy to second-order phase transitions in thermodynamics (Haken, 1978). We will show later that in the stochastic description the sharp transition is replaced by a smooth transition. This is due to the influence of fluctuations. This is

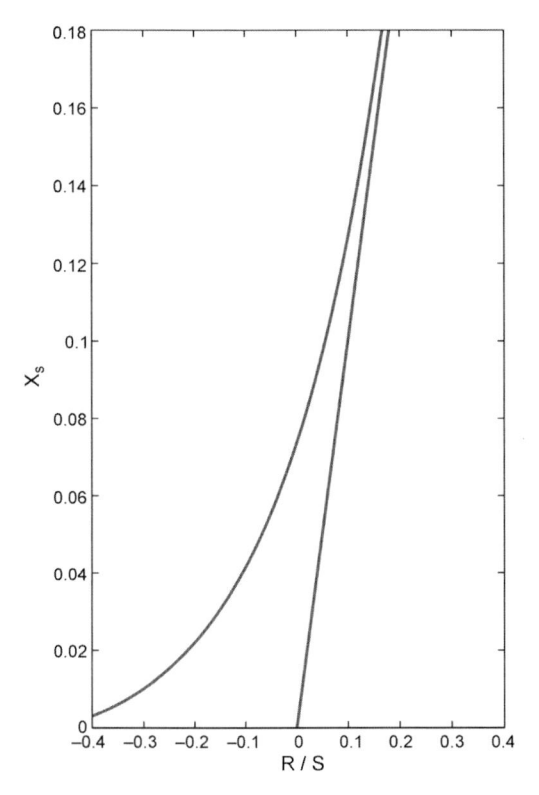

Figure 5.3 The attractors of the first-order autocatalytic reaction system. The abscissa $x = 0$ is stable only for $R < 0$; the line $x = R/S$ is the stable attractor for $R > 0$. The point $R = 0$ marks the transition between two branches of the reaction; a kind of second-order kinetic transition occurs. The increasing curve shows the prediction of maximal probability in a stochastic description, discussed later.

demonstrated by the increasing curve in Figure 5.3, which shows the prediction of the concentration that has the maximal probability in the stochastic description; this is discussed later. In the critical point, we observe the largest deviation of the maximum of the stochastic distribution from the deterministic curve. In other words, the fluctuations play a big role near to the point of the transition. This is another analogy to the theory of thermodynamic phase transitions (Haken, 1978).

Let us now study the system (5.3) with higher order autocatalysis,

$$A + 2X \leftrightarrow 3X, \quad X \leftrightarrow F$$

The corresponding differential equation is

$$\frac{d}{dt} X(t) = k_1 A X^2(t) - k'_1 X^3(t) - k_2 X(t) + k_2 F \qquad (5.12)$$

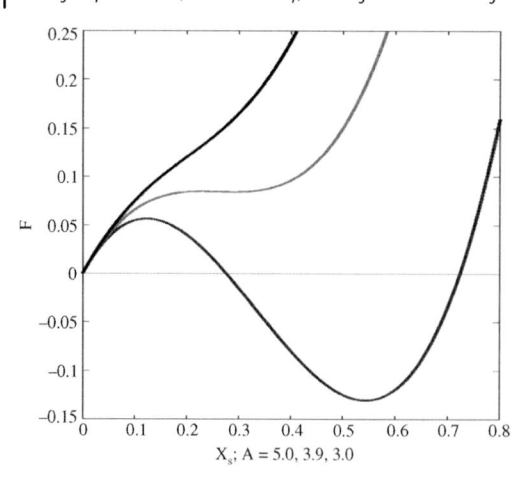

Figure 5.4 The stationary states of a second-order autocatalytic process. In our example, all rates equal unity, and the coefficient A takes three different values (5.0, 3.9, and 3.0). In the first case, $A = 5$, there exist three solutions for constant F. For $-0.13 < F < 0.06$, the outer solutions are stable, the solution in between is unstable. In the second case, $A = 3.9$, we observe a double solution for $F = 0.8$. This is a point of a kinetic transition between bistability and monostability. For larger values of A, the system is monostable.

This system has two stable attractors, the outer roots $X^{(1)}$ and $X^{(3)}$ of the quadratic equation

$$0 = k_1 A X^2 - k'_1 X^3 - k_2 X + k_2 F$$

Note that the solution in between, $X^{(2)}$, is unstable. Our system is bistable. We note that bistable systems are necessarily nonlinear. We demonstrate an example in Figure 5.4, assuming that all rates are one, the coefficient A has three different values (5.0, 3.9, and 3.0). In the first case $A = 5$, there exist three solutions for any constant in the range $-0.13 < F \leq 0.06$, the outer solutions are stable, and the solution in between is unstable. In the second case $A = 3.9$, we observe a double solution for $F = 0.8$, this is a point of a kinetic transition between bistability and monostability. For larger values of A, the system is monostable. We observe an analogy to the thermodynamic phase transitions of first order (Haken, 1978; Ebeling and Feistel, 1982).

The time-dependent solution can be given in the implicit form

$$t = -\int_{X(0)}^{X(t)} \frac{dX}{(X - X^{(1)})(X - X^{(2)})(X - X^{(3)})} \tag{5.13}$$

We see that the outcome of this game depends on the initial conditions. If the initial state, $X(0)$, is below the threshold value $X^{(2)}$, then the final state will be $X^{(1)}$, otherwise, if the initial concentration is higher than $X^{(2)}$, the final state will $X^{(3)}$. Our system is multistable; we repeat that multistable systems are necessarily nonlinear. We will show later that bistable or multistable systems may be used as an information storage, but in the present "century of bits and bytes," this is trivially known already to small children.

A case which is also of some interest for evolutionary processes is the hyperbolic growth, which denotes a very fast growth leading to an explosion already at finite time. This case is typically described by the following type of kinetic equations without contributions from backward reactions. A special case is obtained from Equation 5.12 with $k'_1 = 0, F = 0$:

$$\frac{d}{dt}X(t) = kAX^v(t) - k'X(t) \tag{5.14}$$

with a growth power law (the exponent, v,) that is at least quadratic. The solution can be found in analytical form after introducing the variable $z(t) = [X(t)]^{1-v}$ (Ebeling and Feistel, 1982). Typically, these solutions possess a singularity at finite time. In the special case $k' = 0$, $A = 1$, we get, for example, the formula

$$X(t) = \frac{X(0)}{1 - X(0)kt}$$

As shown in Figure 5.5, the growth is very fast and approaches infinity at the finite time $t = 1/(X(0) k)$. As postulated by Eigen and Schuster (1978), hypercyclic growth processes played an important role in early evolution. The singular behavior of solutions is part of that theory of hypercycles.

The competition between two species with quadratic growth rates leads to rather surprising results as we show for one example. We consider two coupled deterministic equations for two concentrations competing with each other in a way that the sum has the constant value C:

$$\frac{dx_1}{dt} = b_1 x_1^2 - \varphi x_1 \quad \frac{dx_2}{dt} = b_2 x_2^2 - \varphi x_2 \quad x_1 + x_2 = C$$

$$\varphi \equiv \frac{b_1 x_1^2 + b_2 x_2^2}{C} \tag{5.15}$$

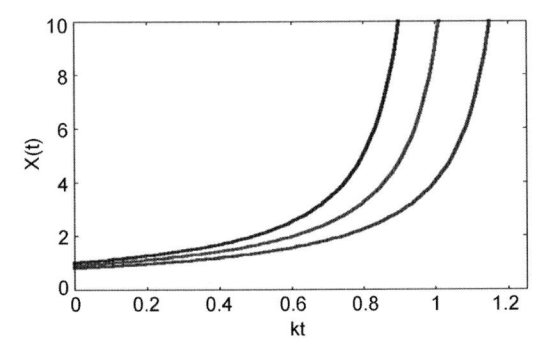

Figure 5.5 Hyperbolic growth of a concentration for three different initial conditions.

$$x_i = 0 \qquad S_i \qquad x_i = C$$

Figure 5.6 Phase space of Equation 5.15 representing the competition of two species with hyperbolic growth.

This dynamical system has one degree of freedom due to the conservation of the total concentration. It has two stable attractors $X_1 = 0$, $X_2 = C$ and $X_2 = 0$, $X_1 = C$. This bistability corresponds to winning of either species 1 or species 2. In between is an unstable state. Which attractor is approached depends not only on the growth rates but also on the initial conditions. This leads to a very peculiar situation. Even if the growth rate of species 2 is much larger than the growth rate of species 1, it cannot win the competition, if the initial state is near to attractor one, and of course, also the opposite is true. In a bistable system, the outcome of the game is determined by the initial condition. Section 5.4 shows that such a strict alternative is weakened in a stochastic description.

As a result of the quadratic terms, the phase space is split into two regions by a separatrix, S_i, as shown in Figure 5.6.

The selection behavior depends on the initial conditions, $x_i (t = 0)$, of the system and the threshold values:

$$S_1 = \frac{Cb_2}{b_1 + b_2} \qquad S_2 = \frac{Cb_1}{b_1 + b_2}$$

The species 1 can win in the long run only if $x_1(0) > S_1$ and correspondingly species 2 will win for $t \to \infty$ if $x_2(0) > S_2$ is fulfilled.

Note that oscillatory chemical reactions may be treated in a similar way by using two coupled differential equations for two concentrations (Feistel and Ebeling, 1978a, 1989; Ebeling and Feistel, 1982).

5.3
Stochastic Theory of Birth-and-Death Processes

Following pioneers such as *Verhulst* and *Pearl* and many later extensions of the theory, we described so far all processes of growth and decay by real numbers and ordinary differential equations, neglecting all aspects of discreteness. In general, this is not a problem since population numbers are very large and therefore the discreteness does not play any role. However, there are also processes where the discreteness of the number of atoms, molecules, or individuals plays an important role. In particular this is the case where only a few units (atoms, molecules, or individuals) are present in the system but nevertheless their role is of much relevance. In this case, a stochastic description is indispensable. This type of dynamic model was also analyzed already in some detail in Chapter 2. In the class of stochastic processes in discrete systems, there is a special stochastic process that was termed *birth-and-death process*. This process is of particular importance for our study of stochastic effects in evolution processes, in particular in processes where a new molecule or a new species appears for the first time and will take over the control in subsequent dynamic growth processes. From technology and economy, innovation processes are an important example. In this sense, a birth process is the random appearance of a new element in a system, and a death process is the disappearance of a formerly existing element. Processes of this kind play a big role in biology, ecology, and sociology. Technological and economic processes are another field where processes of this type are relevant. Since this might be less evident, let us add a few remarks regarding this point (see, e.g. Pyka and Scharnhorst, 2009). Economic growth is characterized by structural changes based on the introduction of new technologies in the economic world. To describe the technological evolution problem in an appropriate theoretical model, one has to specify the system, the elements, and their interactions. Here we may consider plants (production units, firms) and technologies. The plants are introduced as elementary units that play the role of decision carriers according to market conditions (choosing a new technology or not). Plants play also the role of users of a particular technology. Technological change is usually considered as a macro-economic change process. But to describe it as an evolutionary process, one has to consider the microscopic level that means we have to identify the microeconomic carriers of technological changes.

We will get the basic ideas for the modeling of birth-and-death processes by the investigation of so-called urn models. In 1907, the physicists Paul and Tatjana Ehrenfest developed a simple model for the diffusion of N molecules (Ehrenfest and Ehrenfest, 1907). In 1926, Kohlrausch and Schrödinger gave a continuous diffusion approximation for birth-and-death processes (Kohlrausch and Schrödinger, 1926). The Ehrenfests studied two urns A and B which are isolated with respect to exchange with the surrounding. With respect to mutual exchange, there are permeable connections between A and B. Because of the isolation of the two urns, the total number of molecules in A and B remains constant. After regular time intervals a molecule is randomly (that means with the probability of $1/N$) choosen and moved from its urn to the other urn. The basic mechanism of this model is that the probability to move in the next step – a molecule from one urn to the other – is directly proportional to the number of molecules that are contained in the urn at that moment of time.

After the general and historical remarks given above, let us turn to the formal description. The state space which we consider is the set of integers

$$N(t) = \{0, 1, 2, 3, \ldots\}$$

For the class of processes considered, the elementary stochastic act is the so-called one-step process, the change of the particle number by one

$$\begin{pmatrix} N \\ N \end{pmatrix} \to \begin{pmatrix} N+1 \\ N-1 \end{pmatrix}$$

which describes that the number of units increases or decreases by one. We denote the transition probabilities by $W^+(N)$ and $W^-(N)$, respectively.

The transition probabilities may depend on the occupation number N. This may be a more or less complicated function that depends on the concrete process which we want to describe.

In the simplest case we may assume that the transition probability is proportional to the number of the individuals that are performing self-reproduction or decay

$$W^+(N) = kN \quad W^-(N) = k'N \tag{5.16}$$

For the choice of transition probabilities of birth-and-death processes, it is a necessary boundary condition that transitions to negative particle numbers are impossible, that is, $W^-(0) = 0$. We can formulate the master equation for this discrete process in the form of a balance of probabilities. This equation describes the time behavior of the probability distribution of the occupation numbers, $P(N;t)$. Considering the balance between four different possible processes of increase and decrease with the corresponding transition probabilities we infer,

$$\frac{\partial}{\partial t} P(N;t) = W^+(N-1)P(N-1;t) + W^-(N+1)P(N+1;t)$$
$$- W^+(N)P(N;t) - W^-(N)P(N;t) \tag{5.17}$$

In the simplest case of linear transition probabilities as given by Equations 5.13 and 5.14, we get

$$\frac{\partial}{\partial t} P(N;t) = k(N-1)P(N-1;t) + k'(N+1)P(N+1;t)$$
$$- kNP(N;t) - k'NP(N;t) \tag{5.18}$$

In the stationary case, Equations 5.15 and 5.16 reduce to a recursive difference equation

$$W^-(N+1) P^0(N+1) = W^+(N) P^0(N) \tag{5.19}$$

which we may solve by iteration assuming some starting value, $P^0(0)$,

$$P^0(N) = \frac{W^+(N-1) \cdot W^+(N-2) \cdots W^+(0)}{W^-(N) \cdot W^-(N-1) \cdots W^-(1)} P^0(0) \tag{5.20}$$

This solution is known as a *detailed balance* because the four terms on the right-hand side of Equation 5.16 cancel pair wise due to Equation 5.18. The value of $P^0(0)$ is determined from the normalization condition

$$1 = \sum_{N=0}^{\infty} P^0(N) = P^0(0) \sum_{N=0}^{\infty} \frac{P^0(N)}{P^0(0)} \tag{5.21}$$

where the terms of the sum are known from the recursion relation (5.19). Examples for the distribution functions of monostable or bistable systems are demonstrated in Figure 5.7. For large particle numbers N, the result of the formula (5.20) can be

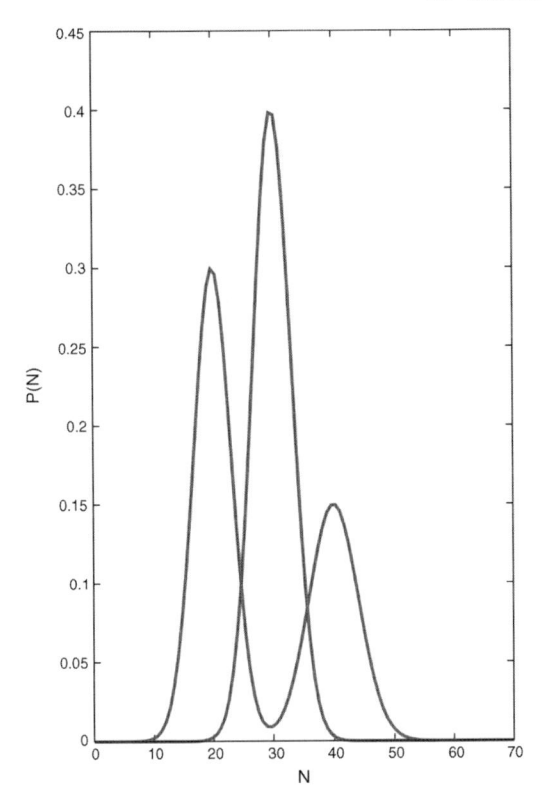

Figure 5.7 Typical shapes of probability distributions for monostable and for bistable systems.

estimated by the integral

$$\ln\frac{P^0(N)}{P^0(0)} \approx \int_0^{N-1}\ln W^+(N')dN' - \int_1^N \ln W^-(N')dN \approx \ln\frac{W^+(0)}{W^-(N)} + \int_1^{N-1}\ln\frac{W^+(N')}{W^-(N')}dN'$$

For polynomial functions W, this integral can be evaluated analytically after decomposition into partial fractions.

For the case of the reaction rates given by Equation 5.16 with $W^+(0) = 0$, the unique stationary solution (5.20) reads

$$P^0(N) = \delta_{N,0} \tag{5.22}$$

that is, the asymptotic extinction of stochastically self-reproducing individuals is statistically certain in the form a *fluctuation catastrophe* because $N=0$ is an absorber state that can be approached but not be left by the system. Such catastrophes may apply frequently to small populations; we know that continuously many biological species disappear from our planet without a chance for recreation, for example, as a result of overfishing or ruthless poaching. For sufficiently large populations, the extinction probability within a reasonable number of generations is found to be an extremely small number. Moreover, new species (or technologies) appear at a slow rate due to mutations, inventions, and so on, such that there is always a small probability for the transition process $0 \to 1$ which, by some tiny chance for spontaneous creation, $W^+(0) = \varepsilon > 0$, leads to a nontrivial stationary solution for Equation 5.20,

$$\frac{P^0(N)}{P^0(0)} = \frac{\varepsilon}{W^-(N)} \prod_{i=1}^{N-1} \frac{W^+(i)}{W^-(i)} = \frac{\varepsilon}{k'N} \left(\frac{k}{k'}\right)^{N-1} \tag{5.23}$$

This distribution can be normalized only if $k < k'$. We find for the value of $P^0(0)$, Equation 5.21,

$$\frac{1}{P^0(0)} = \sum_{N=0}^{\infty} \frac{P^0(N)}{P^0(0)} = 1 + \sum_{N=1}^{\infty} \frac{\varepsilon}{k'N} \left(\frac{k}{k'}\right)^{N-1} = 1 - \frac{\varepsilon}{k} \ln\left(1 - \frac{k}{k'}\right) \tag{5.24}$$

as easily verified by series expansion of the logarithmic function.

Even the time-dependent analytical solution of Equation 5.17 can be found in the linear case (Bartholomay, 1958a, 1958b, 1959; Eigen, 1971; Allen and Ebeling, 1983) by the method of generating functions (see Section 6.4). In particular, the value of $P(0, t)$ is the extinction probability of the population at the time t. A special case of this solution is considered in the Section 5.4.

The stochastic theory of oscillatory reactions may be developed in a similar way (Feistel, 1977a, 1979; Feistel and Ebeling, 1977, 1979a, 1989; Ebeling and Feistel, 1982). The master equations for the probabilities depend then on two particle numbers, which gives rise to several complications. The stationary probability distributions have a crater-like shape. Of particular interest in this context is the relation between neutrally stable phase fluctuations and asymptotically stable amplitude fluctuations.

Finally, let us make a few remarks about simple applications of the stochastic models discussed here. Reviews of biological applications of birth-and-death-type processes were given by Bartholomay (1958a, 1958b, 1959) and Eigen (1971). Many applications to social processes were reviewed by Weidlich (2000). These types of applications are discussed in later chapters. Further, we mention that the model of Ehrenfest represents the prototype for the investigation of decision processes between two possibilities A and B, or respectively, between yes and no (Ebeling, Molgedey, and Reimann, 2000). For example, the decision may be to accept a new solution of a problem or refuse it, or accept a new technology or refuse it. Of special interest for evolutionary processes are stochastic effects in small and sensitive networks (Hartmann-Sonntag, Scharnhorst, and Ebeling, 2009), which have been discussed already in Chapter 4.

5.4
Stochastic Analysis of the Survival of the New

This section considers a question of special relevance for evolutionary processes, the emergence of the one or several exemplars of a new species (the NEW) for the case with linear growth rates. A useful instrument to model such processes is the theory of stochastic transitions between urns with constant overall number established already by Ehrenfest. Between the urnes, symbolic balls are exchanged that stand for few individuals trying to multiply in competition with other species that are present in the system with large numbers. We start with a two-species system. Let us denote with N_i – the number of individuals, which belong to the species i, (here, $i = 1, 2$), that means they belong to the urn i. These numbers are termed occupation numbers

$$N_i(t) = \{0, 1, 2, \ldots\}$$

Following Ehrenfest, we assume the total number of individuals to be constant. For the new individual to increase its occupation number, it has to oust an old one. This is the simplest form of competition.

We formulate the stochastic formalisms for our simple model for a binary competition process considered for the first time already in Section 5.2 with complementary occupation assuming a constant total number

$$N = N_1 + N_2 = \text{const} \tag{5.25}$$

We assume that during elementary transition processes, the occupation number can only change by ± 1. This is the so-called one-step process. During a transition process, at most two occupation numbers can change simultaneously in such a way that

$$\begin{pmatrix} N_1 \\ N_2 \end{pmatrix} \rightarrow \begin{pmatrix} N_1 - 1 \\ N_2 + 1 \end{pmatrix} \tag{5.26}$$

For instance, we assume that E_1 is the growth rate of the individuals of species 1. For a new species introduced into the system 2, the growth rate is E_2. We assume that the new species is somehow "better" and has a larger growth rate:

$$E_2 > E_1$$

The elementary stochastic process

$$\begin{pmatrix} N_1 \\ N_2 \end{pmatrix} \rightarrow \begin{pmatrix} N_1 - 1 \\ N_2 + 1 \end{pmatrix} \tag{5.27}$$

describes how much the new species can perform better than the old one. In a more general case, we may assume that the transition probability has the following structure:

$$W^+\left(N_2 + 1 | N_2\right) = E_2 N_1 \left(\frac{N_2}{N}\right) + E_{21} \equiv W^+_{N_2} \tag{5.28}$$

This means that at first the transition probability is proportional to the number of the individuals that belong to the old species

$$W^+(N_2) \sim N_1$$

and, furthermore, that it is proportional to the relative number of individuals that belong to the new species,

$$W^+(N_2) \sim \frac{N_2}{N}$$

Here, E_{21} is the rate of any spontaneous change from 1 to 2, and E_{12} from 2 to 1. For the opposite case, we can assume:

$$W^-(N_2-1 \mid N_2) = E_1 N_2 \left(\frac{N_1}{N}\right) + E_{12} \equiv W^-_{N_2} \qquad (5.29)$$

Now, we formulate the master equation for this discrete process that describes the time behavior of the probability distribution of the two occupation numbers, $P(N_1, N_2; t)$. With the transition probabilities assumed above, it follows that

$$\frac{\partial}{\partial t} P(N_1, N_2; t) = W^+(N_2 \mid N_2-1) P(N_2-1; t)$$

$$+ W^-(N_2 \mid N_2+1) P(N_2+1; t) \qquad (5.30)$$

$$- W^+(N_2+1 \mid N_2) P(N_2; t)$$

$$- W^-(N_2-1 \mid N_2) P(N_2; t)$$

and, inserting the transition rates, from Equations 5.27 and 5.28, we find

$$\frac{\partial}{\partial t} P(N_1, N_2; t) = \left[E_{21} + \frac{E_2}{N}(N_2-1)(N_1+1) \right] P(N_1+1, N_2-1; t)$$

$$+ \left[E_{12} + \frac{E_1}{N}(N_2+1)(N_1-1) \right] P(N_1-1, N_2+1; t) \qquad (5.31)$$

$$- \left[E_{21} + E_{12} + \frac{1}{N}(E_1+E_2) N_2 (N-N_2) \right] P(N_1, N_2; t)$$

If we use the relation $N_1 + N_2 = N$, (i.e., $N_1 = N-N_2$), we can write

$$\frac{\partial}{\partial t} P(N_1, N_2; t) = \left[E_{21} + \frac{E_2}{N}(N_2-1)(N-N_2+1) \right] P(N_2-1; t)$$

$$+ \left[E_{12} + \frac{E_1}{N}(N_2+1)(N-N_2-1) \right] P(N_2+1; t) \qquad (5.32)$$

$$- \left[E_{21} + E_{12} + \frac{1}{N}(E_1+E_2) N_2 (N-N_2) \right] P(N_2; t)$$

5.4 Stochastic Analysis of the Survival of the New

To obtain statements for the deterministic case, we define the mean values of the stochastic variable, N_2, by

$$\langle N_2(t) \rangle = \sum_{N_2=0}^{\infty} N_2 \, P(N_2;t) \tag{5.33}$$

By multiplication of the master equation with N_2 and subsequent summation, we get

$$\frac{d}{dt}\langle N_2(t) \rangle = \frac{1}{N}(E_2 - E_1)\langle N_2(N - N_2) \rangle + (E_{21} - E_{12}) \tag{5.34}$$

Using the approximation $\langle (N_2)^2 \rangle \approx \langle N_2 \rangle^2$ and the abbreviations

$$x_2 = \frac{\langle N_2 \rangle}{N} \qquad \alpha = E_2 - E_1 \qquad \beta = (E_{21} - E_{12})\frac{1}{N}$$

we derive the deterministic equation

$$\frac{dx_2}{dt} = \alpha\, x_2(1 - x_2) + \beta \tag{5.35}$$

Following the method described in Section 5.3, Equation 5.19, we can also find the stationary solution of Equation 5.32, in the form

$$P^0(N_2) = \frac{W^+(N_2) \cdot W^+(N_2 + 1) \cdots W^+(N-1)}{W^-(N_2 + 1) \cdot W^-(N_2 + 2) \cdots W^-(N)} P^0(N) \tag{5.36}$$

In the deterministic case, after some relaxation time, the final state

$$x_2 = 1, \text{ that is, } N_2 = N$$

will be approached asymptotically.

On the contrary, for the stochastic case we get a probability distribution $P(N_2) \neq 0$ for $N_2 \neq N$. That means, in the stochastic case, the old and new species can coexist (at least if $E_2 - E_1$ is not too large). The most interesting case occurs if the spontaneous transition rates are zero, that is, $E_{21} = E_{12} = 0$. The states $N_1 = 0$, $N_2 = 0$ are so-called absorber states, which means a system cannot leave such states again once it is "trapped" there. For the stationary solution of the master equation, we get (Ebeling and Feistel, 1982):

$$P^0(N_2) = \sigma_1 \delta_{0N_2} + \sigma_2 \delta_{NN_2}; \qquad \sigma_1 + \sigma_2 = 1 \tag{5.37}$$

The absorption probabilities σ_i are real numbers between zero and one. Starting from the absorber states $N_2 = 0$ and $N_1 = 0$ $(N_2 = N)$, the other states cannot be reached.

The initial state determines which of the two alternative stationary solutions will be occupied by the system. After some calculations, we can obtain the following formula for σ_2 in the limit $N \gg 1$ and $N_2(t=0) \ll N$ (Eigen, 1971; Allen and Ebeling,

1983; Ebeling et al., 2006; Hartmann-Sonntag, Scharnhorst, and Ebeling, 2009),

$$\sigma_2 = \frac{1-(E_2/E_1)^{N_2(0)}}{1-(E_2/E_1)^N} \tag{5.38}$$

where $N_2(0) = N_2(t=0)$ is the initial state of the system. Another detailed derivation and discussion of this important formula is given in Chapter 6. If $N_2(0)$ is the number of individuals of the species 2 at the time $t = 0$, then σ_2 is the probability that for $t \to \infty$ $N_2 = N$, individuals will have changed to the species 2. Similarly, σ_1 is the probability that no individuals, $N_2 = 0$, that is, $(N_1 = N)$, will have changed to the species 2 for $t \to \infty$. That means that σ_i is the probability for the species i to be the winner of the competition.

If a small number of individuals of the new species, $N_2(0)$, starts the game in a big population, then this species will for sure disappear if its growth rate E_2 is less than that of the old one. However, if the growth rate E_2 of the new species is greater, the new species will be successful with the survival probability

$$\sigma_2 = 1 - \left(\frac{E_1}{E_2}\right)^{N_2(0)} \tag{5.39}$$

Introducing the selection advantage $\delta = (E_2 - E_1)/E_1$, we find the more convenient form for the survival probability as a function of the relative advantage in a finite population of N members of species 1 in which $N_2(0)$ members of species 2 are imbedded

$$\sigma_2(\delta) = \frac{1-(1+\delta)^{-N_2(0)}}{1-(1+\delta)^{-N}} \tag{5.40}$$

We see in Figure 5.8 that the survival probability increases with the advantage, but the advantage is not a guaranty of the survival and in small populations the advantage of the better species does not necessarily result in winning the competition. In other words, a species having some advantage with respect to the growth rates may fail with some probability and disappear just for stochastic reasons, in spite of the fact that it is actually "better." In the deterministic case, the new species with $E_2 > E_1$ is always successful. In the stochastic situation, this is the case only with a certain chance less than 100%.

5.5
Survival of the New in Bistable Systems

Similar to the dynamical system (5.31), we consider now the case with absorber states, that is, we omit the terms E_{21}, E_{12}, and take instead nonlinear reactions of the linear

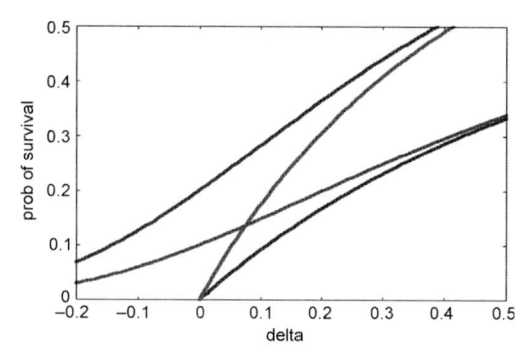

Figure 5.8 The probability that a new better species (relative advantage δ is positive) or a worse species (with negative δ) assuming a linear growth rate will succeed after infecting a big population. Here δ is the relative advantage of growth rates. The two lower curves starting at zero correspond to the case that one or two specimen were introduced in an infinite host population. The two upper curves extending to negative values of δ correspond to the case that one or two specimen of species 2 are infecting a host population consisting of ($N = 10 - 2 = 8$) specimen. Note that in the deterministic case the new better species will always succeed to grow and to win the competition, the corresponding curve would switch from zero (for negative δ) to one (for positive δ).

terms (E_1, E_2) with the rates b_1, b_2, and the volume, V. So we can write the transition probabilities as

$$W^+(N_1) = \frac{b_1}{NV} N_2 N_1^2$$

and (5.41)

$$W^-(N_1) = \frac{b_2}{NV} N_2^2 N_1$$

This case is analyzed in detail for biochemical systems in Ebeling, Sonntag, and Schimansky-Geier (1981) and Sonntag, Feistel, and Ebeling (1981), and for more general cases in Ebeling *et al.* (2006) and in Hartmann-Sonntag, Scharnhorst, and Ebeling (2009)). Again we assume the sum of the occupation numbers to be constant,

$$N_1 + N_2 = N = \text{const} \tag{5.42}$$

The distribution of the occupation numbers is $P(N_1, N_2; t)$, but because of the condition (5.42), the probability can be expressed in terms of a single variable. Similar to Section 5.3, we can formulate the master equation in the form

$$\frac{\partial}{\partial t} P(N_1; t) = W^+(N_1 - 1) P(N_1 - 1; t)$$
$$+ W^-(N_1 + 1) P(N_1 + 1; t) \tag{5.43}$$
$$- [W^+(N_1) + W^-(N_1)] P(N_1; t)$$

By multiplying the master equation with N_k/V, summing up over all occupation numbers, and after factorizing of the mean values, we get

$$\langle N_k \rangle / V = x_k$$
$$x_1 + x_2 = \frac{N}{V} = C = \text{const} \tag{5.44}$$

again the deterministic equation (5.15), which we considered as an example for hyperbolic growth

$$\frac{dx_i}{dt} = b_i x_i^2 - \varphi x_i \quad i = 1, 2, \quad \varphi \equiv \frac{b_1 x_1^2 + b_2 x_2^2}{C} \tag{5.45}$$

This way, we are back at the example of a competition in systems with quadratic growth rate (see Section 5.2). We will later show that these equations play a basic role for the model of hypercyclic evolution (Eigen and Schuster, 1978).

As a result of the quadratic terms in Equation 5.45, the phase space is split into two regions by a separatrix, S_i, as was shown in Figure 5.6. The outcome of the selection depends on the initial conditions, $x_i(t=0)$, of the system. The species i can win, (i.e., $x_i = C$ for $t \to \infty$), only if the conditions

$$x_i(0) > S_i \quad S_i = \frac{Cb_j}{b_i + b_j} \tag{5.46}$$

hold. It is possible that the species i has a large growth rate but it does not win because the starting number of individuals is too small

$$x_i(t=0) < S_i \tag{5.47}$$

This is referred to as "once-forever" behavior of the winner species.

In the stochastic case, the phase space looks as shown in Figure 5.9.

In the stochastic picture of two attractors, we have instead now two absorber states which are again $N_1 = 0$ and $N_1 = N$. Let us assume the initial state $N_1(t=0)$ or in different form:

$$P(N_1; 0) = \delta_{N_1, N_1(t=0)} \tag{5.48}$$

The final state after reaching the absorbers of probability is then

$$P(N_1; t = \infty) = \sigma_1 \delta_{N\,N_1} + \sigma_2 \delta_{0\,N_1} \tag{5.49}$$

After some calculations we get for the case $N_1(0) = 1$ (that means we have at initial time $t = 0$ one exemplear of a "better" species 1 imbedded in a population of species 2

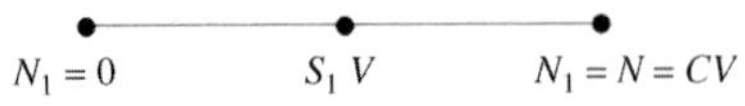

Figure 5.9 Phase space of the stochastic dynamics, Equations 5.41–543.

the following expression for the probability that the better species survives:

$$\sigma_1 = \frac{1}{\left(1 + b_2/b_1\right)^{N-1}} \tag{5.50}$$

where σ_1 is the probability that the new (better) species will win the competition process.

In this model, the separatrix S is penetrated with a well-defined probability; the "once-forever" behavior has disappeared. A "better" species can win the selection process against the established species with a certain probability even if at the beginning, $t = 0$, only a small number of the new species, or if even only one exemplar is present, that is, if $N_1(t=0) = 1$. This probability increases with decreasing size of the population (Ebeling and Sonntag, 1986; Hartmann-Sonntag, Scharnhorst, and Ebeling, 2009), as shown in Figure 5.10.

We now consider the more general case in which the growth rates contain linear as well as nonlinear terms. The transition probabilities are defined by

$$W^+(N_1) = \frac{E_1}{N} N_1 N_2 + \frac{b_1}{NV} N_2 N_1^2$$
$$W^-(N_1) = \frac{E_2}{N} N_1 N_2 + \frac{b_2}{NV} N_1 N_2^2 \tag{5.51}$$

Again we assumed that $E_{12} = E_{21} = 0$, that is, the absorbing boundaries of the positive cone, as well as constant overall number, $N_1 + N_2 = N = $ const. The master equation takes the form of Equation 5.43:

$$\frac{\partial}{\partial t} P(N_1; t) = W^+(N_1-1) P(N_1-1; t)$$
$$+ W^-(N_1 + 1) P(N_1 + 1; t) \tag{5.52}$$
$$- [W^+(N_1) + W^-(N_1)] P(N_1; t)$$

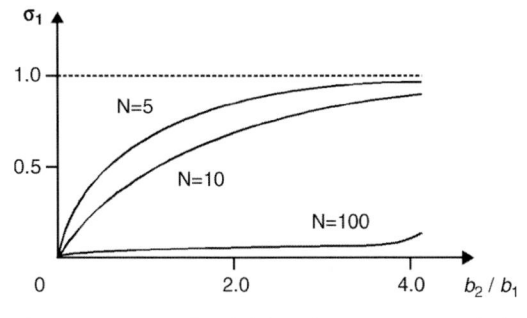

Figure 5.10 Survival probability of a new species in the bistable stochastic model given by Equations 5.41–5.43.

For this more general case, the survival probability $\sigma_{N_2(0),N}$ is found to be

$$\sigma_{N_2(0),N} = \frac{1 + \sum_{j=1}^{N_2(0)-1} \prod_{i=1}^{j} [E_1 + b_1\{(N-i)/V\}]/[E_2 + b_2(i/V)]}{1 + \sum_{j=1}^{N-1} \prod_{i=1}^{j} [E_1 + b_1\{(N-i)/V\}]/[E_2 + b_2(i/V)]}, \quad \text{where} \quad i = N_2$$

(5.53)

(Ebeling et al., 2006; Hartmann-Sonntag, Scharnhorst, and Ebeling, 2009). The value of the survival probability σ does not only depend on the selective advantage (the relation of the parameters E_i, b_i between the competing species) but also on N (the overall user number) and on the initial occupation numbers at $t = 0$.

The stochastic analysis of bistable systems shows that there are severe differences in comparison to the deterministic description. Here we first analyzed the special case of two-dimensional systems with constant total particle numbers (urn model), $N_1 + N_2 = N = \text{const}$, and with two absorber states, $N_2 = 0$ and $N_2 = N$. Here, absorber state means that once either of the states $N_2 = 0$ or $N_2 = N$ is reached, it cannot be left by the system anymore.

Generally, the survival probabilities depend on the system size, the system parameters, and the initial conditions. In the stochastic case, we obviously find a niche effect. In the niche, the sharpness of selection is diminished. In the linear case, "good" and "bad" species can coexist temporarily. In the quadratic case, the "once-forever" effect can be overcome by the niche effect. This effect is the only possibility for a "better" species to win the competition against an "old master species", if only very few exemplars of the new species exist at the beginning. Locally developed niches may play a constructive role in the evolution by overcoming the hyperselection. Large complex systems such as biological, ecological, and social systems usually form local domains (niches). The description with the help of the master equation includes processes that play a decisive role in those small domains. Within a niche, the new species is protected against extinction for some limited time. After winning the competition in the close vicinity, the local new species can infect the whole system and may be established globally.

5.6
Multistability, Information Storage, and Information Transfer

In Section 5.1, we considered the synthesis of polynucleotides by four types of nucleotides as an example of a multistable reaction. A remarkable fact is that the number of different outcomes of such reaction is astronomical. The arbitrary formation of a polynucleotide of the length v with

$$v = v_1 + v_2 + v_3 + v_4 \tag{5.54}$$

may have a total number

$$W = \frac{v!}{v_1! v_2! v_3! v_4!} \tag{5.55}$$

of different outcomes. For a chain length of $v = 100$ and equally frequent nucleotides, $v_1 = \cdots v_4 = 25$, this is an astronomical number of possibilities, $W \approx 1.6 \times 10^{57}$. The alternative ways to structure RNA or DNA chains may be the basis for the storage of a large amount of information. Rather than just $v = 100$, the longest human chromosome has a length of $v \approx 220\,000\,000$ base pairs (Gregory et al., 2006). Here is the physical basis for the storage of information in the human genome. The information aspect is discussed in Chapters 8 and 9 of this book. Here, let us consider multistable reactions in a more formal way.

We take the reaction–kinetic equations as a typical example (Chernavsky, 2001)

$$\frac{dx_i}{dt} = a_i x_i - \sum_{i,j=1}^{W} b_{ij} x_i x_j \quad i = 1, 2, \ldots, W \tag{5.56}$$

This is the structure of a so-called Lotka–Volterra system (Peschel and Mende, 1986). For simplicity, we assume that all coefficients are equal to one, except the main-diagonal elements, b, of the matrix $\mathbf{b}$

$$\frac{dx_i}{dt} = x_i - b x_i^2 - \sum_{i \neq j} x_i x_j \quad 1 = 1, 2, \ldots, W \tag{5.57}$$

In the case that $b > 1$, this systems has just one attractor and is of no interest in our context.

However, for $b < 1$, we find many attractors located at the corner of the cone (the simplex) in the points

$$x_i = 1/a; \quad x_i = 0 \quad i = 1, \ldots, W \quad j = 1, \ldots, i-1, i+1, W \tag{5.58}$$

This way we found W stable attractors, which have the character of stable nodes (see Chapter 4). Inside the cone, a saddle point is located

$$x_i = \frac{1}{1+a}; \quad i = 1, \ldots, W \tag{5.59}$$

The phase space has a quite complicated structure with W basins belonging to the attractors. Inside each of the basins, the state points converge to the stable node in this basin. Following Chernavsky (2001), we may consider this system as the prototype of a chemical reaction system, which is able to store information. More complicated systems of Lotka–Volterra type are discussed in Chapter 6. A mathematical analysis of such systems and ecological applications were discussed in the literature (Svireshev and Logofet, 1978; Peschel and Mende, 1986; Ebeling and Peschel, 1986).

We have shown that multistable systems are potentially able to develop information processing. Information storage and transfer is a key element for the evolution of complex systems. In the 1970s, Rolf Landauer, Mikhail Volkenstein, and Dmitri Chernavsky already presented some deep analyses of the problem which are the physical conditions for the information storage and transfer (Landauer, 1973, 1976; Volkenstein,1979, 1986, 1994; Volkenstein and Chernavsky, 1979; Chernavsky, 2001). These researchers clearly showed that multistability is a *conditio sine qua non* for information storage. The system should have many attractors that are alternative states and may form a memory. A prototype is the reaction studied in the previous section. For biological evolution, multistable reactions of the type we discussed in Sections 5.1, 5.2, and 5.6 are of large relevance. For technological memory systems, many other types of multistable systems are now being used, as electronic, magnetic, and optic states. A second condition mentioned by these researchers is dissipativity. If the system is in some of the alternative states, it should stay there for a sufficiently long time in order to keep a memory about the state (Landauer, 1973, 1976). Other relevant conditions are a sufficiently low level of natural fluctuations and the existence of regions in the phase space from where the system can reach the alternative attractor states. Information is connected with probability and entropy (Stratonovich, 1975; Volkenstein, 1986, 2009; Haken, 1988; Chernavsky, 2001). In the simplest case that all alternative W states are of equal probability, the information capacity or in other terms the uncertainty of the system according to Hartley is equal to

$$I = \log_2 W \qquad (5.60)$$

Up to the different constant in front and the (possibly) different basis of the logarithm, this is identical with the Boltzmann–Planck entropy of a system with equal probabilities

$$S_{BP} = k_B \ln W \qquad (5.61)$$

This way, the Boltzmann–Planck entropy may be understood as the uncertainty that is removed after getting full information on the actual state. In the case that the microstates have probabilities p_i different from $1/W$, the formula (5.61) is to be replaced by the Shannon information (Shannon entropy)

$$H = \sum_{i=1}^{W} p_i \log\left(\frac{1}{p_i}\right) \qquad (5.62)$$

This is to be understood as the quantity of information which is obtained, or in other words, the uncertainty which is removed, after exploring the actual state. One can easily prove that H is minimal (zero) for the case that only one state is occupied and maximal ($\log W$) if the states are populated with equal probability. There exists a closely related function, named the Kullback–Leibler entropy, that depends on two probability distributions

$$K(p_i, p_i^0) = \sum_{i=1}^{W} p_i \log\left(\frac{p_i}{p_i^0}\right) \qquad (5.63)$$

The Kullback–Leibler entropy has several interesting mathematical and physical properties (Kullback and Leibler, 1951; Stratonovich, 1975, 1982; Ebeling and Engel-Herbert, 1989; Uhlmann, 1977) and may be considered as a kind of distance between probability distributions. A similar entropy measure was already introduced by Gibbs (1902, Chapter XI). This entropy is a measure for the gain of information, that is, the loss of uncertainty, by changing from the probability distribution p_i^0 to the distribution p_i. The most interesting question, however, in this context is what is the value of transmitted information. This most important question about the *value of information* is so to say the weak point of information theory. As far as we can see the most convincing answers to this extremely important question was given by two great Russian scientists, the pioneer of modern stochastic theory, Ruslan Stratonovich (1930–1997), and the pioneer of modern biophysics, Mikhail Volkenstein (1912–1992). In the following text, we will try to follow their views as close as possible. However, we have to say that in spite of the fact that we had long personal discussions with both of them, we are not absolutely sure that we understood everything correctly. So let us develop our view on the concept of information value following the spirits of Stratonovich and Volkenstein.

The first important statement is: The value of a transmitted information depends on the user of it, so there is no general measure. The value can be expressed in amounts of basic values such as money, food, shelter, welfare, and biological fitness, and has the same dimension as the basic values. The basic value depends on the state which we assume here for simplicity as discrete and from some possible action u to increase welfare $C_i(u)$. In general, the system has only some small knowledge about the state, expressed in probabilities p_i^0, which Stratonovich calls the a priori probabilities. By the choice of one of the possibilities, we get in average over the *a priori* probabilities the mean welfare

$$\langle C(u) \rangle = \sum_i C_i(u) p_i^0 \tag{5.64}$$

This is the welfare that we get without having any information on the state. The problem is now how the value for welfare changes – improves – by a transition $p_i^0 \to p_i$. In the mean we have a value which is the advantage in welfare coming from the information

$$V = \langle C(u) \rangle = \sum_i C_i(u) p_i - \sum_i C_i(u) p_i^0 \tag{5.65}$$

In the case that the possible actions can be optimized independently before and after the transmission of information, according to Stratonovich (1985) we have

$$V = \langle C(u) \rangle = \max_u \sum_i C_i(u) p_i - \max_u \sum_i C_i(u) p_i^0 \tag{5.66}$$

In order to make clearer what we mean let us consider two simple examples, one connected with the value "money" and one with the value "food". In both cases, we assume that the message is "true."

1) Let us imagine a man leaving a pub after paying his beer. In his hands, he is carrying his purse with 100€ left. Walking 1 km home, after arrival he misses his purse. He decides to make a search and divides the distance into 20 intervals, corresponding to 20 streetlights along the way. To find the purse in an arbitrary sector is 1/20, so the average result for the value obtained from a search in one sector will be 5€. Now he remembers that only 400 m from his house he passed a bus stop where a woman was waiting for the next bus. So he decides to go back and ask. The woman tells him that she indeed saw a purse in his hands. The operating procedure u is simple: Pick up the purse. Assuming that the waiting woman told him the truth, we ask: What is the value of the information transmitted to our man? He is now sure that he lost the purse on the last 400 meters, that is, in one of the last eight sectors. Assuming the same probability of 1/8 for each of the sectors, he finds that the value of the information he got from the waiting women is

$$V = \langle C(u) \rangle = \frac{100}{8} - \frac{100}{20} € = 12.5\ € - 5\ € = 7.5\ €. \tag{5.67}$$

Clearly, the 5€ have to be deducted since they are not due to the information that the woman passed to him. We see that the recipe leads to a clear answer on a concrete question and the resulting value is given in the units of money.

Let us now consider a different task out of the field of biology:

2) A hungry fox smells that in a distance of less than 100 m is a hare. In which direction should it make a search? It divides the possible search directions into 12 sectors with each a priori probability 1/12. The expected mean food value of a search in just one of the sectors is a priori 1/12 hare. Suddenly, wind is coming up and gives it a signal which tells it that the hare should be in direction northwest in one of the corresponding three sectors. The operating procedure is simply: Catch the hare. The value of the information which it obtained by the wind signal is according to our recipe

$$V = \langle C(u) \rangle = \frac{1}{3} \text{hare} - \frac{1}{12} \text{hare} = 0.25\ \text{hare} \tag{5.68}$$

There is again a clear answer in units of the food which the fox needs.

Finally, let us give a more general form of the Stratonovich value of information. We assume that the observable x with the a priori probability $p^0(x)$ has a continuous spectrum of values and that the message y is a function of x which we denote as $y = f(x)$. Assuming that the message can be erroneous in the general case, y is a random quantity with the distribution $p(y,x)$. Then according to Stratonovich, the following generalization of the formula for the value of a message holds (Stratonovich, 1985)

$$V = \int dy \, \max_u \int dx \, C(x, u) p(y, x) p^0(x) - \max_u \int dx C(x, u) p^0(x)$$

A remarkable point is that this procedure of the calculation is quite general; however, the results are in each case specific for the needs of the users of the information.

6
Competition and Selection Processes

Preservation of favorable individual differences and variations, and the destruction of those which are injurious, I have called Natural Selection, or the Survival of the Fittest.

Charles Darwin: The Origin of Species

6.1
Discussion of Basic Terms

Competition and selection are fundamental concepts in the context of self-organization and evolution processes in systems which consist of many similar subsystems. Competition in such systems means rivalry between partial systems or dynamical modes that possess equal qualitative but different quantitative properties and attempt to reach a joint goal which, by the rules of the game, can be achieved by just one or a few of the participants. Mathematically, this behavior is often reflected in models by an optimization of certain parameters, functions, or functionals. The term "competition" is used in various scientific disciplines, so it is, for example, fundamental to political economy. In ecology, competition can be understood as the common exploitation of a vital resource by two or more organisms (Wilson and Bossert, 1973).

As a rule, competition is accompanied by selection, that is, a process which chooses the "best" among the competing components with respect to a certain target criterion. The term "Natural Selection" was coined by Charles Darwin in 1859 in order to mark its relation to man's power of selection, while "Survival of the Fittest" often used by Herbert Spencer was considered even more accurate and sometimes equally convenient by Darwin (1911, ch. III). The term "fittest," its mathematical definition and practical measurability is not rigorously clear by now and remains subject to intense discussions, in particular among biologists, population and molecular geneticists. "Non-Darwinian Evolution" is just one alternative concept here (Kimura, 1968; King and Jukes, 1969; Barrick *et al.*, 2009). Selection occurs among neurons of the brain (Shors *et al.*, 2001) and even in human cancer genomes (Pleasance *et al.*, 2010). A particularly interesting and complicated process is the competition between human

Physics of Self-Organization and Evolution, First Edition. Rainer Feistel and Werner Ebeling.
© 2011 Wiley-VCH Verlag GmbH & Co. KGaA. Published 2011 by Wiley-VCH Verlag GmbH & Co. KGaA.

languages (Haarmann, 2002; Janson, 2006; Hamel, 2007; Atkinson et al., 2008; Hull, 2010), see the motto of the Preface and the origin of language discussed in Chapter 8. Languages compete by social competence of their users rather than by formal properties such as syntax or grammar (Ebeling and Feistel, 1982; Janson, 2006).

In the scientific literature, we find different explanations for the term "selection." Population biologists refer to selection as *the different change of relative genotype frequencies caused by different abilities of their phenotypes to be present in the subsequent generation* (Wilson and Bossert, 1973), or *a process which controls the probability that certain individuals reach their age of reproduction* (Timofeeff-Ressovsky, Voroncov, and Jablokov, 1975). Here, the genotype of a living being is the entity of hereditary dispositions stored in the cell nucleus, and the phenotype is its external kind of appearance (Hagemann, 1976). Following different authors such as Haldane, Wright, Schmalhausen, Timofeeff-Ressovsky, and others, selection processes are classified in various ways, for example, selection by survival or by reproduction, dynamical, displacing, directed, stabilizing selection, selection within or between species, territories, and so on. These definitions reduce selection processes to occur among objects which necessarily possess genotypes, phenotypes, or reproduction, and thus exclude selection that may occur in purely physical or chemical processes. Moreover, selection is often considered as merely assessable by its results; in other words, from this perspective selection is the survival of the survivors (Timofeeff-Ressovsky et al., 1969; Eigen, 1971; Eigen and Winkler, 1975; Eigen and Schuster, 1979). This would turn the Darwin Principle into a tautology. However, following Eigen (1971), at least under specific circumstances there exists an optimal "selective value" which can be calculated in advance from the system's elementary properties and can then be used to predict the result of a selection process. In special, simple cases, fitness can even be measured in a reaction vessel (Küppers, 1975, 1983: Schuster, 2009). Consequently, Eigen and Schuster (1977) distinguish Darwinian behavior (survival by selective advantage) from non-Darwinian behavior (survival without selective advantage).

Our concept, which was explained already in Chapter 1, is that *fitness*, or *selective value*, belongs to the *emergent irreducible properties*. We consider fitness as a special kind of *value* (Feistel, 1976a, 1977b, 1991; Ebeling, 2006). Competition and selection are always based on some kind of valuation. In the simplest case, the valuation occurs on the basis of reproduction rates and may be reduced this way to an order relation between real numbers. On the other hand, as we have already shown in Chapter 5, survival may not simply be based on a comparison of reproduction rates but may also depend on initial conditions, stochastic influences, and so on. Some people expressed their view that the term fitness is a tautology, that is, that winning a competition is just the definition of fittest. We like to emphasize that our view is in strict contrast. We are convinced that fitness is an emergent property comparable to rigidity, entropy, and other concepts in physics which are always based on concrete properties. The argument that we do not have a rigorous mathematical definition or measure of fitness is not relevant in our opinion. In physics we do not have rigorous definitions of most important terms such as mass, space, time, energy, force, entropy, rigidity, and so on, and not even for physics itself, but nevertheless we successfully work with them, and using those terms we even construct functioning machines.

Physicists have accepted that emergent properties are difficult to define but nevertheless reflect physical reality. Practically, all terms we are using to describe processes in biology or society are of emergent character, not reducible to fundamental physical laws but nevertheless consistent with them.

So, in brief our view is that quantities such as fitness exists as emergent, collective properties and are the basis for biological valuation expressed in the result of competition and selection processes. Sewall Wright developed the idea of a fitness landscape (value landscape) which was subsequently worked out by many authors, in particular by Conrad (1983). We shall consider fitness landscapes in more detail in Chapter 7. In the past decade many new results on the structure of landscapes were obtained by the group of Peter Schuster, Peter Stadler, and collaborators (Schuster, 2009). The concepts of values and fitness landscapes are rather abstract and qualitative. Our point of view is that these values are abstract collective properties of systems (species) in a certain dynamical context. Values express the essence of biological, ecological, economic or social properties, and relations with respect to the dynamics of the system (Feistel, 1986, 1990, 1991; Ebeling, 2006). From the point of view of modeling and simulations, values are emergent properties. At the same time values appear as collective properties, they belong to certain ensemble (Harmer et al., 2010). The valuation process is an essential element of the self-organization cycles of evolution. We shall return to the evolution in phenotypic landscapes in the next chapter; here, we focus on the "elementary act" of evolution, the dynamics of selection processes among the members of a given set of similar but unequal items.

The application of "evolution strategies" to the solution of practical problems or to computer experiments (Bremermann, 1970; Rechenberg, 1973; Schwefel, 1977; Voigt et al., 1996) led to a mathematical definition of selection (Papentin, 1973). In this sense, selection means that only the subset of privileged species with a maximum value of some "fitness functional" Φ is permitted to reproduce; Φ is computed from a well-defined algorithm that evaluates each species of the given population. In other words, selection is the subsequent substitution of certain species by ones with a higher fitness, in the competition for a given finite number of survivor positions. The basic mechanism of selection is the competition between genetically different members of a population due to their interaction with some environment which consists in particular of members of the other species (Kamshilov, 1977). Thus, loosely spoken, selection is survival of some and extinction of others. Note that associating selection with a nondecreasing fitness functional is a significant a-priory restriction that is not justified in general. We shall return repeatedly to the existence problem of such extremum principles in this and the subsequent chapters.

When we ask for the possibility of selection in general dynamic systems which may be given by the set of differential equations

$$\dot{x}_i = f_i(x_1, \ldots, x_n)$$

a definition seems obvious which simply requires that some of the x_i disappear for $t \to \infty$ (Ebeling and Feistel, 1976a). But, this trivial approach is too unspecific; decaying radioactive isotopes, for instance, would then be an example for a selection process. It is more reasonable to include here the aspect of competition as an essential element of selection (Ebeling, 1976a). We regard selection as a special form of cooperative behavior of subsystems (Feistel and Ebeling, 1976). The restriction of selection to processes with competition properties appears inappropriate only at the first look. Can we speak of selection if fertile individuals are dying or parts of a population disappear under changing environmental conditions (temperature, humidity, light, and so on)? From the perspective of population dynamics this is selection even though there is no competition involved. From the perspective of system theory, the disappearance in the example mentioned above is similar to the case of radioactive decay; in both situations certain objects vanish irrespective of the presence of other objects. Therefore, we consider here as targets of selection only such objects which are capable of survival under given boundary (environmental) conditions and which are threatened only by the competition with similar present objects of the same quality. A simple choice between suitable and unsuitable candidates will not be regarded as selection in a stricter sense here.

After these introductory considerations we attempt the following *definition* of selection: selection is a special form of coherent behavior of subsystems (species) in complex systems. It occurs if a competition process between self-sustained subsystems (species), which are principally able to survive, leads to the extinction of at least one or of a class of subsystems (species).

A more rigorous selection *criterion* can be derived from the behavior of the system with respect to the addition of new subsystems (species). We can speak of selection if the addition of new subsystems (species) to a set of established subsystems (permanent population) causes the extinction of at least one subsystem or a class of the established subsystems (permanent population) for $t \to \infty$.

Upon their consideration of prebiotic evolution processes, Romanovsky, Stepanova, and Chernavsky (1975) discuss two other forms of selection, besides the "simple" selection (survival of advantageous variants):

1) An object appears earlier than others so that the latter ones are unable to interfere (following Kastner).
2) The choice between equally capable candidates (in the sense of bistability, discussed in Chapter 5).

We classify these special forms of selection here under the aspect that their result depends on the initial conditions (initial concentrations) and that nonlinear growth laws play an important role (hyperbolic growth). With reference to Decker (1975) we regard these forms of selection as *hyperselection*. In contrast, the result of the selection processes considered previously does not depend on the particular quantitative initial conditions, rather on certain sets of properties that characterize a given species (its abstract phenotype). This kind of selection will be denoted as "plain selection" and will play a predominant role in the following because it permits an unlimited sequence of "better and better" species. The plainness of those systems guarantees by no means the existence of a selective value, the approach of an optimum (Ward, 2009), the action of a generalized extremum principle, as we shall see later on.

Plain selection and hyperselection will prove to be fundamental principles of self-organization in the following. They played a key role, too, in the evolution of life on planet Earth.

6.2
Extremum Principles

In the theory of selection and evolution processes, extremum principles play a central role. Mathematically, the term "extremum principle" applies to a variety of quantities, processes, or conditions even if we consider only isothermal, isobaric, and spatially homogeneous systems (see Section 2.3). Here, we are going to discuss a rather special kind of extremum principles. For the discussion of the following sections it is helpful to classify the most important quantities that can approach an extremum during a selection or evolution process. For simplicity we focus on systems which possess isolated stationary states; a generalization to quasistationary states such as limit cycles or strange attractors is straightforward if we apply the terminology to stationary time averages over the attractor. The dynamics in occupation number spaces which we are going to consider now are characterized by nonnegative state vectors, $x(t) = \{x_i(t)\}$, with $x_i(t) \geq 0$, since concentrations or species numbers cannot take negative values. Consequently, the system's dynamics must be governed by equations which keep the state vector $\mathbf{x}$ inside the so-called positive cone, Figure 6.1.

Let R be a simply connected region in the positive cone of the n-dimensional vector space E^n. Let a vector function $\mathbf{f}$ be defined on R which describes the dynamics of the state vector and possesses the properties described above:

$$\dot{\mathbf{x}} = \mathbf{f}(\mathbf{x}, \mathbf{u}) \tag{6.1}$$

Here, $\mathbf{u}$ is a set of parameters. Some singular points $\mathbf{x}^{(i)}$, $i = 1, 2, \ldots, m$ of which exactly one, $\mathbf{x}^{(s)}$, is asymptotically stable in the sense of Lyapunov:

$$\mathbf{f}\left(\mathbf{x}^{(i)}, \mathbf{u}\right) = 0 \quad \text{for } i = 1, 2, \ldots, m \tag{6.2}$$

To classify different extremum principles relevant for selection processes we suggest the following convenient terminology:

Let $G(\mathbf{x}, \mathbf{x}^0, \mathbf{u})$ be a function that is continuously differentiable with respect to $\mathbf{x}$, defined in such a way that

$$L\left(\mathbf{x}, \mathbf{x}^{(s)}, \mathbf{u}\right) \equiv G\left(\mathbf{x}, \mathbf{x}^{(s)}, \mathbf{u}\right) - G\left(\mathbf{x}^{(s)}, \mathbf{x}^{(s)}, \mathbf{u}\right) < 0 \, \forall \mathbf{x} \neq \mathbf{x}^{(s)} \tag{6.3}$$

and

$$\frac{d}{dt} L\left(\mathbf{x}, \mathbf{x}^{(s)}, \mathbf{u}\right) > 0 \, \forall \mathbf{x} \neq \mathbf{x}^{(1)}, \mathbf{x}^{(2)}, \ldots, \mathbf{x}^{(m)} \tag{6.4}$$

holds, then G is regarded as a *dynamical extremum principle* with respect to the dynamical system (6.1). The function L is known as a Lyapunov function.

If a function $F(\mathbf{x}^{(i)}, \mathbf{u})$, $i = 1, 2, \ldots, m$, is defined on R for each of the singular points $\mathbf{x}^{(i)}$ in such a way that

$$F\left(\mathbf{x}^{(s)}, \mathbf{u}\right) > F\left(\mathbf{x}^{(i)}, \mathbf{u}\right) \, \forall i \neq s \tag{6.5}$$

holds, then F is regarded as a *static extremum principle* with respect to the dynamical system (6.1).

If there exists a dynamical extremum principle G such that

$$F\left(\mathbf{x}^{(i)}, \mathbf{u}\right) = G\left(\mathbf{x}^{(i)}, \dot{\mathbf{x}}^{(i)}, \mathbf{u}\right) \tag{6.6}$$

holds, then G is regarded as a *complete extremum principle* with respect to the dynamical system (6.1).

It is of crucial importance for the investigation of selection processes whether or not the concentration of a particular species is either zero or greater than zero, that is, whether the species exists in the system or not. Mathematically, this question is whether the state vector $\mathbf{x}$ is located inside the positive cone or on its boundary, Figure 6.1.

From the theory of sets it is well known that a finite set with m elements possesses 2^m different subsets (Görke, 1970). Correspondingly, the m-dimensional positive cone $R_m^>$ possesses 2^m *positive subspaces* $R_p^>$, $\dim(R_p^>) = p$, in which certain components $x_1, x_2, \ldots, x_p > 0$ are positive and the remaining ones are considered as nonexisting. The consideration of this kind of subspaces is relevant and of significant interest whenever the dynamical system (6.1) has the property

$$f_i(\mathbf{x}, \mathbf{u})|_{x_{p+1}=\ldots=x_m=0} = 0 \quad \text{for } i = p+1, \ldots, m \tag{6.7}$$

In this case, $\dot{\mathbf{x}} = \mathbf{f}(\mathbf{x}, \mathbf{u})$ forms a closed dynamical system within the subspace $R_p^>$, depending on the set $\mathbf{u}$ of external control parameters.

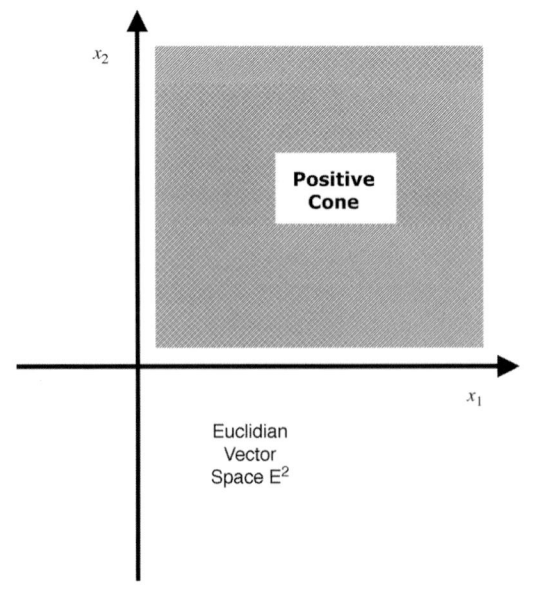

Figure 6.1 The spaces of chemical concentrations or species numbers correspond to the positive cone of the Euclidian vector space.

Similar reasoning can be applied to the *nonnegative subspaces* $R_p^{\geq}$, $\dim(R_p^{\geq}) = p$, in which certain components $x_1, x_2, \ldots, x_p \geq 0$ are nonnegative and the remaining ones are considered as nonexisting.

On this basis, an extremum principle is regarded as

a *segregation principle*, if it is defined on each relevant positive subspace $R_p^{>}$, (6.8)

a *selection principle*, if it is defined on each relevant positive subspace $R_p^{>}$ and its conjunction with the nonnegative subspaces $R_p^{\geq}$, corresponding to the extinct, no-longer existing species, and (6.9)

an *evolution principle*, if it is defined on any relevant nonnegative subspaces $R_p^{\geq}$ corresponding to the no-longer existing as well as the not-yet-existing species. (6.10)

The definitions (6.3)–(6.6) and (6.8)–(6.10) can freely be combined with each other to distinguish nine different principles, from the dynamical segregation principle to the complete evolution principle. Some of them imply others, see Figure 6.2.

The existence of one of those displayed in Figure 6.2 ensures the existence of each of its successors in the directed graph; vice versa, the proof of the nonexistence of a certain principle implies the nonexistence of each of its predecessor principles. Once a complete evolution principle has been discovered, the qualitative properties of any selection and evolution processes possible in the related model can be concluded from the principle.

In some cases, thermodynamics can provide a complete evolution principle. The force-related change $d_X P$ of the entropy production in homogeneous systems is a

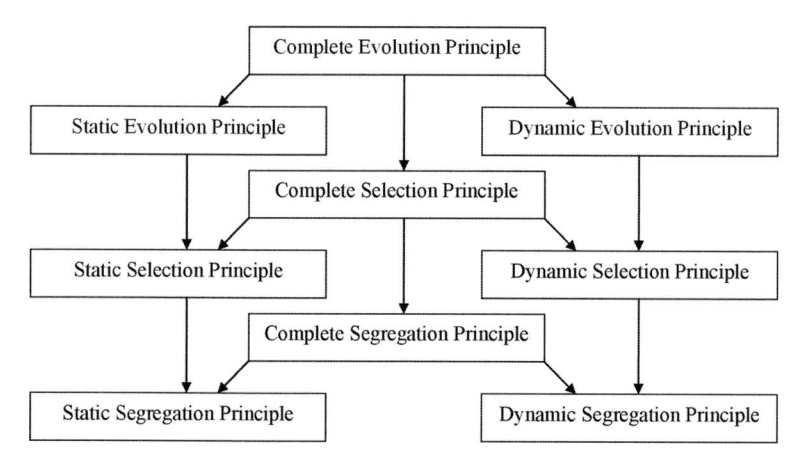

Figure 6.2 Hierarchy of extremum principles, definitions (6.3)–(6.6) and (6.8)–(6.10), that can be used for the classification of selection processes.

Pfaffian form (Margenau and Murphy, 1943; Smirnow, 1968) with respect to its variables (order parameters),

$$d_x P = \sum_i A_i(\mathbf{x}) dx_i \tag{6.11}$$

If this Pfaffian form represents a total differential

$$A_i = -\frac{\partial G}{\partial x_i} \tag{6.12}$$

then $G(\mathbf{x})$ is at least a dynamical but often even a complete evolution principle due to the Glansdorff-Prigogine theorem, see Chapters 2 and 4.

6.3
Dynamical Models with Simple Competition

As a simple standard model for competition and selection we consider in this section competing parallel chemical reactions. The simplest conditions for competition in chemical reactions are given by the following boundary conditions (Eigen, 1971; Schuster, 1972; Eigen and Winkler, 1975; Küppers, 1975; Ebeling and Feistel, 1974, 1976a; Ebeling, 1976a):

(a) limited input of jointly consumed raw materials, or
(b) constrained total particle number of the competing species.

As a model for the case (a) we consider the parallel autocatalytic production of species X_i from a raw material A, and their decay to a final product F (Ebeling, 1976a; Feistel, 1976b; Ebeling and Feistel, 1977):

$$\begin{aligned} A + X_1 &\xrightarrow{k_1} 2X_1; \quad X_1 \xrightarrow{k'_1} F \\ A + X_2 &\xrightarrow{k_2} 2X_2; \quad X_2 \xrightarrow{k'_2} F \\ &\cdots \\ A + X_n &\xrightarrow{k_n} 2X_n; \quad X_n \xrightarrow{k'_n} F \end{aligned} \tag{6.13}$$

An equivalent ecological model exhibits reproduction of the species X_i (with the lifetime $1/k'_i$) that is feeding on A, and its natural death. The related formal kinetic equations read

$$\begin{aligned} \dot{A} &= \Phi - \sum_i k_i A X_i \\ \dot{X}_i &= k_i A X_i - k'_i X_i \end{aligned} \tag{6.14}$$

Here, $\Phi > 0$ is a constant external raw material supply. Analytical solutions of the system (6.14) are not known, but one can integrate the temporal behavior numerically. Figure 6.3 shows an example with $\Phi = 0.5$ and the other parameters as given in Table 6.1.

In the long-term limit, the system (6.14) converges asymptotically to a stable stationary solution. The properties of the stationary solutions of (6.14) can be

Table 6.1 Selected parameter values for the system (6.14) used for the numerical integration displayed in Figure 6.3.

Species	Initial value	k	k'
X_1	0.01	1	0.5
X_2	0.01	1.5	1
X_3	0.01	2.0	1.5
A	0		

investigated analytically, in particular their dynamical stability in the sense of Lyapunov (Ebeling and Feistel, 1974, 1976; Ebeling, 1976a). Denoting the replication rate by

$$E_i = k_i A - k'_i \qquad (6.15)$$

we can write Equation 6.14 in the form

$$\dot{A} = \Phi - \sum_i k_i A X_i k, \quad \dot{X}_i = E_i X_i \qquad (6.16)$$

For the stationary solutions $\dot{A} = \dot{X}_1 = \ldots = \dot{X}_n = 0$ must hold. The condition $\dot{X}_i = 0$ implies either $X_i = 0$ or $E_i = 0$ for each species i. Moreover, $\dot{A} = 0$ requires at least one $X_i > 0$, for which the replication rate E_i must disappear. In general, by a suitable adjustment of A, Equation 6.15 can vanish for only one species, say, $i = s$:

$$\begin{aligned} E_s &= 0, \; X_s > 0 \\ E_i &\neq 0, \; X_i > 0 \; \forall i \neq s \end{aligned} \qquad (6.17)$$

From $\dot{A} = 0$ we infer

$$A^{(s)} = k'_s/k_s \text{ and } X_s^{(s)} = \Phi/k'_s \qquad (6.18)$$

In order to analyze the steady state (s) with the properties (6.17) we compute the eigenvalues of the Jacobian of the system (6.16) at this state:

$$p_{0,s}^{(s)} = -\frac{\Phi}{2} \cdot \frac{k_s}{k'_s} \left\{ 1 \pm \sqrt{1 - \frac{4k'_s}{\Phi} \cdot \frac{k'_s}{k_s}} \right\}$$

and

$$p_i^{(s)} = k_i \left(\frac{k'_s}{k_s} - \frac{k'_i}{k_i} \right) \quad \forall i > 0 \quad i \neq s$$

(6.19)

The stability condition reads

$$p_i^{(s)} = k_i \left(\frac{k'_s}{k_s} - \frac{k'_i}{k_i} \right) < 0 \quad \forall i > 0 \quad i \neq s \qquad (6.20)$$

that is,

$$\frac{k'_s}{k_s} = A^{(s)} < \frac{k'_i}{k_i} = A^{(i)}$$

Thus, the survivor is the species that requires the lowest raw material concentration $A^{(s)}$ of all possible stationary solutions. In other words, the winner of the competition is characterized by a maximum product of the lifetime, $1/k'_s$, and the number of "offsprings" per unit of time and raw material, k_s, that is, it is the species of which each particle produces a maximum number of "offsprings" in the course of its "life."

Such a number that can be computed from the phenotypic properties of the species (here, k_i, k'_i of the species i), and that decides on the result of the selection process between those species, will be regarded as the *selective value* of the species, as suggested by Eigen. As we will see later on, the existence of such a number is by no means granted for selection processes in general.

In the sense of the classification of extremum principles given in Section 6.2, the function

$$F\left(\mathbf{X}^{(i)}, \mathbf{u}\right) = 1/A^{(i)} = \sum_j \frac{k_j}{k'_j} \delta_{ij} \tag{6.21}$$

represents a static evolution principle since this quantity is defined for all stationary states (including those on the boundary of the positive cone) and takes its maximum value at the asymptotically stable steady state. Since the raw material concentration does not change monotonically with time, see Figure 6.3, it is not a dynamical

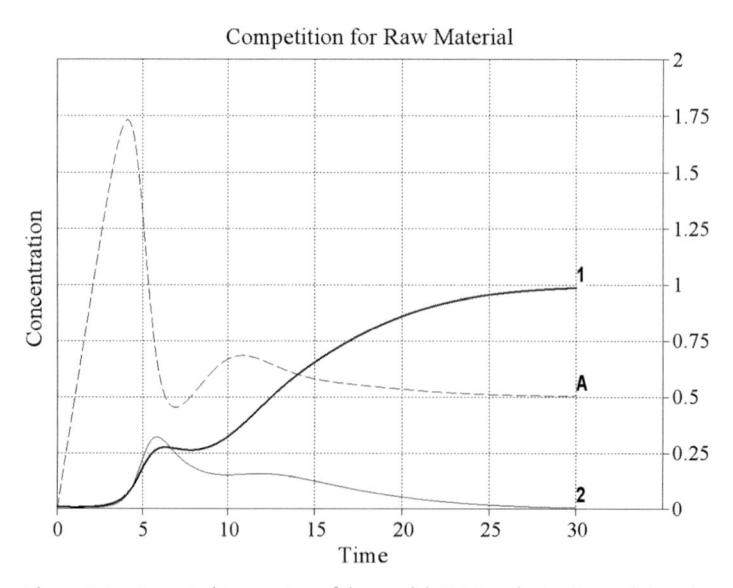

Figure 6.3 Numerical integration of the model (6.14) with $\Phi = 0.5$ and the other parameters as given in Table 6.1. When the raw material A is highly abundant, several species can grow. Asymptotically, only the species X_1 survives the competition.

extremum principle. It should be mentioned, however, that on certain hypersurfaces of the phase space the function $G=1/A$ acts as a dynamical and therefore as a complete evolution principle (Feistel, 1979).

We consider now the system (6.16) with n original species in the steady state (6.17) when a new species μ is added to it with a small initial concentration $X_\mu > 0$. As long as this minor species does not essentially affect the current raw material level $A^{(s)}$, the growth follows the growth law (6.16),

$$\dot{X}_\mu = \left(k_\mu A^{(s)} - k'_\mu\right) X_\mu = k_\mu \left(\frac{k'_s}{k_s} - \frac{k'_\mu}{k_\mu}\right) X_\mu$$

which can be integrated to the approximate solution for short times

$$X_\mu(t) = X_\mu(0) \exp\left\{\left(\frac{k'_s}{k_s} - \frac{k'_\mu}{k_\mu}\right) k_\mu t\right\} = X_\mu(0) \exp\left\{\left(A^{(s)} - A^{(\mu)}\right) k_\mu t\right\}$$

If $(k'_s/k_s) < (k'_\mu/k_\mu)$, then the added species μ is rejected by the system. Provided that $A^{(\mu)} < A^{(s)}$, in agreement with the extremum principle formulated above, the system is unstable with respect to the appearance of the injected species. The selection process will turn the system eventually into a new stable stationary state that has the properties

$$A^{(\mu)} = k'_\mu/k_\mu \quad X^{(\mu)}_\mu = \Phi/k'_\mu \quad X^{(\mu)}_i = 0 \quad \forall i \neq \mu$$

Under the condition $(k'_\mu/k_\mu) < (k'_s/k_s)$, that is,

$$E_\mu = k_\mu A^{(s)} - k'_\mu = k_\mu \left(A^{(s)} - A^{(\mu)}\right) > 0 \tag{6.22}$$

the injected species μ will substitute the previously dominating species s if the initial replication rate E_μ is positive.

From the eigenvalues $p_{0,s}$, Equation 6.19, we can infer that the steady state is approached via an oscillating regime if the raw material supply is subcritically slow, $\Phi < 4(k'_s)^2/k_s$.

Another interesting property of our model is the fact that the result of the competition is independent of the boundary conditions (here, the particular constant value of Φ). No matter whether the raw material supply is plenty or deficient, the same species s is optimal, and the same stationary raw material concentration $A^{(s)}$ is established. As a remarkable consequence, if the "food" supply $\Phi(t)$ is varying slowly on a "seasonal" or "climatological" timescale with a frequency ω, the asymptotic state (6.17) is quantitatively modified only by terms of $O(\omega^2)$, that is, the selection process is almost unaffected with respect to a weak variability of the "environmental" conditions (Feistel, 1976b).

As mentioned in the beginning of this section, there is a second simple choice of possible boundary conditions that can establish competition. We control the flux $\Phi(t)$ now in such a way that the raw material concentration A remains constant, and in addition we dilute the system at a rate $k'(t)$ to keep the total population fixed, that is, maintain a "constant overall organization" (Fisher, 1930; Eigen, 1971):

$$\dot{X}_i = E_i X_i - k'(t) X_i \quad \sum_i X_i = C = \text{const} \tag{6.23}$$

Geometrically this condition means that all trajectories of (6.23) are located on a plane that intersects the axes of the positive cone at $X_i = C$. This state space, a symmetric polyhedron, is mathematically a simplex (Eigen and Schuster, 1977).

To determine an expression for the dilution rate $k'(t)$, we sum up Equations 6.23 over all species

$$\sum_i \dot{X}_i = 0 = \sum_i E_i X_i - k'(t) \sum_i X_i$$

Solving this equation for k', we obtain

$$k'(t) = \sum_i E_i X_i / \sum_i X_i = \langle E \rangle$$

Inserting this result into the dynamical system (6.23), we obtain the Fisher–Eigene equation

$$\dot{X}_i = (E_i - \langle E \rangle) X_i \tag{6.24}$$

Here, $\langle E \rangle = \sum_i E_i X_i / C$ is the mean excess production rate. Note that Equation 6.24 remains unchanged if the same arbitrary constant is added to each E_i. We may therefore require that

$$E_i > 0 \tag{6.25}$$

holds for all species, without loss of generality.

Equations of the form (6.24) appear in various contexts of physical, technical, biological, and social systems (Feistel and Ebeling, 1976). They can be interpreted alternatively as a model for a closed system (with energy exchange but without particle exchange through the boundaries) in which only transformation processes (e.g., chemical reactions),

$$X_i \leftrightarrow X_j \tag{6.26}$$

occur at suitable rates. With this interpretation, a thermodynamic evolution principle of the type (6.11) can be derived that governs the dynamics of the system (6.24). Since we are only interested in particle numbers of the system with instantaneous chemical reaction "collisions," rather than in any details of their interaction forces, we may start from the entropy of mixing of ideal gases/solutions (Landau and Lifschitz, 1966):

$$S = -kV \sum_i X_i \left(\ln \frac{X_i}{X_i^0} - 1 \right) \tag{6.27}$$

The entropy production density is, since no particle exchange is permitted,

$$\sigma = \frac{P}{V} = \frac{1}{V}\frac{dS}{dt} = -k\sum_i \dot{X}_i \ln\frac{X_i}{X_i^0} = -k\sum_i (E_i - \langle E \rangle) X_i \ln\frac{X_i}{X_i^0} \qquad (6.28)$$

Here, $\dot{X}_i$ is the thermodynamic flux and $-k\ln(X_i/X_i^0)$ is the thermodynamic force (the affinity); hence the force-related entropy change is

$$d_X \sigma = -k \sum_{i=1}^{n} (E_i - \langle E \rangle) dX_i \qquad (6.29)$$

To verify that this Pfaffian form is a total differential we consider that the motion is possible only on the $(n-1)$-dimensional simplex $\sum_i X_i = C$ of the phase space, where only $(n-1)$ concentrations can vary independently. By means of

$$dC = 0 = \sum_{i=1}^{n} dX_i \quad dX_n = -\sum_{i=1}^{n-1} dX_i \qquad (6.30)$$

we can eliminate dX_n from (6.29) and obtain

$$d_X \sigma = -k \left\{ \sum_{i=1}^{n-1} (E_i - \langle E \rangle) dX_i - (E_n - \langle E \rangle) \sum_{i=1}^{n-1} dX_i \right\} \qquad (6.31)$$

$$d_X \sigma = -k \sum_{i=1}^{n-1} (E_i - E_n) dX_i$$

Because the integrability condition of the Pfaffian form, Equation 6.31,

$$\frac{\partial}{\partial X_i}(E_j - E_n) = 0 = \frac{\partial}{\partial X_j}(E_i - E_n) \qquad (6.32)$$

is obeyed, a total differential $dG = -d_X\sigma$ is easily found by integration of Equation 6.31, in the form

$$G = kC\langle E \rangle + \text{const} \qquad (6.33)$$

From the Glansdorff–Prigogine principle, $d_X P \leq 0$, we conclude that G represents a dynamical extremum principle. To see this, we study the properties of the system (6.24) in greater detail. Integrating formally Equation 6.24 in the form

$$\int_0^t \frac{dX_i}{X_i} = \int_0^t (E_i - \langle E \rangle) dt \qquad (6.34)$$

and deriving from $\sum_i X_i(t) = C$ an expression for the normalization factor, $\exp\left\{-\int_0^t \langle E \rangle dt\right\}$, which appears in the integral (6.34), we finally obtain the analytical

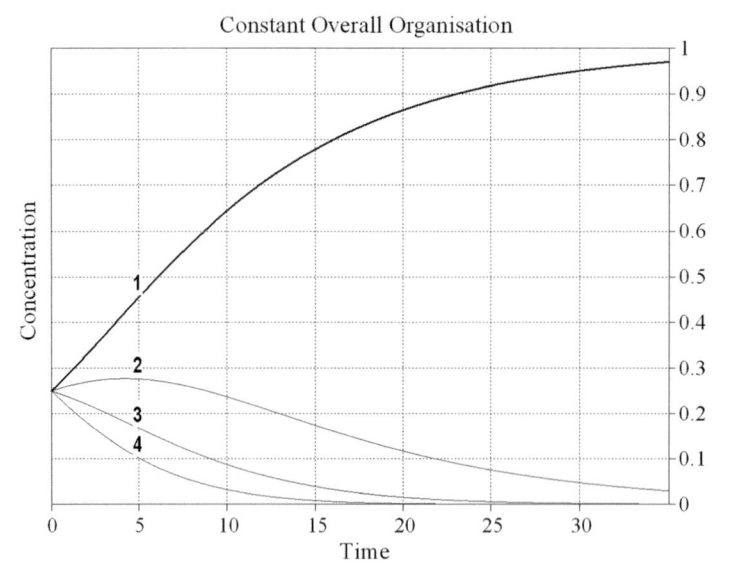

Figure 6.4 Example for the solution (6.35) of selection under the condition of constant total particle number, using Runge–Kutta numerical integration (Sprott, 1991). The replication rates are $E_1=0.4$, $E_2=0.3$, $E_3=0.2$, and $E_4=0.1$, the initial concentrations are $X_i(0)=0.25$.

solution (Eigen, 1971)

$$X_i(t) = C \frac{X_i(0) \exp\{E_i t\}}{\sum_j X_j(0) \exp\{E_j t\}} \qquad (6.35)$$

An example of Equation 6.35 is shown in Figure 6.4.

The n different stationary states $s = 1, 2, \ldots, n$ of the model (6.24) are characterized by the existence of a single species X_s in the system

$$X_i^{(s)} = 0 \quad \forall i \neq s \quad X_s^{(s)} = C \qquad (6.36)$$

Exactly one of those states is asymptotically stable which belongs to the maximum replication rate

$$E_s > E_i \quad \forall i \neq s \qquad (6.37)$$

Consequently, the system (6.24) under consideration possesses a single asymptotically stable steady state with the properties (6.36) and (6.37) which is located at one of the vertices of the simplex. The phase trajectory approaches this stable state in the limit $t \to \infty$.

We verify now that the extremum principle (6.33) is actually a complete one. With respect to Equation 6.3 it is required that

$$L(\mathbf{x}, \mathbf{u}) = \langle E \rangle - E_s < 0 \quad \forall \mathbf{x} \neq \mathbf{x}^{(s)} \qquad (6.38)$$

This is obeyed because of the inequality (6.37), since the mean value cannot exceed the maximum of a given set of numbers.

Next we check whether the condition (6.4) is fulfilled, too:

$$\frac{d}{dt}L = \frac{d}{dt}\langle E \rangle = \frac{1}{C}\sum_i E_i \dot{X}_i = \frac{1}{C}\sum_i (E_i^2 - E_i\langle E \rangle)X_i = \langle E^2 \rangle - \langle E \rangle^2 \geq 0 \quad (6.39)$$

Finally, as follows from the condition (6.37), the numbers

$$F\left(\mathbf{x}^{(i)}, \mathbf{u}\right) = E_i \quad (6.40)$$

constitute a static extremum principle. Since the functions F and G are well defined even on the entire boundary of the positive cone, we conclude that the quantity

$$G(\mathbf{x}, \mathbf{u}) = \langle E \rangle \quad (6.41)$$

as well as the similar quantity defined in Equation 6.33 is a complete evolution principle.

Another interesting property of the Fisher–Eigen model (6.23) is the existence of additional conservation laws for three or more competing species in the form of a mass-action law (Feistel, 1983a):

$$(E_i - E_j)\ln X_k + (E_k - E_i)\ln X_j + (E_j - E_k)\ln X_i = \text{const} \quad (6.42)$$

At the end of this section we like to take a first brief look at an important generalization of the competition situation studied before. Let us assume that the reproduction of the species i is catalyzed by certain other present molecules of the kind j. The formal-kinetic reactions (6.23) then read

$$\dot{X}_i = \left(E_i + \sum_j b_{ij} X_j - k'(t)\right) X_i \quad (6.43)$$

We will consider this model with nonlinear growth again in more detail in different contexts (Chapters 7–9, see also Volterra, 1931; Feistel and Mahnke, 1977; Svireshev and Logofet, 1978; Feistel, 1979; Küppers, 1979; Ebeling et al., 2006). The matrix $\mathbf{B} = \{b_{ij}\}$ is sometimes regarded as the *community matrix*. In certain pre-biological molecular evolution models we are interested in the case $b_{ii} = 0$ (excluding autocatalysis) for all i, and $b_{ij} > 0$ for at least some $i \neq j$. From the condition $\sum_i X_i(t) = C$ we derive the "dilution" function $k'(t)$ in the form

$$k'(t) = \frac{1}{C}\sum_j \left(E_i + \sum_j b_{ij} X_j\right) X_i \quad (6.44)$$

As a result of this constraint, no concentration X_i can grow infinitely in the course of time, but some species may become extinct as a result of the competition. We consider the asymptotic behavior in a positive subspace $R_p^>$ in which the species $X_i > 0$, $i = 1, \ldots, p$, are assumed to survive and the remaining species $X_i = 0$, $i = p + 1, \ldots, n$, to be absent. This state can always be established formally by an appropriate renumbering of the species. The necessary condition, Equation 6.7, for the existence of 2^n different asymptotic states with $p \leq n$ is evidently satisfied by the structure of Equation 6.43.

For the subsystem considered in $R_p^>$, we divide the related Equation 6.43 by the concentration, $X_i > 0$, and take the time average (not to be confused with the population average $\langle \cdots \rangle$ used previously):

$$\left\langle \frac{d}{dt} \ln X_i \right\rangle = E_i + \sum_{j=1}^{p} b_{ij} \langle X_j \rangle - \langle k'(t) \rangle \quad i = 1, \ldots, p \tag{6.45}$$

In the asymptotic limit, $t \to \infty$, the average on the left-hand side vanishes, and the remaining system of p linear equations can be solved for $\langle X_i \rangle$, $i = 1, \ldots, p$, as functions of the constant $\langle k' \rangle$. From linear algebra we know that this solution is unique if the nonnegative quadratic matrix $\mathbf{B}^{(p)} = \{b_{ij}\}$ is regular in $R_p^>$, that is, if $\mathbf{B}^{(p)}$ does not possess a zero eigenvalue:

$$\langle X_i \rangle = -\sum_{j=1}^{p} \left\{\mathbf{B}^{(p)}\right\}_{ij}^{-1} E_j + \langle k'(t) \rangle \sum_{j=1}^{p} \left\{\mathbf{B}^{(p)}\right\}_{ij}^{-1} \tag{6.46}$$

Mathematical properties of nonnegative matrices are subject to the comprehensive theorems of Perron and Frobenius (Gantmacher, 1971; Lancaster, 1969; Feistel and Sändig, 1977; Feistel, 1979; Ebeling and Feistel, 1982, see Section 6.5).

Inserting the solution of Equation 6.45 into $\sum_i \langle X_i \rangle = C$ provides a linear equation for $\langle k'(t) \rangle$. The resulting values for $\langle X_i \rangle$ are required to be positive, otherwise the related state is physically not reasonable and cannot be approached from within the positive cone since no phase trajectories intersect the boundary of the positive cone. If at least one $\langle X_i \rangle$ turns out to be zero or negative, a different positive subspace $R_p^>$ must be considered to host a physically permitted asymptotic state. Note that from stationary asymptotic mean values $\langle X_i(t) \rangle$ one cannot conclude that the attractor state $X_i(t)$ itself is necessarily stationary, too.

For the remaining absent species with $X_i = 0$, $i = p+1, \ldots, n$, the resulting equations read for small concentrations X_i

$$\dot{X}_i = \left(E_i + \sum_{j=1}^{p} b_{ij} \langle X_j \rangle - \langle k' \rangle \right) X_i \tag{6.47}$$

The state $\langle X_j(t) \rangle$, $j = 1, \ldots, p$, considered in $R_p^>$ is asymptotically stable with respect to the injection of an absent species $i > p$ if

$$E_i + \sum_{j=1}^{p} b_{ij} \langle X_j \rangle < \langle k' \rangle \quad i = p+1, \ldots, n \tag{6.48}$$

In dependence on the coupling matrix $\{b_{ij}\}$, the model (6.43) can exhibit an extremely complicated selection behavior. While the system (6.24), that is, the special case $b_{ij} = 0$, has just a single stable steady state located on a vertex of the simplex

$\sum_i X_i(t) = C$, Equation 6.43 in general possesses several stable steady states. The simplex is divided into several attraction basins by corresponding separatrices. The theory of selective systems such as (6.43) was developed in the context of the Eigen–Schuster theory of hypercycles (Eigen and Schuster, 1977, 1979).

6.4
Stochastic of Simple Competition Processes

As a rule, real natural selection processes occur between established master species and an emerging new, possibly advantageous mutant or immigrant. The new species appears first as a single copy, or at most as a few individuals. At such a low population number the usual description by continuous concentrations is inappropriate; the decision whether a single particle will disappear or remain in the system is essentially a stochastic process. The related randomness of the dynamics is entirely a consequence of the discreteness of the integer particle numbers, that is, the fact that the competitors are individuals which can die or multiply in the form of instantaneous stochastic events (shot noise).

Master equations, as they were called later, were used by Kac (1947) in theoretical physics for the description of irreversible kinetic processes (Uhlenbeck, 1959), see also Chapters 4 and 5. They are appropriate mathematical tools to model chemical reactions with small particle numbers, in particular selection and evolution processes such as those considered in the previous section (Feistel, 1976b, 1977a, 1979; Ebeling and Feistel, 1977; Ebeling, Feistel, and Jiménez-Montaño, 1977; Feistel and Mahnke, 1977).

We begin with the stochastic model of the process (6.13) of limited raw material supply (Feistel, 1976b). We use the particle number N_0 for the raw material A in the given volume V, and $N_1, \ldots, N_n$ for the species $X_1, \ldots, X_n$. For the three reaction channels, import of A, reaction $A \to X_i$, and decay of X_i (Figure 6.5), we assume the following transition probabilities W (see also Chapter 4):

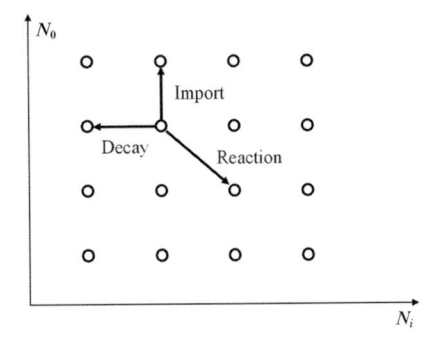

Figure 6.5 Discrete states and stochastic transitions in the model of constant raw material supply, Equation 6.49.

Import, $R \to A$ $N_0 \to N_0 + 1$, $W_1(N_0, \ldots, N_n) = \Phi V$

Reaction, $A \to X_i$ $\begin{pmatrix} N_0 \\ X_i \end{pmatrix} \to \begin{pmatrix} N_0 - 1 \\ X_i + 1 \end{pmatrix}$ $W_2(N_0, \ldots, N_n) = k_i N_0 N_i / V$

Decay, $X_i \to F$ $N_i \to N_i - 1$ $W_3(N_0, \ldots, N_n) = k'_i N_i$

(6.49)

Here, R is a reservoir and F is the discarded final product. With the transition rates (6.49) the master equation for the probability $P(N_0, N_1, \ldots, N_n, t)$ to find at the time t in the system just N_0 particles of A, N_1 particles of X_1, and so on, takes the form

$$\frac{\partial}{\partial t} P(N_0, N_1, \ldots, N_n, t) = \Phi V [P(N_0 - 1, N_1, \ldots, N_n, t) - P(N_0, N_1, \ldots, N_n, t)]$$

$$+ \sum_i \frac{k_i}{V} [(N_0 + 1)(N_i - 1) P(N_0 + 1, N_1, \ldots, N_i - 1, \ldots, N_n, t) - N_0 N_i P(N_0, N_1, \ldots, N_n, t)]$$

$$+ \sum_i k'_i [(N_i + 1) P(N_0 - 1, N_1, \ldots, N_i + 1, \ldots, N_n, t) - N_i P(N_0, N_1, \ldots, N_n, t)]$$

(6.50)

Unfortunately, this master equation is too much complicated to be solved analytically in the general case. We are mainly interested in the asymptotic behavior for $t \to \infty$. If the directed graph of the transition processes possesses sinks, only those absorber states are eventually occupied with nonzero probability (Chapters 4 and 5). This is actually the case for the model (6.50), compare Figure 6.6.

With respect to each species i with $N_i > 0$, the sink is the state $N_i = 0$. In particular, for $t \to \infty$, the stochastic absorber state $N_1 = N_2 = \ldots = N_n = 0$ is approached with probability 1 at which the raw material accumulates boundlessly and all species of interest die out. The time scales related to this "fluctuation catastrophe" are extremely long if reasonable mean particle numbers are considered. On these time scales the model (6.50) is formally correct but may no longer be a reasonable model for real evolution and selection processes. However, one cannot exclude that such extinction

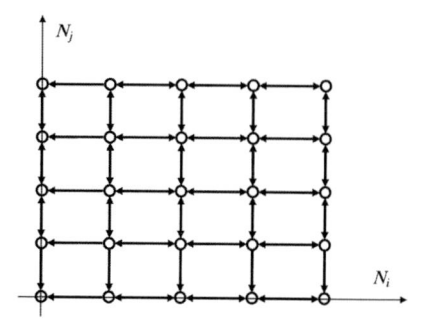

Figure 6.6 Graph of transition processes related to Equation 6.49. Absorber states exist on the boundary of the positive cone.

catastrophes may have occurred in the very early stages of molecular evolution. In the thermodynamic limit, $V \to \infty$, the extinction time becomes infinite.

The typical time scales for selection processes between competing species are much shorter and determined by the chemical reaction rates of the participating particles. Thus, we have to distinguish between stochastic processes on two very different time scales of interest:

1) the approach to a selection equilibrium ($N_s > 0$, $N_i = 0$ $\forall i \neq s$) and
2) the extinction of the only existing species by the fluctuation catastrophe.

Since the second process is rather slow, we consider the probability distribution as approximately stationary as soon as state 1 is established.

Using the generating function

$$G(s_0, \ldots, s_n, t) = \sum_{N_0} \sum_{N_1} \cdots \sum_{N_n} s_0^{N_0}, \ldots, s_n^{N_n} P(N_0, \ldots, N_n, t) \tag{6.51}$$

the master equation (6.50) can be written more compactly in the form of the second-order linear partial differential equation:

$$\frac{\partial G}{\partial t} = \Phi V(s_0-1)G + \sum_i \left[\frac{k_i}{V} s_i(s_i-s_0) \frac{\partial^2 G}{\partial s_0 \partial s_i} + k'_i(1-s_i) \frac{\partial G}{\partial s_i} \right] \tag{6.52}$$

The auxiliary variables $s_0, \ldots, s_i$ can take any values between 0 and 1. In particular, $G(1, 1, \ldots, 1, t) = 1$ is the normalization constant, and $G(1, 0, \ldots, 0, t)$ equals the probability for the fluctuation catastrophe to occur by the time t.

To derive some instructive properties of the stochastic model, we subsequently multiply Equation 6.50 with N_0, N_i, and with the harmonic sum,

$$H(N_0) = \sum_{k=1}^{N_0} \frac{1}{k} \tag{6.53}$$

and sum up over all particle numbers. We obtain the following equations:

$$\frac{\partial}{\partial t} \langle N_0 \rangle = \Phi V - \sum_i \frac{k_i}{V} \langle N_0 N_i \rangle \tag{6.54}$$

$$\frac{\partial}{\partial t} \langle N_i \rangle = \frac{k_i}{V} \langle N_0 N_i \rangle - k'_i \langle N_i \rangle \tag{6.55}$$

$$\frac{\partial}{\partial t} \langle H(N_0) \rangle = \Phi V \left\langle \frac{1}{N_0 + 1} \right\rangle - \frac{k_i}{V} \langle N_i \rangle \tag{6.56}$$

In the stochastic picture, the harmonic sum plays a similar role as the logarithmic function in the deterministic; in the thermodynamic limit they coincide.

We consider now the situation of the selection equilibrium at which only one species N_s populates the system. Setting to zero the left-hand sides of

Equations 6.54–6.56, we obtain

$$\left\langle \frac{V}{N_0+1} \right\rangle = \frac{k_s}{k'_s} \quad \text{and} \quad \left\langle \frac{N_s}{V} \right\rangle = \frac{\Phi}{k'_s} \tag{6.57}$$

similar to the deterministic result (6.17). In contrast, the stochastic model does not imply that a specific, predetermined species is the unambiguous winner of the competition. At the selection equilibrium after some long time t, any of the species i can be the only survivor with the probability σ_i,

$$\sigma_i = \sum_{N_0} \sum_{N_i} P(N_0, 0, \ldots, N_i, \ldots, 0, t) = G(s_0 = 1, 0, \ldots, 0, s_i = 1, 0, \ldots, t) \tag{6.58}$$

The survival does not only depend on the phenotypic properties of the species but also on its initial abundance.

We investigate this question for the most relevant case that a few individuals of a new species i, $N_i \ll N_s$, appear in a system with the established species s. We simplify the problem by neglecting any feedback of the minor species on the current quasistationary state:

$$G(s_0, s_s, s_i, t) \approx G^{(s)}(s_0, s_s) G_i(s_i, t) \tag{6.59}$$

We obtain from Equation 6.52, omitting the index i for simplicity ($G \equiv G_i$, $s \equiv s_i$),

$$G^{(s)} \frac{\partial G}{\partial t} = \left[\Phi V(s_0-1) G^{(s)} + \frac{k_s}{V} s_s(s_s-s_0) \frac{\partial^2 G^{(s)}}{\partial s_0 \partial s_s} + k'_i(1-s_s) \frac{\partial G^{(s)}}{\partial s_s} \right] G \\ + \frac{k_i}{V} \left[s(s-s_0) \frac{\partial G^{(s)}}{\partial s_0} + k'_i(1-s) G^{(s)} \right] \frac{\partial G}{\partial s} \tag{6.60}$$

We set $s_0 = s_s = 1$, consider the normalization $G^{(s)}(1,1,t) = 1$, and obtain a closed equation for the new species (Feistel, 1976b), subject to the only approximation (6.59)

$$\frac{\partial G}{\partial t} = k'_i(s-1)[s(1+v)-1] \frac{\partial G}{\partial s} \tag{6.61}$$

Here, making use of Equations 6.18 and 6.57, the value of v is the selective advantage of the new species over the master species:

$$v = \frac{k_i}{k'_i V} \frac{\partial G^{(s)}}{\partial s_0} \bigg|_{s_0=s_s=1} - 1 = \frac{k_i}{k'_i} \left\langle \frac{N_0}{V} \right\rangle^{(s)} - 1 \approx \frac{k_i/k'_i - k_s/k'_s}{k_s/k'_s} \tag{6.62}$$

The first-order partial differential equation (6.61) can be solved analytically by standard methods (Doob, 1953; Bartholomay, 1958a, 1958b, 1959; Smirnow, 1968; Feistel, 1976b). The ordinary differential equation for the characteristic

6.4 Stochastic of Simple Competition Processes

$x(s, t) = \text{const}$ of Equation 6.61 reads

$$\frac{ds}{(s-1)[s(1+v)-1]} + k_i' dt = 0 \tag{6.63}$$

By decomposition into partial fractions we find the integral of (6.63) in the form

$$x(s, t) = \ln \frac{s-1}{s(1+v)-1} + vk_i't = \text{const} \tag{6.64}$$

which can be inverted analytically to give the function $s(x, t)$:

$$s(x, t) = \frac{\exp(x - vk_i't) - 1}{(1+v)\exp(x - vk_i't) - 1} \tag{6.65}$$

Any function $G(s, t) = G(x(s, t))$ is a solution of Equation 6.61. If N_i^0 particles of the new species are initially present, the boundary condition at $t = 0$ is

$$G(s, 0) = \sum_{N_i} s^{N_i} \delta_{N_i, N_i^0} = s(x, 0)^{N_i^0} = \left[\frac{\exp(x) - 1}{(1+v)\exp(x) - 1}\right]^{N_i^0} \tag{6.66}$$

This is the specific function $G(x)$ we are looking for. The time-dependent solution of Equation 6.61 follows from Equation 6.66 by inserting x from Equation 6.64, in the form

$$G(s, t) = \left[\frac{(s-1)\exp(vk_i't) - s(1+v) + 1}{(1+v)(s-1)\exp(vk_i't) - s(1+v) + 1}\right]^{N_i^0} \tag{6.67}$$

From this function $G(s, t)$, the survival probability for the new species follows to be

$$\sigma_i = 1 - G(0, \infty) = \begin{cases} 1 - (1+v)^{-N_i^0} & \text{if } v \geq 0 \\ 0 & \text{if } v \leq 0 \end{cases} \tag{6.68}$$

Expanding (6.67) into a power series of s, Equation 6.51, we obtain from the starting term the extinction probability by the time t:

$$P(0, t) = G(0, t) = \left[\frac{\exp(vk_i't) - 1}{(1+v)\exp(vk_i't) - 1}\right]^{N_i^0} \tag{6.69}$$

In particular, if the new species emerges as a single copy with $v > 0$, its chance of survival

$$\sigma_i = 1 - \lim_{t \to \infty} P(0, t) = \frac{v}{1+v} \approx \frac{k_i/k_i' - k_s/k_s'}{k_i/k_i'} \quad \text{if} \quad \frac{k_i}{k_i'} > \frac{k_s}{k_s'} \tag{6.70}$$

equals the relative selection disadvantage of the established species. The ratio k_i/k_i' was found as a selection value already in the deterministic model. From this result we can infer for the general case that, as a rule, the species with the highest selective value

$$\frac{k_s}{k_s'} > \frac{k_i}{k_i'} \quad \forall i \neq s \tag{6.71}$$

will possess the highest chance to survive and will most frequently win the competition. This implies that the reciprocal raw material concentration, Equation 6.57, will tend to a maximum value:

$$\frac{1}{A^{(s)}} \approx \left\langle \frac{V}{N_0 + 1} \right\rangle = \frac{k_s}{k'_s} \to \text{Max} \qquad (6.72)$$

Similarly, this is also the extremum principle which is governing the deterministic model (Ebeling, 1976a; Feistel, 1976b).

The analysis of extinction or survival probabilities appears as the stochastic counterpart of the asymptotic stability analysis of stationary states located on the boundary of the positive cone performed in the deterministic case. The results in both pictures are consistent, even though an advantageous new species will always defeat the established one in the deterministic model, while the stochastic dynamics assigns nonzero survival chances to both of them. In the thermodynamic limit this difference disappears.

The model of constant overall organization is mathematically simpler, and it is easier to derive analytical results from its stochastic formulation. If we adopt the interpretation of Equation 6.26, we only need to model the transformation of a particle from one species to the other rather than explicitly counting import and export fluxes (Figure 6.7).

For the transformation process we assume a transition rate of the form (Ebeling and Feistel, 1975, 1977)

$$N_i \to N_j; \quad \begin{pmatrix} N_i \\ N_j \end{pmatrix} \to \begin{pmatrix} N_i - 1 \\ N_j + 1 \end{pmatrix} \quad W(N_1, \ldots, N_n) = A_{ji} N_i + B_{ji} N_i N_j \qquad (6.73)$$

and determine the coefficients A_{ij} and B_{ij} from the condition that the equation for the mean value converges to Equation 6.24 in the thermodynamic limit. It follows that

$$A_{ji} = 0 \quad B_{ji} = \frac{E_i}{N} \quad \text{provided that } E_i > 0 \qquad (6.74)$$

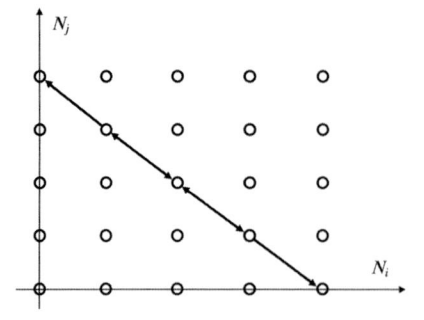

Figure 6.7 Graph of transition processes, Equation 6.75, of the stochastic model with constant total particle number.

We note that, rather than (6.74), the alternative choice, $B_{ji} = R_i/N$, is of interest for certain biological and ecological processes (Ebeling and Feistel, 1976a; Feistel and Ebeling, 1976).

From Equations 6.73 and 6.74 we obtain the master equation:

$$\frac{\partial}{\partial t} P(N_1, \ldots, N_n, t) = \frac{1}{N} \sum_{j \neq i} E_i \left[(N_i - 1)(N_j + 1) P(N_i - 1, N_j + 1) - N_i N_j P \right]$$

(6.75)

where the omitted arguments of P on the right-hand side are the same as on the left-hand side. For the mean values

$$\langle N_i \rangle (t) = \sum_{N_1} \sum_{N_2} \cdots \sum_{N_n} N_i P(N_1, \ldots, N_n, t)$$

(6.76)

we derive from Equation 6.75 the dynamical equation

$$\frac{d}{dt} \langle N_i \rangle = E_i \langle N_i \rangle - \frac{1}{N} \sum_j E_j \langle N_i N_j \rangle$$

(6.77)

In the deterministic model, we know that the mean excess production, $\langle E \rangle$, represents an extremum principle. In the stochastic picture it is defined as

$$\langle E \rangle = \frac{1}{N} \sum_i E_i \langle N_i \rangle = \frac{1}{N} \sum_{N_1} \sum_{N_2} \cdots \sum_{N_n} E_i N_i P(N_1, \ldots, N_n, t)$$

(6.78)

Taking the time derivative of $\langle E \rangle$ with the help of Equation 6.77,

$$\frac{d}{dt} \langle E \rangle = \frac{1}{N} \left\langle \sum_i N_i \left(E_i - \frac{1}{N} \sum_j E_j N_j \right)^2 \right\rangle \geq 0$$

(6.79)

we see that $\langle E \rangle$ is maximized also in the stochastic model. The equality sign is valid only for stationary distributions of the form

$$P^0(N_1, \ldots, N_n) = \sum_s \sigma_s \delta_{N_1 0} \cdots \delta_{N_s, N} \cdots \delta_{N_n 0}$$

(6.80)

Here,

$$\sigma_s \geq 0 \quad \sum_s \sigma_s = 1$$

(6.81)

are the survival probabilities of the species s. Again, in this case the stationary probability distribution is characterized by absorber states (sinks) at $N_i = 0$ which is typical for selection processes (Figure 6.7, see also Section 5.3). They permit the absence of extinct earlier species and leave any future "better" species without influence on the present system. Rather than reaching its maximum value as in the deterministic case, the stochastic quantity $\langle E \rangle$ increases only up to

$$\lim_{t \to \infty} \langle E \rangle = \sum_i \sigma_i E_i \qquad (6.82)$$

where the σ_i depends on the initial conditions.

In the case of two species the survival probabilities can be found exactly without solving the dynamical problem, Equation 6.75. We consider the competition between an established species s and an emerging new species i. Because of Equation 6.81, $\sigma_s + \sigma_i = 1$, we need only a conservation quantity of the stochastic process to compute σ_i. The generating function, Equation 6.51, which corresponds to the master equation (6.75) obeys the following equation:

$$\frac{\partial G}{\partial t} = \frac{1}{N} \sum_{i,j} E_i s_i (s_i - s_j) \frac{\partial G}{\partial s_i \partial s_j} \qquad (6.83)$$

The right-hand side vanishes at the point $s_i = 1/E_i$, that is, the value of

$$G\left(\frac{1}{E_1}, \ldots, \frac{1}{E_n}, t\right) = \langle E_1^{-N_1} E_2^{-N_2}, \ldots, E_n^{-N_n} \rangle \qquad (6.84)$$

is the same at the initial and at the final state (Feistel, 1983a). For the two species i and s, initially present with N_i^0 and $N_s^0 = N - N_i^0$ particles, the conservation of $\langle E_i^{-N_i} E_s^{-N_s} \rangle$ at $t = 0$ and $t \to \infty$ leads to the relation

$$E_i^{-N_i^0} E_s^{-N + N_i^0} = \sigma_i E_i^{-N} + (1 - \sigma_i) E_s^{-N} \qquad (6.85)$$

The solution for the survival probability of the new species is

$$\sigma_i = \frac{1 - (E_s/E_i)^{N_i^0}}{1 - (E_s/E_i)^{N}} \qquad (6.86)$$

and depends on the ratio of the selective values of the two participating species as well as on the initial conditions. These results for the survival probability generalize several formulae given already in Section 5.3. Recall that in the deterministic picture, the differences rather than the ratios of the selective values are most relevant. The stochastic formulation breaks the symmetry with respect to the addition of an arbitrary constant to the E_i, Equation 6.25.

In the case of a single new copy, $N_i^0 = 1$, and large particle numbers, $N \to \infty$, the formula

$$\sigma_i = \begin{cases} 0, & \text{if } E_i < E_s \\ (E_i - E_s)/E_i, & \text{if } E_i > E_s \end{cases} \qquad (6.87)$$

is consistent with the results (6.68), (6.70), that is, the survival probability equals the selective advantage (Chapters 4 and 5).

It is interesting to note that one can formally proceed from the stochastic to the deterministic picture by performing the thermodynamic limit, $V \to \infty$. This limit can be carried out for

$$N_i^0 = c^0 \cdot V \quad N = c \cdot V \quad c > c^0 \tag{6.88}$$

with finite c, c^0 and converts Equation 6.86 into the Boolean decision

$$\sigma_i = \begin{cases} 0, & \text{if } E_i < E_s \\ 1, & \text{if } E_i > E_s \end{cases} \tag{6.89}$$

This shows that the analysis of survival probabilities is the stochastic equivalent of the Lyapunov stability analysis, at least for the model considered here. It is quite interesting to compare the stochastic survival probability with the deterministic results (Allen and Ebeling, 1983). In the deterministic case the analysis given in Section 5.3 says that any competitor which is better will win and will survive therefore with probability 1, and any competitor who is worse will disappear, that is, his survival probability is zero, Equation 6.89. This way, in the stochastic case, the deterministic step-function is replaced by a smooth, increasing function. A better species wins only with certain probability and this probability increases with the selection advantage. If the relative selection advantage is small, the survival probability is not significantly higher than that for worse species. This is a result for infinite populations, and for finite populations the situation is even less clear. The difference between better and worse species is smoothed further as demonstrated already in Figure 5.8. The selection in finite stochastic systems is less decisive than in deterministic systems. This is one essence of "neutral selection" detected by Kimura (1968). We emphasize that the sequence in which the limits $V \to \infty$ and $N \to \infty$ are carried out leads to different results.

To conclude this section, we consider the stochastic generalization of the catalytic enhancement of the selection values, Equation 6.43. Rather than Equations 6.73 and 6.74, we use the ansatz (Ebeling and Jiménez-Montaño, 1980),

$$N_i \to N_j : \begin{pmatrix} N_i \\ N_j \end{pmatrix} \to \begin{pmatrix} N_i - 1 \\ N_j + 1 \end{pmatrix} \quad W(N_1, \ldots, N_n) = \frac{E_j}{N} N_i N_j + \frac{b_{jk}}{NV} N_i N_j N_k \tag{6.90}$$

The related master equation reads

$$\frac{\partial}{\partial t} P(N_1, \ldots, N_n, t) = \frac{1}{N} \sum_{j \neq i} E_i \left[(N_i - 1)(N_j + 1) P(N_i - 1, N_j + 1) - N_i N_j P \right]$$

$$+ \frac{1}{NV} \sum_{j \neq i} \sum_k b_{ij} N_k \left[(N_i - 1)(N_j + 1) P(N_i - 1, N_j + 1) \right.$$

$$\left. - N_i N_j P \right] \tag{6.91}$$

Multiplication with (N_k/V) and summation over all particle numbers leads to the equation for the mean particle numbers:

$$\frac{d}{dt}\frac{\langle N_m\rangle}{V} = E_m\frac{\langle N_m\rangle}{V} - \frac{1}{C}\sum_k E_k\frac{\langle N_m N_k\rangle}{V^2} + \sum_l b_{ml}\frac{\langle N_m N_l\rangle}{V^2} - \frac{1}{C}\sum_{k,l} b_{kl}\frac{\langle N_m N_k N_l\rangle}{V^3} \tag{6.92}$$

Factorization of the momenta turns this equation back into Equation 6.43. The solution of Equation 6.91 shows an extremely complicated and little known temporal behavior. A δ-like initial distribution will initially form a maximum at the deterministic attractor that is located in the same attraction basin. In a later stage, the probability diffuses across the separatrices and fills up the entire simplex relatively homogeneously. For long times, the solution converges toward absorber states on the boundary of the positive cone, that is, selection occurs.

Only a few analytical solutions for this model are currently known (Ebeling et al., 2006; Hartmann-Sonntag, Scharnhorst, and Ebeling, 2009), several of them are demonstrated for a quadratic growth rate in Section 5.4.

Let us summarize the known results for the survival probability of two-dimensional systems with constant total particle numbers, $N_1 + N_2 = N = \text{const}$, and with two absorber states, $N_2 = 0$ and $N_2 = N$. Here, the absorber state means that once either of the states $N_2 = 0$ or $N_2 = N$ is reached, it cannot be left by the system anymore. Formally, this property is expressed in terms of transition probabilities in the form

$$\begin{aligned} W^+(N_2 = 0) = 0 \qquad & W^-(N_2 = 1) > 0 \\ \text{and} & \\ W^+(N_2 = N-1) > 0 \qquad & W^-(N_2 = N) = 0 \end{aligned} \tag{6.93}$$

As demonstrated above, the case of two absorber states can be treated analytically (Schimansky-Geier, 1981; Ebeling, Sonntag, and Schimansky-Geier, 1981; Ebeling and Feistel, 1982). According to the absorber character we may assume that the stationary probability which is the target of evolution has delta character. Accordingly, we have for the stationary solution the formula:

$$P(N_2, t\to\infty) = \sigma\delta_{N\,N_2} + (1-\sigma)\delta_{0\,N_2} \tag{6.94}$$

which is valid with the unknown constant being the survival probability σ of the "new" species 2. An expression for the survival probability σ can be calculated by the help of one conservation quantity. The result is (Ebeling, Sonntag, and Schimansky-Geier, 1981; Ebeling and Feistel, 1982; Hartmann-Sonntag, Scharnhorst, and Ebeling, 2009)

$$\sigma_{N_2(0),N} = \frac{1 + \sum_{j=1}^{N_2(0)-1}\prod_{i=1}^{j}\frac{W_i^-}{W_i^+}}{1 + \sum_{j=1}^{N-1}\prod_{i=1}^{j}\frac{W_i^-}{W_i^+}} \quad \text{for} \quad 0 < N_2(0) < N \tag{6.95}$$

and

$$\sigma_{N_2(0),N} = 1 \quad \text{for} \quad N_2(0) = N \tag{6.96}$$

The transition probabilities W^+ and W^- depend on the parameters, the initial conditions, and the system size. We studied three plausible stochastic models:

1) the linear case:

$$W^+_{N_2} = E_2 \frac{N-N_2}{N} N_2 \qquad W^-_{N_2} = E_1 \frac{N-N_2}{N} N_2 \tag{6.97}$$

2) the quadratic case:

$$W^+_{N_2} = b_2 \frac{N-N_2}{NV} N_2^2 \qquad W^-_{N_2} = b_1 \frac{(N-N_2)^2}{NV} N_2 \tag{6.98}$$

3) and the combined case:

$$W^+_{N_2} = E_2 \frac{N-N_2}{N} N_2 + b_2 \frac{(N-N_2)}{NV} N_2^2$$

$$W^-_{N_2} = E_1 \frac{N-N_2}{N} N_2 + b_1 \frac{(N-N_2)^2}{NV} N_2 \tag{6.99}$$

From Equation 6.95, we derive the survival probabilities for the cases (1) to (3) as follows:

1) For the linear case we find, Equation 6.86,

$$\sigma_{N_2(0),N} = \frac{1-\left(\frac{E_1}{E_2}\right)^{N_2(0)}}{1-\left(\frac{E_1}{E_2}\right)^{N}} \tag{6.100}$$

For large systems, σ can be simplified to

$$\sigma_{N_2(0),N\to\infty} = 0 \quad \text{for} \quad E_2 < E_1 \tag{6.101}$$

and

$$\sigma_{N_2(0);N\to\infty} = 1-\left(\frac{E_1}{E_2}\right)^{N_2(0)} \quad \text{and} \quad E_2 > E_1 \tag{6.102}$$

2) For the quadratic case, we find the survival chance:

$$\sigma_{N_2(0),N} = \frac{1+\sum_{j=1}^{N_2(0)-1}\left(\frac{b_1}{b_2}\right)^j\binom{N-1}{j}}{\left(1+\frac{b_1}{b_2}\right)^{N-1}} \tag{6.103}$$

For a system with an initial number of $N_2(0) = 1$, we obtain in particular

$$\sigma_{N_2(0),N} = \frac{1}{\left(1 + \frac{b_1}{b_2}\right)^{N-1}} \tag{6.104}$$

3) For the combined case we obtain the expression

$$\sigma_{N_2(0),N} = \frac{1 + \sum_{j=1}^{N_2(0)-1} \prod_{i=1}^{j} \frac{E_1 + b_1 \frac{N-i}{V}}{E_2 + b_2 \frac{i}{V}}}{1 + \sum_{j=1}^{N-1} \prod_{i=1}^{j} \frac{E_1 + b_1 \frac{N-i}{V}}{E_2 + b_2 \frac{i}{V}}} \tag{6.105}$$

For a system with an initial occupation number $N_2(0) = 1$ we find in the general case

$$\sigma_{N_2(0),N} = \frac{1}{1 + \sum_{j=1}^{N-1} \prod_{i=1}^{j} \frac{E_1 + b_1 \frac{N-i}{V}}{E_2 + b_2 \frac{i}{V}}} \tag{6.106}$$

As our results show, the survival probabilities depend on the system size, the system parameters, and the initial conditions. In the case of quadratic growth rates, we find niche effects. Due to finite number effects, the sharpness of selection is weakened. In the linear case, "good" and "bad" species can coexist temporarily. In the quadratic case, the "once-forever" effect can be overcome by the niche effect. Niches may play an important role in the evolution by overcoming the hyperselection. Large complex systems – such as biological, ecological, and social systems – usually form niches. Within a niche, the new species is protected against extinction for some limited time. After winning the competition in the close vicinity, the local new species can infect the whole system and may be established globally.

6.5
Competition in Species Networks

The consideration of simple competition in the previous sections was restricted to parallel and independently progressing autocatalytic reactions:

$$A \xrightarrow{X_1} X_1, \ A \xrightarrow{X_2} X_2, \ldots, A \xrightarrow{X_n} X_n$$

The coupling between the reactions which establishes competitive behavior was caused by either constant raw material supply or constant overall organization.

Now we are interested in the more general and more realistic situation that the production of a particular species is cross-catalytic, that is, any species may be produced from the raw material in the presence of certain sets of other species:

$$A \xrightarrow{X_k} X_1, \ A \xrightarrow{X_k} X_2, \ldots, A \xrightarrow{X_k} X_n \quad k = 1, \ldots, n$$

Autocatalysis appears now as just a special case of the catalytic network.

As before, we consider only linear catalysis. We begin with the growth limitation by constant raw material supply, Equation 6.13,

$$A + X_i \xrightarrow{k_{ii}} 2X_i \quad X_i \xrightarrow{k'_i} F \tag{6.107}$$

and permit additionally the catalytic interaction:

$$A + X_i \xrightarrow{k_{ji}} X_i + X_j \tag{6.108}$$

The related generalized form of Equations 6.14,

$$\begin{aligned} \dot{A} &= \Phi - \sum_{ij} k_{ij} A X_j \\ \dot{X}_i &= A \sum_j k_{ij} X_j - k'_i X_i \end{aligned} \tag{6.109}$$

can be conveniently written in the matrix notation:

$$\begin{aligned} \dot{A} &= \Phi - A \cdot \mathbf{1}^T \mathbf{k} \mathbf{X} \\ \dot{\mathbf{X}} &= A \mathbf{k} \mathbf{X} - \mathbf{k}' \mathbf{X} = \mathbf{E} \mathbf{X} \end{aligned} \tag{6.110}$$

Here, the superscript T refers to the transpose matrix:

$$\begin{aligned} \mathbf{X} &= (X_1, X_2, \ldots, X_n)^T \quad \mathbf{1} = (1, 1, \ldots, 1)^T \\ \mathbf{k} &= \{k_{ij}\}, \quad \mathbf{k}' = \{k'_i \delta_{ij}\} \quad \mathbf{E} = A\mathbf{k} - \mathbf{k}' \end{aligned} \tag{6.111}$$

A system of the form (6.110) is also obtained if autocatalytic replication is accompanied by mutations (Eigen, 1971; Ebeling and Feistel, 1975, 1977; Jones, Enns, and Rangnekar, 1976; Eigen and Schuster, 1978). If the cross-catalytic reactions (6.108) are significantly slower than the autocatalysis (6.107),

$$k_{ij} \geq 0 \quad k_{ij} \ll k_{ii} \quad k_{ji} \ll k_{ii} \quad \forall j \neq i \tag{6.112}$$

or, more strictly,

$$\sum_{j \neq i} k_{ij} < k_{ii} \quad \sum_{j \neq i} k_{ji} < k_{ii} \quad k_{ij} \geq 0 \quad \forall i,j \tag{6.113}$$

then the reaction (6.108) is regarded as production of error copies rather than cross-catalysis. Mathematically, the difference between the two interpretations of Equations 6.110 is that the matrix **k** under the condition (6.113) is diagonal-dominant and belongs to the class of monotonous matrices (Varga, 1962) with special spectral properties (Gerschgorin, 1931).

We determine now the stationary solutions of the system (6.110). From $\dot{A} = 0$ and $\dot{\mathbf{X}} = 0$, we infer the following equations (Ebeling and Feistel, 1976a; Feistel, 1976b, 1979):

$$\Phi = \mathbf{1}^T \mathbf{k}' \mathbf{X} \tag{6.114}$$

and

$$\mathbf{k}'^{-1} \mathbf{k} \mathbf{X} = \frac{1}{A} \mathbf{X} \tag{6.115}$$

since $\mathbf{k}'$ is a regular diagonal matrix that can be inverted. The possible stationary values $1/\Lambda$ of the inverse raw material concentration are found to be the eigenvalues of the reproduction matrix $\mathbf{k}'^{-1}\mathbf{k}$, the stationary concentrations of the species are given by the eigenvectors $\mathbf{X}$ of the matrix, and the total population, that is, the norm of the eigenvector, is limited by the raw material supply rate. The discussion of the stationary solutions of the system (6.110) is reduced to the properties of the eigenvalue problem (6.114).

It is a very fortunate circumstance for the analysis of the eigenvalue problem (6.114) that reproduction matrix $\mathbf{k}'^{-1}\mathbf{k}$ does not have any negative elements. For this reason, the well-developed theory of spectral and decomposition properties of nonnegative matrices is available for the investigation of selection processes in catalytic networks (Gantmacher, 1971; Feistel and Ebeling, 1978b; Feistel, 1979). In addition, the qualitative properties of partly populated nonnegative matrices can very instructively be represented by directed graphs (Saaty, 1966; Busacker and Saaty, 1965; Harary, Norman, and Cartwright, 1965; Biess, 1976; Feistel and Sändig, 1977; Jain and Krishna, 2003). Because $\mathbf{k}'$ is a diagonal matrix, the decomposition properties of $\mathbf{k}'^{-1}\mathbf{k}$ and $\mathbf{k}$ are equivalent. For simplicity, it is therefore sufficient to restrict the qualitative analysis to $\mathbf{k}$ as the adjacency matrix of a related directed *production graph*. This graph is closely related to the so-called minor of the system (6.110), (Chapter 4).

The node i of the production graph corresponds to the species X_i, its directed edge from j to i corresponds to a positive matrix element $k_{ij} > 0$, which tells us that the species i is catalytically produced by the species j. The topology of the production graph this way reflects the reaction pathways in the chemical network.

We consider now some selected examples that are representative for the typical qualitatively different structures of the production graph and its related adjacency matrix. The species are numbered in a way consistent with the normal form of the matrix (Gantmacher, 1971).

Our first example is an irreducible matrix $\mathbf{k}$. The matrix elements k_{ij} shown are assumed to be positive. The related graph is shown in Figure 6.8.

$$\mathbf{k} = \begin{Bmatrix} k_{11} & k_{12} & k_{13} & 0 & 0 & 0 & 0 & k_{18} \\ k_{21} & k_{22} & 0 & k_{24} & 0 & 0 & 0 & 0 \\ 0 & 0 & 0 & 0 & 0 & 0 & 0 & 0 \\ 0 & 0 & 0 & 0 & 0 & 0 & 0 & 0 \\ 0 & k_{52} & 0 & 0 & 0 & 0 & 0 & 0 \\ 0 & 0 & 0 & 0 & k_{65} & 0 & k_{67} & 0 \\ 0 & 0 & 0 & 0 & 0 & k_{76} & 0 & 0 \\ 0 & 0 & 0 & 0 & 0 & 0 & k_{87} & 0 \end{Bmatrix} \qquad (6.116)$$

The irreducibility of the production matrix (6.116) is more easily recognized in the graph, Figure 6.8. Following the arrows, it is possible to walk from any node to any other node of the graph. In this case, there exists exactly one stationary state $\mathbf{X}^0$. At this state, no species disappears. A nonnegative irreducible matrix possesses one and only one real, positive and nondegenerate eigenvalue, the modulus of which is not exceeded by any other eigenvalue, and the related eigenvector has positive components (Frobenius

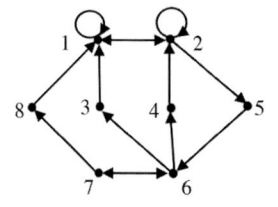

Figure 6.8 Irreducible production graph related to the adjacency matrix **k**, Equation 6.116.

theorem). Since this matrix does not possess two different, linear independent nonnegative eigenvectors, the steady state is the unique solution of Equation 6.115.

A special irreducible matrix is the diagonal-dominant, positive one, which is typically used to describe mutations transforming one species into another. In the related deterministic picture, each present species then continuously produces a "cloud" of all other species with very low abundance. This was sometimes interpreted in a way that "quasispecies" defined by the eigenvectors of the production matrix take over the role of the regular species when mutations are taken into account (Eigen, 1971; Eigen, McCaskill, and Schuster, 1989; Weinberger, 1991; Nowak, 1992; Galluccio, 1997; Feng et al., 2007). As first pointed out by Feistel and Ebeling (1978b) this naive quasispecies picture has to be considered with some care and is a merely formal picture. The Frobenius theorem implies that only one of those "clouds" has positive concentrations, all other "quasispecies" must include "species" with negative particle numbers. The positive eigenvector has one dominating component in comparison to the remaining ones; this expresses the presence of one selectively superior species within the "cloud" of its mutants with low concentrations (Jones, Enns, and Rangnekar, 1976).

Another, particularly simple special case of an irreducible production matrix is the ring (catalytic cycle), Equation 6.116, as shown in Figure 6.9:

$$\mathbf{k} = \begin{Bmatrix} 0 & 0 & 0 & 0 & k_{15} \\ k_{21} & 0 & 0 & 0 & 0 \\ 0 & k_{32} & 0 & 0 & 0 \\ 0 & 0 & k_{43} & 0 & 0 \\ 0 & 0 & 0 & k_{54} & 0 \end{Bmatrix} \quad (6.117)$$

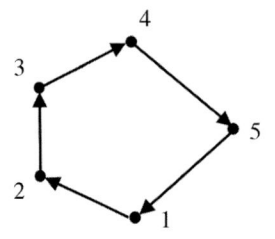

Figure 6.9 Irreducible production graph of a catalytic cycle, related to the adjacency matrix **k**, Equation 6.117.

The fifth power of the matrix (6.117) is the diagonal matrix:

$$\mathbf{k}^5 = k_{15} \cdot k_{54} \cdot k_{43} \cdot k_{32} \cdot k_{21} \mathbf{I} \qquad (6.118)$$

so that the eigenvalues of **k** are the five complex roots of the product of the production rates (**I** is the unity matrix). Only one of the eigenvalues is positive, the geometric mean of the rates. In general, the characteristic polynomial of a ring with n members reads (Eigen, 1971; Feistel, 1976b)

$$\lambda^n = k_{1n} \cdot k_{n,n-1} \cdot \ldots \cdot k_{32} \cdot k_{21} \qquad (6.119)$$

This kind of matrix is regarded as mathematically *imprimitive* or *cyclic* of the index n. The components of the positive eigenvector obey the equation

$$k_{ij} X_j = \lambda X_i \quad \text{or} \quad \frac{X_i}{X_j} = \frac{k_{ij}}{\lambda} \qquad (6.120)$$

for neighboring species i and j of the ring.

We conclude that a system with an irreducible production matrix is not subject to a selection process that leads to the extinction of some of its species.

We consider now the very contrasting structure, a completely reducible matrix **k**, as given in Equation 6.121 with the related production graph displayed in Figure 6.10:

$$\mathbf{k} = \begin{pmatrix} 0 & 0 & 0 & k_{14} & 0 & 0 & 0 & 0 & 0 & 0 & 0 & 0 \\ k_{21} & 0 & 0 & 0 & 0 & 0 & 0 & 0 & 0 & 0 & 0 & 0 \\ 0 & k_{32} & 0 & 0 & 0 & 0 & 0 & 0 & 0 & 0 & 0 & 0 \\ 0 & 0 & k_{43} & 0 & 0 & 0 & 0 & 0 & 0 & 0 & 0 & 0 \\ 0 & 0 & 0 & 0 & k_{55} & k_{56} & 0 & 0 & 0 & 0 & 0 & 0 \\ 0 & 0 & 0 & 0 & k_{65} & 0 & 0 & 0 & 0 & 0 & 0 & 0 \\ 0 & 0 & 0 & 0 & 0 & 0 & 0 & k_{78} & 0 & 0 & 0 & 0 \\ 0 & 0 & 0 & 0 & 0 & 0 & 0 & 0 & 0 & 0 & 0 & k_{8,12} \\ 0 & 0 & 0 & 0 & 0 & 0 & 0 & k_{98} & 0 & 0 & 0 & 0 \\ 0 & 0 & 0 & 0 & 0 & 0 & 0 & 0 & k_{10,9} & 0 & k_{10,11} & 0 \\ 0 & 0 & 0 & 0 & 0 & 0 & 0 & 0 & 0 & 0 & 0 & k_{11,12} \\ 0 & 0 & 0 & 0 & 0 & 0 & k_{12,7} & 0 & 0 & k_{12,10} & 0 & 0 \end{pmatrix}$$

(6.121)

In this matrix, the elements are arranged so that **k** is a diagonal block matrix that consists of three quadratic matrices as its main diagonal elements. In the graph, this structure is reflected by three disconnected irreducible components, Figure 6.10.

The species of such a component form a cluster, that is, a self-contained subnetwork. The previous consideration of the competition in an irreducible catalytic network, Equation 6.116, had shown that the cluster's member species can exist or disappear only jointly.

To the maximum eigenvalue of a completely reducible matrix, a positive eigenvector exits only if all clusters separately have exactly the same maximum eigenvalue.

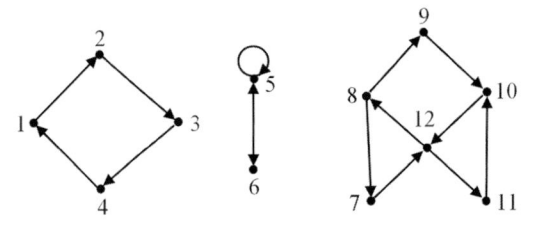

Figure 6.10 Production graph of the adjacency matrix (6.121): a completely reducible catalytic network that consists of three irreducible subnets (clusters).

This case can practically be excluded for the chemical rate constants of different chemicals; then the eigenvector which belongs to the maximum eigenvalue is nonnegative with positive components only for the member species of the cluster with the highest eigenvalue. This means that clusters with different maximum eigenvalues cannot stationarily coexist.

If there exist n irreducible clusters, there are n related different stationary states with one cluster present and any other extinct. Among those, only the state with the maximum eigenvalue $1/A$ is asymptotically stable, Equation 6.115. Hence this situation is very similar to that of competing single species, with the difference that now clusters rather than individual species are the smallest competing entities. Again the stationary reciprocal raw material concentration appears as a static evolution principle. The dependence of stationary values of $1/A$ on the reaction rate is implicitly defined by the positive solutions of

$$\det\left(\mathbf{k}'^{-1}\mathbf{k} - \frac{1}{A}\mathbf{I}\right) = 0 \qquad (6.122)$$

Rather than being just occasional, the obvious similarity between separate autocatalytic species and irreducible clusters expresses the possibility of hierarchical structures to appear in selection processes. This is an extremely simple case for emergent properties – integral properties of several or many cooperating, more elementary entities. For comparison with a rather different example, in statistical physics, elementary particles are the actual subject of the theory. The governing equations offer the possibility that so-called bound states between the particles behave in turn like particles (Ebeling, 1974; Ebeling, Kraeft, and Kremp, 1976). Consider, as an example, a hydrogen plasma that consists of electrons and protons. At lower temperatures, hydrogen atoms H and molecules H_2 can form as bound states which themselves can be treated as autonomous particles with internal degrees of freedom. In statistical physics, this transition between elementary and more integral levels of description of the same physical object is regarded as the transition from the "physical picture" to the "chemical picture." It is apparent that certain aggregates of the molecules in turn may form bigger macroscopic bodies which again play the roles of "mass points," "cells," or "individuals" in a subsequent, higher-level "ecological" theory, and so on.

The situation is very similar in the case of competition between clusters of species. If we consider the species X_i as elementary molecules, clusters represent integral functional aggregates of those molecules, for example, primitive "organisms". As a simple instructive example, we carry out here the formal transition from the "chemical" to the "ecological" level of description, in which new species with internal degrees of freedom are competing with each other, rather than single molecules (Feistel and Ebeling, 1976).

We write for the "organism" l,

$$S_l = \sum_{\text{cluster } l} X_i \tag{6.123}$$

and for its "internal structure,"

$$\varrho_{il} = \begin{cases} X_i/S_l, & \text{if the species } i \text{ belongs to the cluster } l \\ 0, & \text{otherwise} \end{cases} \tag{6.124}$$

Evidently, $\sum_i \varrho_{il} = 1$ holds for each cluster l. Further, we define the cluster averages,

$$\langle k_i \rangle_l = \sum_j k_{ij} \varrho_{jl}, \quad \langle k \rangle_l = \sum_i \langle k_i \rangle_l \quad \text{and} \quad \langle k' \rangle_l = \sum_i k'_i \varrho_{il} \tag{6.125}$$

In these variables, if **k** is completely reducible, Equation 6.109 transforms into the hierarchical system with separate but coupled equations for the internal structure of the cluster l,

$$\dot{\varrho}_{il} = A\left(\langle k_i \rangle_l - \langle k \rangle_l \varrho_{il}\right) - \left(k'_i - \langle k' \rangle_l\right) \varrho_{il} \tag{6.126}$$

for the competition between the clusters,

$$\dot{S}_l = \left(\langle k \rangle_l A - \langle k' \rangle_l\right) S_l \tag{6.127}$$

and for the raw material budget,

$$\dot{A} = \Phi - A \sum_l \langle k \rangle_l S_l \tag{6.128}$$

Obviously, Equations 6.127 and 6.128 are formally very similar to the system (6.14) for the competition between autocatalytic species except that this role has been taken over by the clusters. The reaction rates of the clusters are no longer constant and depend on the equilibration of the internal structure, Equation 6.126.

To elucidate this demonstration example, we consider the case that the decay rates are species independent within each cluster, $k'_i = k'$. In this case Equation 6.126 for the internal structure is simplified to

$$\dot{\varrho}_{il} = A\left(\langle k_i \rangle_l - \langle k \rangle_l \varrho_{il}\right) \tag{6.129}$$

It follows that in this case, the clusters possess a stationary internal structure, ϱ^0_{il}, that is independent of the selection process and is the solution of an apparently nonlinear problem:

$$\langle k_i \rangle_I^0 = \langle k \rangle_I^0 \varrho_{il}^0 \tag{6.130}$$

Making use of the definition (6.5.15), this Equation 6.130 is actually equivalent to the eigenvalue equation:

$$\sum_j k_{ij}\varrho_{jl}^0 = k_i^*\varrho_{il}^0 \tag{6.131}$$

and corresponds to the case of an irreducible matrix such as (6.116) with a unique positive solution, due to Frobenius' theorem.

As soon as the internal balance (6.130) has been reached, the rate coefficients of Equations 6.127 and 6.128 are time-independent constants and the existence of internal degrees of freedom of the clusters is no longer visible in their selection properties.

Similar conclusions can be inferred if the selection process described by Equations 6.127 and 6.128 between the clusters has a significantly longer relaxation time than that of the "metabolism" (6.126) of the clusters themselves. If the time constants of a slow subsystem are sufficiently different from that of the remaining fast subsystems (if expressed in suitable state variables such as ϱ and S rather than X, above), Tikhonov's theorem (Tikhonov, 1952) can be applied. It quantifies the conditions under which the integrals of a dynamical system with and without the application of the quasi-steady-state hypothesis converge to each other (Hahn, 1974). Under these circumstances, we can study the selection behavior between the clusters assuming that those are permanently in their quasistationary internal equilibrium states.

We conclude that the "ecological picture" permits a description of the selection between irreducible clusters of cross-catalytic species similar to that of autocatalytic species.

For the demonstration of this approach we consider the numerical solution of an example for the system of differential equation (6.110). We choose $\Phi = 2$, for $\mathbf{k'}$ the unity matrix, and for $\mathbf{k}$ the completely reducible matrix (see also Figure 6.11):

$$\mathbf{k} = \begin{Bmatrix} 0 & 3/2 & 0 & 0 & 0 & 0 \\ 0 & 0 & 1/2 & 0 & 0 & 0 \\ 4/3 & 0 & 0 & 0 & 0 & 0 \\ 0 & 0 & 0 & 0 & 1 & 0 \\ 0 & 0 & 0 & 0 & 0 & 1/8 \\ 0 & 0 & 0 & 1 & 0 & 0 \end{Bmatrix} \tag{6.132}$$

The system consists of the clusters $S_1 = X_1 + X_2 + X_3$ and $S_2 = X_4 + X_5 + X_6$.

6 Competition and Selection Processes

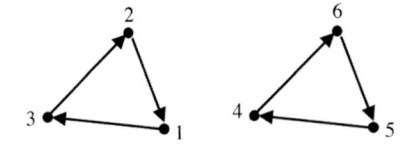

Figure 6.11 Two uncoupled catalytic rings, corresponding to the adjacency matrix (6.132).

Making use of Equations 6.119 and 6.122, the problems (6.131) of the internal cluster structures have the solutions

$$X_1 : X_2 : X_3 = 3 : 2 : 4 \quad \text{and} \quad X_4 : X_5 : X_6 = 2 : 1 : 4$$

$$k_1^* = \sqrt[3]{\frac{4}{3} \cdot \frac{3}{2} \cdot \frac{1}{2}} = 1 \quad \text{and} \quad k_2^* = \sqrt[3]{1 \cdot \frac{1}{8} \cdot 1} = \frac{1}{2} \tag{6.133}$$

In the selection process the ring S_1 with the selective value of 1 is superior over the ring S_2 with $1/2$; for $t \to \infty$ we find the stationary state

$$S_1 = \Phi/k_1' = 2 \quad S_2 = 0 \quad A = 1/k_1^* = 1 \tag{6.134}$$

These analytical results are perfectly confirmed by the numerical integration (Figure 6.12). Note that the internal structure of the ring is stationary already at $t \approx 4$, while the selection process lasts until $t \approx 10$.

We turn our attention now to another import special case, the elementarily reducible catalytic production matrix. An example is given by Equation 6.135 and

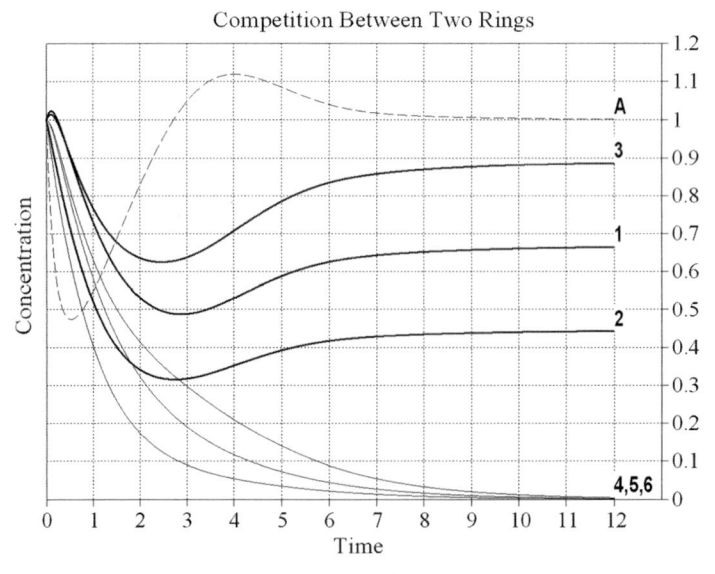

Figure 6.12 Selection between two irreducible clusters (catalytic rings 1-2-3 and 4-5-6), Equation 6.132, under the constraint of limited raw material, A.

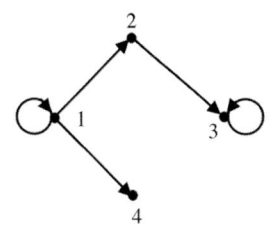

Figure 6.13 Production graph of the adjacency matrix (6.135): an elementarily reducible catalytic network that consists of autocatalytic species and cross-catalytic networks without cycles.

the related graph, Figure 6.13:

$$\mathbf{k} = \left\{ \begin{array}{cccc} k_{11} & 0 & 0 & 0 \\ k_{21} & 0 & 0 & 0 \\ 0 & k_{32} & k_{33} & 0 \\ k_{41} & 0 & 0 & 0 \end{array} \right\} \tag{6.135}$$

The species of an elementarily reducible matrix can be numbered in such a way that the matrix is triangular, that is, all elements above the main diagonal are zero:

$$\mathbf{k} = \left\{ \begin{array}{cccccc} k_{11} & 0 & 0 & . & . & 0 \\ k_{21} & k_{22} & 0 & . & . & 0 \\ k_{31} & k_{32} & k_{33} & . & . & 0 \\ . & . & . & & & . \\ . & . & . & & & . \\ k_{n1} & k_{n2} & k_{n3} & . & . & k_{nn} \end{array} \right\} \quad k_{ij} \geq 0 \tag{6.136}$$

The corresponding production graph contains no catalytic cycles except single loops which are represented by a nonzero diagonal element, $k_{ii} > 0$, of an autocatalytic species.

From Equation 6.115 we infer for the eigenvalues, which coincide with the main diagonal elements of the triangular matrix:

$$\frac{1}{A} = \frac{k_{ii}}{k'_i} \tag{6.137}$$

They define the set of different stationary values of the raw material concentration, A. The analysis of the related eigenvectors is not difficult but requires some effort. The result is that the autocatalytic species with the highest selective value of k_{ss}/k'_s survives the competition. Together with it, all species survive that are successors of the winner in the catalytic production graph, no matter whether or not they are autocatalytic. All other species disappear.

After the consideration of the qualitatively different special cases before, we are now prepared to analyze the general case of a reducible production matrix. An example is the matrix, given here in its normal form (Gantmacher, 1971) with matrix blocks of zeros forming the upper triangle:

$$\mathbf{k} = \begin{Bmatrix} 0 & 0 & k_{13} & 0 & 0 & 0 & 0 \\ k_{21} & 0 & 0 & 0 & 0 & 0 & 0 \\ 0 & k_{32} & 0 & 0 & 0 & 0 & 0 \\ 0 & 0 & k_{43} & k_{44} & 0 & 0 & 0 \\ k_{51} & 0 & 0 & 0 & 0 & k_{56} & 0 \\ 0 & 0 & 0 & 0 & k_{65} & 0 & 0 \\ 0 & 0 & 0 & k_{74} & 0 & k_{76} & k_{77} \end{Bmatrix} \equiv \begin{Bmatrix} k_{11} & 0 & 0 & 0 \\ k_{21} & k_{22} & 0 & 0 \\ k_{31} & 0 & k_{33} & 0 \\ 0 & k_{42} & k_{43} & k_{44} \end{Bmatrix}$$

(6.138)

This matrix can be represented by two different, related graphs, the production graph whose adjacency matrix is **k** and the reduced graph in which the four irreducible diagonal blocks (clusters) of **k** represent the nodes and the nonvanishing nondiagonal blocks represent the directed edges, Figure 6.14.

Similar to Equations 6.123–6.128, we can transform the concentration variables into the ecological picture, taking special care of the nondiagonal blocks here:

$$\dot{\varrho}_{il} = \frac{A}{S_l} \sum_m S_m \left(\langle k_i \rangle_m - \sum_{j \in S_l} \langle k_j \rangle_m \varrho_{il} \right) - \varrho_{il}(k'_i - \langle k' \rangle_l)$$ (6.139)

$$\dot{S}_l = A \sum_m \sum_{j \in S_l} \langle k_j \rangle_m S_m - \langle k' \rangle_l S_l$$ (6.140)

$$\dot{A} = \Phi - A \sum_{l,m} \sum_{j \in S_l} \langle k_j \rangle_m S_m$$ (6.141)

This representation is particularly beneficial if the interaction between the irreducible clusters is materialized by the clusters as a whole, rather than being specific for certain species within each cluster. Under this condition, the internal structure of each cluster is independent of the other clusters, such that in Equation 6.139 the terms with $m \neq l$ disappear as soon as the clusters m and l have reached their internal equilibria, that is, the mutual concentration ratios are controlled by the related eigenvectors of the diagonal matrix block (assuming again a species-unspecific decay, $k'_i = \langle k' \rangle_l$ if $i \in S_l$). Then the model (6.139) can be treated in a similar way as

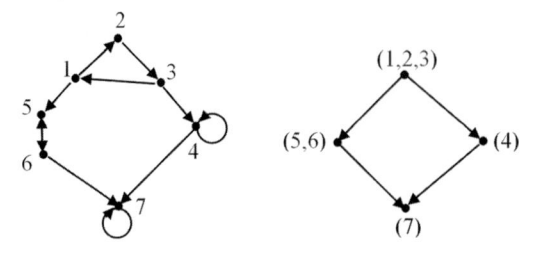

Figure 6.14 Production graph of the reducible adjacency matrix (6.138), and the related reduced graph representing the irreducible clusters as nodes.

Equation 6.129 and the stationary values ϱ_{il} of the internal structure can be inserted into Equations 6.140 and 6.141. The resulting system

$$\dot{S}_l = A \sum_m k^*_{lm} S_m - k'_l S_l \qquad (6.142)$$

$$\dot{A} = \Phi - A \sum_{l,m} k^*_{lm} S_m \qquad (6.143)$$

has the same structure as the original system (6.109), but with the important difference that the matrix $\{k^*_{lm}\}$ is always elementarily reducible, corresponding to the reduced graph as shown in Figure 6.14. The case of an elementarily reducible matrix we considered already earlier. For $t \to \infty$, the survivor is the cluster s with the largest eigenvalue, $1/A = k^*_{ss}/k'_s$, and with it all its successors in the related reduced graph.

This result remains valid even without the simplifications that lead to Equation 6.142, but for its proof one better starts from the eigenvalue problems (6.114) and (6.115) and exploits the spectral properties of nonnegative matrices (Gantmacher, 1971). As an example, we consider a modified version of the model (6.132) that has the production graph shown in Figure 6.15:

$$\mathbf{k} = \begin{pmatrix} 0 & 3/2 & 0 & 0 & 0 & 0 \\ 0 & 0 & 1/2 & 0 & 0 & 0 \\ 4/3 & 0 & 0 & 0 & 0 & 0 \\ 1/3 & 0 & 0 & 0 & 1 & 0 \\ 0 & 0 & 0 & 0 & 0 & 1/8 \\ 0 & 0 & 0 & 1 & 0 & 0 \end{pmatrix} \qquad (6.144)$$

The numerical solution, Figure 6.16, shows the expected behavior with respect to the discussion given before; the ring (1,2,3) is superior with respect to its selective value, but the second ring is its successor in the reduction graph corresponding to Figure 6.15 and cannot disappear. We can refrain here from a more detailed discussion of this model.

We turn our attention now to the model with the alternative constraint of constant overall organization in the generalized version of Equation 6.24 with cross-catalytic growth rates (Eigen, 1971; Ebeling and Feistel, 1977; Feistel, 1979; Bonhoeffer and Stadler, 1993):

$$\dot{X}_i = \sum_j E_{ij} X_j - \langle E \rangle X_i \qquad (6.145)$$

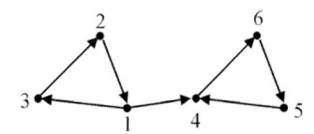

Figure 6.15 Two coupled catalytic rings, corresponding to the reducible adjacency matrix (6.144).

Competition Between Two Coupled Rings

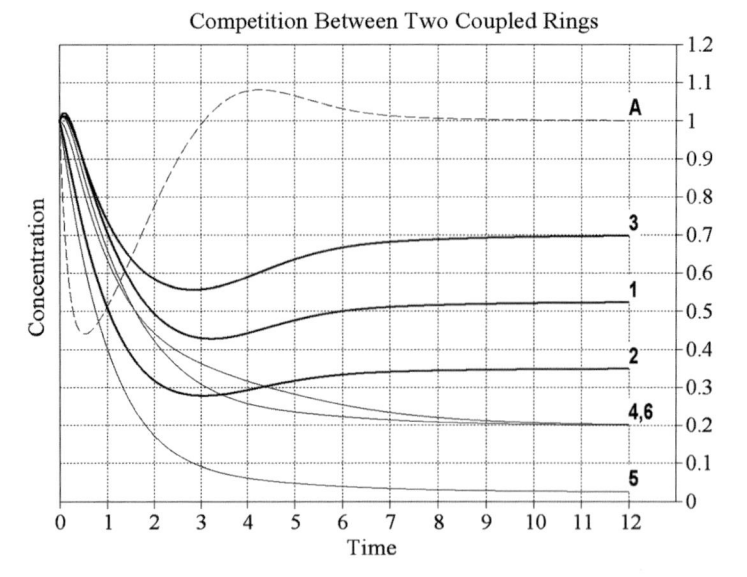

Figure 6.16 Competition between two coupled irreducible clusters (catalytic rings 1-2-3 and 4-5-6), Equation 6.144, under the constraint of limited raw material, A.

where the dilution rate is

$$\langle E \rangle = \frac{1}{C} \sum_{i,j} E_{ij} X_j \quad C = \sum_i X_i \tag{6.146}$$

In matrix notation with the abbreviations (6.111), these equations read

$$\dot{\mathbf{X}} = (\mathbf{E} - \langle E \rangle \mathbf{I})\mathbf{X} \quad \langle E \rangle = \frac{1}{C}\mathbf{1}^T \mathbf{E} \mathbf{X} \quad C = \mathbf{1}^T \mathbf{X} \tag{6.147}$$

It is obvious that the solution of equation $\dot{\mathbf{X}} = 0$ leads to the eigenvalue problem of the matrix $\mathbf{E}$ with the eigenvector $\mathbf{X}$ and the eigenvalue $\langle E \rangle$. In addition to the stationary solution, also the time-dependent integral can be represented analytically. This is achieved by the ansatz (Thompson and McBride, 1974; Jones, Enns, and Rangnekar, 1976; Ebeling and Feistel, 1976a, 1977; Feistel and Ebeling, 1976, 1978b; Jones, 1977a, 1977b; Feistel, 1976b):

$$\mathbf{X} = \mathbf{Z}(t) \exp\left\{ -\int_0^t \langle E \rangle \, dt \right\} \tag{6.148}$$

Inserting this expression into Equation 6.147 yields the linear equation

$$\dot{\mathbf{Z}} = \mathbf{E} \mathbf{Z} \tag{6.149}$$

which has the formal solution

$$\mathbf{Z}(t) = \mathbf{T} \exp(\mathbf{P}t) \mathbf{T}^{-1} \mathbf{Z}(0) \tag{6.150}$$

Here, **T** is the modal matrix (whose columns are the eigenvectors) of **E**, and **P** is the diagonal matrix of the eigenvalues of **E**:

$$\mathbf{ET} = \mathbf{TP} \quad \mathbf{P} = \{\lambda_i \delta_{ij}\} \tag{6.151}$$

Making use of the relation

$$\int_0^t \langle E \rangle dt = \int_0^t \frac{\mathbf{1}^T \mathbf{EX}}{\mathbf{1}^T \mathbf{X}} dt = \int_0^t \frac{\mathbf{1}^T \mathbf{EZ}}{\mathbf{1}^T \mathbf{Z}} dt = \int_0^t \frac{\mathbf{1}^T \dot{\mathbf{Z}}}{\mathbf{1}^T \mathbf{Z}} dt = \ln(\mathbf{1}^T \mathbf{Z}) - \ln C \tag{6.152}$$

Equation 6.148 results in the time-dependent solution for $\mathbf{X}(t)$:

$$\mathbf{X}(t) = \frac{\mathbf{T} \exp(\mathbf{P}t) \mathbf{T}^{-1} \mathbf{X}(0)}{\mathbf{1}^T \mathbf{T} \exp(\mathbf{P}t) \mathbf{T}^{-1} \mathbf{X}(0)} \mathbf{1}^T \mathbf{X}(0) \tag{6.153}$$

For $t \to \infty$, all those species survive which possess a nonzero component in the eigenvector of **E** that belongs to the eigenvalue λ_s with the largest real part.

By the natural condition that phase trajectories of nonnegative variables cannot cross the boundary of the positive cone, the matrix **E** cannot possess negative off-diagonal elements. If ε is any number chosen equal or less than the smallest diagonal element, then the matrix $\mathbf{E}'$

$$\mathbf{E}' = \mathbf{E} - \varepsilon \mathbf{I} \tag{6.154}$$

is nonnegative. Replacing **E** in Equation 6.147 by Equation 6.154, we find

$$\begin{aligned}
\dot{\mathbf{X}} &= \left(\mathbf{E} - \frac{1}{C}(\mathbf{1}^T \mathbf{EX}) \cdot \mathbf{I}\right) \mathbf{X} \\
&= \left(\mathbf{E}' - \frac{1}{C}(\mathbf{1}^T \mathbf{E}' \mathbf{X}) \cdot \mathbf{I}\right) \mathbf{X} + \varepsilon \left(\mathbf{I} - \frac{1}{C}(\mathbf{1}^T \mathbf{IX}) \cdot \mathbf{I}\right) \mathbf{X} \\
&= \left(\mathbf{E}' - \frac{1}{C}(\mathbf{1}^T \mathbf{E}' \mathbf{X}) \cdot \mathbf{I}\right) \mathbf{X}
\end{aligned} \tag{6.155}$$

Since the model is invariant with respect to transformations of the form (6.154), we may assume nonnegative matrices **E** without loss of generality.

In perfect analogy to the matrices $\mathbf{k}'^{-1}\mathbf{k}$ or $\mathbf{k}$ we can investigate the spectral and reducibility properties of the matrix **E**. The introduction of the ecological picture for the description of the competition between irreducible clusters of species is even simpler than under the condition of constant raw material supply because there is not time dependence of the raw material buffer (Feistel and Ebeling, 1978b).

For stationary states, the value of $\langle E \rangle$ equals an eigenvalue of **E**, and for a stable state, this is the maximum eigenvalue. Hence, $\langle E \rangle$ is a static evolution principle:

$$\lim_{t \to \infty} \langle E \rangle = \lambda_{max} \tag{6.156}$$

For the general cross-catalytic system, however, it is neither a dynamical nor a complete evolution principle. A simple example is sufficient to prove this statement:

$$\mathbf{E} = \begin{Bmatrix} 0 & 1/8 \\ 2 & 0 \end{Bmatrix} \quad X_1(0) = C \quad X_2(0) = 0$$

$$\langle E \rangle_{t=0} = 2 \quad \lim_{t \to \infty} \langle E \rangle = 1/2 \tag{6.157}$$

The lowering of $\langle E \rangle$ in this case is a result of the internal equilibration of the cluster; the selection between clusters always results in an increase of $\langle E \rangle$.

6.6
Selection and Coexistence

The reaction equations which we considered in the previous section,

$$A + X_i \xrightarrow{k_i} 2X_i \tag{6.158}$$

chemically represent a simplified, extreme nonequilibrium situation. The neglected backward reaction of (6.158)

$$A + X_i \xleftarrow{k_{-i}} 2X_i \quad i = 1, 2, \ldots, n \tag{6.159}$$

implies relevant consequences for the selection behavior of the system. To investigate this effect, we consider the related formal-kinetic equations under the condition of constant raw material supply:

$$\dot{A} = \Phi - \sum_i k_i A X_i + \sum_i k_{-i} X_i^2 \tag{6.160}$$

$$\dot{X}_i = k_i A X_i - k'_i X_i - k_{-i} X_i^2 \tag{6.161}$$

In comparison to the models studied before, we expect a more complicated behavior as a consequence of the additional nonlinearities. Before we go into those mathematical details, we illustrate the new features of the system (6.160), (6.161) with respect to some special situations.

First, we imagine that the raw material concentration, A, is kept at a constant level by means of a suitable control of the supply, $\Phi(t)$. Provided that $k_{-i} > 0$, Equation 6.161 for the competing species X_i take the form of the so-called logistic growth law:

$$\dot{X}_i = k_{-i}(C_i - X_i)X_i \tag{6.162}$$

where C_i is regarded as the storage capacity:

$$C_i = (k_i A - k'_i)/k_{-i} \tag{6.163}$$

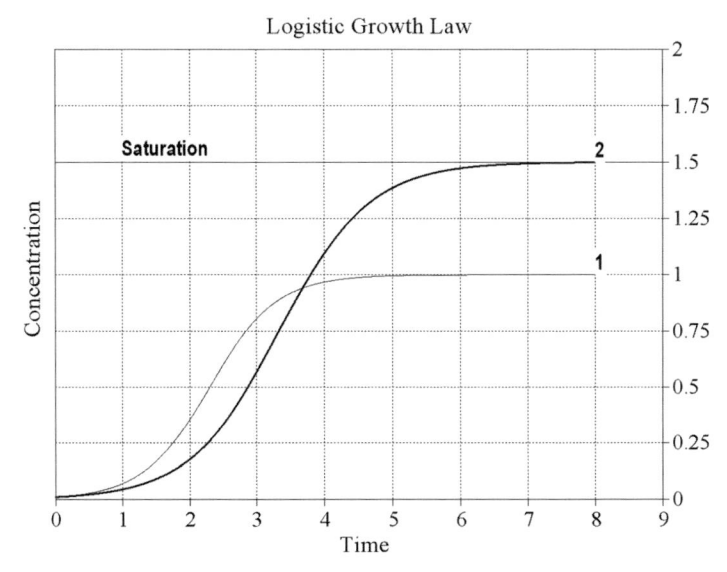

Figure 6.17 Logistic growth of two species, Equations 6.162 and 6.164, with the parameters $k_{-1}=2$, $C_1=1$, $k_{-2}=1$, $C_2=1.5$, until the saturation values C_i are approached asymptotically by the concentrations.

At small concentrations, $X_i \ll C_i$, the species follow an exponential growth (or decay) law which turns into an asymptotic approach to the saturation concentration, $X_i^0 = C_i$, in the long-term limit, Figure 6.17. The analytical solution of the differential equation 6.162 is easily found by separation of variables and decomposition into partial fractions:

$$X_i(t) = C_i \frac{X_i(0)}{X_i(0) + (C_i - X_i(0))\exp(-k_{-i}C_i t)} \quad (6.164)$$

As an aside, we mention that the logistic Equation 6.162 is a fundamental model of mathematical ecology (Wilson and Bossert, 1973), usually written in the form $\dot{x} = rx(1-x/K)$. When ecological niches are newly formed, the species with the highest reproduction rate r have a temporary advantage ("r-selection"), Figure 6.17, while in the long run those with the highest capacity K survive ("K-selection"). In this section we consider only survival in the limit $t \to \infty$ under stationary boundary conditions, that is, selection of the K type in this terminology.

Returning to Equation 6.162, the capacities C_i may have either sign, Equation 6.163, depending on the particular concentration A set by the boundary conditions. Species with negative C_i, Equation 6.164, must disappear; hence, from the condition $C_i > 0$ we infer a critical raw material concentration required as the "margin of subsistence" of this species:

$$A > A_i^c \equiv k_i'/k_i \quad (6.165)$$

When compared to the current value of A, the property A_i^c of a given species considered decides on whether or not it will survive in the system. It appears reasonable to order the species with respect to their margins of subsistence: $A_1^c \leq A_2^c \leq A_3^c \leq \cdots \leq A_n^c$, that is,

$$k_1'/k_1 \leq k_2'/k_2 \leq k_3'/k_3 \leq \cdots \leq k_n'/k_n \tag{6.166}$$

We may imagine these values as ticks on an A-axis, Figure 6.18.

Adding the current value of A to this graph, it is obvious that only species with their ticks on the left of this limiting line can survive. This qualitative behavior remains the same when we release the previously locked value of A and let it freely adjust, subject to the dynamics (6.160). If a species s survives the selection process, then also any species $i < s$ with respect to the order relation (6.166). Whether or not species with different values of A_i^c can coexist depends on the existence of a gap between A_s^c and $A^{(s)}$, the stationary value of A in the presence of the master species s. This value is easily obtained from Equations 6.160 and 6.161:

$$A^{(s)} = \frac{\Phi k_{-s} + k_s'^2}{k_s k_s'} \quad X_s^{(s)} = C_s^{(s)} = \frac{\Phi}{k_s'} \tag{6.167}$$

as well as the asymptotic stability of this state, as long as the species s is the only inhabitant of the system. We compute now the gap width, that is, the excess over the margin of subsistence:

$$\Delta A^{(s)} = A^{(s)} - A_s^c = \frac{\Phi k_{-s}}{k_s k_s'} = \frac{k_{-s}}{k_s} C_i^{(s)} = \frac{k_{-s}}{k_s} X_i^{(s)} > 0 \tag{6.168}$$

As a result of the backward reaction, Equation 6.159, the species builds up a buffer of raw material $\Delta A^{(s)}$, Figure 6.19, proportional to its stationary concentration and the backward reaction rate, $k_{-s} > 0$.

The gap $\Delta A^{(s)}$ forms something formally similar to an "ecological niche" for "worse" species i if their margin of subsistence, A_i^c, fits into the interval (Ebeling and Feistel, 1976a):

$$A_s^c < A_i^c < A^{(s)} \tag{6.169}$$

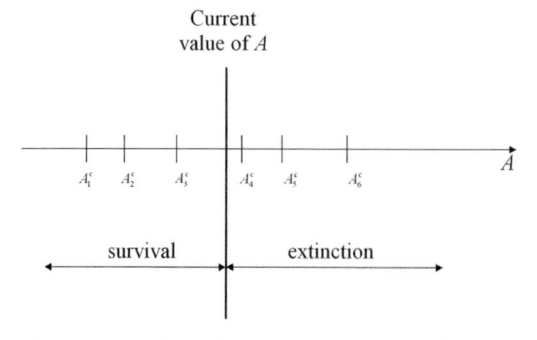

Figure 6.18 Ordering the species in the A-axis with respect their margins of subsistence relative to the current level A of the raw material illustrated the condition for survival of the species.

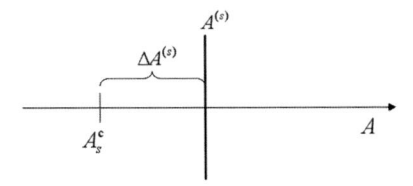

Figure 6.19 Position of the stationary raw material level, $A^{(s)}$, relative to the margin of subsistence, A_s^c.

since it is evident from Equation 6.162 that some added small X_i with the property (6.169) will grow exponentially, that is, the parent system is unstable with respect to the emergence of this species. The backward reaction thus appears as an "altruistic behavior" of the master species that creates the "survival buffer" (6.168) and supports "weaker" species to survive (recall that added superior species, $A_i^c < A_s^c$, will survive even without the backward reaction, Section 6.3). Another interpretation is that the backward reaction permits "parasites" to survive in the wake of the master species.

To answer the question of whether the second species i can coexist with the master species in a stable stationary regime, we investigate now the general case of the system (6.160), (6.161) with n species. From Figure 6.18 it is obvious that of the species ordered their margin of subsistence only a subset of m species $i \leq m$ with the lowest values of $A_i^c \leq A_m^c$ can survive. The number m is implicitly given by the related stationary raw material concentration, $A^{(\leq m)}$, that is, by the inequality (Ebeling and Schmelzer, 1980):

$$A_m^c < A^{(\leq m)} < A_{m+1}^c \tag{6.170}$$

The stationary concentration $A^{(\leq m)}$ follows from Equation 6.6.3 as

$$A^{(\leq m)} = \left(\Phi + \sum_{i=1}^{m} \frac{k_i'^2}{k_{-i}} \right) \bigg/ \sum_{i=1}^{m} \frac{k_i'}{k_i k_{-i}} \tag{6.171}$$

and determines in turn the stationary concentrations of the species as

$$X_i^{(\leq m)} = C_i^{(\leq m)} = \frac{k_i A^{(\leq m)} - k_i'}{k_{-i}} \quad \forall i \leq m \tag{6.172}$$

as well as $X_i^{(\leq m)} = 0 \, \forall i > m$, for the extinct subset of species. For the stability analysis of the solution (6.171), (6.172), we calculate the Jacobian J of the system (6.160), (6.161) at this point:

$$J = \begin{Bmatrix} J_{mm} & J_{mn} \\ 0 & J_{nn} \end{Bmatrix} \tag{6.173}$$

The elements of this block matrix are

$$J_{mm} = \begin{Bmatrix} -\sum_{i=1}^{m} k_i X_i^{(\leq m)} & -k_1' & -k_2' & \cdots & -k_m' \\ k_1 X_1^{(\leq m)} & -k_{-1} X_1^{(\leq m)} & 0 & \cdots & 0 \\ k_2 X_2^{(\leq m)} & 0 & -k_{-2} X_2^{(\leq m)} & \cdots & 0 \\ \cdots & \cdots & \cdots & \cdots & \cdots \\ k_m X_m^{(\leq m)} & 0 & 0 & \cdots & -k_{-m} X_m^{(\leq m)} \end{Bmatrix}$$

$$\tag{6.174}$$

and

$$J_{nn} = \left\{ \left(k_i A^{(\leq m)} - k_i'\right) \delta_{ij} \right\} \quad \forall i,j > m \tag{6.175}$$

The matrix J_{mn} is irrelevant for the stability analysis.

The eigenvalues of J_{nn}, $k_i(A^{(\leq m)} - A_i^c)$, are negative because of the inequality (6.170). The characteristic polynomial for the eigenvalues p of J_{mm} takes the form (Schmelzer, 1979; Ebeling and Schmelzer, 1980)

$$p + \sum_{i=1}^{m} \frac{(k_i' + p) k_i X_i^{(\leq m)}}{k_{-i} X_i^{(\leq m)} + p} = 0 \tag{6.176}$$

When Equation 6.176 is separated into real and imaginary parts after inserting $p = \lambda + i\omega$, each coefficient of the polynomial for λ can be shown to be positive, that is, all eigenvalues of J_{mm} possess negative real parts.

We summarize the results of our analysis. With respect to the raw material concentration A required for survival, the margins of subsistence, $A_i^c = k'_i / k_i$, play the role of selective values for the species i. The asymptotic stationary concentration of A permits the coexistence of m species with the lowest values of A_i^c, where m is determined by the values of Φ and the set of parameters k_i, k_{-i}, and k_i' of the m species. In each positive subspace of the concentrations $X_1, \ldots, X_n$ there exists at most one asymptotically stable stationary state. In each m-dimensional nonnegative there are at most $2^m - 1$ stationary states, of which only the one with the lowest stationary concentration $A^{(\leq m)}$ is asymptotically stable.

Consequently, the function $1/A^{(\leq m)}$ is a static evolution principle in the sense of Section 6.2. With its help we can describe the solution of the coexistence problem as follows.

For the 2^n possible subsets of species we calculate the partial sums

$$S_1 = \sum_i \frac{k_i'}{k_i k_{-i}} \quad S_2 = \sum_i \frac{k_i'^2}{k_{-i}} \tag{6.177}$$

The particular subset of species associated with the maximum value of

$$\frac{1}{A} = \frac{S_1}{\Phi + S_2} \tag{6.178}$$

forms a stable state of coexistence.

The previous statements must be slightly modified if certain properties k_i, A_i^c, and so on, take pairwise exactly equal values. Here we refrain from the analysis of the various related special cases that are practically improbable and structurally unstable.

The possibility of coexistence can be illustrated by a numerical example. Figure 6.20 shows that the number of coexisting species increases with the available raw material supply.

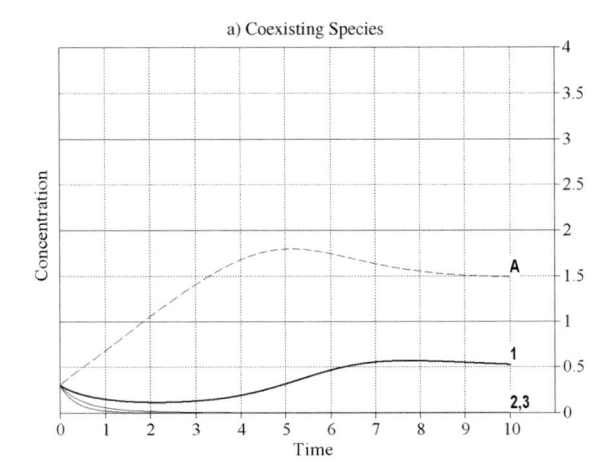

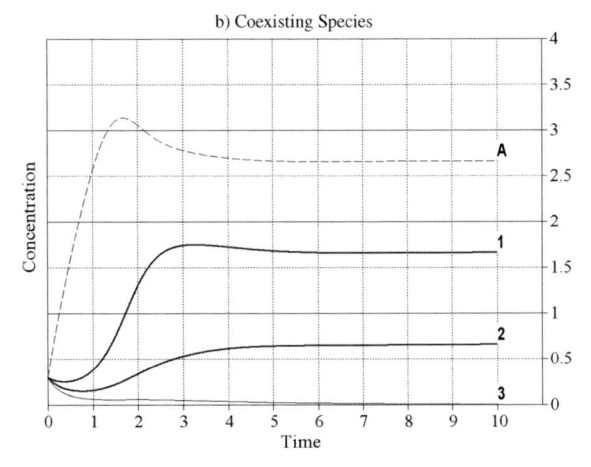

Figure 6.20 Coexisting species in the selection model (6.160), (6.161) with the parameters $k_i = k_{-i} = 1$, $k'_1 = 1$, $k'_2 = 2$, $k'_3 = 3$, that is, the margins of subsistence of $A^c_1 = 1$, $A^c_2 = 2$, $A^c_3 = 3$, indicated by horizontal lines. Panels: (a) $\Phi = 0.5$, (b) $\Phi = 3.0$, (c) $\Phi = 5.0$.

A generalized, ecologically motivated version of the model (6.160), (6.161) with the kinetic equations

$$\dot{A} = \Phi - \sum_i \frac{1}{Y_i} \frac{\mu_i A X_i}{k_i + A} \tag{6.179}$$

$$\dot{X}_i = \frac{\mu_i A X_i}{k_i + A} - k'_i X_i - k_{-i} X_i^2 \tag{6.180}$$

was investigated by Ebeling and Schmelzer (1980). Here, Y_i^{-1} is an efficiency factor which measures how much raw material is consumed for the reproduction of X_i. The growth law is of the Michaelis–Menton type and shows saturation effects at high raw

c) Coexisting Species

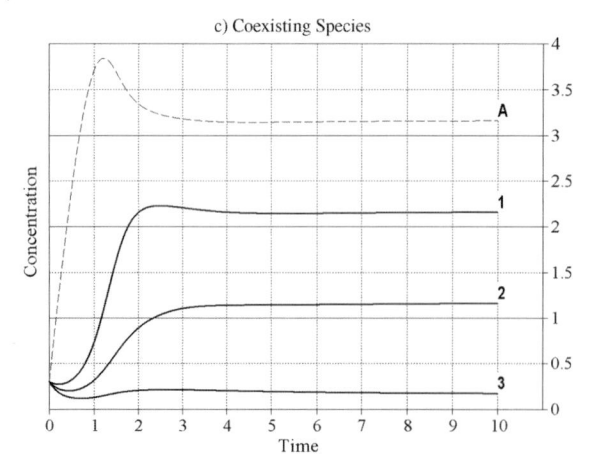

Figure 6.20 *(Continued)*

material concentrations. The quadratic decay term in Equation 6.180 simulates self-inhibition at high densities such as overpopulation, epidemic diseases, and so on. In this case, the necessary and sufficient condition for coexistence of m species reads

$$\frac{k_m k'_m}{\mu_m - k'_m} < A^{(\leq m)} < \frac{k_{m+1} k'_{m+1}}{\mu_{m+1} - k'_{m+1}} \tag{6.181}$$

This result suggests that the qualitative conclusions drawn in this section are of general importance rather than being restricted to the specific model analyzed here.

As another modification of the model (6.160), (6.161) was obtained by its embedding in a flow reactor with an unspecific dilution rate D (Feistel, 1979):

$$\dot{A} = \Phi - \sum_i k_i A X_i + \sum_i k_{-i} X_i^2 - DA \tag{6.182}$$

$$\dot{X}_i = k_i A X_i - k'_i X_i - k_{-i} X_i^2 - D X_i \tag{6.183}$$

For this model it was shown that the reciprocal raw material concentration is a local Lyapunov function and a static evolution principle for the coexisting states, that is, this function possesses almost all properties required for a dynamic evolution principle (Section 6.2).

6.7
Hyperselection

Among organic molecules in the primordial "soup", simple autocatalysis was probably an exceptional case while cross-catalysis, that is, catalytic acceleration of the production of a certain species in the presence of another species is a more plausible hypothetical scenario. Under nonequilibrium conditions, if such mutual catalytic relations form a closed loop among the present species, the members of the reaction cycle may increase their abundance significantly in comparison to the occasional, noncatalytic formation (Ebeling and Feistel, 1982; Ebeling et al., 2006). Mutual catalysis between nucleic acids and proteins in the form of the "hypercycle" was suggested by Eigen (1971) as a model reaction of that type; its properties were investigated mathematically in great detail (Eigen and Schuster, 1977, 1978, 1979). The essential qualitative difference of such systems in comparison to the models studied earlier in this chapter is the appearance of a formation rate that is a nonlinear function of the species concentration. This nonlinearity has two important consequences for selection and evolution processes: (i) there is a "nucleation" effect, that is, species with subcritical abundance must always disappear, and, (ii) there is an explosive, faster-than-exponential growth of supercritical concentrations. None of the both is favorable for gradual improvement of the reproduction rate by natural selection between established parent species and single new mutants.

As a tutorial example, we consider a simple nonlinear autocatalytic model in the form of the quadratic Schlögl reaction (Schlögl, 1972) in a flow reactor with a flushing rate D

$$A + 2X \xrightarrow{k} 3X \tag{6.184}$$

More general cases of cross-catalytic growth rates will be investigated in the subsequent sections. In Equation 6.184, the concentration of the buffer species A is held constant and backward reactions are assumed to be negligible. The differential equation

$$\frac{dX}{dt} = kAX^2 - DX \tag{6.185}$$

can be solved analytically by separation of the variables (Ebeling and Feistel, 1976a),

$$\int \frac{dX}{X\left(X - \frac{D}{kA}\right)} = \int kA \, dt \tag{6.186}$$

for the initial condition, $X(t=0) = X_0$, with the result

$$X(t) = \frac{DX_0}{kAX_0 + (D - kAX_0)\exp(Dt)} \tag{6.187}$$

In dependence on the initial abundance, there are three qualitatively different branches of the solution:

1) if $kAX_0 < D$, the species X disappears exponentially from the system,
2) if $kAX_0 = D$, the species X remains constant (unstable solution), or
3) if $kAX_0 > D$, the species X grows beyond all limits.

Interestingly, the singularity in the case (3) occurs already after a finite latency period of

$$t_\infty = -\frac{1}{D}\ln\left(1 - \frac{D}{kAX_0}\right) \tag{6.188}$$

This "explosion" requires an infinitely fast supply with the raw material A to keep its concentration constant, which is physically unreasonable. For more realistic conditions, we add to Equation 6.185 an equation for constant supply:

$$\frac{dA}{dt} = \Phi - kAX^2 - DA \tag{6.189}$$

Since the concentration $A(t)$ is now a function of time, Equation 6.185 can no longer be solved analytically in full generality. Nevertheless, we can obtain analytical results if we introduce the "total biomass," $M = A + X$. From the addition of Equations 6.185 and 6.189 we infer the linear differential equation

$$\frac{dM}{dt} = \Phi - DM \tag{6.190}$$

which is easily solved as

$$M(t) = \frac{\Phi}{D} + \left(M_0 - \frac{\Phi}{D}\right)\exp(-Dt) \tag{6.191}$$

Independent of the initial condition for X, the value of M is always approaching a unique stable steady state:

$$\lim_{t\to\infty} M(t) = \frac{\Phi}{D} \tag{6.192}$$

Since any initial deviation from the solution (6.192) will fade away with the relaxation time $1/D$, it is sufficient to consider initial conditions consistent with Equation 6.192. Thus, we may dynamically eliminate the "slaved" variable $A(t)$ from Equation 6.185 by means of

$$A(t) = \frac{\Phi}{D} - X(t) \tag{6.193}$$

The resulting equation is of the form of the cubic Schlögl reaction (Schlögl, 1972):

$$\frac{dX}{dt} = k\left(\frac{\Phi}{D} - X\right)X^2 - DX \equiv f(X) \tag{6.194}$$

The properties of this equation are very well known; it is a prototype for bistable dynamical systems (Chapter 5). The stationary solutions are

$$X_1^0 = 0 \quad X_{2/3}^0 = \frac{\Phi}{2D} \pm \sqrt{\left(\frac{\Phi}{2D}\right)^2 - \frac{D}{k}} \tag{6.195}$$

Nontrivial stationary solutions exist only if the condition

$$k\Phi^2 \geq 4D^3 \tag{6.196}$$

is satisfied. The solution X_1^0 is always stable because of $f'(0) = -D < 0$. Of the pair, $X_{2/3}^0$, the larger value, X_3^0, corresponds to a stable solution, the smaller one, X_2^0, to an unstable solution and to a critical threshold. Initial values of X with $X < X_2^0$ will result in the trivial final state, $X_1^0 = 0$, that is, self-replicating species with initially infinitesimally low concentrations have no chance to survive. The solutions (6.195) are similar to those of Equation 6.187 but the singularity is now replaced by a finite value.

After those preparatory considerations, we may now look at the competition among species with nonlinear autocatalysis subject to dynamical equations similar to Equation 6.185:

$$\frac{dX_i}{dt} = k_i A X_i^2 - D X_i \tag{6.197}$$

and with a raw-material supply, similar to Equation 6.189:

$$\frac{dA}{dt} = \Phi - \sum_i k_i A X_i^2 - DA \tag{6.198}$$

The "total biomass", $M = A + \sum_i X_i$, is controlled by the same Equation 6.190 as before. Any asymptotic states are therefore restricted to the condition

$$\lim_{t \to \infty} A(t) \leq \lim_{t \to \infty} M(t) = \frac{\Phi}{D} \tag{6.199}$$

From Equation 6.197 we infer that any new species X_μ that emerges in a given asymptotic state can only grow if its initial concentration is supercritical:

$$k_\mu X_\mu > \frac{D}{A} \geq \frac{D^2}{\Phi} \tag{6.200}$$

Evidently, a higher reproduction coefficient k_μ alone is insufficient here to replace a parent species with a lower value of that coefficient but with much higher concentration. Since new mutants appear with very low concentrations (single molecules), we conclude that deterministic, spatially homogeneous models for nonlinear catalysis of the type (6.197) are not suited for gradual improvement of their properties by Darwinian selection.

For completeness, we derive here some additional general properties of the system (6.197), (6.198) regarding the competition between its species. For this purpose, it is usually helpful to study the dynamics of the relative fractions of the competitors, as it was done in the preceding sections when complete analytical solutions of the differential equations are unavailable. We introduce the fraction variables

$$\varrho_i = \frac{X_i}{B} \quad B \equiv \sum_j X_j \tag{6.201}$$

and obtain from Equation 6.197 the set of differential equations:

$$\frac{d\varrho_i}{dt} = \varrho_i AB(k_i\varrho_i - \langle k\varrho \rangle) \tag{6.202}$$

Fractions grow as long as their individual reproduction, $E_i \equiv k_i\varrho_i$, exceeds the average rate

$$\langle k\varrho \rangle \equiv \langle E \rangle \equiv \sum_i k_i\varrho_i^2 \tag{6.203}$$

for which we find the dynamical evolution principle (Feistel, 1979),

$$\frac{d}{dt}\langle E \rangle = 2\sum_i k_i\varrho_i \frac{d\varrho_i}{dt} = 2AB\{\langle E^2 \rangle - \langle E \rangle^2\} \geq 0 \tag{6.204}$$

We find that the competition maximizes the mean reproduction rate, $\langle E \rangle = \langle k\varrho \rangle$, of the community. Members with a supercritical rate will grow and increase their rate further. Equation 6.202 can be written in the form

$$\frac{dE_i}{dt} = AB(E_i - \langle E \rangle)E_i \tag{6.205}$$

The species with the highest selective value E_i wins the competition, but in addition to the reaction constants, this value depends also on the initial concentration, and the value changes during the competition process according to Equation 6.205, and eventually approaches the winner's k_i value, according to Equation 6.203.

New species starting with infinitesimal initial concentrations have negligible selective values and cannot overturn the established dynamical regime. This kind of once-for-ever selection inhibits evolutionary progress; this is, for example, well known from social systems where dictators, family clans, or other sovereigns may use their political, economic, criminal, or military power to nonlinearly reinforce their leading position independent of their personal qualification, or of the quality of their established social structure. In models, hyperselection can be overcome when stochastic effects and spatially inhomogeneous structures are included, which in social systems have their analogy in the form of riots or revolutions. There is some similarity with phase transitions of the first kind, where in a metastable state thermal fluctuations are responsible for the spontaneous formation of spatial nuclei which may grow and finally convert the entire volume into a new, stable phase (Schlögl, 1972; Ebeling, Feistel, and Jiménez-Montaño, 1977; Poston and Stuart, 1978; Feistel and Ebeling, 1989, 2005).

6.8
Selection in Ecological Systems

Selection is a fundamental mechanism of change in ecological systems; in this field of science the theory of selection processes is evidently highly advanced (Lotka 1910;

Volterra, 1931; Kolomogorov, 1936; Goel, Maitra, and Montroll, 1971; May, 1973; Biswas, Gupta, and Karmakar, 1977; Bojadziev, 1978; Svireshev and Logofet, 1978). We restrict our consideration here to so-called Lotka–Volterra systems:

$$\frac{dN_i}{dt} = N_i\left(\varepsilon_i - \sum_{j=1}^{n} \gamma_{ij} N_j\right) \quad i = 1, \ldots, n \tag{6.206}$$

the standard predator-prey model of mathematical ecology. Here, N_i is the number of individuals of a species i, and ε_i describes the multiplication rates of autotrophic species as well as mortalities,

$$\text{autotrophic species}: \quad \varepsilon_i > 0, \; \gamma_{ij} \geq 0 \text{ for } j = 1, \ldots, n \tag{6.207}$$

$$\text{heterotrophic species}: \quad \varepsilon_i < 0, \; \gamma_{ij} < 0 \text{ for at least one } j \tag{6.208}$$

and γ_{ij} is the so-called community matrix that describes the population change as a result of binary interaction processes between the members of two species (predator–prey interaction, mutual or self-inhibition, competition, and so on).

Despite of the simple form of Equation 6.206, this system represents a wealth of special models, such as the model of plain competition at constant overall organization (Section 6.3). It may be helpful to replace the particle numbers of Equation 6.206 by partial Gibbs energies:

$$g_i = \mu_i N_i \tag{6.209}$$

where μ_i is the chemical potential of the species i (Feistel and Ebeling, 1981; Mauersberger, 1981; Orlob, 1983; Jørgensen and Svireshev, 2004)

$$\mu_i = kT \ln N_i \tag{6.210}$$

In this simplified picture, an organism is considered as a single "macroscopic molecule." The related system reads

$$\frac{dg_i}{dt} = N_i \frac{d\mu_i}{dt} + \mu_i \frac{dN_i}{dt} = g_i\left(a_i - \sum_{j=1}^{n} b_{ij} g_j\right) \tag{6.211}$$

where

$$a_i = \left(1 + \frac{kT}{\mu_i}\right)\varepsilon_i, \; b_{ij} = \left(1 + \frac{kT}{\mu_i}\right)\frac{\gamma_{ij}}{\mu_j} \tag{6.212}$$

As a rule, the factors in front of (6.212) can be set to unity because $kT \ll |\mu_i|$ as long as $N_i \gg 1$, Equation 6.210. In ecology, the term "biomass" is often used rather than g_i; we refrain from this use since this property is physically not well defined and somewhat questionable. Since free energy is not produced by binary interaction between the species, we have

$$g_i g_j (b_{ij} + b_{ji}) \geq 0 \tag{6.213}$$

that is, the symmetric part of the community matrix **b** is nonnegative:

$$\mathbf{b}^S = \frac{1}{2}\left(\mathbf{b} + \mathbf{b}^T\right) \geq 0 \tag{6.214}$$

and describes the rate of collective dissipation (entropy production) of the population. The alternative, antisymmetric part,

$$\mathbf{b}^A = \frac{1}{2}\left(\mathbf{b} - \mathbf{b}^T\right) \tag{6.215}$$

describes the flow of Gibbs energy in the food web, that is, the trophic structure of the ecosystem. More precisely, if the negative elements of $\left(-\mathbf{b}^A\right)$ are replaced by zeros, the remaining nonnegative matrix is an adjacency matrix of a directed graph that represents the transfer of Gibbs energy between the species, that is, the "food web" (Figure 6.21).

This graph is usually considered as cycle free and represents a semi order in this case, permitting a classification of the species into trophic levels. The lowest level, sources of the graph, is formed by autotrophic organisms (plants, phytoplankton in marine ecosystems), the sources of Gibbs energy for the entire community. The next level consists of herbivores, or zooplankton, up to the highest level represented by the sinks of the graph, the top predators (Wilson and Bossert, 1973; Stugren, 1978). There may be several intermediate levels of carnivores; their particular level is not necessarily unique if their precursors and successors in the graph are more than two levels apart. If the matrix $\mathbf{b}^A$ belongs to a cycle-free food web, its elements can numbered in such a way that below the main diagonal are only nonnegative and above it only nonpositive elements. Then, herbs, herbivores, and carnivores are each located in groups of similar indices, and their associated trophic levels are reflected by the values of their indices (Figure 6.21).

It is a natural requirement that an ecological system cannot develop infinite energy g_i in its temporal development. This can, for instance, be achieved by a positive definite matrix **b** and the resulting logistic growth of the species. If the values of the g_i

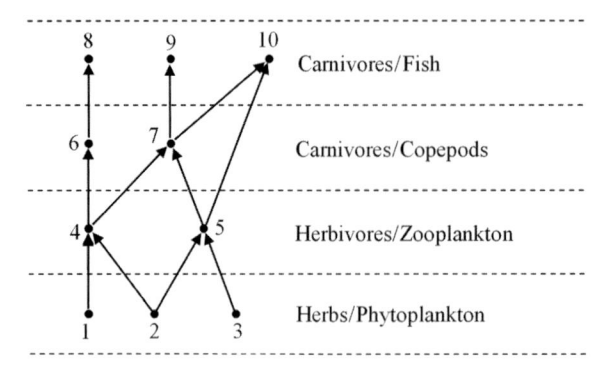

Figure 6.21 Schematic of the trophic structure of an ecosystem.

remain finite, we may investigate the asymptotic behavior of the time averages

$$\langle g_i \rangle = \lim_{t \to \infty} \frac{1}{t} \int_0^t g_i(t') dt' \tag{6.216}$$

If the system (6.211) is started with positive initial values,

$$g_i(0) > 0 \tag{6.217}$$

Equations 6.211 can be divided by g_i and the time average taken:

$$\left\langle \frac{d}{dt} \ln g_i \right\rangle = a_i - \sum_j b_{ij} \langle g_j \rangle \tag{6.218}$$

Asymptotically, any time average, Equation 6.216, of a time derivative converges to zero if the function itself remains finite (Landau and Lifschitz, 1967a):

$$\lim_{T \to \infty} \frac{1}{T} \int_0^T f'(t) dt = \lim_{T \to \infty} \frac{f(T) - f(0)}{T} = 0 \tag{6.219}$$

Hence, there are three possibilities for Equation 6.218. First, the case,

$$\left\langle \frac{d}{dt} \ln g_i \right\rangle > 0 \tag{6.220}$$

corresponds to an unlimited exponential growth of g_i which is excluded here since it contradicts the initial assumption of finite energies. Second, the case,

$$\left\langle \frac{d}{dt} \ln g_i \right\rangle = 0 \tag{6.221}$$

holds for the regular species $1, \ldots, s$ whose abundances remain finite for $t \to \infty$. The third case

$$\left\langle \frac{d}{dt} \ln g_i \right\rangle < 0 \tag{6.222}$$

applies to the species $s+1, \ldots, n$ which disappear exponentially, that is, $\langle g_i \rangle = 0$.

This way we have received the following information. If a_i and b_{ij} describe a Lotka–Volterra scheme with finite particle numbers of n species, then there exists a set of positive numbers $x_1 > 0, x_2 > 0, \ldots, x_s > 0, s \leq n$ (with suitably numbered species) that obeys the properties

$$\sum_{j=1}^s b_{ij} x_j = a_i \quad i = 1, \ldots, s \tag{6.223}$$

and

$$\sum_{j=1}^{s} b_{ij}x_j > a_i \quad i = s+1,\ldots,n \tag{6.224}$$

If the matrix $\mathbf{b}^S$ is positive definite, Equations 6.213 and (6.211) possess a positive stationary state, $g_i^0 > 0$, $i = 1,\ldots, n$, then a Lyapunov function L can be found in the literature (similar to the relative entropy suggested by Gibbs, 1902):

$$L(\mathbf{g}, \mathbf{g}^0) = \sum_i \left\{ g_i - g_i^0 \left(\ln \frac{g_i}{g_i^0} + 1 \right) \right\} \tag{6.225}$$

which proves the asymptotic stability of the steady state $\mathbf{g}^0 = \{g_i^0\}$. A similar function is available for the variables N_i rather than g_i if the related community matrix γ^S is positive definite. In the terminology of Section 6.2, the function

$$G(\mathbf{g}, \mathbf{g}^0) = -\sum_i (g_i - g_i^0 \ln g_i) \tag{6.226}$$

is an incomplete dynamic segregation principle because it is defined only on the positive cone, $g_i > 0$, under the additional condition that this cone contains the stationary state, $\mathbf{g}^0$.

We are going to show, however, that Equation 6.226 is even a dynamic selection principle rather than just a less general segregation principle. For a nonnegative stationary state given by

$$g_i^0 = 0 \quad i = s+1,\ldots, n \tag{6.227}$$

and

$$g_i^0 = x_i > 0 \quad i = 1,\ldots, s \tag{6.228}$$

where x_i obeys Equations 6.223 and 6.224, the function G from Equation 6.226 takes the form

$$G(\mathbf{g}, \mathbf{g}^0) = -\sum_{i=1}^{s} (g_i - g_i^0 \ln g_i) - \sum_{i=s+1}^{n} g_i \tag{6.229}$$

From Equation 6.227 and $\ln x \leq (x-1)$ we infer

$$G(\mathbf{g}, \mathbf{g}^0) - G(\mathbf{g}^0, \mathbf{g}^0) < -\sum_{i=s+1}^{n} g_i < 0 \quad \text{if} \quad \mathbf{g} \neq \mathbf{g}^0 \tag{6.230}$$

For the time derivative of Equation 6.229,

$$\frac{dG}{dt} = -\sum_{i=1}^{s} \left(1 - \frac{g_i^0}{g_i}\right) \frac{dg_i}{dt} - \sum_{i=s+1}^{n} \frac{dg_i}{dt} \tag{6.231}$$

we rearrange (6.211) by means of (6.223):

$$\frac{dg_i}{dt} = g_i \left\{ \sum_{j=1}^{s} b_{ij} g_j^0 - \sum_{j=1}^{n} b_{ij} g_j \right\} = g_i \sum_{j=1}^{n} b_{ij} \left(g_j^0 - g_j \right) \quad i = 1, \ldots, s \tag{6.232}$$

Inserting into Equation 6.231 yields

$$\begin{aligned}\frac{dG}{dt} &= -\sum_{i=1}^{s} (g_i - g_i^0) \sum_{j=1}^{n} b_{ij} \left(g_j^0 - g_j \right) - \sum_{i=s+1}^{n} g_i \left(a_i - \sum_{j=1}^{n} b_{ij} g_j \right) \\ &= \sum_{i,j=1}^{n} (g_i - g_i^0) b_{ij} (g_j - g_j^0) - \sum_{i=s+1}^{n} g_i \left(a_i - \sum_{j=1}^{n} b_{ij} g_j^0 \right) \geq 0\end{aligned} \tag{6.233}$$

Here, we made use of the positive definite matrix $\mathbf{b}^S$ and the inequalities (6.224). The equality sign in Equation 6.233 holds only if $\mathbf{g} = \mathbf{g}^0$. Because of the proof of existence, Equations 6.223 and 6.224, the function G represents a dynamic evolution principle that is defined on each positive subspace and on those boundaries of the positive cone that correspond to extinct species. On the remaining boundaries where $g_i = 0, g_i^0 > 0$, the function G is invalid and is not an evolution principle, that is, it cannot be used to predict the selective advantage of a successful new species.

The principle (6.229) proves that exactly one solution of the type (6.223), (6.224) exits which corresponds to an asymptotically stable state. To find this state, for each of the 2^n subspaces R one has to solve the system

$$\sum_{i,j} b_{ij} x_j = a_i \quad i,j \in R \tag{6.234}$$

and to check the conditions (6.223) and (6.224). If in some subspace exactly one x_j is negative as a solution of Equation 6.234, then the species j is subject to extinction (Strobeck theorem, Strobeck, 1973).

It is important to note that the proof of existence, Equations 6.223 and 6.224, and of the uniqueness and stability by means of the G function for positive definite $\mathbf{b}^S$ demonstrates that these ecological systems are subject to plain selection. The subset of survivors for $t \to \infty$ depends only on the presence of certain species at $t = 0$ rather than on their initial particle numbers. A sufficient condition for the positive definiteness of $\mathbf{b}^S$ is diagonal dominance with positive main diagonal elements. Thus, beyond a critical dissipation of the system as a result of self-inhibition the Lotka–Volterra scheme always has a unique stable stationary solution as described above, independent of the structure of the food web given by the antisymmetric part of the community matrix, $\mathbf{b}^A$. Remember that the famous quasistationary predator–prey oscillations (Lotka, 1910; Volterra 1931) appear under the structurally unstable condition of vanishing cross-dissipation, $\mathbf{b}^S = 0$, similar to those of canonical–dissipative systems (Feistel and Ebeling, 1989; Ebeling, Engel, and Herzel, 1990b, see also Chapter 4).

Unfortunately, general necessary or sufficient conditions for the existence of evolution principles in the sense of Darwin's "fittest" are not known for Lotka–Volterra systems. It is certain that the structure of the food web is of crucial importance in this case. As an illustration, we consider a simple hierarchical trophic

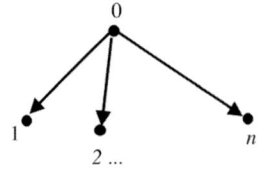

Figure 6.22 Simple trophic tree with one autotrophic species and competing herbivores, Equations 6.235 and 6.236.

network, Figure 6.22, as a special case of Equation 6.206, with the related equations

$$\dot{N}_0 = \left(a_0 - \sum_{j=1}^{n} b_j N_j - c_0 N_0\right) N_0 \tag{6.235}$$

$$\dot{N}_i = (-k'_i + b_i N_i - c_i N_i) N_i \quad i = 1, \ldots, n \tag{6.236}$$

Similar to Section 6.6, it is easy to show that this system permits the coexistence of species. The stable state is given by the condition of a minimum food supply, N_0, for the competing species. This quantity represents a static evolution principle in this case. In the limit $c_i \to 0$, the Gause's principle is valid and stable coexistence is impossible (Allen, 1975).

Alternatively, we consider a cyclic food web, Figure 6.23, with the equations

$$\dot{N}_0 = \left(a_0 - b_0 \sum_{j=1}^{3} N_j - c_0 N_0\right) N_0 \tag{6.237}$$

$$\dot{N}_i = (-a + b_0 N_0 + b N_{i_2} - b N_{i_1} - c N_i) N_i \quad i = 1, 2, 3 \tag{6.238}$$

Here, $i_1 = 1 + (i \bmod 3)$ is the successor of i in the cycle, and $i_2 = 1 + ((i+1) \bmod 3)$ is the precursor of i. This system permits cyclic evolution. Species 1 succeeds over species 3, species 2 over species 1, and in turn species 3 over species 2. Obviously, there cannot exits an evolution principle which assigns a higher fitness value to the winner and lower one to the loser. In ecology, this behavior is known as the

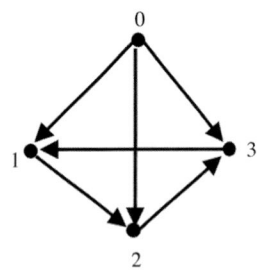

Figure 6.23 Simple cyclic predator–prey interaction with one autotrophic species, Equations 6.237, 6.238.

Red-Queen hypothesis (Fisher, 1930; Van Valen, 1973; Feistel and Ebeling, 1982, 1984; Paterson et al., 2009; see Chapter 7) which expresses the fact that selective advantage may appear only relative to a coevolving background rather than with respect to some fictitious absolute reference frame. Cyclic predator–prey interaction is not as exceptional as it immediately may seem; a well-known example from the Baltic Sea is cod that feeds on sprat which in turn feed of cod larvae (Feistel, Nausch, and Wasmund, 2008a). Deeper investigation of this behavior is evidently of fundamental interest with respect to the qualitative evolution and the optimization properties of ecological systems (Feistel, 1979; Chapter 7).

Another interesting aspect is the physical or biological interpretation of the functions G or L, Equations 6.225 and 6.226. Some authors discuss the similarity between Equation 6.225 and the mixing entropy of ideal gases or solutions; this interpretation has probably only formal character. An interesting attempt is the probability approach to L (Kerner, 1957; Alekseyev, 1976). If only so-called conservative systems with $\mathbf{b}^S = 0$ are considered and no selection occurs, L proves to be a conservation quantity, an integral of motion. With the knowledge of an additive conservation law, the Khinchin (1943) formalisms permit the construction of formal statistical mechanics. In this context, the expression

$$P \propto \exp(-\beta L) \tag{6.239}$$

represents an occupation probability of states with the "ecological energy" L and the "ecological temperature" $T \propto 1/\beta$. However, the actual interpretation of those "thermodynamic functions" turned out to be difficult. In addition, the condition $\mathbf{b}^S = 0$ applies to a special, structurally unstable situation that is coined by a number of unrealistic peculiarities (May, 1973). In general, L will change with time, and occupation probabilities are governed by master equations rather than a formal mathematical equilibrium condition. Below, using this technique, we shall derive a more rigorous interpretation of L.

Related master equations were formulated by Alekseyev and Kostin (1974) and Romanovsky, Stepanova, and Chernavsky (1975) for two-species systems in order to calculate extinction probabilities in a Langevin approximation. Here we generalize those equations for the case of n species.

We consider the following set of discrete transition processes, Table 6.2.

Table 6.2 Stochastic Lotka–Volterra process related to Equation 6.240.

Process	Transition	Rate
Autotrophic growth	$N_i \to N_i + 1$	$W = A_i N_i$
Heterotrophic growth	$N_i \to N_i + 1$	$W = \sum_{i \neq j} B_{ij} N_i N_j$
Self-inhibition	$N_i \to N_i - 1$	$W = C_{ii} N_i (N_i - 1)$
Prey death	$N_i \to N_i - 1$	$W = \sum_{i \neq j} C_{ij} N_i N_j$
Mortality	$N_i \to N_i - 1$	$W = k_i N_i$

The master equation corresponding to the processes of Table 6.2 reads

$$\frac{\partial P}{\partial t} = \sum_i A_i[(N_i-1)P(N_i-1) - N_i P(N_i)]$$
$$+ \sum_i k'_i[(N_i+1)P(N_i+1) - N_i P(N_i)]$$
$$+ \sum_i C_{ii}[(N_i+1)N_i P(N_i+1) - N_i(N_i-1)P(N_i)] \quad (6.240)$$
$$+ \sum_i \sum_{j \neq i} C_{ij} N_j [(N_i+1)P(N_i+1, N_j) - N_i P(N_i, N_j)]$$
$$+ \sum_i \sum_{j \neq i} B_{ij} N_j [(N_i-1)P(N_i-1, N_j) - N_i P(N_i, N_j)]$$

In terms of the generating function,

$$G(s_1, \ldots, s_n, t) = \sum_{N_1} \cdots \sum_{N_n} s_1^{N_1} \cdots s_n^{N_n} P(N_1, \ldots, N_n, t) \quad (6.241)$$

Equation 6.240 is transformed into the partial differential equation:

$$\frac{\partial G}{\partial t} = \sum_i (s_i - 1) \left\{ (A_i s_i - k'_i) \frac{\partial G}{\partial s_i} + \sum_j s_j (B_{ij} s_i - C_{ij}) \frac{\partial^2 G}{\partial s_i \partial s_j} \right\} \quad (6.242)$$

From the formula (6.241) it is obvious that in the region $s_i \ll 1$ the function G is dominated by probabilities for small particle numbers; probabilities of large particle numbers influence G only if values of s very close to unity are considered. These two special cases of significant interest can therefore be estimated by series expansions with respect to either s or $(1-s)$. Looking at large particle numbers, we expand the coefficients of Equation 6.242 with respect to $(1-s)$,

$$\frac{\partial G}{\partial t} \approx \sum_i (s_i - 1) \left\{ \varepsilon_i \frac{\partial G}{\partial s_i} + \sum_j \gamma_{ij} \frac{\partial^2 G}{\partial s_i \partial s_j} \right\} \quad (6.243)$$

where $\varepsilon_i = A_i - k'_i$, $\gamma_{ij} = C_{ij} - B_{ij}$.

In Equation 6.243, the ansatz,

$$G(s_1, \ldots, s_n, t) = \exp\left\{ \sum_i (s_i - 1) f_i(t) \right\} \quad (6.244)$$

leads to the deterministic Lotka–Volterra scheme for the unknown functions f_i,

$$\frac{df_i}{dt} = f_i \left(\varepsilon_i - \sum_{j=1}^n \gamma_{ij} f_j \right) \quad (6.245)$$

that is, $f_i(t) = N_i(t)$ is just the solution of the analytical problem (6.206). Moreover, Equation 6.244 represents a sufficiently general solution of Equation 6.243 since a superposition of functions of the form (6.244) can satisfy any reasonable initial conditions (Gardiner and Chaturvedi, 1977). The probability distribution related to the solution (6.244) is a multi-Poisson distribution:

$$P(N_1, \ldots, N_n, t) = \prod_{i=1}^{n} \frac{f_i(t)^{N_i}}{N_i!} \exp\{-f_i(t)\} \tag{6.246}$$

The probability to encounter the system in its deterministic stationary state is

$$P(N_1^0, \ldots, N_n^0, t) = \prod_{i=1}^{n} \frac{f_i(t)^{N_i^0}}{N_i^0!} \exp\{-f_i(t)\} \tag{6.247}$$

Making use of the Stirling formula for large N, Equation 6.247 can be written by means of the function L, Equation 6.225, in the form

$$P(N_1^0, \ldots, N_n^0, t) = \exp\{-L(f_i(t), N_i^0)\} \tag{6.248}$$

We conclude that the lowering of L reflects the increasing probability of a stochastic ensemble to occupy the deterministic stationary state. For the stochastic perspective it is also reasonable that the function L possesses singularities on certain boundaries of the positive cone: the transition probability is zero to jump from the absorber state on the boundary into the interior of the positive cone.

Finally, we mention a relation between L and the Kullback entropy (Kullback, 1951; Kullback and Leibler, 1951; Uhlmann, 1977; Ebeling and Feistel, 1982):

$$K = \sum_{N_1} \ldots \sum_{N_n} P(N_1, \ldots, N_n, t) \ln \frac{P(N_1, \ldots, N_n, t)}{P^0(N_1, \ldots, N_n)} \tag{6.249}$$

where P^0 is the stationary solution of the master Equation 6.240. In the approximation (6.248), we obtain for K the excess mixing entropy (Gibbs, 1902),

$$K = \sum_i \left\{ N_i \left(\ln \frac{N_i}{N_i^0} - 1 \right) + N_i^0 \right\} \tag{6.250}$$

In contrast to the exact Kullback entropy, this approximation does not necessarily increase monotonically, but reaches its maximum at $N_i = N_i^0$. Moreover, K and L are related by the inequality:

$$K(N_i, N_i^0) + L(N_i, N_i^0) = -(N_i - N_i^0) \ln \frac{N_i}{N_i^0} \geq 0 \tag{6.251}$$

6.9
Selection with Sexual Replication

Sexual recombination is a very special type of cross-catalytic reproduction used by many organisms. Since the related formal-kinetic growth law is proportional to the encounter probability, that is, to the product of the concentrations of the male and the female individuals, and thus to the square of the total population, the question appears whether this may lead to hyperselection of the form discussed in Section 6.8. As described by Darwin (1911), sexual selection is a widespread and very powerful tool for the evolution of special and sometimes even exotic traits (Rubenstein and Lovette, 2009; van Doorn, Edelaar, and Weissing, 2009). For example, colorful fur of certain dinosaurs may have appeared originally to increase sexual attraction and developed later into the phylogenetic root of all feathers that enable today's birds to fly (Zhang et al., 2008; Choiniere et al., 2010; Xu, Zheng, and You, 2010). Also, sexual selection may have been important for the development of human speech and music (Levithin, 2007; Sell et al., 2010). It is the aim of this section to see how mutants with initially minor concentrations but superior properties can take the role of fluctuations which lead to an instability of the parental community, in contrast to the characteristics of hypercycle systems, Section 6.7, which are insensitive against the emergence of any new species, may it be "fitter" or not.

In the regular case, sexually reproducing organisms are equipped with a set of *diploid* genes, each consisting of two *alleles* (Schrödinger, 1944; Böhme, Hagemann, and Löther, 1976; Günther, 1978). If the phenotype of the organism is determined by only one of the two alleles, that allele is regarded as *dominant* and the other as *recessive*. If both alleles influence the phenotype to some extent, the alleles are termed *combinant*. For simplicity, we consider here only individuals, X_{ab}, with a single diploid gene, (a, b), where the alleles a and b may take the labels $1, \ldots, N_A$. A related formal-kinetic dynamical system is (Feistel, 1977c)

$$\frac{d}{dt} X_{ab} = \sum_{efgh} R^{ab}_{ef;gh} X_{ef} X_{gh} - k_{ab} X_{ab} \tag{6.252}$$

Here, k_{ab} is the mortality of organisms with the alleles (a, b). The recombination matrix, $R^{ab}_{ef;gh}$, describes the rate at which parents with the alleles (e, f) and (g, h) produce offsprings carrying the alleles (a, b):

$$R^{ab}_{ef;gh} = E_{ef} E_{gh} C_{ef;gh} \sum_{c,d} W^{ab}_{cd} M^{cd}_{ef;gh} \tag{6.253}$$

The selective value E_{ab} is the chance that an individual of the species (a, b) reaches adulthood. The matrix $C_{ef;gh}$ is the copulation rate between individuals equipped with the genes (e, f) and (g, h); this is the key quantity that enforces sexual selection. The transition matrix, W^{ab}_{cd}, describes the mutation frequency that the alleles (c, d) that were expected to be deployed by the parents in the absence of mutations, actually appear as the alleles (a, b) found in the offspring,

$$W^{ab}_{cd} = \frac{1}{2} \left(A^a_c A^b_d + A^b_c A^a_d \right) \tag{6.254}$$

where the single-allele conversion rate is

$$A^a_b = \mu^a_b (1 - \delta_{ab}) + \delta_{ab} \left[1 - \sum_c \mu^c_a (1 - \delta_{ac}) \right] \tag{6.255}$$

The small number μ_b^a is the mutation rate at which an allele b is changed into a by some random influence, and δ_{ab} is the Kronecker delta. Finally, in Equation 6.253 the Mendelian matrix has the form

$$M^{ab}_{cd;ef} = \frac{1}{8}(1+\delta_{ab})\sum_{v=0}^{1}\left(\frac{\delta_{ac}+\delta_{ad}}{v}\right)\left(\frac{\delta_{ac}+\delta_{af}}{1-v+\delta_{ab}}\right)\left(\frac{\delta_{bc}+\delta_{bd}}{1-v}\right)\left(\frac{\delta_{be}+\delta_{bf}}{v-\delta_{ab}}\right) \tag{6.256}$$

which counts the combinatorial possibilities to randomly choose a certain pair of alleles (a, b) out of the parental genetic pool (c, d) and (e, f). All quantities are assumed to be symmetric in the allele occurrence, that is, an organism with the gene (a, b) is the same as with (b, a).

We are interested in the frequency of a certain gene,

$$\varrho_{ab} = X_{ab}/B \tag{6.257}$$

within the population, rather than in the growth of the population as a whole:

$$B = \sum_{a,b} X_{ab} \tag{6.258}$$

For the growth of the frequency we infer from the derivative of Equation 6.257, that is,

$$\frac{d\varrho_{ab}}{dt} = \varrho_{ab}\left(\frac{1}{X_{ab}}\frac{dX_{ab}}{dt} - \frac{1}{B}\frac{dB}{dt}\right) \tag{6.259}$$

the relation

$$\frac{d\varrho_{ab}}{dt} = B\sum_{efgh}\varrho_{ef}\varrho_{gh}E_{ef}E_{gh}C_{ef;gh}\left(\sum_{c,d}W^{ab}_{cd}M^{cd}_{ef;gh}-\varrho_{ab}\right) + \varrho_{ab}\sum_{c,d}\varrho_{cd}(k_{cd}-k_{ab}) \tag{6.260}$$

First we consider the dynamics of neutral genetic diversity, that is, of a set of different alleles with equal phenotypic properties, without broken sexual symmetry (i.e., without distinct males and females) and without mutations. From

$$W^{ab}_{cd} = \frac{1}{2}(\delta_{ac}\delta_{db} + \delta_{ad}\delta_{cb}) \tag{6.261}$$

we obtain the simplified relation

$$\frac{d\varrho_{ab}}{dt} = BE^2 C\sum_{efgh}\varrho_{ef}\varrho_{gh}\left(M^{ab}_{ef;gh}-\varrho_{ab}\right) \tag{6.262}$$

For any symmetric matrix, g_{ab}, the Mendel matrix obeys the identity

$$\sum_{efgh}M^{ab}_{ef;gh}g_{ef}g_{gh} = \left(\sum_{c}g_{ac}\right)\left(\sum_{d}g_{bd}\right) \tag{6.263}$$

6 Competition and Selection Processes

so that we obtain a "master equation" for the allele drift:

$$\frac{d\varrho_{ab}}{dt} = BE^2 C \sum_{c,d} (\varrho_{ac}\varrho_{bd} - \varrho_{ab}\varrho_{cd}) \tag{6.264}$$

Note some formal similarity to the collision integral of the Boltzmann equation. If we consider B as a constant for simplicity and introduce a rescaled time variable, $\tau = BE^2 Ct$, we find the solution of Equation 6.264 in the form (Feistel, 1977c)

$$\varrho_{ab}(t) = \left(\sum_c \varrho_{ac}(0)\right)\left(\sum_d \varrho_{bd}(0)\right)(1 - e^{-\tau}) + \varrho_{ab}(0)e^{-\tau} \tag{6.265}$$

The unique final stationary distribution is known as the genetic equilibrium, or Hardy–Weinberg Law, in population genetics (Hardy, 1908; Weinberg, 1908).

It is instructive to introduce allele frequencies, F_i, as new variables in the form

$$F_i(t) = \frac{1}{2}\sum_{a,b} (\delta_{ia} + \delta_{ib})\varrho_{ab}(t) \tag{6.266}$$

to describe the reproduction of "selfish alleles." Then, the dynamics (6.264) reduces to a conservation law separately for each allele found in the population

$$\frac{dF_i}{dt} = 0 \tag{6.267}$$

and the Hardy–Weinberg Law takes the form

$$\lim_{t \to \infty} \varrho_{ab}(t) = F_a(0) F_b(0) \tag{6.268}$$

We see that we can apply two alternative description levels here, the dynamics of alleles in the genotype space, and the dynamics of individuals in the phenotype space. We shall return to this aspect in greater detail in the following chapter.

To investigate the selection between different alleles in a population, we transform Equation 6.260 for the fractions of individuals into an equation for the allele frequencies, by means of Equation 6.266

$$\frac{dF_i}{dt} = B\sum_{efgh} \varrho_{ef}\varrho_{gh} E_{ef} E_{gh} C_{ef;gh} \left(\sum_{b,c,d} W^{ib}_{cd} M^{cd}_{ef;gh} - F_i\right) + F_i \sum_{c,d} \varrho_{cd} k_{cd} - \sum_d \varrho_{id} k_{id} \tag{6.269}$$

Here, the product $E_{ab} E_{cd} C_{ab;cd}$ evaluates the balance of the tolerable loss in individual fitness, E_{ab}, against higher performance of the copulation success, $C_{ab;cd}$, which is the actual target coefficient of sexual selection.

In order to return to our initial problem of whether once-for-ever selection may be a relevant problem in sexually replicating populations, we simplify the general relation (6.269) for the case that the mortalities are the same for all alleles, $k_{ab} = k$, that the pairing rates are all the same, $C_{ab;cd} = C$, and that mutations can be neglected, $A^a_b = \delta_{ab}$. Exploiting Equation 6.261 and the identity

$$\sum_d M^{id}_{ef;gh} = \frac{1}{4}(\delta_{ie} + \delta_{if} + \delta_{ig} + \delta_{ih}) \qquad (6.270)$$

we obtain from Equation 6.269 after some mathematical manipulations the selection equation for the allele i, in the form

$$\frac{dF_i}{dt} = BC\langle E \rangle (E_i - \langle E \rangle) F_i \qquad (6.271)$$

Here,

$$E_i = \frac{\sum_a \varrho_{ia} E_{ia}}{\sum_a \varrho_{ia}} \qquad (6.272)$$

is the average fitness of the individuals equipped with the allele i, and

$$\langle E \rangle = \sum_{a,b} \varrho_{ab} E_{ab} \qquad (6.273)$$

is the mean fitness of the community. Thus, the allele frequency, F_i, will increase if the carriers of that particular allele possess superior mean phenotypic selective values, E_i, Equation 6.272, even though the allele may initially be present at only minor abundance.

We conclude that the selection for genotypic traits in sexually reproducing populations, realized via the related phenotypic traits of diploid individuals, follows a dynamics of the Fisher–Eigen type, Equation 6.271, compare Section 6.3, without hyperselection behavior, notwithstanding the quadratic growth law, Equation 6.252. This theoretical result is not trivial but is of course consistent with biological observation.

6.10
Selection between Microreactors

The selection models studied in the previous sections apply to spatially homogeneous situations such as a restricted territory or a stirred test tube. They are hardly suitable for a vast volume such as the primordial ocean, or populations scattered over different continents. In lab experiments, droplets of hydrophobic substances are frequently formed in aqueous systems, such as "microspheres" or "coacervates" (Oparin, 1924; Fox and Dose, 1972). Such droplets may be enclosed by self-assembled membranes or liquid–liquid phase boundaries with high surface tension which keep the droplet's shape spherical as long as the volume is small, and provide some internal chemical milieu different and somewhat protected from the environment (White, 1980).

Here, we want to elucidate the differences between selection processes that occur in a homogeneous volume, and those confined to separate units. As a simple mathematical approach, we imagine that the droplets are internally homogeneous, that their growth is controlled by chemical reactions running in the interior, and that growing droplets are occasionally split up into smaller fractions by some mechanical forces, such as waves in the ocean (Feistel, 1983c). This way, autocatalytic or cross-catalytic reaction systems may result in primitive self-replicating spatial entities, and in fact any biological organism is of similar nature, even if much more complicated and functionally well organized. Such systems were termed "Sysers" (systems of self-reproduction) by Redko (1986).

We consider the general case of some reaction kinetics, Equation 6.1:

$$\frac{d\mathbf{N}}{dt} = V\mathbf{f}(\mathbf{c}) \tag{6.274}$$

where the vector $\mathbf{c} = \mathbf{N}/V$ represents the concentrations of the chemical species with the particle numbers $\mathbf{N}$ in the volume V of the droplet, and the vector $\mathbf{f}$ the related chemical reaction rates. Here, Equation 6.274 is written in a way that permits the reactor volume being a function of time, in contrast to Equation 6.1 where the volume is considered fixed. Let

$$c = \sum_k c_k = \frac{1}{V}\sum_k N_k = \frac{N}{V} \tag{6.275}$$

be the total number of particles per volume, and let the vector of mole fractions, $\mathbf{x}$, be defined by

$$\mathbf{x} = \frac{\mathbf{c}}{c} = \frac{\mathbf{N}}{N} \tag{6.276}$$

In terms of the molar fractions, the kinetics of Equation 6.274 reads (Ebeling and Feistel, 1979, 1982; Feistel, 1983d, 1985)

$$\frac{dx_i}{dt} = \frac{d}{dt}\left(\frac{c_i}{c}\right) = \frac{d}{dt}\left(\frac{N_i}{\sum_k N_k}\right) = \frac{1}{c}\left\{f_i(\mathbf{xc}) - x_i \sum_k f_k(\mathbf{xc})\right\} \tag{6.277}$$

We imagine that the particles, similar to the plasma of a living cell, form a dense pack of molecules. Then, the particles of the kind i, each having the molecular volume v_i, take the partial volume

$$V_i = N_i v_i \tag{6.278}$$

The total volume of the droplet is obtained by summation of all chemical species

$$V = \sum_i V_i \tag{6.279}$$

Hence, the concentrations c_i can be expressed by the molar fraction in the form

$$c_i = \frac{N_i}{V} = \frac{N_i}{\sum_k N_k v_k} = \frac{x_i}{\sum_k x_k v_k} \tag{6.280}$$

and, because of $\sum_k x_k = 1$,

$$c = \frac{1}{\sum_k x_k v_k} \qquad (6.281)$$

We shall assume for simplicity that the molecular volumes, v_i, are "rigid," that is, independent of the particular size or composition of the droplet. Then, Equation 6.277 is a closed equation in the molar fractions, $\mathbf{x}$, independent of the total volume or particle number. The solution of that equation may of course converge to periodic or chaotic attractors, but for simplicity we restrict our discussion here to stationary states, that is, to solutions $\mathbf{x}^{(s)}$ of the following equation:

$$f_i\left(\mathbf{x}^{(s)} c^{(s)}\right) = x_i^{(s)} \sum_k f_k\left(\mathbf{x}^{(s)} c^{(s)}\right) \qquad (6.282)$$

We explicitly permit chemical constituents with zero concentration, $x_i = 0$, and consider the vector $\mathbf{x}$ to have as its components any possible chemical compound. Then, a very large number of different solutions of Equation 6.282 describe any possible stationary microreactor composition, even if its total particle number may be zero, that is, even if that type of microreactor does not yet or may no longer exist in our system. In other words, the numbers $s = 1, 2, \ldots$ are labels of all the different "biological species" that may exist in the form of microreactors or coacervates. Those species are characterized by a certain chemical composition, independent of their size or their number of existing individuals. Different species (s) correspond to different solutions of Equation 6.282, that is, to different composition vectors.

This chemical definition of a simple model for biological species enables us to formally link the selection process between different species to the reaction chemistry of their metabolism (Ebeling and Feistel, 1979, 1982; Feistel, Romanovsky, and Vasiliev, 1980; Feistel, 1983d, 1985). To do this, we introduce the "biomass" $m(\mathbf{x}, t)$ of a biological species with the composition $\mathbf{x}$ at the time t:

$$m(\mathbf{x}, t) = N \sum_k \mu_k x_k \qquad (6.283)$$

Here, μ_k is the molecular mass of the chemical species k. Making use of Equation 6.274, we obtain for the time derivative

$$\frac{d}{dt} m(\mathbf{x}, t) = \sum_k \mu_k \frac{d}{dt} N_k = V \sum_k \mu_k f_k(\mathbf{x} c) \qquad (6.284)$$

We consider now a species (s) which is in internal chemical balance, Equation 6.282. The related growth law reads

$$\frac{d}{dt} m\left(\mathbf{x}^{(s)}, t\right) \equiv \frac{dm_s}{dt} = \frac{1}{c^{(s)}} \sum_k f_k\left(\mathbf{x}^{(s)} c^{(s)}\right) m_s = E_s m_s \qquad (6.285)$$

This equation is the important result derived for the growth of coacervates. No matter of which complicated and nonlinear form the internal chemical kinetics of the droplet may be, it obeys a linear autocatalytic growth law with the replication rate:

$$E_s = \frac{1}{c^{(s)}} \sum_k f_k\left(x^{(s)} c^{(s)}\right) \qquad (6.286)$$

This rate is determined from the particular metabolic steady state, Equation 6.282, in the form

$$\frac{1}{c^{(s)}} \mathbf{f}\left(\mathbf{x}^{(s)} c^{(s)}\right) = E_s \mathbf{x}^{(s)} \qquad (6.287)$$

When different kinds of coacervates exist and multiply in a system under consideration, we may ask for the time evolution of the mass fractions of those species:

$$\varrho_s(t) = \frac{m_s(t)}{\sum_k m_k(t)} \qquad (6.288)$$

and derive from Equation 6.285 the equation

$$\frac{d}{dt} \varrho_s = \varrho_s (E_s - \langle E \rangle) \qquad (6.289)$$

where

$$\langle E \rangle = \frac{\sum_k E_k m_k}{\sum_k m_k} \qquad (6.290)$$

is the average reproduction rate of the population. Regardless of the complexity of the internal chemistry, the competition between coacervates belongs to the category of simple selection, such as the Fisher–Eigen equation(6.24). There exists a dynamic evolution principle, Equation 6.39, that controls the system's behavior. The selective value of a coacervate can be computed from the reaction rate of the chemicals it is consisting of, Equation 6.287.

To demonstrate the latter, we consider three simple examples for the chemical kinetics within the droplet. First, let the chemical reactions be autocatalytic:

$$f_i(\mathbf{c}) = k_i c_i \qquad (6.291)$$

where $k_i \neq k_j$ is assumed if $i \neq j$. Then, Equation 6.287 takes the form

$$k_i x_i^{(s)} = E_s x_i^{(s)} \qquad (6.292)$$

This is an eigenvalue problem for a diagonal matrix with the elements k_i on the main diagonal, the eigenvalues E_s and the related eigenvector $\mathbf{x}^{(s)}$. The solution is $E_s = k_s$ for the eigenvalues and $x_i^{(s)} = \delta_{is}$ for the eigenvectors. Hence, each type of coacervate can contain only one of the autocatalytic chemicals, and the reaction rate k equals the selective value E of the coacervate. In other words, for linear autocatalytic species there is no difference in the selection behavior, regardless of whether they are either confined to droplets or homogeneously distributed solutes.

As a second example, let the chemical reactions be cross-catalytic:

$$f_i(\mathbf{c}) = \sum_j k_{ij} c_{ij} \tag{6.293a}$$

Then, Equation 6.287 takes the form

$$\sum_j k_{ij} x_j^{(s)} = E_s x_i^{(s)} \tag{6.293b}$$

This is an eigenvalue problem for a nonnegative matrix with the elements k_{ij}, the eigenvalues E_s, and the related eigenvector $\mathbf{x}^{(s)}$. The solutions for the eigenvalues and the eigenvectors depend on the reducibility of the matrix $\mathbf{k}$. If the matrix is irreducible, there is only one nonnegative eigenvector. Hence, each type of coacervate can contain only one of the autocatalytic chemicals which belong to separate irreducible clusters of the catalytic network, and their "parasitic" side chains. The selective value E of the coacervate is the maximum eigenvalue of that cluster matrix, corresponding to the theorems of Perron and Frobenius (Gantmacher, 1971), see Section 6.5. We infer that for linear cross-catalytic species there is no difference in the selection behavior, regardless of whether the chemicals are confined to droplets or homogeneously distributed solutes.

The third example is quadratic autocatalysis, Equation 6.184:

$$f_i(\mathbf{c}) = k_i c_i^2 \tag{6.294}$$

Then, Equation 6.287 takes the form

$$k_i \left(x_i^{(s)}\right)^2 c^{(s)} = E_s x_i^{(s)} \tag{6.295}$$

The solution is $E_s = k_s c^{(s)}$ for the "eigenvalues" and $x_i^{(s)} = \delta_{is}$ for the "eigenvectors" of the nonlinear problem, Equation 6.295. Each coacervate type consists only of a single chemical species. The selective value is given by the autocatalytic reaction rate times the concentration in the droplet, which is determined by the molecular volume of the densely packed particles, $c^{(s)} = 1/v_s$, Equation 6.280.

We have found that even coacervates with nonlinear catalytic couplings of the internal chemistry follow linear competition laws of the Fisher–Eigen type. This model avoids hyperselective once-for-ever behavior of the homogeneous solution and permits the evolutionary improvement of complicated chemical networks confined to droplets. The mutual selective relevance of individuals with very different metabolic chemistry becomes commensurable in the form of their selective values, similar to the mutual comparability between qualitatively different commodities by their prices on a market. The formation of self-replicating coacervates already at an early stage of the pre-biological development was indispensible for the molecular evolution of catalytic networks such as the hypercycle (Eigen, 1971) or "autogenes" (White, 1980). For various model-reaction schemes the related selective values of coacervates can be derived from solving the nonlinear "eigenvalue" problem given by

Equation 6.287, in the form of certain functions, $E(\mathbf{k}, \mathbf{u})$, where $\mathbf{k}$ is the set of reaction constants and $\mathbf{u}$ of external conditions. The results for the derivatives $\partial E/\partial \mathbf{k}$ show that with increasing complexity, only fine-tuned balances between the chemical constituents lead to maximum selective values of the competing "individuals," rather than simply maximum rates (Ebeling and Feistel, 1982).

6.11
Selection in Social Systems

Competition and selection processes occur in various physical, chemical, biological, social, or technical systems where certain mechanisms choose a preferred mode or subsystem out of several alternatives under constrained resources of the system as a whole (Feistel and Ebeling, 1976; Haken, 1978). In social and in particular in economic systems, competition and selection processes play a central role with specific inherent aspects (Dosi et al., 1988; Faber and Proops, 1990). On the other hand, there are many common problems and methods in natural and social sciences of evolutionary processes that special branches such as "econophysics" and "evolutionary economics" are currently developing (Jiménez-Montaño and Ebeling, 1980; Dosi et al., 1988; Bruckner et al., 1996; Saviotti and Mani, 1995; Schweitzer, 2002; Pyka and Scharnhorst, 2009; Schweitzer et al., 2009; Kaldasch, 2011). This is a wide field; we restrict ourselves here to just a few remarks regarding some general viewpoints and tendencies. In economy, competition between different producers on the market is the central driving force for the development of new technologies and the regulator of proportionality between different branches of industry. Marx (1951b) emphasized this as the differentiating, stimulating, and regulating function of the "value law," which states that commodities are exchanged in the market by their values which can be measured quantitatively in the form of prices. Market competition has a similarly fundamental role for the social evolution of humans as Darwin's law of natural selection in biology, and it can be modeled mathematically in similar terms (Feistel, 1976a, 1977b; Ebeling, 1976b; 1978b; Feistel and Ebeling, 1976; Ebeling and Feistel, 1976b, 1982; Bruckner et al., 1996; Pyka and Scharnhorst, 2009; Kaldasch, 2011). Darwin's selective value "fitness" of species is taken over here by Marx' exchange value "price" of commodities. We note that the exchange of goods between owners is not a physical process; the change of possession is a change of a person's right of disposal; this is a social category rather than a physical property of the commodity, see also Sections 8.11 and 8.13. Possession and price belong to the emergent properties; they essentially control the dynamics of societies but are not reducible to elementary physical laws.

Let p be the market price that results from the balance between supply and demand of a certain amount of a commodity, x. If a producer i can provide the market with that good and sell it at the rate

$$y_i = \frac{dx_i}{dt} \qquad (6.296)$$

his profit rate is, after covering the production costs of κ_i,

$$\frac{dm_i}{dt} = (p - \kappa_i) y_i \tag{6.297}$$

Of that profit, a certain fraction, α, may be invested to extend the production:

$$\frac{dy_i}{dt} = \alpha \frac{dm_i}{dt} \tag{6.298}$$

For the model's simplicity, we have taken equal fractions for all producers. The producer's market share,

$$\varrho_i = \frac{y_i}{\sum_k y_k} \tag{6.299}$$

will then develop according to the competition equation

$$\frac{d\varrho_i}{dt} = \varrho_i \left(\frac{1}{y_i} \frac{dy_i}{dt} - \frac{1}{\sum_k y_k} \sum_k \frac{dy_k}{dt} \right) = \alpha \varrho_i (\langle \kappa \rangle - \kappa_i) \tag{6.300}$$

Here, the mean cost price is defined as

$$\langle \kappa \rangle = \sum_k \kappa_i \varrho_i \tag{6.301}$$

Note that this relation is independent of the change of the joint market price as a function of the supply. As a result of the competition, the average cost price will fall

$$\frac{d\langle \kappa \rangle}{dt} = \sum_k \kappa_i \frac{d\varrho_i}{dt} = \alpha(\langle \kappa \rangle^2 - \langle \kappa^2 \rangle) \leq 0 \tag{6.302}$$

that is, the cost price acts as a dynamical evolution principle. In the long run, falling production costs can only be realized by higher productivity of labor,

$$\omega_i = 1/\kappa_i$$

and more advanced technology. Thus, the evolution principle(6.302) can also be formulated as a law of increasing productivity, in the form

$$\frac{d}{dt}\left\langle \frac{1}{\omega} \right\rangle = \alpha\left(\left\langle \frac{1}{\omega} \right\rangle^2 - \left\langle \frac{1}{\omega^2} \right\rangle \right) \leq 0 \tag{6.303}$$

The establishment of a common market price is a stochastic process and can be modeled accordingly (Ebeling, 1978). Of fundamental interest is the question what are the responsible mechanism for the empirical fact that exchange values which are originally realized by exchange quantities virtually appear as state quantities associated with certain objects (Feistel, 1991). The difference between the two aspects is reflected in the violation of value conservation, that is, objects in possession of a certain owner may gain or lose value even if those objects remain physically unchanged. This difference between exchange and state quantity is particularly important if possession consists of paper money or other symbols for values. The change of currency exchange rates or inflation permanently generates or destroys immense values in the property of private persons, companies, or states, see also Section 8.13.

Similar to most of the competition processes discussed before in this chapter, economic competition is of course a stochastic process in which discrete numbers of producers or owners participate. Let us briefly repeat here the schematic of occupation numbers for the simplest case that the overall number of competitors is fixed. Under this boundary condition, competition and selection appear as substitution processes (Ebeling and Feistel, 1974, 1976a; Jiménez-Montaño and Ebeling, 1980). For our model we introduce a set of producers that are using certain technologies for the production of goods, enumerated by $i = 1,2,\ldots, s$. By $N_i(t)$ we denote the number of producers which use the technology i. These occupation numbers are either positive or zero and depend on time. The state of the system at the time t is described by the probability distribution for the occupation numbers:

$$P(N_1, N_2, \ldots, N_s; t) = P(\mathbf{N}; t) \tag{6.304}$$

As elementary processes we consider only transition processes during which two occupation numbers change simultaneously:

$$\begin{pmatrix} N_i \\ N_j \end{pmatrix} \rightarrow \begin{pmatrix} N_i - 1 \\ N_j + 1 \end{pmatrix}.$$

If we assume that all substitutions which lead to a change of occupation numbers depend mainly on the current state rather than on its history, we can apply the concept of Markov processes. The microscopic economic process of substitution may be considered as such a kind of elementary processes. Then the dynamics of the system can be described by the help of the master equation. This equation is a balance equation between formation and decay processes,

$$\frac{\partial P(\mathbf{N}; t)}{\partial t} = \sum_{\mathbf{N}'} \{ W(\mathbf{N}|\mathbf{N}') P(\mathbf{N}'; t) - W(\mathbf{N}'|\mathbf{N}) P(\mathbf{N}; t) \} \tag{6.305}$$

where $\mathbf{N}$ is the vector $\mathbf{N} = \{N_1, N_2, \ldots, N_s\}$.

Per unit time, the system's transition probability to change from the state $\mathbf{N}$ to the state $\mathbf{N}'$ is expressed by the rate $W(\mathbf{N}'|\mathbf{N})$. The transition probabilities reflect underlying economic processes. Here we consider only the simplest case that the transition probabilities can be modeled as polynomials (Ebeling and Feistel, 1974, 1976; Jiménez-Montaño and Ebeling, 1980; Heinrich and Sonntag, 1981; Ebeling and Feistel, 1982; Ebeling and Sonntag, 1986). Examples for the transition probabilities of special processes are given in Section 6.4. An ansatz that is sufficiently general for this model reads (Ebeling and Feistel, 1974, 1976; Bruckner et al., 1996)

$$W(\ldots, N_i + 1, \ldots, N_j - 1, \ldots, N_k, \ldots | , N_i, \ldots, N_j, \ldots, N_k, \ldots) = A_{ij} N_j + B_{ij} N_i N_j \tag{6.306}$$

The simplest nontrivial choice of the coefficients is

$$A_{ij} = 0 \quad B_{ij} = \frac{\alpha}{N}(\kappa_0 - \kappa_i) \tag{6.307}$$

where κ_0 denotes some fictive maximal production cost using the most primitive technology. A producer which reaches this maximal cost does not reproduce. In the statistical mean, the stochastic process defined by Equation 6.307 is equivalent to Equation 6.300, as easily found by averaging. In addition the stochastic version includes also the effect of fluctuations about the mean values. Note that the term κ_0 cancels in averaging procedures. Generalizations may be obtained by assuming (Jiménez-Montaño and Ebeling, 1980; Bruckner et al., 1996)

$$A_{ij} = IN a_{ij} \geq 0 \quad B_{ij} = \frac{\alpha}{N}(\kappa_0 - \kappa_i) + IM \frac{\alpha}{N} \Theta(\kappa_j - \kappa_i) \tag{6.308}$$

Here we introduced the Heaviside step function, Θ, which equals one for positive arguments and zero for negative arguments. In other words, the term IM is different from zero only for transitions to lower production costs. What is the meaning of the two new additional terms? The IN term describes a tendency to change from technology j to technology i quite independently of the present use of technology i. This rate is in general small but extremely important since it describes the possibility of innovations. Innovations are transitions to a state (a way of production) unused so far. It introduces a new competitor into the game. This is the analogy to mutations in biology. We see that innovations are stochastic elements in the technological evolution. The stochastic transitions lead to a novel state, to a way of production which was not used in the system before. There is no need to explain here in detail how important innovation is for social systems; it may suffice to remind the reader to the statement (Ebeling and Feistel, 1982, see the Preface to this book): "Refusal of innovation is virtually an optimal tactic, but ultimately it is a fatal strategy."

The second new contribution in Equation 6.308 is even more plausible than the first. A producer j makes a transition in dependence on the quality of the market success. In other words, the own production costs are compared with the costs of a competitor, and if the latter is doing better, then that way of production is simply copied. In many case it is prohibited by patents or by moral laws to copy the technology of the competitor i if it is better; nevertheless, copying and imitating is a standard strategy in technological evolution. Alternative technological transition processes were studied by Bruckner et al. (1996). Further we mention applications of the model described here to the control of search dynamics (Ebeling and Reimann, 2002).

Innovation and imitation are characteristic for human activities and are driving forces in social evolution. In comparable form, the targeted search for better genes and the purposeful copying of good genes from others is impossible in biological evolution, except to some extent by sexual selection (Section 8.7). In social evolution this strategy belongs to the basic driving forces. Needless to say that imitation is also quite characteristic for some other modes of human behavior such as dressing, opinion formation, and decision making (Ebeling, Molgedey, and Reimann, 2000), and is even quite typical for the behavior of scientists in making their choice for the preferred field of work (Ebeling and Scharnhorst, 1986; Bruckner, Ebeling, and Scharnhorst, 1990). Imitation is an amplifying nonlinear feedback mechanism that may result in positive Lyaponov coefficients and chaotic behavior of related mathematical models.

7
Models of Evolution Processes

The ultimate result is that each creature tends to become more and more improved in relation to its conditions.
Charles Darwin: The Origin of Species

Now, here, you see, it takes all the running you can do, to keep in the same place.
Lewis Carroll: Through the Looking-Glass

The discoveries of Clausius (1876) that there is a physical state quantity, entropy, with a natural tendency to grow, and of Darwin (1911), that biological evolution is characterized by the survival of the fittest organism or species, had created the lingering impression of a systematic inconsistency between the laws of physics and the realm of life. This feeling was theoretically underpinned by the work of Schrödinger (1944), who argued that living beings maintain a stable state of lowered entropy. On the other hand, already Szilard (1929) had shown that the entropy lowering by conscious activity of humans can only be achieved at the cost of additional entropy production in the environment, similar to a refrigerator which owes its cold inside to an additional warming outside. The virtually divergent physical theories of life and matter were eventually merged into a satisfactory consistent picture by the work of Glansdorff and Prigogine (1971) and Eigen (1971). It became obvious that the second law of thermodynamics requires that entropy can only be produced rather than destroyed but does not prevent an open system, such as an organism or a refrigerator, from exporting more entropy than it is actually producing and this way, from lowering its internal entropy and maintaining its ordered state and function, as discussed in more detailed already in Chapters 1 and 2.

Those findings inspired various studies on properties on nonequilibrium dissipative structures and processes of self-organization, in particular in the 1970s and 1980s. Prigogine's principle on the minimum entropy production of stationary states in the vicinity of the thermodynamic equilibrium served as a paradigm for the search of extremum principles valid under more general circumstances (Paltridge, 1975, 2001), often vividly debated on conferences between different schools or research

Physics of Self-Organization and Evolution, First Edition. Rainer Feistel and Werner Ebeling.
© 2011 Wiley-VCH Verlag GmbH & Co. KGaA. Published 2011 by Wiley-VCH Verlag GmbH & Co. KGaA.

teams, sometimes based on suspicious foundations such as Lovelock's Gaia world (Lovelock, 2000; Ward, 2009).

It is suggestive to believe that Darwin's survival of the "fittest" species or, say, the maximized heat transport by the onset of thermal convection may perhaps be examples for a universal extremum principle that governs the formation of and the transition between self-organized structures, similar to the second law applied to the world as a whole (Clausius, 1865; Greene, 2004) or the Kullback–Leibler entropy for a wide class of stochastic processes (Kullback and Leibler, 1951; van Kampen, 1981; Ebeling and Feistel, 1982).

With increasing distance from equilibrium, there are three qualitatively different regimes of dynamic behavior. For small deviations, thermodynamic fluxes are linear functions of the thermodynamic forces. In this linear thermodynamic branch, Prigogine's theorem of minimum energy production holds; instabilities and dissipative structures are excluded. Very far from equilibrium, in contrast, instabilities, symmetry-breaking, and self-organized structures such as thermal convection or turbulence occur. No general extremum principle is known under those conditions. As an example which we considered in Chapter 2 (see in particular Section 2.4 and Figure 2.2), entropy production as a function of the Rayleigh number in the bifurcation diagram of kinetic transitions between different convective regimes exhibits deviations to higher and to lower values of the entropy production (Linde, 1978; Ebeling and Feistel, 1982), thus demonstrating that in the general case entropy production does not follow any definite rule. In between Onsager's linear regime close to equilibrium and that of self-organized structures far from equilibrium, a nonlinear thermodynamic branch can often be observed where despite of fluxes being nonlinear functions of forces, dissipation still dominates the dynamics and prevents any instability. In this regime, the qualitative behavior of the system is very similar to that in the linear branch, and it is sometimes (silently) assumed that Prigogine's principle of minimum entropy principle may be extrapolated and still be valid under these circumstances. Unfortunately, already the physically simple case of stationary radiation between black bodies violates the putative principle of minimum entropy production as soon as the relation between fluxes and forces is nonlinear, see Section 3.2. This result suggests that the validity of Prigogine's principle is definitely restricted to the linear branch.

As already shown in Section 2.4, we emphasize here again that the principle of minimum entropy production is not an appropriate candidate for a general evolution principle. The derivation of Prigogine's principle given there is valid for linear processes only. In contrast, relevant physical processes of interest for evolution are essentially nonlinear. This does not exclude, though, that for special evolutionary processes extremum principles may exist and can be found. The search for such principles is an essential part of the investigation in this chapter.

In this book, as pointed out in the introduction, evolution is understood as a potentially unlimited succession of self-organization steps, each beginning with an instability (mutation, innovation, invention) of the previously established parent regime. In Chapter 6, a state function depending on the order parameters was regarded as (i) a *static evolution principle*, if its value associated to a given regime is lower that that to the subsequent one, (ii) a *dynamic evolution principle*, if its value is increased by the initial fluctuation and further grows during the establishment of the new regime, and finally (iii) a *complete evolution principle*, if it obeys both properties. We have seen that such a principle can be found for Fisher–Eigen models, but it is unclear whether this is just a special property of a particular class of models, or may be of more general relevance.

In the special case that such a principle can be derived from a system of dynamic equations, its function value is a real number that can only increase with time and can be termed the "selective value" or "fitness" of a given state or species for that reason. If in such a way a real number can be associated with each state of the system, those states are subject to an order relation (see Section 4.1); they can be ordered by their "fitness," "goodness," "complexity," "efficiency," or whatever their qualification may be.

Independent of whether such a function is known or can at least be formally defined, the necessary condition for its existence is the validity of an order relation. If an order relation (or in some cases a semiorder) exists between the states, an appropriate selective value can be assigned to each of them in such a way that a given state can only be replaced by a "better" state. If the existence of a semiorder or order relation can be disproved, the existence of a properly defined fitness value is impossible in that case. The question is how large the class of dynamical systems may be for which order relations exist? If a state A consists of a number of species, and the addition of a new species turns A unstable and leads to a new state B, then the reverse process is normally excluded. That means, as a rule, that the binary relation "successor" between the states A and B is asymmetric (Section 4.1). Asymmetry is a necessary property for irreflexive semiorder relations, but it is not a sufficient one. While we may interpret the asymmetry as a selective advantage of the state B over the state A, we may not conclude that this selective advantage is sufficient for the proper definition of a selective value. The main reason is that asymmetry is not sufficient for transitivity. If B has a selective advantage over A, according to the particular fitness measure we use, and C has an advantage over B by the same measure, it is not always granted that by this measure also C has an advantage over A. While there is a large class of selection processes that are asymmetric (although hyperselection is not, for example), the required transitivity is much more restrictive and in general more difficult to prove for a model given by, say, a set of differential equations. While selective advantage is a differential property as it compares states that differ from each other only infinitesimally, selective values are integral properties that permit a universal comparison between any two given states, no matter how distant they may be. How to compare a bacterium with the Internet? We emphasize that integrals (i.e., solutions) of differential equations may possess qualitatively different symmetries and properties than the equations themselves. Asymmetry between subsequent states is expressing the irreversibility of the process, but only in special situations such as closed systems, irreversibility in fact implies the existence of an integral extremum principle such as entropy. We shall return to this relevant point.

The models and examples of this chapter are intended to illustrate these properties using simple dynamical models, but unfortunately, they will not provide ultimate answers to our questions, perhaps, just because evolution is an open process that has no predetermined goal and does not like ultimate answers of any kind. Consequently, the idea that evolution has a certain aim or target may only apply to special situations. In other words, evolution is associated with local optimization and progress, and often with local search for extrema of some functions, but we may have to give up hope for a global goal of evolution to exist. While evolution proceeds always forward from one state to a subsequent one, one should be careful in using words such as better, higher, fitter, and so on in so far as those attributes imply the existence of order relation which is not granted in general.

Closely related to the question of the evolution to putatively "higher" or "fitter" states are those of the evolution velocity, that is, the rate at which a fitness value may increase, and of neutral evolution, that is, the reasons and the effects of changes without direct selective advantage. In order to illustrate the related problems, methods, and conclusions, we address evolution as a succession of physical states here with three different tools. The first is sequence evolution, where a pool of sequences is repeatedly randomly modified and evaluated by certain selection criteria. One of the central concepts which will be developed here as the second option is that evolution can be modeled by a dynamics on valuation landscapes. Finally we formulate the evolution problem in terms of abstract relations between sets in a Boolean reaction system (BRS).

7.1
Sequence-Evolution Models

Availability, speed, and convenience of use of today's personal computers makes it very easy to play simple "evolution games" by numerical simulation (Eigen and Winkler, 1975; Sprott, Bolliger, and Mladenoff, 2002). For instance, one may implement a Runge–Kutta algorithm (Sprott, 1991) to solve the system of differential equations of the form (see Section 6.3)

$$\dot{X}_i = (E_i - \langle E \rangle) X_i \qquad (7.1)$$

where $\langle E \rangle = \sum_i E_i X_i / C$ is the mean excess production rate and $\sum_i X_i = C = $ const is the constant total population. At random times, a "mutation" is carried out on the computer by replacing one copy of a randomly chosen species i with the selective value E_i by another species j with an associated selective value E_j. This mixed model of deterministic growth and stochastic mutations can be designed more consistently in the form of an entirely stochastic simulation (Ebeling and Feistel, 1974; Feistel, 1976b, 1977a). For that purpose, we write down a master equation of the form (Sections 4.5 and 6.4)

$$\frac{\partial}{\partial t} P(N_1, \ldots, N_n, t) = \sum_{j \neq i} \left[W_{ij}(N_i - 1, N_j + 1) P(N_i - 1, N_j + 1) - W_{ij}(N_i, N_j) P(N_i, N_j) \right]$$

$$(7.2)$$

For the probabilities of the transitions,

$$\begin{pmatrix} N_i \\ N_j \end{pmatrix} \rightarrow \begin{pmatrix} N_i + 1 \\ N_j - 1 \end{pmatrix} \tag{7.3}$$

we assume the function

$$W_{ij}(N_1, \ldots, N_i, N_j, \ldots, N_n) = E_i \frac{N_i N_j}{\sum_k N_k} + \mu_{ij} N_j \tag{7.4}$$

where $E_i > 0$ is the selective value of the species i, N_i is the number of individuals of that species at the time t, and μ_{ij} is the (small) mutation rate for the spontaneous transformation of an individual of the species j into one of the species i. Rather than attempting to compute the probability P as a function of time, it is often more instructive (and entertaining) to watch a random process that is controlled by Equation 7.2. To do so, we start from an arbitrary initial state given by some particle numbers; in particular, there may be just one species with N individuals in the beginning.

At any intermediate time t, the current state of the system is given by a certain set of occupation numbers for the species, $N_1, N_2, \ldots$. From that given state, we can compute the waiting time (Feistel, 1977a; Feistel and Ebeling, 1978a; Ebeling and Feistel, 1982)

$$\tau = \frac{1}{\sum_{i \neq j} W_{ij}(N_1, \ldots, N_i, N_j, \ldots, N_n)} \tag{7.5}$$

which is the average time the system remains in that state before a stochastic event will happen to change the state. To simulate this process, we compute a stochastic waiting time, Δt, which must be exponentially distributed with the mean value of τ. Computer languages usually provide random numbers r equally distributed in the interval $0 < r \leq 1$. In that case

$$\Delta t = -\tau \ln r \tag{7.6}$$

is a suitable time step by which we update the system age variable, $t \rightarrow t + \Delta t$. Next, we have to decide which of the possible events should occur. After summing up the values of $\tau \times W_{ij}$ for each pair i, j, and storing away the partial sums in a list of numbers growing from 0 to 1, we can use another random number r in (0, 1] to pick the particular interval of the length $\tau \times W_{ij}$ into which the current value of r is pointing. Then we increment the state variable N_i by 1 and decrement the variable N_j, according the Equation 7.3. This new state completes the iteration step that may be repeated arbitrarily often.

This raises the question of how to choose suitable values for E_i and μ_{ij} for the species i and j. To imitate biological evolution, each species i can be represented by a sequence S_i of symbols from a specified alphabet. As an illustration, Eigen (1976) chose strings of 20 Latin characters, starting from the sequence

$$S_1 = \text{"KORN AUS DEN FELDERN"} \tag{7.7}$$

For the selective value of any given sequence S_i, the deviation

$$E_i = 21 - d_H(S_i, S_0) \tag{7.8}$$

from a target sequence, such as

$$S_0 = \text{"LERN AUS DEN FEHLERN"} \tag{7.9}$$

can be used (see also Försterling and Kuhn, 1971; Försterling, Kuhn, and Tews, 1972). Here, $d_H(S_i, S_0)$ is the Hamming distance between two sequences which counts the number of positions at which their characters disagree. We may specify the mutation rate by

$$\mu_{ij} = \mu \times \delta_{d_H(S_i, S_j), 1} \tag{7.10}$$

Then, transitions between sequences are possible by point mutations, that is, substitution of one letter by another.

This model is useful to study the evolution speed. Depending on the value chosen for the mutation rate μ, the value of

$$\langle E \rangle = \frac{\sum_k E_k N_k}{\sum_k N_k} \tag{7.11}$$

may approach some vicinity of the maximum, $\langle E \rangle_{max} = 21$, after different times when starting from the same initial sequence. At very small μ, mutations occur very seldom and the growth of $\langle E \rangle$ is sluggish. At some medium μ, the evolution proceeds most rapidly. Beyond that optimum, too frequent mutations destroy the advantage of higher selective values and a random population of sequences appears, termed the "mutation catastrophe." Evidently, best evolutionary progress is made in the immediate vicinity of the critical threshold, in a state of self-organized criticality (SOC), see Chapter 3.

Clearly, this evolution model presents only little creativity of the evolution process; it rather functions as a problem solver to reach a definite target state sooner or later. It resembles relaxation of an initial perturbation to a unique equilibrium state rather than evolutionary progress and innovation. In particular, the rule (7.8) completely impedes any historicity which is so typical for evolution; the final dominating state is completely independent of any fluctuations that may have happened in the past. It is obvious from this simple example that even though the far-from-equilibrium dynamics, Equation 7.1, permits evolution, the rules of the game, Equations 7.8 and 7.10, play a decisive role, too. Rules like this determine in reality which particular modification of a phenotype represents an advantage, or which idea or invention in detail is responsible for the success of a certain commodity or technology on the market. It seems evident that we cannot derive those rules from general physical laws; the governing selective or exchange values are emergent properties. Physically, the state far from equilibrium is a necessary rather than a sufficient condition for evolution.

A model with some modified, arbitrarily chosen, intentionally less regular rules can demonstrate that the same dynamics as before may enable nontrivial, potentially unlimited evolutionary progress and historicity (Ebeling and Feistel, 1974; Feistel, 1976b). Let the sequence S_i consist of the four letters $z = A, B, C, D$, associated with the values $w(z) = 1, 2, 3, 4$, respectively. To compute the value $W_L(S_i)$ for a given sequence

Table 7.1 Function values of $f_{ab}(w, p)$, Equation 7.13, to evaluate pairs of characters.

$f_{ab}(w, p)$	$a = A$	$a = B$	$a = C$	$a = D$
$b = A$	w	w	$w+1$	w/p
$b = B$	$w+p/2$	w	$w/2$	$w+2$
$b = C$	$w/3$	$w+p/3$	w	$w/3$
$b = D$	$w+4/p$	$w/4$	$W+p/4$	w

of the length L, we read the sequence, starting with the value of the letter at the leftmost position, $p = 1$,

$$W_1(S_i) = w(z_1) \tag{7.12}$$

Then, the sequence is read and for each overlapping pair at the position $(p, p+1)$, a pair value is added by the recursion formula

$$W_{p+1}(S_i) = f_{ab}[W_p(S_i), p^*] \tag{7.13}$$

Here, the subscripts refer to the letters of the pair, $a \equiv z_p$, $b \equiv z_{p+1}$, at the position p of the sequence; the modified position number p^* is defined by

$$p^* = \begin{cases} p, & \text{if } p < 8 \\ (p-6)/2, & \text{if } p \geq 8 \end{cases} \tag{7.14}$$

and the function f_{ab} is given by Table 7.1. Finally, the selective value is computed from

$$E_i = \ln W_L(S_i) \tag{7.15}$$

The mutation rate is taken proportional to the length of the sequence. Permitted mutations are

- substitution of a letter by another letter,
- addition of a letter on the left or on the right end, and
- truncation of the sequence at an arbitrary position, wasting the rest.

As a typical evolution sequence of that model, we obtained the regular history:

$$\begin{array}{r} \text{D} \\ \text{DC} \\ \text{BDC} \\ \text{BDA} \\ \text{BDAC} \\ \text{BDACB} \\ \text{CBDACB} \\ \text{CBDACBA} \\ \text{DCBDACBA} \\ \text{DABDACBA} \\ \text{DABDACBAC} \\ \ldots \end{array} \tag{7.16}$$

Here, progress occurred by point mutations along a certain branch of the evolution tree. Each step was associated with an increase in the selective value of the particular dominant species.

An alternative realization was found which takes its route to a separate branch. Soon after the bifurcation point, the change from DACBDCBA to DACBACBDA, any return to the previous branch (7.16) is highly improbable:

$$
\begin{array}{l}
\text{D} \\
\text{DA} \\
\text{DAC} \\
\text{DACB} \\
\text{DACBD} \\
\text{DACBDC} \\
\text{DACBDCB} \\
\text{DACBDCBD} \\
\text{DACBDCBA} \\
\text{DACBACBDA} \\
\ldots
\end{array}
\qquad (7.17)
$$

The bifurcation fluctuation that led from DA to either BDA or to DAC is preserved throughout the later history. Note that despite the low probability, the last step was a triple mutation rewarded with a substantial selective advantage. But even continual progress of selective values is not mandatory in a stochastic evolution model, as observed in a third realization:

$$
\begin{array}{lll}
\ldots & & \\
\text{CBD} & E = & 1.67 \\
\text{CBDD} & E = & 1.67 \\
\text{CBDDC} & E = & 1.85 \\
\text{CBBDC} & E = & 1.85 \\
\text{CABDC} & E = & 1.39 \\
\text{DABDC} & E = & 2.40 \\
\ldots & &
\end{array}
\qquad (7.18)
$$

The examples given above clearly demonstrate that the simple competition dynamics, Equation 7.1, in combination with suitable rules for mutations and selective values, is capable of showing typical features of evolution such as historicity, branching, neutral changes, or hopping. The models are less suitable, though, for a more systematic study of the conditions and properties of such qualitative characteristics. Those aspects will be addressed more thoroughly in the models investigated in the following sections.

Evolution processes of the kind considered here proceed in physical state spaces that are vastly underoccupied; the number of possible states is astronomically large and exceeds significantly the number of particles or possible practical realizations. This situation implies a severe and systematic but inevitable difficulty; the selection rules (selective values or other properties associated with each state) can neither be derived from general physical laws (because they are emergent properties) nor from observation (because the observed states form just a negligible minority of the manifold of possible states). Consequently, these evolution models remain inherently hypothetical or even fictitious to a large extent. They are of qualitative and tutorial rather than of descriptive nature. This assessment applies similarly to the models presented in the subsequent sections.

7.2
Evolution on Fitness Landscapes

The idea that evolution may be modeled or at least visualized as a search process on landscapes was first suggested by Wright (1931, 1932) and was further developed by Eigen (1971), Conrad (1978, 1983), Kauffman (1995), Stadler (1996), Schuster (2009), and other workers (see also Feistel and Ebeling, 1982, 1984, 1989; Dawkins, 1996b; Gavrilets, 2004; Pigliucci, 2008).

Roughly speaking, the idea is that genes or phenotypes or whatever is valuated in the course of evolution forms a discrete or continuous space in which the dynamics of a population is defined by some dynamical model. Having in mind the high dimensionality of all models based on sequence and phenotype descriptions, the state space of the evolution dynamics is in realistic cases high-dimensional (not just two-dimensional which we used in our examples in Figures 7.1 and 7.2 for illustration). The dynamics in the assumed state space is controlled (in the widest sense) by a valuation function which is defined by some kind of landscape over the space in which the dynamics occurs. In the simplest case, the dynamics is just "hill-climbing," looking for the highest mountains of the valuation landscape, but the dynamics might be much more complicated, including, for example, interactions in the population and deformation of the landscape itself as a result of the evolution process.

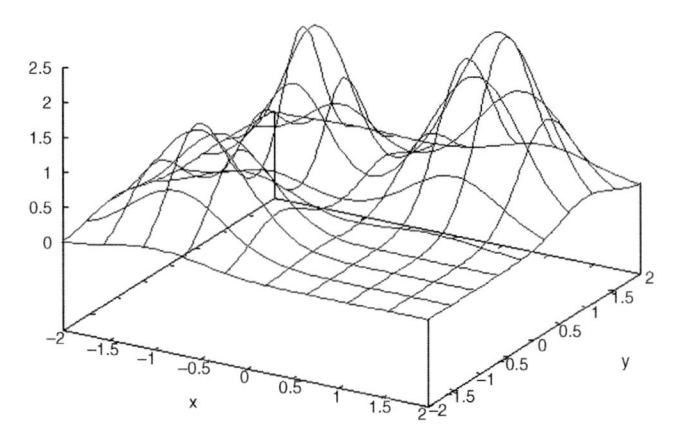

Figure 7.1 Schematic of a valuation landscape with many hills, controlling a population dynamics in the underlying state space, which here is a plane (see also the schematic of Figure 7.2).

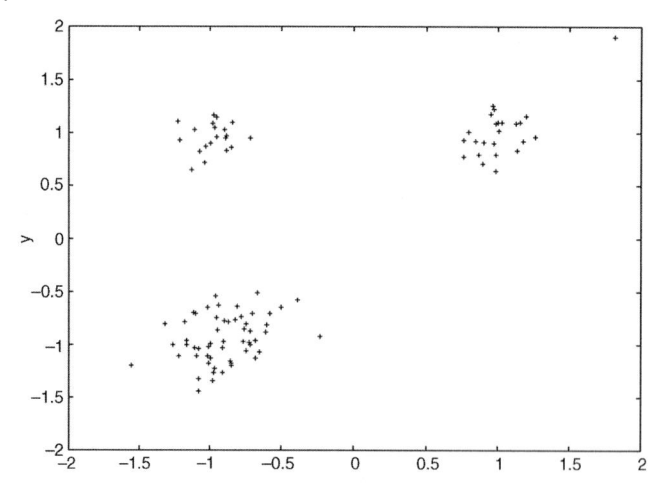

Figure 7.2 Schematic of a hill-climbing dynamics of a population controlled by the valuation landscape shown in Figure 7.1.

The schematics, Figures 7.1 and 7.2, are certainly oversimplifications but, on the other hand, are already surprisingly rich, in particular if the extended manifold of possible valuation landscapes in high-dimensional spaces is taken into account. The potential richness of this picture led to a new branch in modeling social and technological phenomena which Andrea Scharnhorst called G_O_E_THE (geometrically oriented evolution theories, see Scharnhorst,1998, 2001; Scharnhorst and Ebeling, 2005).

If evolutionary processes are controlled by a complex landscape, then the structure of this landscape is of primary importance. In a completely uncorrelated stochastic landscape, no search strategy can succeed. In order to be able to develop good search strategies, some kind of *smoothness postulate* is adopted (Rechenberg, 1973; Conrad, 1978, 1983; Conrad and Rizki, 1980; Voigt et al., 1996; Schuster, 2009), which describes the property that species with small distances in the state space are assumed to also possess similar selective values (Feistel, 1976b; Feistel and Ebeling, 1982, 1989). The high dimensionality of the valuation landscape plays a special role which guarantees that the number of saddle points is high in comparison to the number of extrema (Conrad and Ebeling, 1992). It is known that in particular the distribution of local extrema and of saddle points separating them in the landscape is dominating the dynamics of evolution processes, such as of biopolymer folding (Klemm, Flamm, and Stadler, 2008; Schuster, 2009).

In the following three sections, three qualitatively different regimes of evolution processes are modeled, on Fisher–Eigen landscapes which possess extremum principles and on Lotka–Volterra landscapes which in general do not. We distinguish between the extremes of smooth and random landscapes; in the first case, the correlation length of the landscape is much greater than the correlation length of the population distribution, while for random landscapes the opposite is assumed. The correlation length of the landscape is also related to the species diversity problem (Beardmore et al., 2011).

7.3
Evolution on Smooth Fisher–Eigen Landscapes

Selection equations of the type (7.1) are denoted here as *Fisher–Eigen equations*

$$\dot{X}_i = (E_i - \langle E \rangle) X_i \tag{7.19}$$

They were first introduced by Fisher (1930) to study biological populations, and later used by Eigen (1971) for the description of molecular evolution. As shown in Chapter 6, the related dynamical model can represent an extended class of very different competition systems (Feistel and Ebeling, 1976). In the previous section, it turned out that beyond the structure of Equation 7.19, the way by which different values E_i are associated with the species i, as well as the specification of a suitable similarity measure between two species i and j, their "genetic distance" $d(i, j)$, are highly relevant for qualitative features of the resulting evolution process.

It is usually termed as a *smoothness postulate* (Rechenberg, 1973; Conrad, 1978, 1983; Conrad and Rizki, 1980; Schuster, 2009) if pairs of species with small genetic distances are assumed to also possess similar phenotypic properties, including their selective values. Under this condition, we can formally replace the picture of discrete species with the index i by a species with a continuously variable phenotype $\mathbf{q}$ (Feistel, 1976b; Feistel and Ebeling, 1982, 1989). This leads to the continuous form of the Fisher–Eigen equation

$$\frac{\partial}{\partial t} x(\mathbf{q}, t) = (E(\mathbf{q}) - \langle E \rangle) x(\mathbf{q}, t) \tag{7.20}$$

where $x(\mathbf{q}, t)$ is now the *population density* in the *phenotype space*, Q, the scalar valuation function $E(\mathbf{q})$ forms a *landscape* over Q. The detailed structure of this landscape is unknown, except several general properties.

We assume in the following properties such as smoothness and a quasistochastic structure similar as in physics assumed for the energy landscape of disordered semiconductors (Ebeling et al., 1984; Engel and Ebeling, 1987b; Ebeling, Engel, and Feistel, 1990a; Klemm, Flamm, and Stadler, 2008; Schuster, 2009). The mean value of the valuation function depends on the distribution and is defined by

$$\langle E \rangle(t) = \frac{\int d\mathbf{q}\, E(\mathbf{q}) x(\mathbf{q}, t)}{\int d\mathbf{q}\, x(\mathbf{q}, t)} \tag{7.21}$$

This functional has the meaning of the average fitness of the particular population, $x(\mathbf{q}, t)$. The system of ordinary differential equation (7.19) is now replaced by a nonlinear partial integral-differential equation (7.20). For the beginning, let us consider a one-dimensional space Q for simplicity. Typical properties of the original discrete version of Equation 7.19 are maintained in the continuous case. The total population

$$N(t) = \int d\,qx(q,t) \tag{7.22}$$

is constant within the present model

$$\frac{\partial N}{\partial t} = \frac{\partial}{\partial t}\int d q x(q,t) = \int d q E(q) x(q,t) - \langle E \rangle \int d q x(q,t) = 0 \tag{7.23}$$

and the mean fitness is increasing monotonously due to Jensen's inequality (Jensen, 1906)

$$\frac{\partial}{\partial t}\langle E \rangle = \frac{1}{N}\int dq E(q) \frac{\partial}{\partial t} x(q,t) = \langle E^2 \rangle - \langle E \rangle^2 \geq 0 \tag{7.24}$$

The ansatz

$$x(q,t) = y(q,t)\exp\left(-\int_0^t \langle E \rangle(t')dt'\right) \tag{7.25}$$

transforms Equation 7.20 into the linear equation

$$\frac{\partial}{\partial t} y(q,t) = E(q) y(q,t) \tag{7.26}$$

with the solution

$$y(q,t) = y(q,0)\exp(E(q)t) = x(q,0)\exp(E(q)t) \tag{7.27}$$

For the mean fitness, Equation 7.21, we obtain

$$\langle E \rangle(t) = \frac{\int d\,q E(q) x(q,t)}{\int d\,q x(q,t)} = \frac{\int d\,q E(q) y(q,t)}{\int d\,q y(q,t)} = \frac{\partial}{\partial t}\ln\int d q y\,(q,t) \tag{7.28}$$

with the time integral

$$\int_0^t \langle E \rangle(t')dt' = \ln\frac{\int d q y(q,t)}{\int d q y(q,0)} = \ln\frac{\int d q y(q,t)}{\int d q x(q,0)} \tag{7.29}$$

From Equation 7.25 we infer the general solution in the form

$$x(q,t) = y(q,t)\exp(-\int \langle E \rangle dt) = y(q,t)\frac{\int d\,q x(q,0)}{\int d\,q y(q,t)}$$
$$= N\frac{x(q,0)\exp(E(q)t)}{\int dq' x(q',0)\exp(E(q')t)} \tag{7.30}$$

7.3 Evolution on Smooth Fisher–Eigen Landscapes

For the displacement of the phenotypic barycenter of the population,

$$\langle q \rangle(t) = \frac{1}{N} \int q x(q,t) \, dq \tag{7.31}$$

we derive from Equation 7.20 the correlation expression

$$\frac{d}{dt}\langle q \rangle = \langle Eq \rangle - \langle E \rangle \langle q \rangle = \langle (E - \langle E \rangle)(q - \langle q \rangle) \rangle \tag{7.32}$$

that is, the mean phenotype of a species will drift as long as there is a correlation between phenotype change and reproduction rate. Similarly, for the phenotypic diversity of the species we infer

$$\frac{d}{dt}(\langle q^2 \rangle - \langle q \rangle^2) = \left\langle (E - \langle E \rangle)(q - \langle q \rangle)^2 \right\rangle \tag{7.33}$$

If the function $E(q)$ varies only slightly over the population's spreading range, we may expand E locally into a power series

$$E(q) = E(\langle q \rangle) + (q - \langle q \rangle)\frac{\partial E}{\partial q} + \frac{1}{2}(q - \langle q \rangle)^2 \frac{\partial^2 E}{\partial q^2} + \dots \tag{7.33}$$

For the leading term of the displacement rate, Equation 7.31, we get the equation

$$\frac{d}{dt}\langle q \rangle = \left\langle (q - \langle q \rangle)^2 \right\rangle \frac{\partial E}{\partial q} + \dots \tag{7.34}$$

This relation is known in population biology as *Fisher's fundamental law of natural selection* (Wilson and Bossert, 1973), for short, Fisher's law. For the diversity we infer

$$\frac{d}{dt}(\langle q^2 \rangle - \langle q \rangle^2) = \left\langle (q - \langle q \rangle)^3 \right\rangle \frac{\partial E}{\partial q} + \frac{1}{2}\left\langle (q - \langle q \rangle)^4 \right\rangle \frac{\partial^2 E}{\partial q^2} + \dots \tag{7.35}$$

Hence, the width of the phenotypic pool will change with the skewness (asymmetry) of the distribution if the selective value has a gradient, and with the kurtosis of the distribution if the selective value is either a convex or a concave function.

Equipped with this set of general relations, we may consider some Gaussian distribution of a population with the phenotypic diversity σ localized at q_0,

$$x(q, 0) = A \exp\left\{ -\frac{(q - q_0)^2}{2\sigma^2} \right\} \tag{7.36}$$

If the diversity is small in comparison to the scales on which the landscape $E(q)$ is changing, we may approximate the latter by Equation 7.33. The average fitness, Equation 7.21, is then

$$\langle E \rangle \approx E(\langle q \rangle) + \frac{1}{2}\sigma^2 \frac{\partial^2 E}{\partial q^2} \tag{7.37}$$

The center of the phenotypic pool will be displaced by Fisher's law (Feistel, 1976b; Feistel and Ebeling, 1982, 1989) with a velocity, Equation 7.34,

$$\frac{d}{dt}\langle q \rangle \approx \sigma^2 \frac{\partial E}{\partial q} \tag{7.38}$$

proportional to the phenotypic diversity and in direction of the phenotypic gradient of the selective value. For the time development of the phenotypic diversity, we infer from the vanishing higher cumulants of the Gaussian distribution

$$\frac{d}{dt}\sigma^2 = 0 \tag{7.39}$$

Hence, Equation 7.38 can be integrated to the trajectory in the phenotype space, to give

$$\langle q \rangle(t) = q_0 + \sigma^2 t \frac{\partial E}{\partial q} \tag{7.40}$$

The traveling Gaussian distribution,

$$x(q,t) = A \exp\left\{-\frac{1}{2\sigma^2}\left(q - q_0 - \sigma^2 t \frac{\partial E}{\partial q}\right)^2\right\} \tag{7.41}$$

is actually a time-dependent solution of Equation 7.20 for the case of a landscape with a constant gradient of the selective value. The variance σ is determined by the initial state and remains unchanged at later times. This process is known as the "shifting form" of selection in evolution theory (Timofeeff-Ressovsky, Voroncov, and Jablokov, 1975; Feistel and Ebeling, 1984).

So far, the continuous model, Equation 7.20, describes only self-reproduction. If a phenotype q is not occupied by individuals at the beginning, $x(q,0) = 0$, it will never become occupied in the future. Evolution is impossible here in the sense of creating something previously nonexistent. This was not obvious in the example of the initial Gaussian distribution because the state at $t=0$ included already every possible state, even though with a vanishingly small density. Alternatively, if we had begun with a point distribution, $x(q,0) = \delta(q-q_0)$, nothing at all had happened in the model since any Dirac delta function is a stationary solution of Equation 7.20.

To overcome this unrealistic limitation, mutations must be included which create species with the phenotype $\mathbf{q}'$ from present species with the phenotype $\mathbf{q}$:

$$\frac{\partial}{\partial t} x(\mathbf{q},t) = (E(\mathbf{q}) - \langle E \rangle)x(\mathbf{q},t) + \int d\mathbf{q}'[A(\mathbf{q},\mathbf{q}')x(\mathbf{q}',t) - A(\mathbf{q}',\mathbf{q})x(\mathbf{q},t)] \tag{7.42}$$

The additional term is similar to the description of transitions between different states by the master equation formalism, Equation 4.63 in Section 4.3. The nonnegative matrix element $A(\mathbf{q},\mathbf{q}')$ is the transition rate from a phenotype $\mathbf{q}'$ to a phenotype $\mathbf{q}$.

Here, we have returned to a higher-dimensional phenotype space, Q, which is spanned, at least in principle, by any quantitative property of any species that may be relevant in the context of the model. If we think of human populations, we know that a large number of different, very detailed properties are important for successful social interaction, such as the body mass, the color of the hair, of the skin and of the eyes, the length and breadth of the nose, and many other measures that we inherit and sometimes even try to correct by technical means to get a higher "selective value." Alternatively, for primitive organisms, we may think of counting the numbers of all the chemical species that are present in a prokaryote cell. Thus, we shall imagine Q to possess many dimensions, similar to state spaces studied in statistical thermodynamics.

Either by genetic mutations or by sexual recombination (compare Section 6.9), offsprings may slightly deviate in their phenotypic appearance from that of their parents. High dimensionality is relevant here; in a high-dimensional space much more points are located within a short distance of a given one – there are many directions into which small deviations may occur. When a function $E(\mathbf{q})$ describes the selective values of species with the phenotype $\mathbf{q}$, apart from the minima or maxima of the landscape, in a one-dimensional state Q all neighbors of a given point q necessarily possess either higher or lower selective values. In an N-dimensional space Q, in contrast, the condition $E(\mathbf{q}) = \text{const}$ specifies an $(N-1)$-dimensional subspace rather than a point, which means that almost every mutation is necessarily a strictly or nearly neutral mutation, and only exceptional changes will lead to elevated or lowered selective values. The high dimension of the phenotype space is fundamental for understanding the dynamical role of "neutral evolution" frequently stressed in the literature (Conrad, 1982; Kimura, 1983; Conrad and Ebeling, 1992; Stoltzfus, 1999; Speijer, 2010). Here, with neutral mutations we refer to modifications which in fact change the phenotype $\mathbf{q}$ but in such a way that leaves the function value of $E(\mathbf{q})$ unaltered, rather than "homologous substitutions," or "silent mutations," which change the genetic sequence without affecting the phenotype $\mathbf{q}$ at all (Povolotskaya and Kondrashov, 2010).

In terms of the model (7.42), we may assume the related transition matrix to be rapidly (exponentially) decreasing with the distance between the points $\mathbf{q}$ and $\mathbf{q}'$ in Q, that is,

$$A(\mathbf{q}, \mathbf{q}') = A(\mathbf{q}-\mathbf{q}'), \quad \lim_{|\mathbf{r}| \to \infty} A(\mathbf{r}) = 0 \tag{7.43}$$

If we make the reasonable assumption of symmetry,

$$A(\mathbf{q}-\mathbf{q}') = A(\mathbf{q}'-\mathbf{q}) \tag{7.44}$$

the transition integral can be written in the form

$$\int d\mathbf{q}' [A(\mathbf{q}, \mathbf{q}')x(\mathbf{q}', t) - A(\mathbf{q}', \mathbf{q})x(\mathbf{q}, t)] = \int d\mathbf{r} A(\mathbf{r})[x(\mathbf{q}+\mathbf{r}, t) - x(\mathbf{q}, t)] \tag{7.45}$$

where the new integration variable is the transition distance vector, $\mathbf{r} = \mathbf{q}'-\mathbf{q}$. If the population density is a smooth function over the short mutation range, we may expand it into a local power series with respect to $\mathbf{r}$,

$$x(\mathbf{q}+\mathbf{r}, t) = x(\mathbf{q}, t) + \mathbf{r}\frac{\partial}{\partial \mathbf{q}} x(\mathbf{q}, t) + \frac{1}{2}\left(\mathbf{r}\frac{\partial}{\partial \mathbf{q}}\right)^2 x(\mathbf{q}, t) + \cdots \tag{7.46}$$

Under the integral (7.45), the linear term of Equation 7.46 disappears for the symmetry of A, Equation 7.44, and as the first nonvanishing term there remains the diffusion approximation of the transition integral

$$\int d\mathbf{q}'[A(\mathbf{q},\mathbf{q}')x(\mathbf{q}',t) - A(\mathbf{q}',\mathbf{q})x(\mathbf{q},t)] \approx \frac{\partial}{\partial \mathbf{q}} \mathbf{D} \frac{\partial}{\partial \mathbf{q}} x(\mathbf{q},t) \qquad (7.47)$$

where the diffusion matrix

$$\mathbf{D} = \frac{1}{2}\int d\mathbf{r} A(\mathbf{r})(\mathbf{r}\otimes\mathbf{r}) \qquad (7.48)$$

is independent of $\mathbf{q}$. The symbol $\otimes$ denotes the dyadic vector product. The coordinates $\mathbf{q}$ in the phenotype space are not specified yet in any way; we may carry out a linear (rotation and scaling) transformation such that the matrix $\mathbf{D}$ becomes diagonal with equal elements on the main diagonal, $\mathbf{D} = D\mathbf{I}$, where $\mathbf{I}$ is the unity matrix. We shall regard this set of rearranged phenotypic properties as the *normal form*. The value of the scalar diffusion constant D depends on the way the phenotypic coordinate transformation is specified. In the normal form, Equation 7.42 reads (Feistel and Ebeling, 1982, 1989)

$$\frac{\partial}{\partial t} x(\mathbf{q},t) = (E(\mathbf{q}) - \langle E \rangle) x(\mathbf{q},t) + D\frac{\partial^2}{\partial \mathbf{q}^2} x(\mathbf{q},t) \qquad (7.49)$$

For this general selection–mutation equation, in comparison to the case of pure selection, Equations 7.155, we may derive relations of the temporal development of some statistical moments. For the mean selective value, $\langle E \rangle$, we obtain by partial integration and application of the Gauss theorem

$$\frac{\partial}{\partial t} \langle E \rangle = \langle E^2 \rangle - \langle E \rangle^2 + D\left\langle \frac{\partial^2 E}{\partial \mathbf{q}^2} \right\rangle \qquad (7.50)$$

In particular in the vicinity of local maxima of the landscape $E(\mathbf{q})$ where E is concave and the second derivative of E is negative, the universal growth of the mean fitness is no longer granted; mutations permanently generate a cloud of less fit copies which lower the average.

For the population average, $\langle \mathbf{q} \rangle$, we get again after partial integration still Fisher's law

$$\frac{d}{dt}\langle \mathbf{q} \rangle = \langle E\mathbf{q} \rangle - \langle E \rangle\langle \mathbf{q} \rangle \qquad (7.51)$$

which is virtually unaffected by mutations. Finally, for the diversity

$$\frac{d}{dt}(\langle \mathbf{q}^2 \rangle - \langle \mathbf{q} \rangle^2) = \left\langle (E - \langle E \rangle)(\mathbf{q} - \langle \mathbf{q} \rangle)^2 \right\rangle + 2D \qquad (7.52)$$

we expectedly find an additional spreading of properties due to mutations. With respect to neutral mutations mentioned above, the diffusion term will necessarily tend to increase the phenotypic diversity along the hypersurface $E(\mathbf{q}) = \text{const}$, that is, into all dimensions except one, namely the direction of the gradient of E, Equation 7.51. This is a rather general and unexpected result obtained from this fairly simple evolution model; it seriously raises the question why at all the properties of biological species are confined to some relatively narrow scatter range of their traits. From the viewpoint of this theory, two answers can be given. First, if a species is in evolutionary progress according to Fisher's law, neutral mutations (perpendicular in Q to the gradient of the selective value) cannot keep pace with the driving force of competition; they will either form some "Mach cone" or "bow wave" left behind that finally gets extinct if the gradient is locally weaker, or they will take over the lead if the local gradient is steeper there. Second, if a species is located at a local maximum of the fitness landscape, there will be no strictly neutral mutations but, due to the missing gradient, a practically neutral vicinity with a radius given by the local curvature of the maximum. This case will be considered in more detail below.

The transformation, Equation 7.25, to a new variable $y(\mathbf{q}, t)$ is also possible in the presence of mutations, with the result

$$\frac{\partial}{\partial t} y(\mathbf{q}, t) = E(\mathbf{q}) y(\mathbf{q}, t) + D \frac{\partial^2}{\partial \mathbf{q}^2} y(\mathbf{q}, t) \tag{7.53}$$

This equation can be written in the form of a Fréchet derivative

$$\frac{\partial}{\partial t} y = -\frac{\delta W}{\delta y(\mathbf{q})} \tag{7.54}$$

where W is a suitably defined Landau–Ginzburg functional (Landau and Lifshitz, 1966; Haken, 1978; Feistel and Ebeling, 1982)

$$W = \frac{1}{2} \int d\mathbf{q} \left\{ D \left(\frac{\partial y}{\partial \mathbf{q}} \right)^2 - E(\mathbf{q}) y(\mathbf{q})^2 \right\} \tag{7.55}$$

Here, mutations result in a "negative surface-tension" effect in the phenotype space. From Equation 7.54, we derive the evolution principle for Fisher–Eigen models of selection–mutation processes

$$\frac{d}{dt} W = \int d\mathbf{q} \frac{\delta W}{\delta y(\mathbf{q})} \frac{\partial y(\mathbf{q}, t)}{\partial t} = -\int d\mathbf{q} \left(\frac{\delta W}{\delta y(\mathbf{q})} \right)^2 \leq 0 \tag{7.56}$$

It is instructive to study the special case of selection–mutation behavior in the vicinity of a local maximum of the landscape $E(\mathbf{q})$, when the population has approached some optimum and short-range mutations do not provide any selective advantages,

$$E(q) = E_{max} - \frac{1}{2} \kappa q^2 \tag{7.57}$$

For simplicity of writing, we have returned here to a one-dimensional phenotype space. Note that the value of E_{max} can be chosen arbitrarily because it cancels in Equation 7.49 for $x(q, t)$. For convenience, we shall use $E_{max} = 0$ here.

For the particular landscape given by Equation 7.57, the selection–mutation equation (7.53) has a helpful similarity to the stationary Schrödinger equation of a harmonic oscillator

$$E_n \psi_n = -\frac{\hbar^2}{2m} \frac{\partial^2 \psi_n}{\partial q^2} + \frac{1}{2} \kappa q^2 \psi_n \qquad (7.58)$$

where for the original quantum constants the relation $\hbar^2/2m \equiv D$ holds in this formal analogy. Integrable solutions of Equation 7.58 are known to be the eigenfunctions

$$\psi_n(q) = \exp\left\{-\frac{q^2}{2}\sqrt{\frac{\kappa}{2D}}\right\} H_n\left[q\left(\frac{\kappa}{2D}\right)^{1/4}\right] \qquad (7.59)$$

associated with the discrete eigenvalues

$$E_n = \left(n + \frac{1}{2}\right)\sqrt{2\kappa D} \qquad (7.60)$$

Here, H_n are the Hermite polynomials (Abramowitz and Stegun, 1968). The time-dependent solution can be composed of a superposition of eigenfunctions

$$x(q,t) = \sum_{n=0}^{\infty} A_n \psi_n(q) \exp\left\{-\int_0^t [\langle E(t')\rangle + E_n] dt'\right\} \qquad (7.61)$$

where the coefficients A_n result from the initial conditions. The final stationary state is given by

$$\lim_{t \to \infty} \langle E(t) \rangle = -E_0 \qquad (7.62)$$

which is below the maximum fitness, and

$$\lim_{t \to \infty} x(q,t) = A_0 \psi_0(q) \qquad (7.63)$$

that is, a Gaussian distribution with a phenotypic diversity of $\sigma^2 = 2D/\kappa$. This result is known as the "stabilizing form" of selection in evolution theory (Timofeeff-Ressovsky, Voroncov, and Jablokov, 1975; Feistel and Ebeling, 1984).

7.4
Evolution on Random Fisher–Eigen Landscapes

Many results from quantum mechanics may be translated to the evolutionary problem considered in the previous section. A more realistic landscape $E(q)$ will possess various separate maxima which correspond to local oscillator states in our picture. If initially only one or a few of those maxima are populated, the first quick process will consist in some "falling down" to the local oscillator ground states (remember, minima in quantum mechanical potential energy correspond to maxima of the selective value in the evolution picture). The second, usually much slower

7.4 Evolution on Random Fisher–Eigen Landscapes

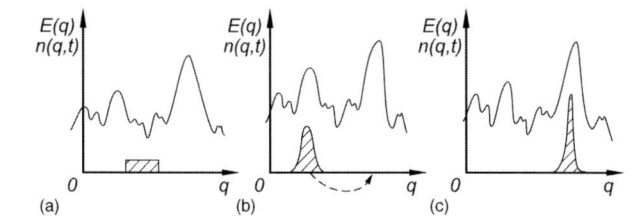

Figure 7.3 Schematic of the time development of a localized population density, $n(q, t)$, as a function of the phenotype, q, on a random fitness landscape, $E(q)$ (adapted from Ebeling, Engel, and Feistel, 1990a).

process is the "quantum tunneling" through the barriers of unfavorable states to another, not yet populated maximum in the neighborhood, as shown schematically in Figure 7.3. If the new maximum has a higher selective value, the local population will grow and the previous maximum will remain unoccupied, the related phenotype extinct. Thus, evolution consists of a hopping process from certain maxima to higher maxima somewhere nearby. This scenario resembles the quantum theory of disordered systems where the Schrödinger equation is solved for very complicated potential functions. This problem is subject to the modern theory of condensed matter (Mott and Davis, 1971). We will focus on such aspects in this section (Ebeling et al., 1984; Ebeling, Engel, and Feistel, 1990a).

The general selection–mutation problem on a Fisher–Eigen landscape, $E(\mathbf{q})$, Equation 7.53, leads to an eigenvalue problem similar to Equation 7.58, in the form of the partial differential equation

$$E_n \psi_n(\mathbf{q}) = D \frac{\partial^2}{\partial \mathbf{q}^2} \psi_n(\mathbf{q}) + E(\mathbf{q}) \psi_n(\mathbf{q}) \tag{7.64}$$

which equals the stationary Schrödinger equation for a potential $U(\mathbf{q}) = -E(\mathbf{q})$ and eigenvalues E_n, which correspond to negative energy values. The functions $\psi_n(\mathbf{q})$ are the natural generalizations of the "quasispecies" suggested by Eigen and Schuster (1977, 1979). Note, however, that only nonnegative solutions of Equation 7.64 can be considered as proper density distributions of some species. As a rule, this condition is obeyed by only one of the infinitely many solutions of Equation 7.64, due to their pairwise orthogonality. Any other "quasispecies" possess zeros where the function changes its sign. While those functions can act mathematically as formal modes in series expansions such as Equation 7.61, they can never appear as some autonomous observable entity. This situation is very similar for the original discrete "quasispecies" defined as the eigenvectors of certain cross-catalytic matrices. If, for example, such a matrix is nonnegative and irreducible (see Sections 4.4 and 6.5), only one of its eigenvectors can be nonnegative and hence be understood as a combination of population numbers (Feistel and Ebeling, 1978b).

As a simple example for the analysis by means of Equation 7.64 we consider the case of two different maxima of the function $E(q)$, let the higher one, $E^{(1)}$, be narrower,

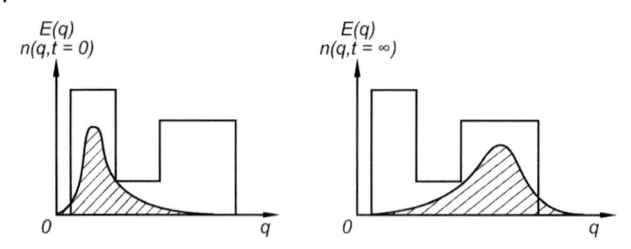

Figure 7.4 Schematic of the transition of a localized population density, $n(q, t)$, as a function of the phenotype, q, from a high but narrow maximum fitness landscape, $E(q)$, to a lower but wider one (adapted from Ebeling, Engel, and Feistel, 1990a).

$\Delta q^{(1)}$ the lower one, $E^{(1)} > E^{(2)}$, be wider, $\Delta q^{(1)} < \Delta q^{(2)}$, see Figure 7.4. Without mutations, $D=0$, it is clear that the preferred final state is the maximum 1 with the higher selective value. For positive D, the eigenvalue E_0 of the ground state in a potential well is known from quantum mechanics (Landau and Lifschitz, 1988),

$$E_0^{(i)} = E^{(i)} - \frac{D\pi}{\Delta q^{(i)}} \tag{7.65}$$

where we have inverted the sign of E in comparison to the quantum result. The fitness maximum with the higher "effective fitness" $E_0^{(i)}$ will eventually be populated; the steeper the peak, the more its top value is lowered by diffusion (mutations). This effect becomes even more important in phenotype spaces of high dimension, $d_Q \gg 1$, because the generalization of Equation 7.65 for that case reads

$$E_0^{(i)} = E^{(i)} - d_Q \frac{D\pi}{|\Delta \mathbf{q}^{(i)}|} \tag{7.66}$$

It is evident already from this simple example that in high-dimensional phenotype spaces, the structure of the fitness function in the vicinity of the maxima may be more important than the peak value itself, quite in contrast to selection in the absence of mutations, that is, the solution of Equation 7.64 may pose a nontrivial theoretical problem. This is in particular true for populations on random Fisher–Eigen landscapes that are mathematically equivalent to electron wave functions in a random potential. Such results are known from systematic theoretical studies of disordered solid states (Thouless, 1974; Anderson, 1978; Lee and Ramakrishnan, 1985). Here we report three general findings that are also important for the case of evolution (Ebeling, Engel, and Feistel, 1990a).

The density of eigenvalues of Equation 7.64 is similar to that of many quantum systems such as a hydrogen atom. There is a so-called mobility edge E_0 (for the H atom, the ionization energy) such that below that fitness (above that energy) a continuous spectrum exists which corresponds to spatially extended scattering states. Above the threshold value of E_0 (below the mobility energy level), the eigenvalues are discrete (spectral lines of the H atom), the related eigenfunctions are localized (Mott, 1967; Kimball, 1978), and the number of possible states

7.4 Evolution on Random Fisher–Eigen Landscapes

decreases rapidly with higher fitness (stronger binding energy). In the evolution picture, localized states correspond to species with some diversity in their phenotypic appearance, and with high fitness. Asymptotic expressions for the density of eigenvalues in the limit $E \to \infty$ can be estimated from the statistical properties of the landscape (Lifshitz, Gredeskul, and Pastur, 1982).

Eigenfunctions $\psi_n(\mathbf{q})$ which belong to eigenvalues $E_n > E_0$ are localized. They possess a center phenotype $\mathbf{q}_n$ and a *localization length* L_n such that the population density away from the center decreases asymptotically in the form (Anderson, 1978)

$$\psi_n(\mathbf{q}) \propto \exp\left\{-\frac{|\mathbf{q}-\mathbf{q}_n|}{L_n}\right\} \tag{7.67}$$

For a Gaussian random landscape $E(\mathbf{q})$ with vanishing correlation length, that is,

$$\langle E(\mathbf{q}) E(\mathbf{q}') \rangle \propto \delta(\mathbf{q}-\mathbf{q}') \tag{7.68}$$

there exist no localized states for $d_Q \geq 4$, that is, the mobility threshold E_0 is infinitely high. In contrast, stochastic landscapes with finite correlation lengths possess localized states for any dimension of the phenotype space, d_Q.

If we denote the integral of the eigenfunction $\psi_n(\mathbf{q})$ by a_n

$$a_n = \int d\mathbf{q}\, \psi_n(\mathbf{q}) \tag{7.69}$$

we can write the population density by means of Equations 7.30 and 7.61 in the form

$$x(\mathbf{q}, t) = N \frac{\sum_n A_n \exp(E_n t) \psi_n(\mathbf{q})}{\sum_n A_n \exp(E_n t) a_n} \tag{7.70}$$

For increasing t, this distribution will be dominated by the eigenfunctions, $\psi_n(\mathbf{q})$, that belong to the highest selective values, E_n. If we start from, say, a homogeneous initial situation, $x(\mathbf{q}, 0) = $ const, then only the localized eigenfunctions will contribute substantially after the time $t > 1/E_0$. This formation of populated islands in the phenotype space can be interpreted as the emergence of separate species with specific phenotypes, including some diversity, out of a diffuse variety of individual properties. Under the increasing selective pressure, only individuals with specific, suitably adjusted combinations of traits will survive. A similar formation of population structures without mutations was discussed by Zeldovich (1983) in the context of percolation theory. We may conclude quite generally that reasonable evolution processes in the form of successions of fitter and fitter well-defined species can only be expected if localized eigenfunctions exist. For this condition, because the dimension of the phenotype space is usually much higher than four, the landscape must exhibit some smoothness with appropriate correlation lengths rather than being too rough (Conrad, 1978, 1983; Conrad and Ebeling, 1992; Voigt et al., 1996; Rose, Ebeling, and Asselmeyer, 1996). This smoothness postulate (Conrad, 1978,

1983) is plausible; typical realizations of ensembles of random functions with correlations similar to Equation 7.67 may be continuous but not differentiable and show abrupt changes from one point to the next. The occupation of some locus with a high selective value is not of any benefit for the future progress; the vicinity of the given maximum includes as many unfavorable states as any other region of the phenotype space. As a result, mutations will spread the population all over the rugged landscape with occasional peaks of only infinitesimal width. Landscapes of this type are obviously not very suitable as models for real evolution processes, where small property changes of a favorable species are usually accompanied with only small selective disadvantages rather than deadly deficiencies. An example for the rugged landscape may be the development of computer code. For a functioning code, already the change of a single bit is usually disastrous, and any method to systematically improve the code by random exchange of certain bits is a rather hopeless enterprise. It is the correlation length of the phenotype landscape that makes the difference between genetic and computer code. A sensitive measure for the *difficulty of optimization* in a random landscape is the density of states (Rose, Ebeling, and Asselmeyer, 1996).

Provided that localized eigenfunctions exist, they dominate the later stages of evolution as a result of their high selective values. Consequently, later evolution steps consist of transition processes between different localized states similar to the example discussed in the beginning of this section. For illustration, we consider a system with two localized eigenfunctions, $\psi_1(\mathbf{q})$ and $\psi_2(\mathbf{q})$, and the eigenvalues E_1 and $E_2 = E_1 + \Delta E$, respectively. Then, the solution (7.70) takes the form

$$x(\mathbf{q}, t) = N \frac{A_1 \psi_1(\mathbf{q}) + A_2 \psi_2(\mathbf{q}) \exp(\Delta E t)}{A_1 a_1 + A_2 a_2 \exp(\Delta E t)} \tag{7.71}$$

Starting at $t=0$ from a situation at which one eigenfunction is dominating, say, $|A_2| \ll |A_1|$, to establish a balance between the states, that is,

$$|A_1| \approx |A_2| \exp(\Delta E t) \tag{7.72}$$

it takes the time

$$\Delta t \approx \frac{1}{\Delta E} \ln \left| \frac{A_1}{A_2} \right| \tag{7.73}$$

which can be regarded as a typical transition time between the two states. Making use of the estimate for the amplitudes, Equation 7.67,

$$\left| \frac{A_2}{A_1} \right| \approx \exp\left(-\frac{|\Delta \mathbf{q}|}{L} \right) \tag{7.74}$$

where $\Delta \mathbf{q}$ is the distance between the localization centers and L the correlation length of the landscape, we get for the transition time

$$\Delta t \approx \frac{|\Delta \mathbf{q}|}{L \Delta E} \tag{7.75}$$

between two maxima with the distance $\Delta\mathbf{q}$ and the selective advantage ΔE on a landscape with the correlation length L. Hence, the value of $L\Delta E$ estimates some effective evolution speed on the phenotype landscape between separate maxima. A more detailed statistical analysis (Engel, 1983, 1985; Ebeling, Engel, and Feistel, 1990a) shows that in random landscapes, an optimum advantage ΔE can be defined by the condition that the mean transition time Δt has a minimum. For much smaller values of ΔE, the time takes longer because of the minor selective advantage; for much higher gains of ΔE, the typical distance $|\Delta\mathbf{q}|$ is quite long, depending on the statistics of the landscape. This result leads to the conclusion that evolution is of a hopping nature; it always takes a typical time to "discover" a new, distant attractor, followed by a typical, relatively fast cross-over process to the new advantageous state. These mathematical findings may shed some new light on the lasting empirical controversy between phyletic gradualism and punctuated equilibrium theory in cladogenesis (Gould and Eldridge, 1977; Volkenstein, 1984; Engel and Ebeling, 1987b; Dawkins, 1996a). Similar problems are also discussed for the evolution of languages (Dixon, 1997).

7.5
Evolution on Lotka–Volterra Landscapes

The selection–mutation equation for Fisher–Eigen landscapes, Equation 7.49, can be written in the form

$$\frac{\partial}{\partial t}x(\mathbf{q},t) = w\{\mathbf{q}|x(\mathbf{q},t)\}x(\mathbf{q},t) + D\frac{\partial^2}{\partial \mathbf{q}^2}x(\mathbf{q},t) \tag{7.76}$$

where the functional w of the population density, $x(\mathbf{q},t)$, is defined by

$$w\{\mathbf{q}|x(\mathbf{q},t)\} \equiv E(\mathbf{q}) - \langle E \rangle = E(\mathbf{q}) - \int d\mathbf{q}' \frac{E(\mathbf{q}')}{N} x(\mathbf{q}',t) \tag{7.77}$$

In the previous sections, we have considered $E(\mathbf{q})$ as a fixed landscape given in the phenotype space Q, where according to Fisher's law, Equation 7.34, the population center $\langle \mathbf{q} \rangle$ is moving uphill to higher selective values, $E(\mathbf{q})$. Rather than by those absolute fitness values, $E(\mathbf{q})$, the reproduction rate or effective fertility of an individual with the phenotype $\mathbf{q}$ is actually given by the functional $w\{\mathbf{q}|x(\mathbf{q},t)\}$ in Equation 7.76, which expresses the selective advantage a phenotype $\mathbf{q}$ has in the presence of the current population, $x(\mathbf{q},t)$. Therefore it is reasonable to consider as a proper landscape the selective advantage

$$p(\mathbf{q},t) \equiv w\{\mathbf{q}|x(\mathbf{q},t)\} \tag{7.78}$$

Species $\mathbf{q}$ with $p(\mathbf{q}) < 0$ will decay, those with $p(\mathbf{q}) > 0$ grow, as long as the deviation of p from zero exceeds the mutation effects. In contrast to the rigid landscape $E(\mathbf{q})$, the landscape $p(\mathbf{q},t)$ will change in time as a result of the joint feedback of population scattered over the phenotype space. This collective effect on the rigid landscape can be computed from Equation 7.77, to be

$$\frac{\partial}{\partial t}p(\mathbf{q},t) = \int d\mathbf{q}' \frac{\delta w}{\delta x(\mathbf{q}')} \frac{\partial x(\mathbf{q}')}{\partial t} = -\langle E^2 \rangle + \langle E \rangle^2 - \frac{D}{N}\left\langle \frac{\partial^2 E}{\partial \mathbf{q}^2} \right\rangle \leq -\frac{D}{N}\left\langle \frac{\partial^2 E}{\partial \mathbf{q}^2} \right\rangle \quad (7.79)$$

Except for situations when the population is confined to the vicinity of a peak and mutations dominate, the rigid landscape $p(\mathbf{q},t)$ is permanently "sinking down" as a result of the competition. Since the right-hand side of Equation 7.79 does not depend on the particular location $\mathbf{q}$, the effect is similar to "flooding" of a landscape by rising sea level; the landscape is submerged without being otherwise deformed. Since the average elevation of occupied phenotype regions is zero by definition, Equation 7.77,

$$\langle p(\mathbf{q},t) \rangle \equiv 0 \quad (7.80)$$

individuals will populate the vicinity of the "shorelines," $p(\mathbf{q},t) = 0$, while the drowned ones disappear, the survivors are climbing uphill and this way pushing the landscape further down.

The occupation of "islands" in this picture was discussed in the form of localized eigenfunctions in the previous section; evolution finally takes the form of an "island-hopping" process.

The Fisher–Eigen landscape, Equation 7.77, is a special case of a linear fitness functional, w. In general, the integral kernel may also depend on the particular location $\mathbf{q}$, that is,

$$w\{\mathbf{q}|x(\mathbf{q},t)\} = a(\mathbf{q}) + \int d\mathbf{q}' b(\mathbf{q},\mathbf{q}')x(\mathbf{q}',t) \quad (7.81)$$

This more general landscape, $p(\mathbf{q},t) \equiv w\{\mathbf{q}|x(\mathbf{q},t)\}$, will be regarded here as a *Lotka–Volterra landscape*. To avoid subtle but irrelevant mathematical difficulties, we imagine the phenotype space here as extended but finite; it is not reasonable to permit phenotypic properties with infinite values. From the theory of integral equations (Smirnow, 1968) and quantum mechanics (Landau and Lifschitz, 1988), it is known that under relatively general conditions integral kernels such as the *community matrix* of Equation 7.81, $b(\mathbf{q},\mathbf{q}')$, can be represented in the form of a Hilbert–Schmidt expansion,

$$b(\mathbf{q},\mathbf{q}') = \sum_k \lambda_k \Phi_k(\mathbf{q}) \Psi_k(\mathbf{q}') \quad (7.82)$$

For the definition of $\Psi_k(\mathbf{q})$ and $\Phi_k(\mathbf{q})$, we consider the symmetric kernel

$$B(\mathbf{q},\mathbf{q}') = \int b(\mathbf{q}'',\mathbf{q})b(\mathbf{q}'',\mathbf{q}')d\mathbf{q}'' \quad (7.83)$$

which has real eigenvalues, μ_k, and real orthogonal eigenfunctions, $\Psi_k(\mathbf{q})$, as defined by the eigenvalue equation

$$\int B(\mathbf{q},\mathbf{q}') \Psi_k(\mathbf{q}') d\mathbf{q}' = \mu_k \Psi_k(\mathbf{q}) \quad (7.84)$$

Because the quadratic form

$$\int B(\mathbf{q},\mathbf{q}')f(\mathbf{q})f(\mathbf{q}')d\mathbf{q}d\mathbf{q}' = \int \left[\int b(\mathbf{q}'',\mathbf{q})f(\mathbf{q})d\mathbf{q}\right]^2 d\mathbf{q}'' > 0 \qquad (7.85)$$

is positive for any nonvanishing function f, all eigenvalues μ_k must be positive due to Rayleigh's theorem. If we define the functions $\Phi_k(\mathbf{q})$ by the relation

$$\sqrt{\mu_k}\Phi_k(\mathbf{q}) = \int b(\mathbf{q},\mathbf{q}')\Psi_k(\mathbf{q}')d\mathbf{q}' \qquad (7.86)$$

we infer from (7.84) the conjugate relation

$$\int b(\mathbf{q}',\mathbf{q})\Phi_k(\mathbf{q}')d\mathbf{q}' = \sqrt{\mu_k}\Psi_k(\mathbf{q}) \qquad (7.87)$$

After inserting the Hilbert–Schmidt expansion, Equation 7.82, into Equation 7.86 we obtain for the expansion coefficients, λ_k, the relation

$$\lambda_k = \frac{\sqrt{\mu_k}}{\int \Psi_k(\mathbf{q})^2 d\mathbf{q}} = \frac{\sqrt{\mu_k}}{\int \Phi_k(\mathbf{q})^2 d\mathbf{q}} \qquad (7.88)$$

Thus, for a given community matrix, $b(\mathbf{q},\mathbf{q}')$, the expansion terms of Equation 7.82 can uniquely be calculated from Equations 7.84 and 7.88.

If the Hilbert–Schmidt expansion, Equation 7.82, consists of a finite number of terms, the kernel is termed *degenerate*. If in particular the sum has just one term, the kernel is termed a *product kernel*

$$b(\mathbf{q},\mathbf{q}') = \Phi(\mathbf{q})\Psi(\mathbf{q}') \qquad (7.89)$$

In the Fisher–Eigen model, the community matrix, $b(\mathbf{q},\mathbf{q}')$, has an exceptionally simple structure in the form of a product kernel with $\Phi(\mathbf{q}) = \text{const}$

$$b_{FE}(\mathbf{q},\mathbf{q}') = -\frac{E(\mathbf{q}')}{N} \qquad (7.90)$$

If we ignore the diffusion term of Equation 7.76 for a moment, we see that the equation

$$\frac{\partial}{\partial t}x(\mathbf{q},t) \approx x(\mathbf{q},t)p(\mathbf{q},t) = x(\mathbf{q},t)\left[a(\mathbf{q}) + \int d\mathbf{q}'b(\mathbf{q},\mathbf{q}')x(\mathbf{q}',t)\right] \qquad (7.91)$$

has stationary solutions $x(\mathbf{q})$ only if for any value of $\mathbf{q}$ either the population vanishes, $x(\mathbf{q}) = 0$, or the landscape is zero, $p(\mathbf{q}) = 0$. Hence, if we assume that selection between existing individuals is a much faster process than the emergence of new phenotypes, the populated regions of the phenotype space are the "shorelines" $p(\mathbf{q},t) = 0$. Adjacent individuals with $p < 0$ will disappear, while those with $p > 0$ will grow. For stability against mutations in the immediate vicinity, $p(\mathbf{q}) < 0$ must hold in the accessible surrounding of an occupied locus; in other words, those loci must be relative maxima of the fitness landscape, Equation 7.92, where the peak value is zero.

We consider a distribution $x(\mathbf{q}, t)$ that consists of n localized, evolving species of a form similar to the traveling Gaussian distribution of Equation 7.41

$$x(\mathbf{q}, t) = \sum_{k=1}^{n} A_k(t) \exp\left\{-\frac{1}{2\sigma_k^2(t)} [\mathbf{q} - \mathbf{q}_k(t)]^2\right\} \tag{7.92}$$

We assume that the species are well separated

$$\left(\mathbf{q}_j - \mathbf{q}_k\right)^2 \gg \sigma_j^2 + \sigma_k^2 \tag{7.93}$$

for any pair j, k. For the abundance of the species k, we get by integration over its phenotypic diversity range, (k),

$$x_k(t) = \int_{(k)} x(\mathbf{q}, t) \, d\mathbf{q} \tag{7.94}$$

the temporal change

$$\frac{d}{dt} x_k(t) = \int_{(k)} \left\{ x(\mathbf{q}, t) p(\mathbf{q}, t) + D \frac{\partial^2}{\partial \mathbf{q}^2} x(\mathbf{q}, t) \right\} d\mathbf{q} = \langle p \rangle_k x_k \tag{7.95}$$

Here, the average with subscript refers to the phenotypic mean value of a species

$$\langle p \rangle_k \equiv \frac{1}{x_k} \int_{(k)} p(\mathbf{q}, t) x(\mathbf{q}, t) d\mathbf{q} \tag{7.96}$$

The integral of p can also be split into those over the separate occupied regions

$$p(\mathbf{q}, t) = a(\mathbf{q}) + \sum_j \int_{(j)} d\mathbf{q}' b(\mathbf{q}, \mathbf{q}') x(\mathbf{q}', t) \tag{7.97}$$

We make use of the spectral decomposition, Equation 7.82,

$$\begin{aligned} p(\mathbf{q}, t) &= a(\mathbf{q}) + \sum_i \lambda_i \Phi_i(\mathbf{q}) \sum_j \int_{(j)} d\mathbf{q}' \Psi_i(\mathbf{q}') x(\mathbf{q}', t) \\ &= a(\mathbf{q}) + \sum_i \lambda_i \Phi_i(\mathbf{q}) \sum_j \langle \Psi_i \rangle_j x_j \end{aligned} \tag{7.98}$$

and can then carry out the average of Equation 7.96,

$$\langle p \rangle_k = \langle a \rangle_k + \sum_i \lambda_i \langle \Phi_i \rangle_k \sum_j \langle \Psi_i \rangle_j x_j \tag{7.99}$$

We define now the matrix

$$b_{kj} = \sum_i \lambda_i \langle \Phi_i \rangle_k \langle \Psi_i \rangle_j \tag{7.100}$$

and rewrite Equation 7.95 in the form

$$\frac{d}{dt} x_k(t) = \left(\langle a \rangle_k + \sum_j b_{kj} x_j \right) x_k \qquad (7.101)$$

This is the usual Lotka–Volterra model as discussed in Section 6.8. Thus we have shown that the continuous Lotka–Volterra landscape, Equation 7.81, is equivalent to the classical ecological model for the case of species with separate phenotypic appearances.

For a steady-state solution of Equation 7.101, either in the form of stationary concentrations or of stationary averages over Lotka–Volterra oscillations, see Section 6.8, the equations

$$\langle a \rangle_k + \sum_{j=1}^{s} b_{kj} x_j = 0 \quad k = 1, \ldots, s \qquad (7.102)$$

must permit a positive solution for the abundances of the s different species x_j. When we decompose the matrix, Equation 7.100, and introduce new formal variables, y_i, by

$$y_i = \sum_{j=1}^{s} \langle \Psi_i \rangle_j x_j, \quad i = 1, \ldots, n \qquad (7.103)$$

we obtain Equation 7.102 in the alternative form

$$\langle a \rangle_k + \sum_{i=1}^{n} \lambda_i \langle \Phi_i \rangle_k y_i = 0, \quad k = 1, \ldots, s \qquad (7.104)$$

This is now a system of linear equations for the n variables y_i. If $n < s$, this system of s equations is overdetermined and has in general no solution for the n unknowns. We infer that the number of expansion terms of Equation 7.82 provides an upper bound for number of discrete species that may exist in the system. In the sense of Gause's Law (Gause and Witt, 1935), we may denote n as the number of ecological niches of the Lotka–Volterra landscape; the number of existing species cannot exceed the number of existing niches. As an aside, we mention that Gause's coauthor Witt was a member of the famous Russian school on nonlinear oscillations (Andronow, Chaikin, and Witt, 1965). We may interpret the terms of Equation 7.82 in such a way that each ecological niche k is modeled mathematically by a positive number λ_k which describes the relative strength or importance of that niche, by a function $\Psi_k(\mathbf{q})$ which tells us how much a phenotype $\mathbf{q}$ contributes to the resource k, and a function $\Phi_k(\mathbf{q})$ that says how fast multiplication with a phenotype $\mathbf{q}$ is influenced by that niche. For example, in the Fisher–Eigen case, Equation 7.90, the individual growth rate $\Psi_1(\mathbf{q}) = E(\mathbf{q})$ is the factor by which a species $\mathbf{q}$ contributes to the dilution flux $\langle E \rangle$, and $\Phi_1(\mathbf{q}) = $ const indicates that all phenotypes are equally affected by the dilution. This example also shows that niches may be of supporting or of inhibiting nature. We conclude that for a degenerate community matrix, $b(\mathbf{q}, \mathbf{q}')$, the maximum number of potential species is finite, while in general there is no mathematical upper limit to the number of existing niches that may subsequently be conquered by the evolving species. In the terminology of Eigen and Schuster, the occupation number y_i of a particular niche, Equation 7.103, may be seen as the population number of a related "quasispecies."

While the discrete Lotka–Volterra system cannot make any predictions for the phenotypic change and diversity of the system, we can derive equations for the drift and spreading of properties in the phenotype space. For the species phenotype we get by integration

$$\mathbf{q}_k(t) = \langle \mathbf{q} \rangle_k \equiv \frac{1}{x_k} \int_{(k)} \mathbf{q} x(\mathbf{q}, t) \mathrm{d}\mathbf{q} \qquad (7.105)$$

similar to Equation 7.95 the temporal phenotypic change

$$\frac{\mathrm{d}}{\mathrm{d}t} \mathbf{q}_k(t) = \langle p\mathbf{q} \rangle_k - \langle p \rangle_k \langle \mathbf{q} \rangle_k \qquad (7.106)$$

Phenotypic properties change in directions that are positive correlated with the phenotypic landscape. For localized species, we may expand the landscape with respect to the small phenotypic scatter of the species

$$p(\mathbf{q}, t) = p(\mathbf{q}_k, t) + (\mathbf{q} - \mathbf{q}_k) \frac{\partial}{\partial \mathbf{q}_k} p(\mathbf{q}_k, t) + \cdots \qquad (7.107)$$

and get Fisher's Law in the form

$$\frac{\mathrm{d}}{\mathrm{d}t} \mathbf{q}_k(t) \approx (\langle \mathbf{q}^2 \rangle_k - \mathbf{q}_k^2) \frac{\partial}{\partial \mathbf{q}_k} p(\mathbf{q}_k, t) = \sigma_k^2 \frac{\partial}{\partial \mathbf{q}_k} p(\mathbf{q}_k, t) \qquad (7.108)$$

Thus, the species are "climbing uphill" the Lotka–Volterra landscape. Although this system of equations looks rather similar to a dynamical gradient system, see Section 4.2, its behavior is qualitatively different from its Fisher–Eigen counterpart, Equation 7.38. Using Equation 7.98, we get the formula

$$\frac{\partial}{\partial \mathbf{q}_k} p(\mathbf{q}_k, t) = \frac{\partial a(\mathbf{q}_k)}{\partial \mathbf{q}_k} + \sum_i \lambda_i \frac{\partial \Phi_i(\mathbf{q}_k)}{\partial \mathbf{q}_k} \sum_j \int_{(j)} \mathrm{d}\mathbf{q}' \Psi_i(\mathbf{q}') x(\mathbf{q}', t)$$

$$\approx \frac{\partial \langle a \rangle_k}{\partial \mathbf{q}_k} + \sum_i \lambda_i \frac{\partial \langle \Phi_i \rangle_k}{\partial \mathbf{q}_k} \sum_j \langle \Psi_i \rangle_j x_j(t) \qquad (7.109)$$

Here, the derivative with respect to species phenotype vector, $\mathbf{q}_k$, is taken at constant abundances x_j. It shows that the landscape is not rigidly lowered, as in the Fisher–Eigen model, but will be deformed as a result of the associated population numbers of the species involved. At the slow evolution time scale, we may consider the selection process as fast and the conditions (7.102) permanently obeyed

$$p(\mathbf{q}_k, t) = 0 \qquad (7.110)$$

for each species, k. We infer from the vanishing total differential

$$\frac{d}{dt}p(\mathbf{q}_k, t) = 0 = \frac{\partial}{\partial t}p(\mathbf{q}_k, t) + \frac{d\mathbf{q}_k}{dt}\frac{\partial}{\partial \mathbf{q}_k}p(\mathbf{q}_k, t) \qquad (7.111)$$

together with Fisher's law, Equation 7.108, that,

$$\frac{\partial}{\partial t}p(\mathbf{q}_k, t) = -\frac{1}{\sigma_k^2}\left(\frac{d\mathbf{q}_k}{dt}\right)^2 < 0 \qquad (7.112)$$

holds for the evolution process. This inequality is the general evolution principle for localized species on a Lotka–Volterra landscapes (Feistel and Ebeling, 1984, 1989). The species are located at the shorelines of the landscape, Equation 7.110. While the phenotypes evolve in uphill direction, Equation 7.108, the landscape is lowering and deforming, Equation 7.108 such that the species remain at the shoreline during their displacements. The lateral motion of the species is, therefore, combined with a lowering at the current location, Equation 7.103.

If the community matrix is degenerate, only a finite number of ecological niches is available. This number equals the number of degrees of freedom of deformation the landscape has. Consequently, if the number of present species exceeds the degrees of freedom, the landscape cannot continually adjust to the independent trajectories of the species; some will get extinct. As an example, in the case of Eigen–Fisher landscapes, Equation 7.90, the community matrix is a product kernel and possesses only one niche; consequently, in general only one species can survive, consistent with Gause's Law. In the continuous Lotka–Volterra model, we have transferred the qualitative problems of phenotypic change, extinction, and emergence of species to the analysis of geometric properties of smooth flexible hypersurfaces in a high-dimensional space.

The most exciting question here is whether those equations for the evolution of the species follow some general extremum principle, that is, whether the term "fitter" can be associated with a proper mathematical expression as in the case of Fisher–Eigen landscapes, or in other words, whether an order relation between subsequent evolution stages can be found. The general evolution criterion, Equation 7.103, indicates that the evolution process is irreversible and locally directed to increasing fitness.

The inequality (7.112) is derived for the phenotypic change of a particular species while all other species are considered as fixed, Equation 7.109. As soon as we consider the simultaneous evolution of all species, the occupation numbers (or their time averages over predator-prey cycles) satisfy Equation 7.102, that is, the x_j become functions of their changing phenotypes. The implicit system of linear equations for $x_j(\mathbf{q}_1, \ldots \mathbf{q}_s)$ is

$$a(\mathbf{q}_k) + \sum_{j=1}^{s} b(\mathbf{q}_k, \mathbf{q}_j) x_j = 0 \quad k = 1, \ldots, s \qquad (7.113)$$

Similar to Equation 7.109, we consider the functions

$$A_k(\mathbf{q}_1,\ldots,\mathbf{q}_s) = \frac{\partial a(\mathbf{q}_k)}{\partial \mathbf{q}_k} + \sum_i \lambda_i \frac{\partial \Phi_i(\mathbf{q}_k)}{\partial \mathbf{q}_k} \sum_j \Psi_i(\mathbf{q}_j) x_j(\mathbf{q}_1,\ldots,\mathbf{q}_s) \quad (7.114)$$

and the related Pfaffian form

$$d'W(t) = \sum_k A_k(\mathbf{q}_1,\ldots,\mathbf{q}_s) d\mathbf{q}_k \quad (7.115)$$

For the differential change, we find from Equations 7.109 and 7.112 the local evolution criterion

$$d'W = -\sum_k \frac{1}{\sigma_k^2}\left(\frac{d\mathbf{q}_k}{dt}\right)^2 dt \leq 0 \quad (7.116)$$

Nevertheless, because the integrability condition

$$\frac{\partial A_i}{\partial \mathbf{q}_k} = \frac{\partial A_k}{\partial \mathbf{q}_i} \quad (7.117)$$

is in general not satisfied for more than one species, the Pfaffian form, Equation 7.115, is not a total differential and is not integrable to provide a scalar function W that decreases with time. Subject to *Caratheodory's theorem* (Margenau and Murphy, 1943; Buchdahl, 1949; Smirnow, 1968; Stepanow, 1982; Feistel and Hagen, 1994), the integral of the differential form corresponds to a manifold of curves, rather than forming a surface. The local tendency, Equation 7.116, does not necessarily imply a systematic decrease in some function on the global scale.

We conclude that the evolution dynamics of localized species on a Lotka–Volterra landscape, given by Equations 7.108 and 7.110, is irreversible as described by the local evolution principle, Equation 7.116. No integral evolution criterion could be found that quantifies the term "fitter" for more than one species; the evolution process for many interacting species may be cyclic or chaotic without a definite, finite, or asymptotic target state. This qualitative dynamic behavior is known as the *Red Queen hypothesis* in evolution theory (Fisher, 1930; van Valen, 1973; Darlington, 1977; Ebeling and Feistel, 1982; Zimmer, 2009a; Venditti, Meade, and Pagel, 2010). We shall address to this problem again from rather different mathematical approach developed in the following sections.

7.6
Axiomatic Evolution Models

Evolution among organic molecules, either in the form of competing chemical reactions, or as "selfish genes" (Dawkins, 1976) that are equipped with sophisticated survival apparatuses known as organisms, takes place in a vastly underoccupied state space. A DNA chain with 10^9 base pairs has $d = 4^{1000\,000\,000} \approx 10^{600\,000\,000}$ different possible configurations, this is roughly the dimension of the phase space we are talking about here (Wright, 1932). If the species are taken as discrete rather than continuous as done in the previous sections, each axis of the space represents the occupation (or population) number of a certain species, that is, the number of existing copies of the particular molecule. Already from the comparatively "small" number of atoms available in the entire universe, sometimes estimated as "only" 10^{80}, and from the even "smaller" one of only those which form our planet Earth, perhaps about 10^{50}, it is clear that the overwhelming number of those potential molecules or species must be permanently absent. The vector of occupation numbers has many, many zeros and comparatively few positive components; the related representative point in the state space is located on the surface of the positive cone (see Chapters 4 and 6).

Evolution in the sense discussed here is a process in which by mutation, genetic recombination, or spatial immigration a new species appears, that is, the state vector gets another nonzero component, and the related representative point in the state space moves from the particular boundary of the positive cone into the interior. If this fluctuation represents an instability, the newcomer will grow and perhaps produce additional new chemicals as its direct or indirect reaction products, that is, in the initial phase of the self-organization process in response to an instability, the number of nonzero components of the state vector may further increase. Asymptotically, when the self-organization phase approaches its mature stage, some species may disappear as a result of the competition, and the number of nonzero components may decrease again.

In this Boolean picture, only existence or absence of certain species is relevant, rather than any dynamical details of attracter states and so on. Let C_n be the set of species present at the evolution step n (the "time" variable) in the mature parent state. Fluctuations may not delete any of those directly but may add new species at minor abundances, say, just one copy each, forming a new set of species, A_{n+1}. Depending on the survival probability (Sections 5.4 and 6.4), this state may either return to the parent state, C_n, or develop into an intermediate state, B_{n+1}, by self-organization. In the final phase, a new parent set of species, C_{n+1}, will be established by selection, while some previously dominating members may disappear. In this picture, evolution takes the form of a sequence of sets and binary relations between those sets.

$$\cdots \to C_n \leftrightarrow A_{n+1} \to B_{n+1} \to C_{n+1} \leftrightarrow \cdots \qquad (7.118)$$

In this abstract form, evolution is easily expressed in terms of sets and relations, that is, in terms of general structure models (Sections 4.4 and 6.5). Between the sets of (7.118), the inclusion relations (Feistel, 1979; Ebeling and Feistel, 1982)

$$C_n \subset A_{n+1}, \quad A_{n+1} \subseteq B_{n+1}, \quad B_{n+1} \supseteq C_{n+1} \qquad (7.119)$$

are obeyed for the reasons explained above, as illustrated in Figure 7.5. Similar to the landscape models in the previous sections, the mutation–selection step results in some displacement of the occupied region of the appropriate space.

The overall evolution step

$$C_n \Rightarrow C_{n+1} \qquad (7.120)$$

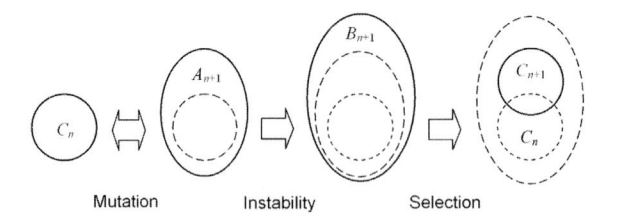

Figure 7.5 Transition from a Boolean evolution state, C_n, to its successor, C_{n+1}, via some critical fluctuation which lead to A_{n+1}, the related, triggered instability to B_{n+1}, and the asymptotic relaxation to the new mature state C_{n+1}, Equation 7.118. The enclosed regions indicate the subsets of species present in the subsequent phases of the self-organization step, Equation 7.119.

is irreversible, and any direct backward transition $C_{n+1} \Rightarrow C_n$ is definitely excluded. The central question here is whether or not, or under which conditions, an indirect backward transition

$$C_{n+1} \Rightarrow \cdots \Rightarrow \cdots \Rightarrow C_n \tag{7.121}$$

is permitted or prohibited, in other words, whether the binary relation "$\Rightarrow$" between the elements C_n is irreducible, reducible, or elementarily reducible (Sections 4.4 and 6.5). Only in the last case, an order relation or a semiorder between the elements can be constructed and a value $E(C_n)$ can be assigned to each evolution stage in such a way that

$$C_n \Rightarrow C_{n+1} \text{ implies } E(C_n) < E(C_{n+1}) \tag{7.122}$$

If so, the value E is a measure of "fitter," "better," or "higher" associated with the evolution process. For classes of evolution dynamics whose evaluation in the form of Equation 7.122 is impossible, talking about evolution to "higher states" makes no sense at all, even though the single step (7.120) is irreversible also in this case. For such systems, any search for certain general extremum principles obeyed by evolution is evidently in vain. The fundamental difference is that Equation 7.120 describes a local evolution criterion, while Equation 7.122 represents a global one. In the previous sections, we have seen that Fisher–Eigen models are of the type (7.122) while for Lotka–Volterra models in general only the relation (7.120) is granted.

In the following sections, we shall look in more detail at the problems of which dynamical models for chemical or ecological systems permit a Boolean description in the form displayed in Figure 7.5, and it turn, which of those may be consistent with a global evaluation criterion of the form (7.122).

7.7
Boolean Behavior in the Positive Cone

In particular in Chapter 6 we analyzed a number of different dynamical models with selection properties. Among those is a competition model in which different species,

7.7 Boolean Behavior in the Positive Cone

X_i, struggle for a common raw material, A, with restricted supply (Section 6.3),

$$\dot{A} = \Phi - A \sum_i f_i(X_i)$$
$$\dot{X}_i = A f_i(X_i) - k'_i X_i \qquad (7.123)$$

It is a characteristic feature of this model that two different classes of species are involved. The nutrient A does not need to be present at the beginning, $t = 0$; nevertheless it will be present in the system at any later time, $t > 0$, because it is supplied to the system at a constant rate. Provided that $f_i(0) = 0$, the competitors, X_i, on the other hand, can only exist if they are present from the beginning on. Consequently, X_i can be subject to a selection process while A cannot.

Models can also be classified under the aspect of behavior in the asymptotic limits of very small concentrations, such as

(a) $\lim_{X \to 0} f(X) \propto X$,

(b) $\lim_{X \to 0} f(X) \propto X^\nu$, $\quad \nu < 1$ $\qquad (7.124)$

(c) $\lim_{X \to 0} f(X) \propto X^\nu$, $\quad \nu > 1$

or of very high concentrations, such as

(a) $\lim_{X \to \infty} f(X) = -\infty$

(b) $\lim_{X \to \infty} f(X) = \text{const}$ $\qquad (7.125)$

(c) $\lim_{X \to \infty} f(X) = 0$

Moreover, species may interact in various ways, for example, in cross-catalytic reactions (Section 6.5)

$$\dot{X}_i = A \sum_j k_{ij} X_j - k'_i X_i \qquad (7.126)$$

or in food webs (Section 6.8)

$$\frac{dX_i}{dt} = X_i \left(a_i - \sum_{j=1}^n b_{ij} X_j \right) \qquad (7.127)$$

In a qualitative theory of selection processes, the aim is a suitable classification of the variety of such models.

Quite generally, we will consider only spatially homogeneous processes (e.g., in a stirred tank reactor) at constant temperature. Fluctuations will be ignored except those which lead to the emergence of new species. The description makes use of the thermodynamic limes of an infinite volume that contains infinitely many particles; the number of different species is large but finite.

The description of such systems is usually done in the form of systems of first-order ordinary differential equations

$$\dot{X}_i = f_i(X_1, \ldots, X_n) \qquad (7.128)$$

where the functions are continuously differentiable and may be nonlinear. Such an unspecific dynamical system is subject to only a few general requirements such as existence and uniqueness of its solutions, nonnegativity and finiteness of the concentrations X_i, as well as some additional conditions:

(a) The functions $f_i(X_1, \ldots, X_n)$, $i = 1, \ldots, n$, are defined in a certain region of the positive cone, $X_i \geq 0$, and are continuously differentiable functions of their independent variables.
(b) The functions $f_i(X_1, \ldots, X_n)$, $i = 1, \ldots, n$, ensure finiteness of the variables for any times t, that is,

$$X_i(t) < \infty \text{ for } 0 \leq t \leq \infty, \quad i = 1, \ldots, n \tag{7.129}$$

A sufficient condition for Equation 7.129 may take the form

$$\frac{d}{dt}|\mathbf{X}| < 0 \text{ if } |\mathbf{X}| > M \tag{7.130}$$

where M is some finite upper bound and $\|\cdot\|$ the Euclidian norm.
(c) The functions $f_i(X_1, \ldots, X_n)$, $i = 1, \ldots, n$, ensure nonnegativity of the variables for any times t, that is,

$$X_i(t) \geq 0 \text{ for } 0 \leq t \leq \infty, \quad i = 1, \ldots, n \tag{7.131}$$

A necessary condition for Equation 7.131 takes the form

$$f_i(X_1, \ldots, X_{i-1}, 0, X_{i+1}, \ldots, X_n) \geq 0, \quad i = 1, \ldots, n \tag{7.132}$$

The permitted part of the state space is the positive cone, $E_n^+ \subset E_n$, of the Euclidian vector space E_n (Section 6.2). The condition (c) keeps the trajectory inside the positive cone

$$\mathbf{X}(0) \in E_n^+ \rightarrow \mathbf{X}(t) \in E_n^+, \quad t > 0 \tag{7.133}$$

Substantial contributions to such an axiomatic approach to chemical kinetics were provided by Aris (1965, 1968), Gavalas (1968), Czajkowski (1975), and Jiménez-Montaño (1975).

Selection implies the disappearance of certain species, that is, the convergence of the state vector toward the boundary of the positive cone. The properties of the functions $f_i(X_1, \ldots, X_n)$ in the vicinity of that boundary play a central role in this approach to chemical kinetics (Prigogine, Nicolis, and Babloyantz, 1972; Allen, 1976; Feistel, 1979; Ebeling and Feistel, 1982).

The specifics of selection systems are reflected in the topology of the field of trajectories near the edge of the positive cone. Loci on that boundary can be distinguished as being either regular or singular points. If

$$\mathbf{X}^{(s)} = \left\{ X_1^{(s)}, \ldots, X_n^{(s)} \right\} \tag{7.134}$$

is a singular point where $X_i^{(s)} = 0$ for at least one i, then the definition of singular points implies that for the related function f_i

$$f_i\left(X_1^{(s)}, \ldots, X_n^{(s)}\right) = 0 \tag{7.135}$$

must hold. Singular points can be approached by a dynamical system only asymptotically in the limit $t \to \infty$ (Section 4.2). This also applies to other attractors such as limit cycles. Selection means that certain attractors of the system are located on the boundary of the positive cone. We consider the positive cone of an r-dimensional subspace

$$E_r^+ \subset E_n^+, \quad \text{where} \quad r << n \tag{7.136}$$

and a positive initial state in E_r^+

$$X_1(0) > 0, \ldots, X_r(0) > 0, \quad X_{r+1}(0) = 0, \ldots, X_n(0) = 0 \tag{7.137}$$

If for any initial state of the form (7.137) there is a unique attractor located on the boundary of the positive cone of E_r^+, the topology of trajectories, that is, the selection process, is qualified as *simple competition* (Section 6.3). If there exist several attractors and related separatrices between the trajectories, the result of the competition will depend on the particular initial values of the positive components, Equation 7.137. This type of topology is referred to as *hyperselection* (Section 6.7), a term introduced by Decker (1975), see Figure 7.6.

In the later text, we will focus on simple competition because it is the most interesting case of evolution; in particular, it occurs between multiplying spatial compartments independent of the complexity of the internal chemical kinetics (Section 6.10).

In addition to the conditions (a)–(c) above, another requirement imposed on selection models is *structural stability*. A dynamical system is structurally stable if under any minor modification

$$\frac{d}{dt}\tilde{\mathbf{X}} = \mathbf{f}(\tilde{\mathbf{X}}) + \delta \mathbf{f}(\tilde{\mathbf{X}}) \tag{7.138}$$

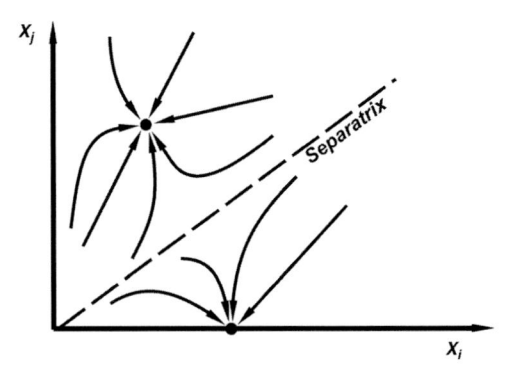

Figure 7.6 In the case of hyperselection, the result of the competition depends on the initial conditions. Separatrices divide the phase space into complementary regions associated with the different stable attractor states.

which is bounded in the form of the inequality

$$\left|\delta f(\tilde{X})\right| + \left|\frac{\partial}{\partial \tilde{X}} \delta f(\tilde{X})\right| < \varepsilon \tag{7.139}$$

the topology of trajectories is not qualitatively changed. In particular, the number of attractors, their quality, and the topology of the region of attraction must be invariant with respect to those modifications. For selection systems, however, the definition of structural stability must be formulated in a less general way (Feistel, 1979; Ebeling and Feistel, 1982). Attractors located on the boundary of the positive cone may be displaced to regions outside the positive cone by suitable perturbations of the form (7.138). In particular, the condition (7.132) may easily be violated. Thus, attractors on the boundary are structurally unstable in the general, formal sense. In contrast, as we have seen in the beginning of this chapter, such attractor states are inevitable in state spaces of extremely high dimensions, as it is the case for evolution processes, and those states must explicitly be permitted. We conclude that the general form of Equation 7.138 is inappropriate for evolution models; specific constraints must be applied to the choice of the perturbations, δf, in addition to Equation 7.139. On the other hand, we cannot entirely abstain from the condition of structural stability; for instance, we need to exclude models where at some finite time $t_1 > 0$ the trajectory touches the boundary of the positive cone and later returns to the interior for $t > t_1$, see Figure 7.7.

Let us take a closer look at the structure of the trajectory field in the immediate vicinity of the boundary of positive cone. As a result of the conditions above, only singular points (attractors) on the boundary can be approached by trajectories from the interior, rather than regular points. Attractor points can be reached only in the limit $t \to \infty$, while any other, regular points can either be occupied at $t = 0$ exclusively, or the trajectory remains on the particular boundary for all times. In fact all evolution processes run on certain boundaries since "truly internal" points with positive

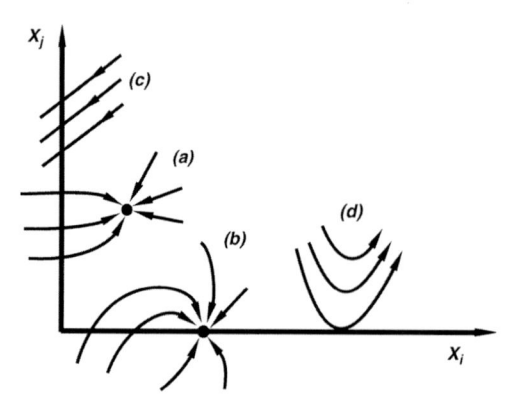

Figure 7.7 Structural stability of the dynamics in the positive cone. Cases (a) and (b) are permitted situations, (c) and (d) are forbidden.

particle numbers of all theoretically possible molecules are practically impossible by the sheer astronomical multitude of those species. Touching a boundary temporarily at some finite time is excluded. In summary, for the models we are interested in here, there are only two qualitatively different possibilities for the abundance of a selected species, either

$$X_i(t) = 0 \quad \text{for} \quad 0 \leq t \leq \infty \tag{7.140}$$

or

$$X_i(0) \geq 0, \quad X_i(t) > 0 \quad \text{for} \quad 0 < t < \infty, \quad X_i(\infty) \geq 0 \tag{7.141}$$

For a certain subset of species, the option (7.140) is excluded; those will be regarded as *mandatory* species, in contrast to *optional* species which may or may not be absent in a given system. For a mandatory species i, there is no singular point on the related hyperplane, $X_i = 0$, that is, the inequalities

$$f_i(X_1, \ldots, X_{i-1}, 0, X_{i+1}, \ldots, X_n) > 0 \tag{7.142}$$

$$X_i(0) \geq 0, \quad X_i(t) > 0 \quad \text{for} \quad 0 < t \leq \infty \tag{7.143}$$

hold for arbitrary values of the other variables. Obviously, only optional species may be subject to selection.

When the system's stability with respect to fluctuations is analyzed using Lyapunov's method, the response to arbitrary small perturbations of a given state is calculated from a local approximation to the dynamical equations. Also in this case, it is clear that for its application to selection systems additional restrictions must be considered with respect to the choice of permitted perturbations. Fluctuations near or on the boundary of the positive cone are not entirely arbitrary; they must not violate the condition of nonnegativity, that is, they cannot exit from the positive cone. We distinguish two groups of fluctuations, δX_i, either

$$\textit{regular fluctuations}, \delta X_i, \text{ about variables } X_i > 0 \tag{7.144}$$

or

$$\textit{singular fluctuations}, \delta X_i \geq 0, \text{ about variables } X_i = 0 \tag{7.145}$$

While regular fluctuations are arbitrary with respect to their sign but sufficiently small to remain within the positive cone, singular fluctuations are nonnegative. The latter may describe mutations of a system and are by orders of magnitude less probable than regular fluctuations. In particular, of the nonnegative state vector belonging to the high-dimensional state space of evolution, only a small number of its zero components may undergo singular fluctuations at the same time. Otherwise singular fluctuations would create a fully positive state vector, which is practically impossible. Fluctuations of the form which, say, would represent the sudden appearance of small number of elephants in the primordial soup, or of a couple of dinosaurs in the present world, can be taken as irrelevant and should be disregarded. Reasonable singular fluctuations are those which belong to a suitably defined neighborhood of the established optional species, as described in the earlier sections of this chapter. Regular fluctuations such as minor concentration variations, on the contrary, are permanently observed and occur with high probability.

Let the system

$$\frac{d\mathbf{X}}{dt} = \mathbf{f}(\mathbf{X}) \tag{7.146}$$

possess a stationary point at $\mathbf{X} = \mathbf{X}^{(s)}$, that is, $\mathbf{f}(\mathbf{X}^{(s)}) = 0$, with the properties

$$X_1^{(s)}(0) > 0, \ldots, X_r^{(s)}(0) > 0, \quad X_{r+1}^{(s)}(0) = 0, \ldots, X_n^{(s)}(0) = 0 \tag{7.147}$$

The linearized dynamical equations read

$$\frac{d}{dt}\delta X_i = \sum_{j=1}^{n} \left.\frac{\partial f_i}{\partial X_j}\right|_{\mathbf{X}^{(s)}} \delta X_j \tag{7.148}$$

The singular fluctuations are subject to the conditions

$$\delta X_j \geq 0 \quad \text{for} \quad r+1 \leq j \leq n \tag{7.149}$$

as well as

$$\frac{d}{dt}\delta X_j \geq 0, \text{ if } \delta X_i = 0 \text{ for } r+1 \leq i \leq n \tag{7.150}$$

Fluctuations are independent and arbitrary within the constraints mentioned before; hence we infer for the partial derivatives, Equation 7.148, the conditions

$$\left.\frac{\partial f_i}{\partial X_j}\right|_{\mathbf{X}^{(s)}} = 0 \text{ for } i = r+1, \ldots, n, \text{ and } j = 1, \ldots, r \tag{7.151}$$

and

$$\left.\frac{\partial f_i}{\partial X_j}\right|_{\mathbf{X}^{(s)}} \geq 0 \text{ for } i = r+1, \ldots, n, \text{ and } j = r+1, \ldots, i-1, i+1, \ldots, n \tag{7.152}$$

In other words, the Jacobian of the system at the state $\mathbf{X} = \mathbf{X}^{(s)}$ is a reducible matrix. Consequently, since the eigenvalue problem of a reducible matrix can be decomposed into separate investigations of the irreducible submatrices, the problems of stability against regular and singular fluctuations can be studied independently. Moreover, the submatrix responsible for the singular fluctuations is nonnegative except for the main-diagonal elements. This finding simplifies the stability analysis; for instance, the theorems of Perron, Frobenius, and Kotelyanski (Gantmacher, 1971) rather than the more demanding criteria of Routh and Hurwitz, or of Gershgorin, can be exploited. In many practical cases, it is even sufficient to merely inspect the values of the main-diagonal elements.

Of particular interest for the evolution behavior is the question whether a new species is accepted or rejected by the established parent species, whether or not a singular fluctuation is attenuated or amplified. It is of minor importance in this context whether the parent attractor is a stationary state, a limit cycle, or a fractal manifold. In contrast, it is of central importance whether separatrices exist which make the decision of acceptance of rejection dependent on the initial abundance of the new species. For the reasons discussed above, we shall restrict ourselves in the following sections to the models with simple selection. In that most relevant case, rather than the initial abundance, only the phenotype of the new species is crucial for its survival.

For those systems, with respect to Equation 7.141, only three qualitatively different states exist in the temporal development of a given self-organization step such as the one shown in Figure 7.7, namely, the situation at the beginning, at $t = 0$, the intermediate one at $t = 1$ (in an arbitrary unit), and the asymptotic one at $t = \infty$. Each of the three stages is described by the presence or absence of certain species, and the mapping from one stage to the next is solely controlled by the phenotypes of the species present, rather than by their actual quantitative abundances in the system. Those circumstances permit the development a Boolean model of evolution, as outlined in the following sections.

7.8
Axiomatic Description of a Boolean Reaction System

As a BRS, we define the triple (Feistel, 1979; Ebeling and Feistel, 1982)

$$U = (M, s, d) \tag{7.153}$$

Here, M is a finite set (the set of possible species), and d and s are mappings of the power set of M (i.e., the set of all subsets of M), **M**, onto itself, with the properties specified in Equations 7.154–7.161.

Let $S \subseteq M$ be the *initial set*, that is, the set of all species that are present at $t = 0$. For all possible S, the *unfolding map* d is defined, which uniquely associates to S an intermediate set P that represents the species present at $t = 1$

$$\forall S \in \mathbf{M} \rightarrow \exists P = d(S) \in \mathbf{P} \subseteq \mathbf{M} \tag{7.154}$$

Here, **P** is the set of all P. For each intermediate set P there is a *selection map* s defined, which uniquely associates to P an *attractor set* V that represent the species present at $t = \infty$

$$\forall P \in \mathbf{M} \rightarrow \exists V = s(P) \in \mathbf{V} \subseteq \mathbf{M}. \tag{7.155}$$

Here, **V** is the set of all V.

The considerations of the previous section are reflected in the formal properties that in the unfolding phase species cannot disappear, that is, the initial set is a subset of the intermediate set, see Figure 7.8,

$$d(S) \supseteq S \text{ for any } S \in \mathbf{M} \tag{7.156}$$

7 Models of Evolution Processes

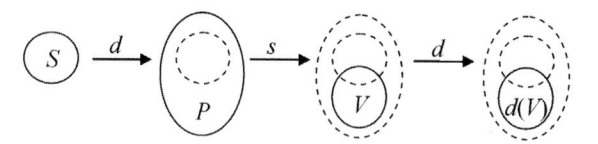

Figure 7.8 Result of the mappings d and s: the unfolding map d cannot reduce the set while the selection map cannot extend the set. Application of d after s leaves the set unchanged, $d(V) = V$.

and that in the selection phase species cannot emerge, that is, the attractor set is a subset of the intermediate set, see Figure 7.8,

$$s(P) \subseteq P \text{ for any } P \in \mathbf{P} \tag{7.157}$$

Any attractor set is a fixpoint of the unfolding map, see Figure 7.8,

$$d(V) = V \text{ for any } V \in \mathbf{V} \tag{7.158}$$

The mapping functions are continuous in the sense that the intermediate set is invariant with respect to addition of its elements to the related initial set, that is, Figure 7.9,

$$d(S_1) = d(S_2) \text{ if } S_1 \subseteq S_2 \subseteq d(S_1) \tag{7.159}$$

and the attractor set is invariant with respect to removal of dispensable elements from the related intermediate set, that is, Figure 7.10,

$$s(P_1) = s(P_2) \text{ if } P_1 \supseteq P_2 \supseteq s(P_1) \tag{7.160}$$

Finally, the unfolding map is monotonous such that addition to the initial set cannot result in a reduction in the intermediate set, Figure 7.11,

$$d(S_1) \subseteq d(S_2) \text{ if } S_1 \subseteq S_2 \tag{7.161}$$

This property concludes the definition of the BRS. The statements (7.156)-(7.161) are consistent and independent.

The general properties of the BRS permit the derivation of some useful general conclusions. The mappings d and s are idempotent. If the initial set is one of the

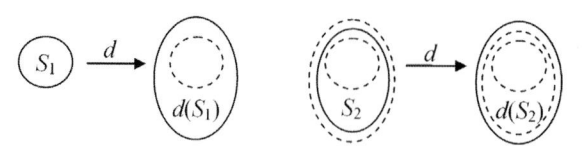

Figure 7.9 Continuity of the mapping d: adding species to S_1, which are elements of $d(S_1)$, leaves the result of the unfolding map unchanged, $d(S_1) = d(S_2)$.

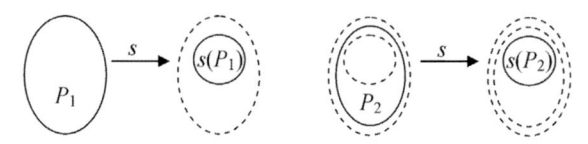

Figure 7.10 Continuity of the mapping s: removing species from P_1, which are not elements of $s(P_1)$, leaves the result of the selection map unchanged, $s(P_1) = s(P_2)$.

intermediate sets, it remains unchanged, that is,

$$d(S) = S \text{ for any } S \in \mathbf{P} \tag{7.162}$$

or equivalently,

$$d(d(S)) = d(S) \text{ for any } S \in \mathbf{M} \tag{7.163}$$

If the intermediate set is one of the victor sets, it remains unchanged, that is,

$$s(P) = P \text{ for any } P \in \mathbf{V} \tag{7.164}$$

or equivalently

$$s(s(P)) = s(P) \text{ for any } P \in \mathbf{P} \tag{7.165}$$

We infer that the set of victor sets is a subset of the set of intermediate sets,

$$\mathbf{V} \subseteq \mathbf{P} \subseteq \mathbf{M} \tag{7.166}$$

Consequently, the mappings s and d are commutable for intermediate and attractor sets, that is,

$$s(d(P)) = d(s(P)) \text{ for any } P \in \mathbf{P} \tag{7.167}$$

and as a special case as well,

$$s(d(V)) = d(s(V)) \text{ for any } V \in \mathbf{V} \tag{7.168}$$

Also, there is the "addition theorem,"

$$d(S_1 \cup S_2) \supseteq S_1 \cup S_2 \cup d(S_1) \cup d(S_2) = d(S_1) \cup d(S_2) \tag{7.169}$$

Additional properties are described by Feistel (1979).

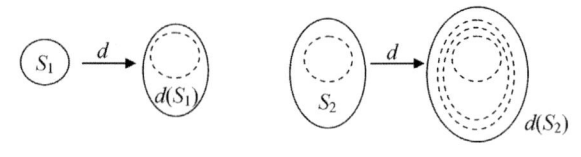

Figure 7.11 Monotony of the mapping d: adding species to S_1 cannot reduce the result of the unfolding map, $d(S_1) \subseteq d(S_2)$.

As an example, we write down the BRS representing the dynamical system

$$\dot{A} = \Phi - A\left(k'_0 + \sum_{i=1}^{2} k_i X_i\right)$$
$$\dot{X}_i = (Ak_i - k'_i)X_i, \ i = 1, 2, \ \frac{k_1}{k'_1} > \frac{k_2}{k'_2} \qquad (7.170)$$

The related BRS, $U = (M, s, d)$, consists of the set of species

$$M \equiv \{A, X_1, X_2\} \qquad (7.171)$$

the unfolding map, d, is defined for any subset S of M, in the form

$$d(S) = S \cup \{A\} \qquad (7.172)$$

such that the power set of intermediate sets, **P**, is

$$\mathbf{P} = \{\{A\}, \{A, X_1\}, \{A, X_2\}, \{A, X_1, X_2\}\} \qquad (7.173)$$

The selection map takes as arguments the elements of **P** and is given by the list

$$\begin{aligned} s(\{A\}) &\equiv \{A\} \\ s(\{A, X_1\}) &\equiv \{A, X_1\} \\ s(\{A, X_2\}) &\equiv \{A, X_2\} \\ s(\{A, X_1, X_2\}) &\equiv \{A, X_1\} \end{aligned} \qquad (7.174)$$

which results in the power set of attractor sets, **V**, of the form

$$\mathbf{V} = \{\{A\}, \{A, X_1\}, \{A, X_2\}\} \qquad (7.175)$$

It is evident that **V** is a subset of **P**, and that **P** is a subset of the power set of M.

7.9
Reducible, Linear, and Ideal Boolean Reaction Systems

The properties (7.153)–(7.169) described in the previous section apply to any BRS by definition. In this section, additional specific properties of certain BRS will be defined on the basis of the axiomatic foundation given before.

First, we shall refer to a set of species, Y, as *isolated*, if the disjunction

$$Y \cap S = 0 \qquad (7.176)$$

implies that also

$$Y \cap d(S) = 0 \qquad (7.177)$$

for the given $Y \subseteq M$, $Y \supset 0$, and any set $S \subseteq M$. Here, 0 denotes the empty set. Thus, a subset of species is isolated if none of its elements can appear in the system unless some of them were already present initially. The union of two isolated sets is also isolated.

7.9 Reducible, Linear, and Ideal Boolean Reaction Systems

The set of *optional* species, B, is defined by

$$B = M - d(0) \tag{7.178}$$

while $d(0)$ is the set of *mandatory* species. It follows that isolated sets contain only optional species, and that B is an isolated set.

The BRS is regarded as *reducible* if its set of optional species contains at least one isolated set Y as a genuine subset, that is, if

$$\exists Y \subset B \tag{7.179}$$

Otherwise the BRS is *irreducible*.

If U is reducible, then the triple $U_1 = (M-Y, s, d)$ is a *subsystem* of U which has the properties of a BRS.

A reducible BRS is termed *disconnected* if two complementary disjoint isolated sets Y_1, Y_2 exist such that

$$Y_1 \cup Y_2 = B = M - d(0), \text{ and } Y_1 \cap Y_2 = 0 \tag{7.180}$$

Otherwise, the BRS is *connected*. The subsystem, $U_1 = (M-Y_1, s, d)$, of a disconnected BRS is termed a *section* of U and is regarded as *complementary* to the subsystem, $U_2 = (M-Y_2, s, d)$, Equation 7.180, see Figure 7.12.

As an example we consider an unfolding map d with the properties

$$
\begin{aligned}
0 &\rightarrow \{A\} \\
\{X_1\} &\rightarrow \{A, X_1, X_2\} \\
\{X_2\} &\rightarrow \{A, X_2\} \\
\{X_3\} &\rightarrow \{A, X_3\} \\
\{X_2, X_3\} &\rightarrow \{A, X_2, X_3\}
\end{aligned}
$$

The results of the application of d on initial sets not mentioned here can be derived from those given in combination with the axiomatic properties. In this BRS,

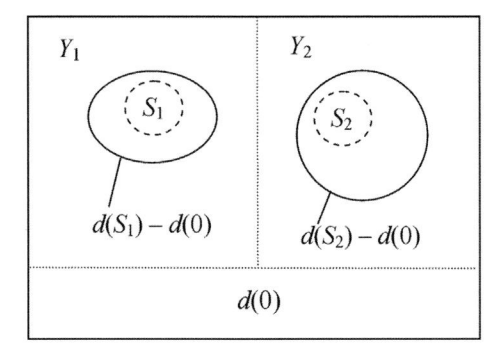

Figure 7.12 Example for a disconnected system that consists of two sections, $U_1 = (M-Y_1, s, d)$ and $U_2 = (M-Y_2, s, d)$, which share the same set of mandatory species, $d(0)$.

subsystems exist which have the alternative sets of species $\{A, X_2\}$, $\{A, X_3\}$, and $\{A, X_1, X_2\}$, and *sections* which have the sets $\{A, X_1, X_2\}$ and $\{A, X_3\}$.

A BRS is *linear* over the set $L \subseteq M$ if its unfolding map d has the property

$$d(S_1 \cup S_2) = d(S_1) \cup d(S_2) \tag{7.181}$$

for any $S_1 \cup S_2 \subseteq L$. Linearity is a special case of the general "convexity" rule expressed by Equation 7.169.

A section is termed *linear* if Y is the optional set of that section, and d is linear over the set $Y \cup d(0)$, plus

$$Y \cap d(S) = Y \cap d(Y \cap S) \tag{7.182}$$

holds for any $S \subseteq M$. The sections of a linear BRS are linear.

In a linear BRS it is impossible that two optional species react to produce a third species. This property is rather restrictive for chemical systems but is not exceptional in ecological systems such prey–predator communities.

Finally, we define an *ideal* BRS by the property that

$$sd(S_1 \cup S_2) = sd(S_1 \cup sd(S_2)) \tag{7.183}$$

holds for any pair, $S_1 \cup S_2 \subseteq M$. Here, the map sd is the shorthand notation

$$sd(X) \equiv s[d(X)] \tag{7.184}$$

As an example, we may think of grouping the species of a given set S into certain portions $S_1, S_2, \ldots$, that is,

$$S = S_1 \cup S_2 \cup \cdots \cup S_n \tag{7.185}$$

and successively adding them to the particular attractor set V of the previous stage such that

$$\begin{aligned} V_1 &= sd(S_1) \\ V_2 &= sd(V_1 \cup S_2) \\ &\cdots \\ V_n &= sd(V_{n-1} \cup S_n) \end{aligned} \tag{7.186}$$

In an ideal BRS, the final attractor set, V_n, does not depend of the way the set S was split into portions, in particular,

$$V_n = V = sd(S) \tag{7.189}$$

Ideality is a plausible BRS property for competitors in one ecological niche but is somewhat restrictive with respect to food webs in general. If we imagine two species such as lynx and hare, adding to an existing system first the hare and later the lynx will result in the same situation as if the two had entered together. But if first the lynx appears, it will lack of prey and will disappear again, and adding hare in a second step will not bring back the extinct lynx. Hence, the portioning of species matters in this case, in contrast to ideal BRS as defined by Equation 7.183.

7.10
Minor and Major of a Boolean Reaction System

The unfolding map d of a BRS, $U = (M, d, s)$, represents a binary relation between the elements of the power set of species, $\mathbf{M}$. If we consider the particular elements of $\mathbf{M}$ which contain either only one or two optional species, or elements of $\mathbf{M}$ which contain all species except one or two, we can transfer the properties of d to binary relations between those one or two optional species. Those relations may serve as certain upper or lower bounds of the unfolding map d.

Let $B \in \mathbf{M}$ be the set of optional species of U. A species X_i is regarded as *produced by a species* X_j if

$$X_i, X_j \in B \quad X_i \in d(\{X_j\}) \tag{7.190}$$

Because of Equation 7.156, each optional species produces itself, that is, "produced" is a reflexive relation (Section 4.1). If X_i is produced by X_j, $X_i \in d(\{X_j\})$, and in turn X_j is produced by some X_k, $X_j \in d(\{X_k\})$, we infer from $\{X_j\} \subseteq d(\{X_k\})$, making use of Equations 7.161 and 7.163, that $X_i \in d(\{X_j\}) \subseteq d(\{X_k\})$, that is, X_i is produced by X_k, in other words, "produced" is a transitive relation (Section 4.1).

The set G_i, which contains all species X_j that produce X_i, is termed the *producing set* of X_i. In addition, if there exists an intermediate set P of U which neither includes X_i nor X_j, that is,

$$\{X_i, X_j\} \cap P = 0 \tag{7.191}$$

and obeys the condition

$$X_i \in d(P \cup \{X_j\}) \tag{7.192}$$

then the species X_i is termed *producible by the species* X_j.

If we consider the particular intermediate set, $P = d(0)$, we infer that the species X_i is producible by the species X_j if it is produced by the species X_j. For a given X_j, the set of its produced species is a subset of its producible species. Species for which the reverse is also true, that is, producible species are always produced species, are termed *simple species*. Simple species that are not produced by X_j are not producible by any $X_j \in B$, $X_j \notin G_i$.

We can draw the important conclusion that a section (or a BRS) is linear if and only if all of its optional species are simple species (for details see Feistel, 1979)

$$\tag{7.193}$$

The binary relations "produced" and "producible" between the species can be represented by directed graphs G_{min} and G_{max}, which we denote as the *Minor* and the *Major*, respectively, related to the unfolding map d. The vertices of the two graphs represent optional species; a directed edge of G_{min} (or, G_{max}) leads from the species X_i to the species X_j if X_j is produced (or, producible) by the species X_j.

For a given set S, by the sets $d_{min}(S)$ and $d_{max}(S)$ we define the set of all vertices that are accessible from S via a path of edges representing the "produced" or "producible" relation, respectively. It follows that

$$d_{min}(S) \subseteq d(S) - d(0) \subseteq d_{max}(S) \tag{7.194}$$

where the equality signs apply for linear BRS.

For a given dynamical system of kinetic equations

$$\dot{X}_i = f_i(X_1, \ldots, X_n), \quad i = 1, \ldots, n \tag{7.195}$$

the related Minor and Major of the BRS can be derived from the functions f_i. If the function value on the hyperplane $X_i = 0$ is always positive, $f_i(X_1, \ldots, X_{i-1}, 0, X_{i+1}, X_n) > 0$, then X_i is a mandatory species. If the function value vanishes always, $f_i(X_1, \ldots, X_{i-1}, 0, X_{i+1}, X_n) = 0$, the species X_i is optional and is neither produced nor producible by any other species.

The components of the graph G_{max} correspond to sections of the BRS. If G_{max} is reducible, the related BRS is reducible. If G_{min} is connected or irreducible, the related BRS is connected or irreducible, respectively. For linear BRS, the graphs G_{min} and G_{max} are identical.

7.11
Selection and Evolution in Boolean Reaction Systems

Following the definition given in previous sections such as in Chapter 6, selection is a process by which established parents species are eliminated as a result of the appearance of certain new species – mutants – in the system. In the BRS, this process is modeled by the addition of new, disjoint species to a given attractor set, and the subsequent mappings by d and s to produce a new attractor set.

Here, we define that a BRS $U = (M, d, s)$ has *selection properties* if a species set, $S \subseteq M$, and an attractor set, $V \subseteq M$, exist such that

$$R = V - sd(V \cup S) \supset 0 \tag{7.196}$$

The set of species R is referred to as *eliminated by S*, as shown in Figure 7.13.

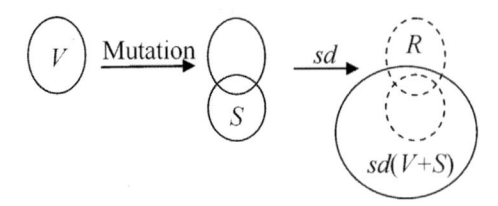

Figure 7.13 Selection properties as defined by Equation 7.196. Mutations add new species, S, to the existing parent set, V. As a result of subsequent unfolding and selection, the new attractor set, $sd(V \cup S)$, does no longer include the subset R of the previous parent species, that is, R is eliminated from the attractor by S.

7.11 Selection and Evolution in Boolean Reaction Systems

For convenience, we define a new binary relation between two attractor sets, $V_2 \prec V_1$, where V_2 is denoted as the *heir* of V_1 if a species set S exists such that

$$V_2 = sd(V_1 \cup S) \neq V_1 \tag{7.197}$$

Two attractor sets are in inheritance relation if the heir emerges as a result of mutations of the parent attractor set. Hence, a BRS has selection properties if two attractor sets, V_1 and V_2, exist with the properties

$$V_2 \prec V_1 \text{ and } V_1 - V_2 \supset 0 \tag{7.198}$$

It is not difficult to show that a BRS *cannot* possess selection properties if its selection mapping, s, is monotonously increasing, that is, if for any pair of intermediate sets with the property, $P_2 \supseteq P_1$, the condition

$$s(P_2) \supseteq s(P_1) \tag{7.199}$$

is obeyed.

For later discussion, we note some properties of the inheritance relation.

1) The relations

$$V_2 \prec V_1 \text{ and } V_2 \subseteq V_1 \text{ exclude one other.} \tag{7.200}$$

2) The inheritance relation is irreflexive:

$$V_1 \prec V_1 \text{ is false} \tag{7.201}$$

3) For linear BRS, inheritance is asymmetric:

$$V_2 \prec V_1 \text{ and } V_1 \prec V_2 \text{ exclude one other} \tag{7.202}$$

4) For ideal BRS, inheritance is transitive:

$$V_1 \prec V_2 \text{ and } V_2 \prec V_3 \text{ implies } V_1 \prec V_3 \tag{7.203}$$

We conclude that for linear, ideal BRS inheritance is an irreflexive semiorder (Section 4.1).

If an attractor set, V_2, is a proper subset of another attractor set, $V_1 \supset V_2$, the difference $N = V_1 - V_2 \supset 0$ is termed a *niche* of V_1, see Figure 7.14. If there is a third attractor set, V_3, with the properties

$$\begin{aligned} V_3 &\subset N \cup V_2 = V_1 \\ V_3 &- V_2 \supset 0 \\ V_2 &- V_3 \supset 0 \\ sd(V_2 &\cup V_3) \supseteq V_2 \cup V_3 \end{aligned} \tag{7.204}$$

then N is termed a *proper niche*, and V_2 and V_3 are *coexistent* attractor sets, see Figure 7.15. If no attractor set of a BRS has a proper niche, the selection mapping s is termed *strict*.

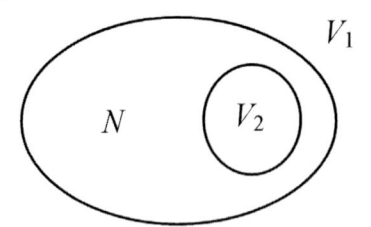

Figure 7.14 A niche N is defined as the part of an attractor set, V_1, which does not belong to another attractor set, V_2, such that $V_1 = V_2 \cup N$.

The context of this axiomatic system permits a reasonable formal definition of an evolution process.

An *evolution process* of an abstract BRS is a sequence of attractor sets, $V_1, \ldots, V_n$, which are successively concatenated by the inheritance relation, Equation 7.197,

$$V_n \prec V_{n-1} \prec \cdots \prec V_2 \prec V_1 \tag{7.205}$$

Here, evolution is modeled as a chain of mutation–selection transitions. For a given BRS, a directed graph can be used to represent the attractor sets, V_i, by its vertices, and the inheritance relations, $V_i \prec V_j$, by its edges from V_j to V_i. This graph is regarded as the *evolution tree* $T(\infty)$ of the BRS. This graph is in general not a tree in the strict sense of graph theory (Section 4.1): it has exactly one root, $V_1 = sd(0)$, which cannot be the heir of another set, and one top, $V_n = sd(M)$, which cannot possess a heir. Each path in this graph corresponds to a possible evolution process, Equation 7.205. Initially the evolution tree is divergent and convergent toward the end, where these terms refer here to the number of heirs rather than to the cardinalities of the different attractor sets, as usual in biology.

So far, arbitrary species were permitted to take the role of mutants in the inheritance relation (7.197). For a more realistic approach, we may generalize the BRS, $U = (M, d, s)$, to a *Metric Boolean Reaction System* (MBRS), $E = (M, d, s, m)$,

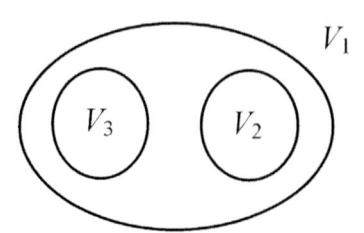

Figure 7.15 A proper niche of the attractor set, V_1, see Figure 7.14, contains two different attractor sets, V_2, V_3, Equation 7.204, which are regarded as coexistent then.

where $m(S_1, S_2)$ is a distance measure defined between any pair of sets of species, $S_1, S_2 \in M$. As a proper metric, m must satisfy the following conditions:

1. nonnegativity : $m(S_1, S_2) \geq 0$, where the equal sign holds only if $S_1 = S_2$
2. symmetry : $m(S_1, S_2) = m(S_2, S_1)$, and the
3. triangle inequality : $m(S_1, S_2) \leq m(S_2, S_3) + m(S_3, S_1)$

(7.206)

By this measure, we redefine for the MBRS the inheritance relation between two attractor sets, $V_2 \prec |_\varepsilon V_1$, if a species set S exists such that

$$V_2 = sd(V_1 \cup S) \neq V_1, \text{ where } m(V_1, S) \leq \varepsilon \quad (7.207)$$

The related graph is regarded as the *evolution tree* $T(\varepsilon)$ of the MBRS. All graphs $T(\varepsilon)$ have the same vertices but with decreasing ε, edges related to larger distances, Equation 7.207, are subsequently omitted.

If a distance is defined between the different species, such as the Hamming distance (Section 7.1), $d_H(X_1, X_2)$, or one of its generalizations (Beyer et al., 1974; Feistel, 1976b; Ebeling, Feistel, and Jiménez-Montaño, 1977; Ebeling and Feistel, 1982; Jiménez-Montaño, Feistel, and Diez-Martínez, 2004), a related distance between sets can be specified as the maximum of the pairwise smallest distances between the elements of the two sets

$$m(S_1, S_2) = \max_{X_1 \in S_1} \left\{ \min_{X_2 \in S_2} \{d_H(X_1, X_2)\} \right\} \quad (7.208)$$

Having formally defined an evolution tree $T(\varepsilon)$ in terms of the axioms of the BRS, we can now return to the core problem of whether general extremum principles may exist which quantify the "fitness" in the form of a scalar measure that takes higher values along each edge of the evolution tree. Note that in general such a value is not reasonably assigned to single species; how would one compare the fitness of a fly to that of a mushroom? Rather, if at all, fitness is associated with an evolutionary stage, with a complete ecosystem which is the predecessor of another, an evolved ecosystem. If all parts of such a system are considered fixed except the competitors for a particular niche, the fitness measure may be "downscaled" to the fitness of single, comparable species; the better exploitation of a niche improves the fitness of the ecosystem as a whole.

The assignment of a real number to each vertex of a graph in a way that the number increases along each directed edge implies that the graph represents an order relation or at least a semiorder. We define, therefore, that

An MBRS is *Darwinian* if its evolution tree $T(\varepsilon)$ is elementarily reducible

(7.209)

Reducibility of binary relations and related graphs is specified in Sections 4.4 and 6.5. If a BRS is Darwinian, any related MBRS is also Darwinian because the graph $T(\varepsilon)$ has lesser edges the smaller the value of ε gets; any removal of edges cannot create additional cycles of the graph.

Table 7.2 Selection map of the example (7.210).

P	s(P)	P	s(P)
0	0	{a, b, c}	{a, b, c}
{a}	{a}	{b, c}	{c}
{b}	{b}	{a, c}	{a}
{c}	{c}	{a, b}	{b}

From the properties (7.201)–(7.203), we conclude that linear ideal BRS are Darwinian because their inheritance relation is asymmetric, transitive, and irreflexive, and hence represents an irreflexive semiorder (Section 4.1, Görke, 1970; Feistel, 1979; Ebeling and Feistel, 1982). Being linear and ideal is a sufficient condition for a Darwinian BRS. There may be Darwinian BRS which are not linear or ideal but we are not aware of any rigorous, necessary criteria for such systems.

As an example for a non-Darwinian BRS, we consider the linear system $U = (M, d, s)$, defined by

- the species set, $M = \{a, b, c\}$,
- the unfolding map, $d(S) = S$, and (7.210)
- the selection map $s(P)$ given in Table 7.2

In this example, it is impossible to assign increasing values to any sequence of $\{a\}$, $\{b\}$, or $\{c\}$ because they possess a cyclic inheritance relation

$$\{b\} \prec \{a\}, \quad \{a\} \prec \{c\}, \quad \{c\} \prec \{b\} \qquad (7.211)$$

Although the evolution tree of the system (7.210) is not elementarily reducible, it is reducible. If we consider the "condensed" graph which is formed of the irreducible clusters of the original evolution tree, that graph is elementarily reducible; reasonable selective values can be assigned only to the irreducible clusters of the evolution tree.

We could demonstrate that evolution processes can be modeled in a rather formal and abstract way by means of axiomatic BRS. Consistent with the landscape models considered earlier in the chapter, it turns out that proper "fitness" values can be assigned to evolution processes only under special, restrictive circumstances. Although evolutionary "progress" is irreversible and locally directed, this does in general not exclude cyclic processes; it cannot be expected that evolution has an overall goal of maximizing some fitness value.

In Darwin's sense, biological evolution is a process of subsequently better adaptation to the environmental conditions. With respect to assumingly fixed "physical" conditions, this process can usually be described by means of a fitness value that is subsequently increased and reflects some distance to an optimum adaptation to the prevailing situation. In contrast, species are actually living within a changing ecosystem, and the continual adaptation to that moving target is not necessarily a process that has some kind of global goal or aim. Rather, this situation is described by the Red Queen hypothesis (Fisher, 1930; van Valen, 1973; Darlington, 1977), referring to the queen in Alice's wonderland. Since biological evolution proceeds within the framework of physical laws, we cannot expect to discover a general, global physical evolution criterion for processes far from equilibrium similar to the role entropy takes for equilibrium states. This somewhat negative conclusion can also be seen in a positive light: physics apparently does not confine evolution to any targets specified by higher, divine beings or forces; evolution is an open process with apparently unlimited capacities to create new and unpredicted forms, structures, and phenomena, just as envisaged by Stanislaw Lem in the citation given in the "Introduction."

8
Self-Organization of Information and Symbols

For my part, I look at the geological record as a history of the world imperfectly kept and written in a changing dialect.
                                                    Charles Darwin: The Origin of Species

Language, more than anything else, is what makes us human.
                                                    Tecumseh Fitch: The Evolution of Language

Similarities between written human languages, computer codes, and genetic sequences are obvious, even though those strings serve very different purposes, use different symbols, and are physically realized by different substances, structures, and processing machineries. What they have in common is usually regarded as "information" (Shannon, 1948; Kämmerer, 1974; Stratonovich, 1975; Ebeling and Feistel, 1982; Jiménez-Montaño, Feistel, and Diez-Martínez, 2004; Marris, 2008). That term, used in various articles and books published in so different disciplines such as physics, neurobiology, computer technology, or philosophy, still eludes a generally accepted and universally valid definition. A related, particularly difficult question is the definition of the value of information (Stratonovich, 1965, 1975, 1985 Volkenstein, 1986), that is, of a suitable quantification of the impact of information on dynamical processes such as competition and evolution. While we do not associate information processing with simple physical systems such as a gas, a pendulum, or a magnetic coil, there is no doubt that very complex systems such as humans are able to store, exchange, and accumulate information. Where is the separation line between those two classes of systems, and what is the physical nature of the transition between them? What is the relation between information and the symbols used to store and transmit it?

Our plain and basic statement is that there is no symbolic information processing without life, and there is no life without symbolic information processing (Ebeling and Feistel, 1982, 1992, 1994; Feistel, 1990, 2008b). In this context, technology is understood as an "honorary living thing" (Dawkins, 1996a), as a part of the human culture that belongs to the realm of life (Donald, 2008). Similarly, Eigen (1994) formulated that *a living entity can be described as a complex adaptive system which differs from any, however complex, chemical structure by its capability of functional self-organization based on processing of information. If one asks where does this information come from and what is its primary semantics the answer is: information generates itself in feed back loops via replication and selection, the objective being "to be or not to be."* This way the origin of information processes is intimately connected with the evolution of life (Ayres, 1994; Avery, 2003; Yockey, 2005). Information processing, we may conclude, is a key process in the struggle for existence of living beings. The search for the origin and the physics of information takes us to the self-organization and evolution of life.

There must have been a point in history at which preinformational processes and structures smoothly transformed into rudimentary genuine information processing for the very first time. That transition had a number of properties in common with later, similar events in the course of biological, social, and technical evolution. Behavior biologists and ethologists were apparently the first who observed and analyzed the general character of that phenomenon. Sir Julian Huxley (1914) described it as *the gradual change of a useful action into a symbol and then into a ritual; or in other words, the change by which the same act which first subserved a definite purpose directly comes later to subserve it only indirectly (symbolically) and then not at all.* Later, the transition process was termed *ritualization* (Lorenz, 1963; Tembrock, 1977; Klix, 1980). In the more general approach taken here, the origin of life is understood as the first ritualization transition that converted an abiogenic complex physicochemical structure into a primitive living organism. From then on, that kind of transition process repeated over and over again, at very different levels of organization, under very different circumstances. In our view, the ritualization transition is the process that refutes the pessimistic hypothesis that *"physicodynamics alone cannot organize itself into formally functional systems requiring algorithmic optimization, computational halting, and circuit integration"* (Abel, 2009).

8.1
Symbolic Information

When a child is born, it is equipped with a minimum bootstrap program for rudimentary forms of behavior – breathing, eating, crying, sleeping, and so on. The first phase of learning is restricted to the correlation analysis of signals coming from the receptor cells for light, sound, tactile sensation, temperature, smell, and taste, and in particular to the analysis of their changes in response to own actions such as moving or crying (Gopnik, Meltzoff, and Kuhl, 2000; Held *et al.*, 2011). Out of the random babbling which a baby can produce with its voice, the parents repeat and imitate selected sounds that resemble simple words in the parents' spoken language, such as "ma-ma," which may again be repeated and imitated by the child. In conjunction with other correlated events and activities, the exchange of simple sounds develops to a primitive acoustic signal system between child and parents. The child discovers that the sound contains a coding system used by the parents. By concatenating simple sounds to short sequences that subsequently extend to words and sentences, a new, additional information channel is opened that works separate from the visual and acoustical signal processing system even though it also uses sound as a carrier and ears as receivers.

Obviously, the same information can be transferred in two qualitatively different ways, I can see a dog or somebody can tell me "there is a dog." What we can extract from structures and processes which we recognize in the world outside, we may refer to as *structural*, bound, analog, concrete, or implicit information. What we learn by listening or reading, we may refer to as *symbolic*, free, digital, abstract, or explicit information, respectively. While symbolic information is "genuine" information, or just "information," structural information is only potential information. Provided that structural information exists in a given system, symbolic information can be obtained, extracted from it.

A process similar to what every baby more or less randomly develops in the first months and years of its life must have taken place when early humans separated from the other primates. In contrast to modern babies, first humans had no teachers and there was no preexisting acoustic coding system. That transition process was self-organized and resulted eventually in the use of a novel symbolic information system, the spoken language (Fitch, 2005, 2010). Ritualization, as we shall understand the term here, is the self-organized emergence of systems capable of processing symbolic information (Feistel, 1990).

Consequently, here we distinguish humans from their precursor species by the self-organized use of a symbolic language, as we distinguish living entities from chemical microprocessors by the self-organized use of a symbolic molecular coding system (Tattersall, 2010). Under this definition, Oparin's coacervates equipped with a genetic translation apparatus such as Eigen's hypercycle count already as living organisms and are subject to Darwinian biological evolution (Feistel, Romanovsky, and Vasiliev, 1980; Feistel, 1983d). The main argument for this terminology is that ritualization exhibits physical properties of kinetic phase transitions of the second kind, and is thus a pronounced qualitative separator between chemical and biological systems on the one hand, and between biological and social systems on the other hand (Ebeling and Feistel, 1992), as shown in Figure 8.1. Ritualization was and is an overly successful trick of mother nature that has been repeated many times after it once happened for the very first time. While written language was the first symbolic information system for primordial cells, in the social evolution of humankind it emerged as a second communication channel with significant delay after spoken language (see Section 8.10 and Table 3.4). The direct digital brain-to-brain interface is an unprecedented, fast, flexible, and extremely powerful tool that enabled the evolutionary triumph of the human species (Donald, 2008; Tattersall, 2010). We shall return to the self-organization of language in Sections 8.10 and 8.12.

In addition to the origin of life and the invention of human language (Figure 8.1), another self-organized system for symbolic information processing is of fundamental importance and unbeatable success, namely the invention of sexual reproduction by the eukaryotes (Butterfield, 2000) (see Section 8.7).

Symbolic information is formally quantified by Shannon's (1948) information theory, as shown in Figure 8.2. In order to transfer information from a source to the

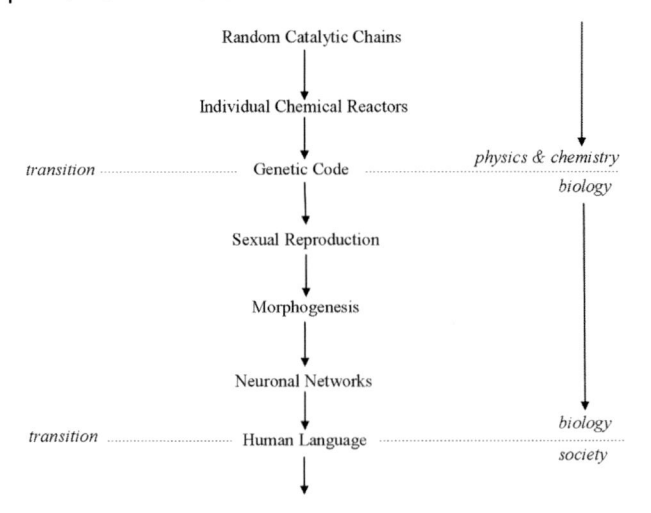

Figure 8.1 Schematic of the two fundamental ritualization transitions during the evolution from prebiological chemistry to the human society. (modified from Ebeling and Feistel, 1992).

destination, a transmitter is used that encodes the information into symbols that are temporally stored in a communication channel from where a receiver can retrieve the information if the same code is used on both sides. From the receiver, the information is passed to the destination. Physically, the process is causal, that is, the information arrives at the destination later than it is sent out, no matter what reference frame is used to observe the process. The process is irreversible; information transferred from A to B may cause transitions in the destination system which cannot be withdrawn, say, by a later backward transfer of the message from B to A.

Evidently, in practice, the components of Shannon's information machine are physical systems that obey physical laws. It is a fundamental aspect of that machine that the particular physical nature of those components is irrelevant as long as they possess a few indispensable general properties. Symbolic information cannot be

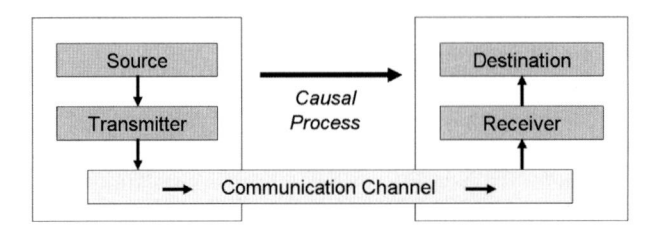

Figure 8.2 Shannon's information concept: A transmitter encodes the information into symbols that are temporarily stored in a communication channel from where a receiver can retrieve the information if the same coding rule is used on both sides.

revealed from investigating, say, the physical properties of the atoms that constitute the machine; the same atoms may convey very different messages, and the same message may be transferred by very different physical carriers. Symbolic information is an emergent property; while it cannot exist outside of or in contradiction to the physical laws which control the machine, it cannot be reduced to those laws.

Shannon's information entropy is a measure for the maximum amount of information that can be transferred by a particular communication channel and coding system. The entropy value does not tell how much information is actually transferred by that system. The entropy formula can also be applied to arbitrary sequences of symbols (Ebeling and Feistel, 1982; Jiménez-Montaño, Feistel, and Diez-Martínez, 2004) and provides an upper bound for the possible information content of that sequence. A trivial string with zero entropy does not provide any information; high entropy applies equally to information-less random sequences or information-rich "compressed" information without redundancy. The Kolmogorov–Solomonov complexity is a similar measure; it estimates the minimum length of a string that a given string can "reversibly," that is, loss-free be converted from and to. With the recent advance of modern computers, of communication networks and genetic sequencing techniques, those classical mathematical concepts fertilized explosively growing research and practical application fields.

Symbolic information has a number of general properties (Feistel,1990, 2008b; Ebeling and Feistel, 1994):

1) Symbolic information systems possess a new symmetry, the *carrier invariance*. Information can loss-free be copied to other carriers or multiplied in the form of an unlimited number of physical instances. The information content is independent of the physical carrier system used.
2) Symbolic information systems possess a new symmetry, the *coding invariance*. The functionality of the processing system is unaffected by substitution of symbols by other symbols as long as unambiguous bidirectional conversion remains possible. In particular, the stock of symbols can be extended by the addition of new symbols or the differentiation of existing symbols. At higher functional levels, code invariance applies similarly also to the substitution of groups of symbols, synonymous words, or equivalent languages.
3) Within the physical relaxation time of the carrier structure, discrete symbols represent quanta of information that do not degrade and can be refreshed unlimitedly.
4) Imperfect functioning or external interference may destroy symbolic information but only life-based processing systems can generate new or recover lost information.
5) Symbolic information systems consist of complementary physical components that are capable of producing the structures of each of the symbols in an arbitrary sequence upon writing, of keeping the structures intact over the duration of transmission or storage, and of detecting each of those structures upon reading the message. If the stock of symbols is subject to evolutionary change, a consistent coevolution of all components is required.

6) Symbolic information is an emergent property; its governing laws are beyond the framework of physics, even though the supporting structures and processes do not violate physical laws.
7) Symbolic information has a meaning or purpose beyond the scope of physics.
8) In their structural information, the constituents of the symbolic information system preserve a frozen history ("fossils") of their evolution pathway.
9) Symbolic information processing is an irreversible, nonequilibrium processes that produces entropy and requires free-energy supply.
10) Symbolic information is encoded in the form of structural information of its carrier system. Source, transmitter, and destination represent and transform physical structures.
11) Symbolic information exists only in the context of life.

Chemical reaction systems, complex organisms, or primate species which by chance were the first to benefit from those properties gained essential selective advantages and prevailed, even at the cost of maintaining the necessary complex machinery.

Several of those properties refer to structural information; this point is subject of the next section.

8.2
Structural Information

Thermodynamic equilibrium is the stable and unique final state of any macroscopic system if it is isolated from the surrounding and left on its own for a sufficiently long time (Chapter 2). The approach to that state is described by the increase of entropy, S, to its maximum value, S^{eq}, under the given condition of constant volume, V, and without import or export of energy, E, and particles, $\mathbf{N}$. The equilibrium state does not provide any clue on its history; it is the same macroscopic state for any initial state the system may have had. As a measure of the information available on its past, the *entropy lowering* (Klimontovich, 1991, see Chapter 2),

$$\Delta S = S^{eq}(E, V, \mathbf{N}) - S(E, V, \mathbf{N}, t) \tag{8.1}$$

decreases with time in a closed system as a consequence of the second law,

$$\frac{d}{dt}\Delta S = -\frac{d_i S}{dt} \leq 0 \tag{8.2}$$

A concept similar to Equation 8.1 is "negentropy" (Schrödinger, 1944; Brillouin, 1953).

Equilibrium states are not necessarily homogeneous; systems exposed to gravity or containing different phases exhibit gradients of density, entropy, and particle numbers.

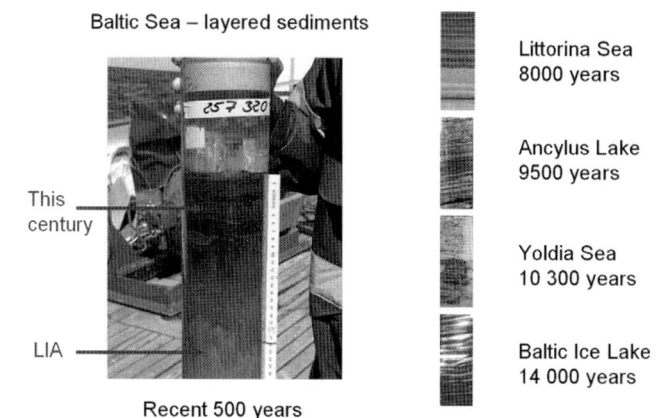

Figure 8.3 Structural information in the form of stratified sediments in drilling cores of the Baltic Sea. "LIA" refers to the Little Ice Age. From the properties of various cores, on the right, different stages could be revealed with alternating brackish seas and freshwater lakes, with subsequent layers of oxic and anoxic conditions in the history of the Baltic Sea from the last glaciation to present. (adapted from Leipe et al., 2008, with permission).

The relaxation time to thermodynamic equilibrium depends on various conditions. Turbulent water in a tea cup comes to rest after a couple of minutes; the thermal equilibration may take many hours. If sugar lumps are in the cup, without stirring it may take even years to reach the equilibrium state, with the speed depending very much on the water temperature. In glaciers and muddy sediments, such as those of the Baltic Sea shown in Figure 8.3, information may be preserved from several hundred years up to 0.4 Myr (Hewitt, 2000), and in petrified sediments and rocks, such as filigree fossils, up to several 1000 Myr (see Chapter 3, Tables 3.1 and 3.2). The historical record of remnant magnetism in the spreading sea-floor rocks belongs to most revolutionary and fascinating geological discoveries (Hohl, 1981); its signature is erased as soon as the material is heated above the Curie point.

Scientists observe and investigate the structures encountered in nature and in lab experiments. Structural information is extracted and converted into symbolic information, in the form of articles and books, data tables, or oral lectures given to students and colleagues. Although we do not possess a suitable general theory for the formulation of conservation laws valid for the research process, we feel intuitively that the amount of symbolic information produced cannot exceed the amount of structural information that is embodied by the research target. If there were a perpetuum mobile of the fourth kind that does nothing but generate useful symbolic information without input of structural information, a lot of money could be saved that is currently spent on satellites, particle accelerators, or geological drilling. The invention of such a machine of universal knowledge is very unlikely even though its existence is not forbidden by any natural law we know by now. Similar to the overwhelming empirical evidence for the validity of the second law, we may infer the physical impossibility of a perpetuum mobile of the fourth kind from the experience that systematic prophecy and fortune telling always failed in the past, without exception. Winners of lotteries are not counted here as prophets. Humans tend to believe in divine beings equipped with the gift of prophecy, but this wishful thinking is not supported by any observational evidence.

Structural information has a number of general properties (Feistel, 1990, 2008b; Ebeling and Feistel, 1994):

1) Structural information is inherent to its carrier substance or process. Information cannot loss-free be copied to any other carrier or identically multiplied in the form of additional physical instances (see Figure 8.4). The physical carrier is an integral constituent of the information. The state of the physical context of the system is an integral part of the information.
2) There is no invariance of information with respect to structure transformations. Different structures represent different information.

Figure 8.4 Structural information cannot be copied identically. The photograph shows a replica of *Archaeopteryx* exhibited in the Solnhofen museum. While that "3D copy" has a surface shape similar to the original fossil shown in Berlin, it is impossible to determine from the replica, say, the geological age as it is done for the original structure by isotopic ratio analysis of the embedding limestone quarries of Solnhofen (Gradstein, Ogg, and Smith, 2005). As printed in this figure, the "copy" of the "copy" has an optical appearance similar to the replica, at least to the human eye, but the structural information carried by this 2D image is further reduced.

3) Structural information emerges and exists on its own, without being produced or supported by any kind of separate information source. No coding rules are involved when the structure is formed by natural processes.
4) Over the relaxation time of the carrier structure, structural information degrades systematically as a consequence of the second law, and disappears when the equilibrium state is approached.
5) Internal physical processes or external interference may destroy structural information; it cannot be regenerated or recovered. Periodic processes can rebuild similar structures but never exactly the same, in particular because the surrounding world is not the same again at any later point of time.
6) Structural information is not represented in the form of codes. No particular coding rule or language is required or distinguished to decipher a structure.
7) Structural information is a physical property; it is represented by the spatial and temporal configuration of matter, and its governing laws are the laws of physics.
8) Structural information is of physical nature and is independent of life.

It is a certain subtlety that symbols are necessarily physical structures und that symbolic information is necessarily accompanied by structural information. The essential difference between the two is that the structural information of the symbol or of the carrier is not relevant for the symbolic information; the relevance of the latter lies in the processes that occur at the source and at the destination of information (Figure 8.2) rather than those acting along the transmission channel.

8.3
Extracting Structural Information

Virtually, the world around us is endlessly wide, diverse, and unsteady; it is full of structures, colors, sounds and scents. To survive in a complex environment, it is an enormous selective advantage for an organism to be able to react suitably on external changes and to develop active behavior, to detect and avoid dangerous circumstances, and to find sources of energy and metabolic substances, such as light or nutrients (Gong et al., 2010). Possibly, the danger of a situation or the appropriateness of an action turns out only with some delay after the activity or inactivity, often too late to subsequently correct the earlier decision. The most rudimentary form of active behavior of a living entity is to grow and to multiply. Step by step, for faster growth and multiplication, chemical, thermal, or optical receptors were exploited to extract structural information from the particular environmental conditions and developed subsequently to extremely high sensitivity and complexity. Organisms that react more quickly, efficiently, and successfully than their competitors are the winners in Darwin's race of natural selection (see Chapter 6). Looking back from today's perspective, we ourselves are the living proof that the evolution of more and more sophisticated information-processing systems was a sufficient – and perhaps also necessary – condition to achieve this goal: no doubt, we do exist.

Contemporary science is a tutorial example for the interaction between an individual, say, a physicist, and the environment. The physicist's physical body is separated from the rest of the world by an interface. The only way a human can recognize the surrounding world is by those physical processes that transfer energy or matter across that interface and activate related processes inside the body, in particular within specialized amplifying receptors, for example, for light, heat, or mechanical pressure. Similar to any other organism, we can learn about the world only by analyzing spatial and temporal relations between the different signals that are exchanged between us and what is "out there." We are principally unable to recognize what the world or any of its parts actually "is"; inevitably watching only shadows in Plato's proverbial cave through Immanuel Kant's yellow glasses, we always see a biased picture. Perhaps, any apparently solid and massive substance around us, including ourselves, as well as all we can see in a starry night, is nothing but excited whitecaps on a Debye ocean or a Higgs ocean of the "true matter" that fills the universe (Greene, 2004), but our senses do not recognize the latter because it has always been irrelevant for our biological survival. Any signal we can receive is a change that happens simultaneously on both sides of the interface between us and the outer world (Figure 8.5). In the terminology of physics, we exclusively recognize exchange quantities rather than state quantities. This physically evident statement is related to but different from philosophical scepticism (Butchvarov, 1998). Virtually, however, our immediate subjective impression is quite a different one: We open our eyes and see clouds in the sky or people walking around.

This deception has its roots in the evolution of life, and it cannot be avoided. Its particular form differs significantly from organism to organism. Bats, for example, consider every smooth horizontal plane as a water surface (Greif and Siemers, 2010). Recent mammals stem from animals that were mostly active during the night over millions of years of dinosaur dominance; their eyes had less color receptors than

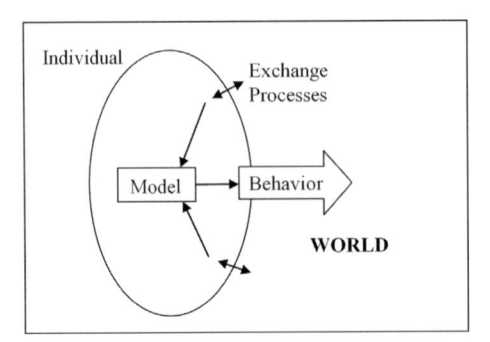

Figure 8.5 To recognize the environment, an individual is restricted to physical exchange processes across the interface. From correlations between the receptor signals, a model is created that represents an "illusion" of the world. That illusion is consistent with sensation experience, may possess predictive capabilities, and is controlling the individual's active behavior.

Figure 8.6 With a little fantasy, the reader may easily discover the grim face of a troll in the bark of this beech tree. Photo taken on the chalk cliff of the Jasmund National Park, Rügen island, Germany, in fall 2010.

ours, in contrast to the tetrachromacy of many fish, birds, and reptiles. It happened only late in the primate evolution, independently in the New World and the Old World, that we began to distinguish red from green, ripe fruits from unripe (Nathans, Thomas, and Hogness, 1986; Hunt *et al.*, 1998; Surridge, Osorio, and Mundy, 2003; Jacobs, 2009; Demb and Brainard, 2010). For social apes and humans, the distinction of a friendly face from others is crucial. Other humans were always our most important friends and enemies in the struggle for survival. The visual preprocessing pipeline of macaques for the recognition of faces consists of six subsequent filter layers (Freiwald and Tsao, 2010). Any visual pattern we receive is apparently first analyzed for face contours before any other objects are identified. This explains naturally why we so often automatically see a face in vague and diffuse patterns, such as Cydonia, the famous "face on Mars" (NASA, 2001), or ghosts and trolls in a dim forest (see Figure 8.6). Caricatures and comics, paintings, and sculptures exploit the human visual preprocessing to trigger certain impressions or emotions by means of highly simplified or faked key signals.

Structural information of the surrounding world is reflected by correlations in the input signals we receive via our receptors. A very strong correlation occurs between the image on our retina and active motion of our eye and our head. An autocorrection filter immediately removes this transformation of the coordinate frame from the image of the world before it is passed to our consciousness. This filter gets in temporary trouble when we use varifocals for the first time. When I touch a tree, my visual observation is consistent with the tactile sensation from my finger tips. The related redundancy of the incoming signals can be exploited to predict missing or disturbed signals. We easily know about the presence of a dog from its noise even if we do not see it, or from a visible piece of fur if the rest of the body is hidden behind an opaque object. Temporal correlations are similarly important. If we are able to store input signals for a while, we can correlate a sudden very pleasant or unpleasant event with our record of its immediate past (Seo and Lee, 2009). That correlation, in turn, permits predictions of future relevant events from the analysis of present, less relevant signals. At an advanced level, this complex processing of memory information in combination with the evaluation of the present internal status and current signal input is subjectively recognized as decision-making, as "free will" (Pauen and Roth, 2008; Resulaj *et al.*, 2009; Desmurget *et al.*, 2009; Springer, 2009; Brembs, 2011).

The entirety of correlations detected in the receptor signals can be termed a *model* of the world. A model is a mapping of the natural objects and laws in a simplified form, observed from a subjective perspective and evaluated under subjective criteria. The ultimate benefit provided by a model is its predictive capability. The basic principle of internal model building from experience has been the same from the very first organisms to modern science. When we observe the world, we do it through inherited models in the form of hidden powerful preprocessors that convert chaotically streaming clouds of photons into colored objects moving in space and time, into lovely faces or scaring enemies. Complex air-pressure fluctuations appear to us as spoken words, singing birds, or rolling thunder. Concentrations of chemicals are projected into the reception of sweet taste or disgusting smell. It is clear that none of those familiar properties actually applies to particles such as electrons or photons of which our world is consisting (according to our currently best models): electrons have neither colors nor smells. The model picture as we consciously recognize our world was simply the most successful one in our evolution history in order to quickly assess the particular environmental situation and to make appropriate decisions on own activities. Our historically developed way of seeing and understanding the world is casting long shadows. We are unable to switch off those preprocessors unless we explore the world by emotionless measurements using artificial technical sensors and receptors. It is a painfully difficult mental process to develop alternative models of space, time, and matter that are fully consistent with all the experiments and observations made in quantum mechanics, particle physics, and astronomy (Callender, 2010). Even the comparatively simple concepts of entropy and information are still imperfect theoretical models. Very likely, we are still watching the physical world of space, time, and matter from a Ptolemaios perspective, we observe complicated, apparently unrelated "epicycles" in the quantum and the cosmic reality, and we are still waiting for a new Copernicus to explain how all the fundamental mysteries may form a logical and self-consistent picture (Greene, 2004).

When humans began to consciously explore the world, there were no teachers and not yet any theoretical models readily developed, tested, and made available by previous generations. Immediate observational evidence over the lifespan of a human is insufficient to explain the complexity and unpredictability of weather, natural catastrophes, or individual fates. Gaps in knowledge were tentatively filled with mysterious invisible forces and divine beings. The main drawback of god models is their total lack of predictive capabilities. Prophetic dogmas, either religious or political, written down as "eternal truth" are just models reflecting the accumulated information up to the time of their formulation; they are only good as "predictors" as long as the world is stationary or periodic; they fail utterly when the world is changing. Creation theory as an alternative model to Darwinian evolution would require a creator empowered with unlimited prediction capabilities, a perpetuum mobile of the fourth kind. The everyday extinction of species on our globe demonstrates that their "creator" obviously failed to foresee the currently prevailing conditions and to ensure their survival. There is overwhelming evidence for "unintelligent design" (Haszprunar, 2009), for imperfect properties of living beings, including us humans, as a result of insufficient predictive capacities of the models encoded in our genes. The ultimate benefit provided by a model, and the final reason for its existence, is an estimate for information that is not or not yet available otherwise. The ultimate verification criterion for any model is the comparison between prediction and observation, but already contradictions between different models may prove that they cannot equally be true.

Models can be built up as analog devices similar to Steinbuch's learning matrix (Kämmerer, 1974). The models controlling the behavior of honey bees and ants are perhaps of that nature. Models created in the form of symbolic information are much more efficient and flexible. They can be stored, transmitted, and copied in a loss-free way. Genetic strains and scientific articles are models of certain parts of the world in the form of symbolic information. Continuous extraction of structural information from the environment combined with gradual accumulation of symbolic information was and is the backbone of Darwinian evolution and is a successful recipe for survival.

8.4
Physical Properties of Symbols

Information and thermodynamic entropy are closely related concepts. Pauling's (1935) residual entropy is an illustrative example for that relation.

When water is cooled down under ambient pressure, it freezes and forms ice Ih, a hexagonal crystal lattice as shown in Figure 8.7. When it is further cooled down to the vicinity of zero temperature, its entropy does not asymptotically approach zero, in contrast to the conventional form of the third law of thermodynamics (Gutzow and Schmelzer, 2011, see Chapter 2). The reason is that the ground state of ice Ih is degenerate; more than one microstate is possible at the same energy and density. The position of the H atom on the hydrogen bound between two O atoms is asymmetric and bistable; the H atom is always closer to one of the O atoms. Of the four H atoms surrounding one O atom in the crystal, exactly two must be located nearby, no matter which two of them. For a given O atom, there are six possible configurations of two H atoms on four hydrogen bounds, 1100, 1010, 1001, 0110, 0101, 0011. For the adjacent O atoms, the number of free choices is restricted by the particular configuration chosen already for the first atom. Rather than 6^N, the number of microstates W of N water molecules in the hexagonal cage was first quickly estimated by Pauling (on a

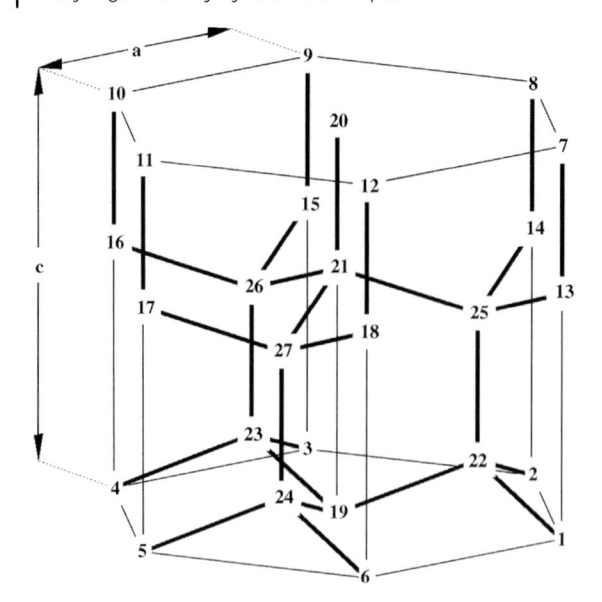

Figure 8.7 Elementary crystal of hexagonal ice I (Penny, 1948; Schulson, 1999; Feistel and Wagner, 2005) with the lattice constants a, c. The numbers 1–27 indicate oxygen (O) atoms. Bold lines between those atoms represent hydrogen (H) bonds. The H atoms are not located half-way between two O atoms, rather, they are closer to one of them. Of the four H atoms connected to one O in the center of the tetrahedron, exactly two are in the nearby position, forming the H_2O molecule, and the other two are in the more distant position. That rule does not uniquely determine the positions of all H atoms. Note that at very low temperatures, in contrast, there is a hypothetical stable proton-ordered phase, ice XI, which has never been observed to form spontaneously without the help of catalysts (Johari and Jones, 1975; Singer et al., 2005).

beer mat, as the anecdote says) to take the value (Bjerrum, 1952; Singer et al., 2005),

$$W \approx 1.5^N \tag{8.3}$$

which was later mathematically perfected by Nagle (1966). Using Boltzmann's famous entropy formula, $S = k_B \ln W$, written in golden letters on his grave in Vienna (see Figure 8.8), we obtain for the residual specific entropy of water at 0 K,

$$\frac{S_0}{m} = \frac{k_B N_A}{N M_W} \ln W = 189 \, \text{J/kg/K} \tag{8.4}$$

where N_A is Avogadro's number, M_W the molar mass of water, and $m = N M_W / N_A$ the mass of the ice sample. The result of Equation 8.4 is consistent with other thermal water properties (Giauque and Stout, 1936; Feistel and Wagner, 2005, 2006). The residual entropy depends only on the crystal symmetry and is therefore independent of density or pressure.

Figure 8.8 Tombstone of Ludwig Boltzmann, 1844–1906, on Vienna's Central Cemetery, with his equation engraved, $S = k \log W$. Photo taken in fall 2010.

Pauling's residual entropy, S_0, has a number of remarkable properties:

1) S_0 cannot be measured by thermodynamic methods.
2) S_0 is a macroscopic state quantity rather than an exchange quantity.
3) S_0 is not equivalent to heat, in contrast to Clausius' thermal entropy, $dS = d'Q/T$.
4) The relation of S_0 to heat is similar to the relation of relativistic rest mass to mechanical energy.
5) The determination of S_0 requires statistical mechanics, that is, mathematical models for particles, probabilities, and ensembles.
6) S_0 represents the perhaps closest link between theoretical physics and information.
7) S_0 was discovered long before information theory was developed.

By thermal fluctuations, any occasional transitions between the microstates of H atoms in ice Ih can practically be neglected at sufficiently low temperatures. If we could use a laser beam to switch between the microstates of ice, and a weaker laser beam to read the current microstate without destroying it, ice could serve as an information storage, similar to quantum memory (Specht *et al.*, 2011). From the thermodynamic residual entropy, we can compute the memory C in bits of such a device as

$$C = \log_2 W = \frac{S_0}{k_B \ln 2} \approx m \frac{N_A \ln(3/2)}{M_W \ln 2} \approx 2 \times 10^{25} \text{ m/kg} \qquad (8.5)$$

Thus, to store one terabyte, $C = 2^{43} \approx 8 \times 10^{12}$, an ice crystal of $m \approx 0.5$ ng is big enough, which is a cube of less than 0.01 mm size, similar to a dust particle. If used to store symbolic information, proton configurations in the ice crystal represent those symbols. It is obvious that neither the physical nature of ice nor that of the symbols is in any way related to the kind or purpose of the stored information, which may be data, texts, pictures, music, or anything else.

There is a close similarity between the thermodynamic entropy of a physical structure and Shannon's entropy of symbolic information, as we already notice from Equations 8.4 and 8.5. The energetic degeneracy of different states that may be used as symbols is reflected in the system's physical entropy and in its symbolic information capacity, such as in the case of ice. On the other hand, there are systems that possess thermodynamic entropy but do not act as symbolic information carriers, such as an ideal gas, and there are systems such as the single-chain molecule of RNA that has a symbolic information capacity but lacks of a well-defined thermodynamic entropy. These examples show that the concepts of thermodynamic entropy and information entropy are not equivalent to each other; they do not exclude, however, the existence of a more general entropy definition which includes the former two as special cases.

Because stored or conveyed symbolic information is independent of the carrier, there is a large variety of possible candidates available from almost any branch of physics to be chosen as a communication medium (Landauer,1973, 1976; Volkenstein and Chernavsky, 1979). Equilibrium systems with degenerate ground states such as ice can be used, or frustrated spin glasses (Ebeling, Engel, and Feistel, 1990a), dissipative structures, as well as acoustic or electromagnetic waves. Traditional media for historical, political, economic, or scientific information storage are "frozen-in" nonequilibrium structures with sluggish relaxation to equilibrium, such as carved stones (Figure 8.8) or printed books (Figure 8.9). They have in common that many alternative structures may exist under similar boundary conditions and persist as such over sufficiently long times, depending on the purpose of the information system in its physical, biological, or social context. For example, books may be readable over 1000 years, sound waves for about 1 s, and RAM capacitors of computer memories for 1 ns.

Of particular interest for symbolic information processes are neutrally stable and critical states that possess so-called Goldstone modes, that is, dynamical modes associated with vanishing Lyapunov coefficients. Once such a system has changed its state, there is no restoring force that might try to reestablish the previous configuration. In linear physical systems such as acoustics or electrodynamics, wave amplitudes and phases are neutrally stable; there is no "equilibrium amplitude" to which any initial value would tend to converge. Amplitudes are conserved by the carrier waves until external noise may dissipate the initial signal. That conservation

Figure 8.9 Printed title of the 1982 German original of this book.

[Shown in figure: **Physik der Selbstorganisation und Evolution**, Von Werner Ebeling und Rainer Feistel]

property was readily exploited during the evolution of the human spoken language or for radio and television transmission. In physically autonomous systems, that is, under time-independent conditions, oscillators possess neutrally stable phase lags. This property is preferably used to store and exchange time information between humans, by our clocks (Feistel 1979, 1982; Ebeling, Engel, and Herzel, 1990b). The accuracy of clocks is principally limited by the phase diffusion rate of those oscillators (Kramer, 1988). Nonlinear, self-sustained oscillators as well as chaotic, strange attractors possess the neutral-phase symmetry, although the amplitudes of those systems are usually controlled by feedback mechanisms. Exceptions from this rule with neutrally stable amplitudes such as Lotka–Volterra (predator-prey) oscillations are usually structurally unstable.

The kinetic phase transition from a stable steady state to a limit cycle is commonly termed a Hopf bifurcation. It is a tutorial physical example for what happens qualitatively during a ritualization transition. A highly simplified model for that transition is given by the dynamical equations for the amplitude, A,

$$\frac{d}{dt} A = A(c - A^2) \tag{8.6}$$

where c is a control parameter, subject to externally imposed conditions, and for the phase, φ,

$$\frac{d}{dt} \varphi = \omega \tag{8.7}$$

where ω is some real number. The bifurcation diagram of this system (Figure 8.10) shows the stable states for the order parameter, A, as a function of the control parameter, c.

We imagine the evolution of the system (Equations 8.6 and 8.7) under the assumption that information storage is a selective advantage for an organism possessing that system, starting from negative values of c which may be subject to small random mutations. In other words, we consider c as a phenotypic property in a valuation landscape (Chapter 7). Such an advantage may be realized by the availability of an internal biological clock to adjust metabolic processes to "predicted" environmental cycles. For $c \ll 0$, no such clock is running since the Lyapunov coefficient, $\lambda = c$, of Equation 8.6 associated with the attractor state $A = 0$ will immediately damp

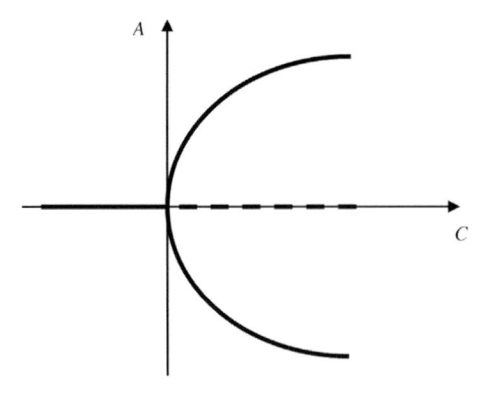

Figure 8.10 Hopf bifurcation diagram for the transition of a steady state to a nonlinear oscillator (Equations 8.6 and 8.7). The dependence of the order parameter A on the control parameter c is shown as bold curves. The steady state with $A=0$ is asymptotically stable for $c<0$, neutrally stable for $c=0$, and unstable (dashed) for $c>0$. In the latter case, oscillations with finite amplitudes occur.

any fluctuations about that state. If by random mutation a negative value of c in the order of the oscillation frequency ω appears, random amplitude fluctuations will relax more slowly and permit the system to perform a few noisy cycles around the stable focus state. From then on, the selective pressure (Fisher's law, see Chapter 7) will gradually increase the mean value of c in an ensemble of competing individuals. When approaching $c=0$, amplitude fluctuations grow macroscopic and form a random oscillator (Feistel and Ebeling, 1978a). The oscillation amplitude gets stabilized for $c>0$, until it sufficiently exceeds the noise level of the system, and special values, such as zero passage or tipping point, may serve as trigger signals for useful metabolic processes controlled by the clock.

Phase transitions of the second kind are characterized by the properties that (a) the two phases possess different symmetries and (b) the two phases are indistinguishable at the transition point (Landau and Lifschitz, 1966). The Hopf bifurcation corresponds to such a kinetic phase transition of the second kind (Feistel, 1979; Feistel and Ebeling, 1989). For $c>0$, the time symmetry is broken; the states the system occupies at different times are no longer identical, in contrast to the steady state at $c<0$. At the transition point, $c=0$, the two phases coincide; an oscillator with zero amplitude cannot be distinguished from a steady state (Figure 8.10).

From the viewpoint of information, the broken time symmetry of an oscillator can be used to transmit information from a source that generates the undulation to a detector that reacts on a threshold value of the amplitude as a receiver for the coded signal. The newly gained symmetry of coding invariance is given by the fact that any trigger value may take the role of the symbolic bit of information, as long as source and receiver agree on it. This property may be particularly important when oscillator and detector become physically separated units at later stages of an evolution process, differentiate subsequently into two or more functions, and use different amplitudes to encode different activities.

8.5
Properties of the Ritualization Transition

The ritualization transition has properties of a kinetic phase transition of the second kind. Before the transition point, the system under consideration possesses structural information. Beyond that point, it exhibits additional symbolic information plus a new symmetry, the coding invariance. At the transition point, the symbolic information is identical with the structural information. A schematic for the qualitative turnover is shown in Figure 8.11.

Consider a causal chain that includes a number of steps that lead from the cause to the effect. If that chain is part of an organism, evolutionary modifications may make the chain gradually more effective, robust of flexible. This may imply that an intermediate process is reduced to a specific aspect of its structural information which then represents the native form of symbolic information. The interface between transmitter and receiver gains the freedom of coding invariance, that is, modifications of the physical details of the symbols in use are possible without affecting the result of the causal cascade. Macroscopic fluctuations emerge as a result of that symmetry. They permit quick development, ripening, optimization, or diversification of the initial coding system.

Because of the coding symmetry, the particular symbol in use for a specific signal is completely arbitrary, but, unlike Buridan's ass, there is always one that is initially chosen out of the set of physically equivalent possibilities. The choice made is obviously the result of the preceding earlier evolution process; in other words, the set of symbols actually used is reflecting the process of emergence of those symbols. The structural information physically contained in the symbols represents a record of

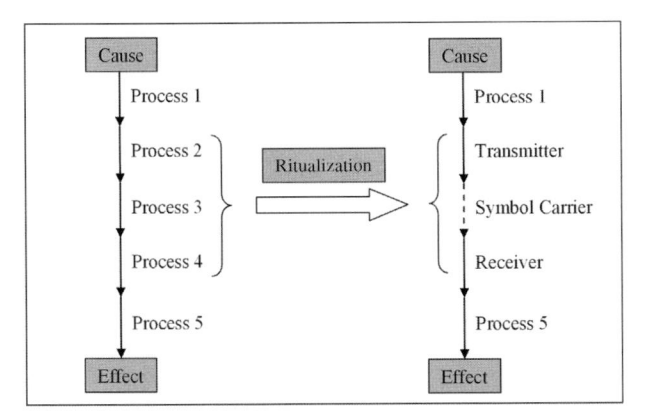

Figure 8.11 Schematic of a ritualization transition: subsequent adjacent segments of a causal chain are converted into transmitter, carrier, and receiver of symbolic information.

the evolution history of the particular symbolic information system. For example, the specific shapes of the black paint that form the printed letters "P," "h," and so on in Figure 8.9 are the product of the long and winding evolutionary road along which written languages in the Mediterranean area were successively perfected; their structural information still preserves some clues on the early beginnings.

Any practical realization of a clock in the form of a periodic process such as a pendulum is subject to phase diffusion as a result of the symmetry of time with respect to translation; different points in time are physically equivalent. Similarly, coding systems are subject to random drift as a result of the coding symmetry, as is well known for any spoken human language. On the other hand, that drift requires simultaneous coevolution of transmitters and receivers that operate the code. With the increasing complexity of the code interpreters, and perhaps their distribution over many independent instances such as biological individuals, freezing mechanisms evolve that slow down the drift of the code to a tolerable rate.

We typically observe three evolution stages of such a transition process. In the initial phase, the original structure is only slowly variable (frozen) to not degrade the system's functionality. Successively, the affected structures are reduced to some "caricatures" of themselves that represent the minimum complexity necessary to maintain the function (Klix, 1980). Irrelevant modes or partial structures are no longer subject to restrictions or restoring forces, and related fluctuations may increase significantly. At the transition point, the caricatures have turned into mere symbols that may be modified arbitrarily (coding invariance) and permit divergent, macroscopic fluctuations. As a result, the kind and pool of symbols can quickly adjust to new requirements. Soon after, in the third phase, the code gets standardized to maintain intrinsic consistency and compatibility of the new-born information-processing system. Fluctuations are suppressed to a necessary minimum, the code is frozen-in and preserves in its arbitrary form a record of its own evolution history.

Although we are mainly interested in the physical self-organization mechanism of the transition shown in Figure 8.11, it may also be instructive to consider illustrative examples from technology or biology, which elucidate the idea behind that general and abstract schematic.

A well-known example is photography. In that case, the "Cause" in Figure 8.11 means a certain visible structure or process of interest. The chain of processes refers to a camera with lenses where a film is exposed to the incoming light. Next, the film material is subject to a series of chemical and technical manipulations, until as the "Effect" an image produced. The purpose of the procedure is that the image can be looked at, where or when the original scene may be gone, out of reach, or inaccessible for a human eye. All steps from taking the photo to the final hardcopy are "physical" ones, that is, no information processing is involved, and no arbitrary symbols are used in between. Substituting intermediate devices of the process chain by others will produce a different result.

In a few decades of technical progress, this widespread technique of the twentieth century has completely changed. Still, in the beginning is a camera and at the end is an image. The incoming light is now directed to a light-sensitive chip. A microprocessor measures the exposure intensity on a discrete matrix of pixels and stores the result in the form of digital numbers in a memory chip, or transmits them to a remote receiver. The receiver converts back the digital numbers to analog signals in the form of the brightness of a screen point, or ink density of a printed dot. In this process, bits are used as symbols to represent numbers. The kind of binary code used or the way the memory chip is technically constructed does not at all influence the final image. The intermediate information-processing steps have made digital photography by far superior over the obsolete analog film technology, in particular because of the ease of storage, loss-free copying, and fast transmission as well as the flexibility in using different image-producing output devices such as mobile phones, projectors, screens, or printers. The digital tools we used to produce figures for this book are amazingly different from those used in 1979 for the original edition.

The intermediate step of symbolic information processing in between a certain scene of physical reality as the input and a physical image of that scene as the output of a digital camera permits cooperation with arbitrary other compatible information-processing devices. This way, the intermediate symbolic representation of the input can be used for other purposes than displaying the image; for instance, for electronic image processing, pattern recognition, or correlation analysis. Vice versa, the output image can be produced from symbolic information which is not necessarily extracted from the real world but may be a product of model computations or human fantasy, such as the movies "Jurassic Park" or "Avatar." In contrast to classical analog records of observations, fictitious objects visible in those outputs virtually do not need to obey physical laws; they may move faster than light, travel through time, and violate the first and the second laws by their sudden appearance or disappearance. It is an important feature of symbolic information that the meaning it represents is no longer confined to natural laws, contrary to structural information.

In the actual crossover phase of the ritualization transition, various types of digital cameras, chips, interfaces, and data formats appear on the market and undergo an "explosive" development of diversification. At some later point, the technical features have adjusted to the customer's needs and wishes. Some saturation is reached; the fluctuations slow down, the wealth of storage and exchange formats is reduced to just one or a few standards. Those standards are usually only one particular random choice out of a practical and theoretical diversity of other technically equivalent options for, say, thickness or width of the memory chip, its number and specification of electric contacts, and so on. The selected version often merely reflects the particular realization used by the winner of the market competition and is a petrified trace of various occasional decisions and random events that happened in the explosive phase before, such as personal preferences of certain engineers, availability of certain materials, friendly or hostile relations between competing persons or firms, and so on.

The transition from the analog to the digital camera is a ritualization process with respect to the self-organization of the human society and in particular its scientific and technological culture. It is by no means a self-organized process of the camera on its own in the sense that, say, under certain physical circumstances the substance that exists in the form of a chemical film material might actually undergo a miraculous "phase transition" to mutate into a digital chip. Most ritualization transitions occur in the context of life, that is, in the presence of preformed information processing systems. This was radically different at the origin of life when no such context existed yet. In that case, the ritualization event in fact required a stepwise conversion of an analog, physical–chemical process into a digital, symbolic, biological one by self-organization of symbolic information processing without the assistance of any external "intelligent designer."

By our definition, ritualization occurs only in the context of life. Hence, the simplest physical example we can imagine for a ritualization process is a system that starts as a physical and ends as a biological one; in other words, this example is the emergence of life. It is of course a tremendous exaggeration to call such a process "simple." Other, better known, and more instructive examples can more easily elucidate the basic phenomena accompanying ritualization, but they inevitably come from biological, social, or technological evolution, such as the digital camera. Nevertheless, the interest in this transition is focused here on the self-organization of information, on the way how a physical system can be enabled to originate symbols, and the related symbol-processing machinery out of "ordinary" prebiological roots.

In the following sections, we present details of fundamental ritualization transitions from the origin of life to the development of symbolic money as an information system on human property. So different the emergence processes are in the cases of the genetic code, morphogenesis, neuronal networks, spoken human language, private property, written language and money, they are described here in terms of simple models and scenarios that illustrate the common features characteristic for the ritualization transition.

8.6
Genetic Code

We will hardly ever know how exactly life began. It may have emerged under the conditions that were rather different from those in today's typical scientific laboratories, it may have required vast areas or volumes, and it probably took millions of years – much too long for a human researcher to watch it in a simulation tank (Chapter 9). It is unlikely that any fossil traces of the prebiological macromolecules may be discovered in ancient rocks to witness the earliest history. Thus, model scenarios for the succession of physicochemical steps on the way to the first living entities can only rely on reasonable assumptions rather than on well-defined initial conditions.

It is plausible that in the beginning randomly synthesized complex molecules formed chemical networks that included catalytic feedback loops and formed precellular spatial units such as membrane-coated droplets. On the other side of the threshold, there are bacteria-like primitive cells that possess a chain molecule such as RNA or DNA that consists of base triplets that symbolically represent certain amino acids. The mapping of the set of triplets onto the set of amino acids, commonly referred to as the genetic code, is physically realized by transfer-RNA (tRNA) molecules that possess chemical affinity to the triplet on the one end and complementary binding structures to the amino acid on the other end of the molecule. The energetic coupling between a certain triplet and a certain binding site for an amino acid is insignificantly weak; any triplet could in principle be combined with any binding structure to form an equally well-functioning tRNA molecule. This neutral stability with respect to permutations of the genetic code permits critical fluctuations in the absence of external restoring forces. It this sense, the genetic code is arbitrary (Crick, 1968), thereby expressing the coding invariance of symbolic information. Because of this arbitrariness, rather than being entirely random, the genetic code is a fossil record of the earliest life, even though imperfectly kept.

Also in the human language, the apparent randomness of structures that represent symbolic information, the question whether symbols have some relation to their meaning or constitute arbitrary and purely conventional coinages, has vividly been discussed already since Plato's *Cratylus* (Fitch, 2010). In the ritualization approach, that old philosophical controversy seems to be merely virtual and artificial; symbols gradually emancipate from the physical incarnation of their meanings until they are completely arbitrary. They preserve frozen traces of their origin as long as those are not erased completely by the immanent sluggish drift due to neutrally stable fluctuations.

The symbolic information kept in a cell is represented by the sequence of triplets which form the genetic chain molecule. Differences in the chemical binding energy of particular adjacent pairs of triplets are weak, so that arbitrary messages can be stored without significant energetic preference of one over the other. Coding invariance is physically established this way. Neutral stability of the chain molecule with respect to the substitution of one amino acid by another permits critical fluctuations (mutations) unless external restoring forces (repair enzymes and redundant, error-correcting code) may prevent that.

The genetic chain molecule is a written recipe for survival. The transmitter of this message is the mature cell that by copying the information from its memory tells the daughter cell how to survive. The receiver of the message is the intracellular apparatus that, by means of the tRNA code table, translates the written text into a chain of amino acids. That assembled chain folds into a catalytic macromolecule that realizes the purpose of the genetic message: it joins the family of proteins that cooperatively make the cell grow and multiply by means of its phenotypic traits.

As indicated in Figures 8.11 and 8.12, the ritualization process is the transition from a catalytic network to a reproduction system controlled by symbolic information in the course of molecular evolution. From the recent system, we may infer plausible precursor states. The current system with 20 amino acids was probably simpler before and may have consisted, in the extreme case, of only two different elements. The present triplets may have been longer sequences originally, corresponding to a larger piece of the tRNA molecule. That larger section may have been energetically correlated with the amino acid it is coding for. Then, coding for the entire precursory tRNA itself, rather than for the amino acid it represents, may have been an earlier way of reading the genetic message, where the associated tRNA acted as a catalyst and did not yet exhibit the independence of code and function. Still before, the RNA strain

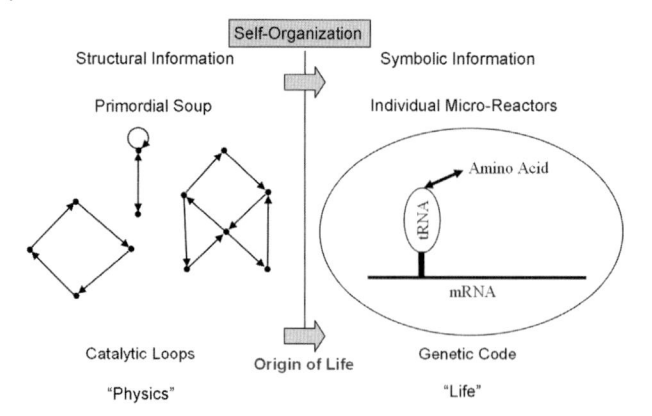

Figure 8.12 The very first ritualization transition was the emergence of the symbolic genetic code out of a random catalytic network, that is, the origin of life.

and its complement may have been the template and its read-out, the message and its meaning in one and the same, perhaps paired molecule. This would then be the stage of a catalytic network without symbolic information, prior to the ritualization transition (Chapter 9). Thus, we can imagine a plausible stepwise process of gradual changes that cross over from the world of physical chemistry to the world of biology. This process was certainly long and complicated in reality, but at least in principle, its various stages can be tested experimentally. From the viewpoint of physical self-organization, the critical point is where the formally inseparable aspects of 1D sequence and 3D structure of the catalyst differentiate into code on the one hand and function on the other, and eventually lose their physical correlation completely in favor of an arbitrary mutual assignment. At that point, fluctuations diverge, a new symmetry appears, and a kinetic phase transition occurs which is considered here to be the emergence of life.

As long as the genetic message is conveyed only from the parent to the siblings, along with molecular instructions how to read it, there is no selection pressure to freeze the code, rather, its modification and differentiation may have offered selective advantages. This situation changed qualitatively as soon as organisms started to exchange records of their individual experience in the form of sexual reproduction (Section 8.7). The benefit from this communication is that the flexibility resulting from the heterogeneous gene pool of a species to survive under unpredictably changed environmental conditions could only be exploited if the symbolic information provided by the male parent was readable by the apparatus inherited from the female parent. This condition implied a strong selective pressure to freeze the genetic code at latest when the eukaryotes appeared on the stage of life, about 2 Gyr ago (see Table 3.1). In contrast to this simplified reasoning, the widespread universality of the genetic code of recent prokaryotes suggests that the freezing must have happened already at an earlier time, whereas the eukaryotes introduced additional coding rules that are still insufficiently understood.

Beginning with Crick's (1966) wobble hypothesis, many attempts were made in the scientific literature to reconstruct the history of the genetic code from its structural information, that is, from its recent redundancy and physicochemical properties (Crick, 1968; Jiménez-Montaño, de la Mora-Basanez, and Pöschel, 1994, 1996; Jiménez-Montaño and He, 2009; Tlusty, 2010). As an illustration, we provide here a rather simplified, schematic example for the extraction of a code history from its structural information (Ebeling and Feistel, 1982), being well aware that the true history was certainly quite different and much more complicated.

The code is redundant; only 20 amino acids, such as Phenylalanine (Phe) or Leucine (Leu), are assigned to 61 RNA codons that consist of the nucleotides uracil (U), cytosine (C), adenine (A), and guanine (G) (Table 8.1). On average, an amino acid is represented by three different codons. It is immediately evident that the assignment of multiple codons to certain amino acids is neither completely random nor perfectly regular. In most cases, the third nucleotide in the triplet is only incompletely or not at all recognized. In the sense of Crick's hypothesis, the table suggests that reading the third position may be the result of a late and unfinished differentiation process, and that in an earlier system that place may have served as an irrelevant placeholder, a "comma" between two codons. At that stage, no more than 16 amino acids could be coded for, each by 4 codons.

Table 8.1 Genetic code table (Pauling and Pauling, 1975), where U, C, A, and G are nucleotides of the RNA chain, and Phe, ..., Gly abbreviate the amino acids associated with the respective triplet(s).

		\multicolumn{8}{c}{Second base}	
		U		C		A		G	
1st base	U	UUU	Phe	UCU	Ser	UAU	Tyr	UGU	Cys	U
		UUC		UCC		UAC		UGC		C
		UUA	Leu	UCA		UAA	*a)	UGA	*	A
		UUG		UCG		UAG		UGG	Try	G
	C	CUU		CCU	Pro	CAU	His	CGU	Arg	U
		CUC		CCC		CAC		CGC		C
		CUA		CCA		CAA	Gln	CGA		A
		CUG		CCG		CAG		CGG		G
	A	AUU	Ile	ACU	Thr	AAU	Asn	AGU	Ser	U
		AUC		ACC		AAC		AGC		C
		AUA		ACA		AAA	Lys	AGA	Arg	A
		AUG	Met	ACG		AAG		AGG		G
	G	GUU	Val	GCU	Ala	GAU	Asp	GGU	Gly	U
		GUC		GCC		GAC		GGC		C
		GUA		GCA		GAA	Glu	GGA		A
		GUG		GCG		GAG		GGG		G

a) The asterisk * represents the termination codon. Amino acids printed in *italic* possess a polar side chain (Vasilenko, 1978; Hausman and Cooper, 2004).

Behind this apparently self-evident conclusion, the plausible assumptions silently made are that (i) reading of codons was less accurate at earlier stages of the evolution, (ii) the pool of used amino acids was smaller at earlier times, and (iii) the set of codons associated with an amino acid was successively shrunken rather than extended with other codons. In order to have a criterion to decide which of the amino acids were more likely in the beginning, we may add as another empirical rule that (iv) on differentiation of an amino acid, the newly added one was similar but more complex. As a convenient measure of simplicity, one may use, for example, the concentration of a certain amino acid in inorganic synthesis experiments (Miller, 1953; Parker et al., 2011). For example, glycine (Gly), alanine (Ala), and serine (Ser) appeared with the highest abundances among Oró's (1965) reaction products. Alternatively, as a simple complexity measure we may just count the number of carbon (C), oxygen (O), nitrogen (N), and sulfur (S) atoms in the molecule. Chemists surely know less trivial measures for the complexity or rarity of organic compounds.

Under such formal construction rules, the earliest nontrivial code table consistent with the present one is the binary code in which Gly belongs to the codons XRX and Ala to XYX, Table 8.2. Here, $X = \{U,C,A,G\}$ represents any of the nucleotides, $R = \{A, G\}$ a purine base and $Y = \{U,C\}$ a pyrimidine base.

A symmetric successor could be the quadruple code as shown in Table 8.3.

Starting from the quadruple code and applying formal construction rules, it is possible to successively introduce new amino acids in such a way that eventually the recent code appears, Table 8.1.

Table 8.2 Formally constructed initial binary-code table[a].

	U		C		A		G		
U	UUU	Ala	UCU	Ala	UAU	Gly	UGU	Gly	U
	UUC		UCC		UAC		UGC		C
	UUA		UCA		UAA		UGA		A
	UUG		UCG		UAG		UGG		G
C	CUU		CCU		CAU		CGU		U
	CUC		CCC		CAC		CGC		C
	CUA		CCA		CAA		CGA		A
	CUG		CCG		CAG		CGG		G
A	AUU		ACU		AAU		AGU		U
	AUC		ACC		AAC		AGC		C
	AUA		ACA		AAA		AGA		A
	AUG		ACG		AAG		AGG		G
G	GUU		GCU		GAU		GGU		U
	GUC		GCC		GAC		GGC		C
	GUA		GCA		GAA		GGA		A
	GUG		GCG		GAG		GGG		G

a) Only the purine–pyrimidine property of the central nucleotide is recognized.

Table 8.3 Formally constructed early quadruple-code table[a].

	U		C		A		G		
U	UUU	Ser	UCU	Ser	UAU	Cys	UGU	Cys	U
	UUC		UCC		UAC		UGC		C
	UUA		UCA		UAA		UGA		A
	UUG		UCG		UAG		UGG		G
C	CUU		CCU		CAU		CGU		U
	CUC		CCC		CAC		CGC		C
	CUA		CCA		CAA		CGA		A
	CUG		CCG		CAG		CGG		G
A	AUU	Ala	ACU	Ala	AAU	Gly	AGU	Gly	U
	AUC		ACC		AAC		AGC		C
	AUA		ACA		AAA		AGA		A
	AUG		ACG		AAG		AGG		G
G	GUU		GCU		GAU		GGU		U
	GUC		GCC		GAC		GGC		C
	GUA		GCA		GAA		GGA		A
	GUG		GCG		GAG		GGG		G

a) Only the purine–pyrimidine properties of the left and the central nucleotide are recognized.

This example should demonstrate that, at least in principle, a reasonable evolution history of the genetic code can be reconstructed from its symbol structure when suitable external, independent information, such as on chemical properties of amino acids, is additionally available. A similar statistical procedure was used by Eigen and Winkler-Oswatitsch (1981) to determine the common tRNA ancestor, where apparently the first triplet base, RXX versus YXX, was distinguished initially, and the central triplet base latest, in contrast to Table 8.2. A frameshift by one base between the two systems could be an explanation. More sophisticated rules may consider smoothness and other requirements (Rechenberg, 1973; Feistel and Ebeling, 1982; Tlusty, 2010), for instance, that the differentiation of an amino acid should result in a chemically similar pair. Otherwise, substitution will likely have fatal consequences for the owner of the new code version. In Table 8.1, the rather regular distribution of, for example, the polarity property indicates that certain smoothness rules must in fact have been in force in the real expansion phase of the genetic code.

Unfortunately, formal reconstruction procedures such as the one applied above are not precise and detailed enough to reveal the true evolutionary tipping point, the ritualization transition act, at which structures were transformed into symbols representing the original structures. A look at the other side of the transition may be helpful, at a time when no symbols in the form of codons existed yet, and the self-reproduction processes was just a catalytic network that probably included short polynucleotices and polypeptides. The selective advantage of using sequences of symbols may have been the capability to control the assembly of bigger molecules and the flexibility in modifying those. If so, a primitive mechanism for the sequential concatenation of certain building blocks such as amino acids should have existed already before the transition (Ebeling and Feistel, 1982). As a primitive model, we may imagine that randomly synthesized simple amino acids condensed along an RNA strand in a sequence that was related to the 3D structure of the RNA. That RNA in turn may have been a copy of another chain molecule such as a complementary RNA or DNA. The latter, say, pre-mRNA, initially played the role of a storage of structural information that catalyzed via the intermediate pre-tRNA a functioning simple protein. In the following development, the sections of the pre-mRNA that corresponded to certain amino acids were subsequently reduced to shorter identifiers. The pre-tRNA had to maintain the originally identical two capabilities of being associated with those identifiers as well as of temporarily binding a certain amino acid. Such a differentiation may happen in the form of an occasional chain doubling and a subsequently different development of the twin parts. This process eventually resulted in a complete decorrelation of the short 1D-identifyer structure from the longer 3D structure of the site for binding the amino acid. The 1D structure represents only a minimized residual "caricature" of the second, the 3D structure. At that point, the pre-mRNA had turned into a symbolic information storage, and the ritualization act was completed. Later, the new coding symmetry provided the freedom of a flexible differentiation of the code as discussed before.

The neutral stability of biochemical chain molecules with respect to changes of the sequence of their elements, that is, the absence of physical restoring forces that would always quickly drive the molecule back toward a unique "thermodynamic equilibrium sequence," is an essential precondition for evolution because it permits the storage of genetic information over millions of years in a liquid, chemically very active medium (Fekry, Tipton, and Gates, 2011). For a molecule to act as an information storage, a reader with periphery must exist that can transform an arbitrary symbolic message into an associated controlled catalytic network. This machinery emerged through the first ritualization transition. The evolutionary benefit gained from this primal information processing system must have outweighed by far its costs in the form of the system's chemical reaction complexity and related vulnerability. In today's retrospect, this spectacular transition appears as if it was intentional or purposeful, but it was a mere act of molecular physico-chemical self-organization under the given conditions. There was nothing and nobody who at that time could have foreseen the amazing and fascinating world of life it kicked off. In Chapter 9, we shall take a closer look at the pre-ritualization stage of the genetic code.

8.7
Sexual Recombination

I can see no good reason to doubt that female birds, by selecting during thousands of generations the most melodious or beautiful males, according to their standard of beauty, might produce a marked effect

Charles Darwin wrote in his "Origin of Species."

Over the decades, many publications appeared on the role of sex in evolution (Darwin, 1911; Huxley, 1914; East, 1918; Feistel, 1977c, 1990; Smith, 1978; Butterfield, 2000; Morran, Parmenter, and Phillips, 2009; Zimmer, 2009a; Becks and Agrawal, 2010) (see also Section 6.9). From the perspective of ritualization and symbolic information considered here, two aspects of sex appear to be most interesting (Feistel, 1990):

i) Sexual reproduction permits *lateral exchange* of genetic information between the individuals of a species. In contrast to asexual reproduction where information is exclusively passed vertically from one parent to the offspring, selective evaluation of diploid progenies exploits the result of an information mixture from numerous phylogenetic lineages, experiences, and genetic recipes inherited from a multitude of necessarily successful survivors under alternative external conditions over a long history of ancestors.

Sexual information exchange within a species is apparently similarly essential as lateral communication is among humans. We, the "chatty apes," are spending most of our time with reading books or websites; watching films or TV; chatting on the street, on the phone, or on the Internet; and talking with friends and colleagues at dates, meetings, or dinner parties. Curiosity and exchange of experience is crucial for survival under changing, unpredictable circumstances, and for accelerated accumulation of information, such as in science. In contrast to imitation, language and sex are means to transfer compact, preprocessed, symbolic information through a specifically designed digital interface from individual to individual.

The second aspect is:

ii) Sexual reproduction introduces *beauty* as a new quality of valuation. Beauty, or sexual attractiveness, is arbitrary within certain limits and relies on convention between individuals ("fashion"), that is, on their particular "standard of beauty."

In contrast, selective values that result from the extraction of structural information from the environment embody a certain model of natural laws; rather than being arbitrary and merely conventional, those traits imply, for example, convergent evolution of phenotypes in the same niches (Lukeš, Leander, and Keeling, 2009). Where natural selection is neutral and offers room for randomness, sexual selection may evaluate additional, arbitrary criteria. There is an interesting similarity between beauty and symbolic information on the one hand, and between fitness and structural information on the other hand. Natural shape and behavior of successful mates developed into sexual symbols recognized by the other gender; a typical ethological ritualization process during which symbols emancipated from their physical root. This aspect of the fitness of survivors may be regarded as the "beauty of the winner."

It is not exactly clear how sex originally came into the world, whether it possibly originated from cannibalism or endosymbiosis among early single-cell organisms. Partial horizontal exchange of genes as practiced by recent pathogens may be an example for the success of this strategy already at a rudimentary level (Gutierrez *et al.*, 2005). In the course of later morphogenesis, the previous symmetry between the genders was broken in favor of distinct male and female counterparts (Bellott *et al.*, 2010; Xiong *et al.*, 2010).

8.8
Morphogenesis

In the problems of morphogenesis, of the controlled formativeness of living matter, center the most important concerns of biological science

(Sinnott, 1946).

The theory of how multicellular organisms, that is, metazoa, may have developed from colonies of single cells, that is, of protozoa, was first proposed by Haeckel (Haeckel, 1874; Christen et al., 1991; Richardson and Jeffrey, 2002). Cells are thermodynamically open systems; they must import energy and matter such as nutrients, and export entropy and metabolic waste products. Cells interact mechanically by their volume need and growth. A spherical layer of cells, similar to *Volvox globator*, separates interior from exterior and may produce physicochemical conditions inside the balloon that are different from those outside. If the internal mechanical pressure differs from that of the outside, the spherical symmetry between the cells may be broken by rupture or invagination of the hull. Different local environmental conditions may cause the cells to switch to different dynamical regimes. It is relatively easy to understand qualitatively that an initially homogeneous group of cells may differentiate into spatial regions of different "tissues," separated by smooth separatrices, as a result of the physicochemical interaction between the cells and their nonlinear response to the conditions they locally perceive (Eiraku et al., 2011). In fact, the 2000-cell *Volvox carteri* and a single-celled green alga, *Chlamydomonas reinhardtii*, are genetically not very different (Pennisi, 2010).

Relatively simple mathematical models support this concept. Mathematical approaches, in particular in the form of chemical reaction–diffusion equations such as "Gierer-Meinhardt models" are able to reproduce various shapes and patterns known from animals or plants (Turing, 1952; Wardlaw, 1953; Gierer and Meinhardt, 1972; Malchow and Feistel, 1982; Malchow, Ebeling, and Feistel 1983a; Malchow et al., 1983b; Meinhardt and Klingler, 1987; Allen et al., 2011). Manifolds of critical conditions under which polynomials possess zeros or extrema resemble the topology of morphogenetic structures, as known from the mathematical "catastrophe theory" (Thom, 1975; Poston and Stuart, 1978).

Formally, let $\mathbf{c}$ be an m-dimensional vector of concentrations. In the phase space $\mathbf{R}\{\mathbf{c}\}$, $\dim(\mathbf{R}\{\mathbf{c}\}) = m$, spanned by $\mathbf{c}$, a given cell may take different dynamical states. If the dynamics of the cell is governed by n order parameters $\mathbf{x}$, in the form of a dynamical system

$$\frac{d\mathbf{x}}{dt} = \mathbf{f}(\mathbf{x}, \mathbf{c}) \tag{8.8}$$

then $\mathbf{x}$ is assumed to approach a d-dimensional stable attractor state the phase space $\mathbf{R}\{\mathbf{x}\}$, $\dim(\mathbf{R}\{\mathbf{x}\}) = n$, with $n > d$,

$$\lim_{t \to \infty} \mathbf{x}(t) = \mathbf{x}^0(\mathbf{c}, t) \tag{8.9}$$

depending on the value of the vector of the control parameters, $\mathbf{c}$. The representation of $\mathbf{x}^0(\mathbf{c}, t)$ in the product space $\mathbf{R}\{\mathbf{c}\} \oplus \mathbf{R}\{\mathbf{x}\}$ is commonly termed the bifurcation diagram of the system (8.8), that is, of the cell considered. By the function

$$C(\mathbf{c}) = 0 \tag{8.10}$$

we describe critical values of $\mathbf{c}$ that are defined by qualitatively different attractor states $\mathbf{x}^0(\mathbf{c},t)$ that exist in an infinitesimal vicinity of the critical point c in $\mathbf{R}\{\mathbf{c}\}$. Usually, the function C is a scalar and piecewise smooth function. If so, the condition $C(\mathbf{c}) = 0$ describes $(m-1)$-dimensional manifolds that separate compact regions in $\mathbf{R}\{\mathbf{c}\}$ from each other, each of those characterized by a different attractor state $\mathbf{x}^0(\mathbf{c},t)$ of the cell. Now we consider a given concentration distribution $\mathbf{c}(\mathbf{r})$ in the "real" physical configuration space, $\mathbf{R}_3$. In this form

$$C\{\mathbf{c}(\mathbf{r})\} = 0 \tag{8.11}$$

the condition (8.10) then defines "phase boundaries" which separate compact regions from each other in $\mathbf{R}_3$, each of those characterized by a different attractor state $\mathbf{x}^0(\mathbf{c},t)$, or "thermodynamic phase," or "tissue," of the "pluripotent" cell.

The growth and differentiation of a cell colony can be modeled in $\mathbf{R}_3$ by a coupled dynamical system for the concentration field

$$\frac{\partial}{\partial t}\mathbf{c}(\mathbf{r},t) = \int [A(\mathbf{r},\mathbf{r}')\mathbf{c}(\mathbf{r}',t) - A(\mathbf{r}',\mathbf{r})\mathbf{c}(\mathbf{r},t)]\,d\mathbf{r}' + \mathbf{P}[\mathbf{x}(\mathbf{r})] - \mathbf{Q}[\mathbf{x}(\mathbf{r})]\mathbf{c}(\mathbf{r},t) \tag{8.12}$$

and for the cell configuration (Equation 8.8)

$$\frac{\partial}{\partial t}\mathbf{x}(\mathbf{r},t) + (\mathbf{V}\nabla)\mathbf{x}(\mathbf{r},t) = \mathbf{f}[\mathbf{x}(\mathbf{r},t),\mathbf{c}(\mathbf{r},t)] \tag{8.13}$$

In Equation 8.12, the transport of the solutes $\mathbf{c}$ by advection and diffusion is considered as a fast, quasistationary process in comparison to the temporal development of the cell accumulation. The nonnegative integral kernel $A(\mathbf{r},\mathbf{r}')$ describes the resulting conservative transport of the substances $\mathbf{c}$ from the position $\mathbf{r}'$ to $\mathbf{r}$ and may depend on integral properties of the cell cluster, such as the surface shape, the existence of channels, openings, and so on. For simplicity, we have assumed here that the transfer rates are the same for all species of $\mathbf{c}$, otherwise the scalar function A may be replaced by a (diagonal) matrix $\mathbf{A}$. Truncated Taylor expansion of $A(\mathbf{r},\mathbf{r}')$ with respect to small distances $|\mathbf{r}-\mathbf{r}'|$ recovers a local advection–diffusion equation (Kramers–Moyal expansion, see Section 4.3). The nonnegative vector $\mathbf{P}(\mathbf{x})$ describes the production rate of the species $\mathbf{c}$ by a cell in the state $\mathbf{x}$ at the position $\mathbf{r}$, and similarly, the nonnegative diagonal matrix $\mathbf{Q}(\mathbf{x})$ describes the consumption or degradation of $\mathbf{c}$ by a cell in the state $\mathbf{x}$. The cell state $\mathbf{x}$ itself is controlled by Equation 8.13. The vector $\mathbf{V}(\mathbf{x},\mathbf{c})$ is the cell's local growth rate, that is, the propagation velocity in $\mathbf{R}_3$ of the interfaces between different cell states. Here, the absence of any cells at the position $\mathbf{r}$ is formally modeled by a distinguished cell state such as $\mathbf{x}=0$. The function $\mathbf{f}(\mathbf{x},\mathbf{c})$ describes the biochemical reaction rates of the cell constituents in the local chemical environment $\mathbf{c}$.

For any given cell cluster, $\mathbf{x}(\mathbf{r})$, that is, for any given related functions $\mathbf{P}(\mathbf{r}) \equiv \mathbf{P}[\mathbf{x}(\mathbf{r})]$ and $\mathbf{Q}(\mathbf{r}) \equiv \mathbf{Q}[\mathbf{x}(\mathbf{r})]$, the stationary limit of Equation 8.12 is a linear integral equation for $c(\mathbf{r})$, which has a unique and stable solution under relatively general circumstances. Hence, the time-dependent solution for the concentration field $c(\mathbf{r}, t)$ is a delayed response to the cell configuration, $\mathbf{x}(\mathbf{r}, t)$. The nonlinear functions $\mathbf{V}(\mathbf{x}, \mathbf{c})$ and $\mathbf{f}(\mathbf{x}, \mathbf{c})$ in Equation 8.13 are responsible for growth, bifurcations, and regional differentiation in the cell cluster.

Initially, the substances $\mathbf{c}$ may be nutrients, metabolic products, or external excretions that represent structural information, that is, those substances permit in principle the detection of the presence or absence of other cells or food resources in the vicinity. Natural selection favors the development of tools for the exploitation of this information in the form of some suitable active response behavior (Klix, 1992). If so, those substances simultaneously serve two different functions, being part of the metabolic chains and representing signals for mutual recognition. In particular, after protection from external influences, such as in the interior of a cell cluster, two cells interacting that way may gradually separate the two functions by using two different substances, a signal substance that slightly deviates from the metabolic substance but is similar enough to be still received as the old signal. This step initiates the ritualization transition; a signal emerges that is merely a symbol of the original metabolic process. That symbol can be modified arbitrarily later on without interrupting the metabolic process, and the metabolism may adapt to external changes without interference with the communication.

The possibility to arbitrarily modify and diversify the control substances in the course of the ritualization transition has an important consequence for morphogenesis and the related selective advantages of differentiated cell composites (Feistel, 1990; Ebeling and Feistel, 1994). Organisms need various filigree and sophisticated tools and tissues. To result in such a complicated differentiation patterns by using a small number m of "morphogenes" $\mathbf{c}$, the dynamics $\mathbf{f}(\mathbf{x}, \mathbf{c})$ must be rather sensitive to small changes of $\mathbf{c}$ in order to generate a complex bifurcation net, $C(\mathbf{c}) = 0$ (Equation 8.10). Such high sensitivity will probably imply high error chances with respect to small perturbations; fatal morphogenetic defects may easily occur under only slightly varied external or internal conditions. This lack of structural stability can be avoided by an enlarged number of morphogenetic substances. If $m \gg 3$, the mapping (8.11) of the bifurcation net from the space $\mathbf{R}\{\mathbf{c}\}$ to $\mathbf{R}_3$ increases significantly the number of qualitatively different regions with complicated shapes that are structurally more stable. To conveniently illustrate this, we consider a fictitious case with $m = 3$ mapped onto a 2D "real" world, $\mathbf{R}_2$. Let the bifurcation net in $\mathbf{R}\{\mathbf{c}\}$ consist of six planes corresponding to the faces of hexahedron, similar to cuboid. This net is rather simple and robust; it is given by a function $C(\mathbf{c})$ which vanishes on those planes, such as

$$C(\mathbf{c}) \equiv (c_1 - A_1)(c_1 - A_2)(c_2 - A_3)(c_2 - A_4)(c_3 - A_5)(c_3 - A_6) \tag{8.14}$$

Here, c_i are the components of $\mathbf{c}$ and A_i are some constant numbers. In Figure 8.13, we see the geometrical transect (Equation 8.11), of such a cuboid with a 2D plane, $c_i(x, y) = a_i x + b_i y$, that is, with the fictitious "real" world with the coordinates x, y. Each black line in Figure 8.13, here in the form of a Star of David, corresponds to an

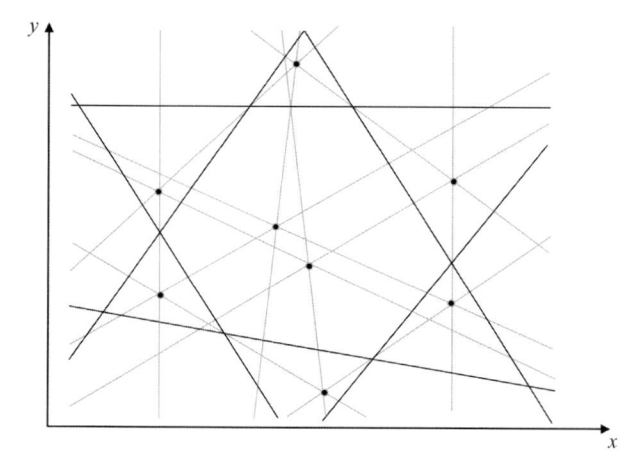

Figure 8.13 2D transect of a simple 3D bifurcation net (Equation 8.14). The resulting "morphogenetic shape" in 2D may look like a Star of David, shown by the bold lines. The eight dots mark the vertices of the 3D hexahedron. The planes associated with the six faces of the distorted cuboid intersect pairwise in 12 edges, shown in gray. A seventh plane, the 2D transect, is identical with the xy-plane of the drawing and cuts the cuboid in the form of the black hexagon. The black lines represent the 2D bifurcation net, $C(c(x, y)) = 0$, that is, the intersections of the xy-plane with the six planes that represent the 3D bifurcation net, $C(c) = 0$.

interface between different cell states. One can easily imagine that the projection $C(c(r)) = 0$ of a high-dimensional bifurcation net into R_3 may result in a large number of different regions separated by interfaces with complicated shapes. This way the ritualization transition and the subsequent diversification of the substances **c** support the development of complex morphogenetic structures. The simple metazoan *Hydra* is estimated to possess about $m = 1000$ regulating peptides (Lohmann and Bosch, 2000). In contrast, no molecule related to morphogenetic control of higher animals has been identified yet experimentally.

Dynamical models such as Equations 8.12 and 8.13 have two important features in common with observed morphogenetic structures. First, deviations from a stable steady state of those equations result in restoring forces, that is, tissues regenerate after small lesions. In particular, the two vector fields **c**(**r**) and **x**(**r**) redundantly represent the same information on the system's spatial structure. If **x**(**r**) is fixed and **c**(**r**) is set to, say, zero everywhere, Equation 8.12 will build up again the original stationary solution for **c**(**r**). Vice versa, if **c**(**r**) is fixed, small perturbations of **x**(**r**) will be damped and the steady state will be retrieved as long as the perturbation remains within the attraction basis. Globally, the system given by Equations 8.12 and 8.13 may possess several different stable solutions, that is, alternative builds of the organism. Perturbations in the gene expression that lead to local instabilities of the morphogenetic structure may serve as a model for tumors or cancer.

Second, if only the symbolic information transmitters, that is, the production functions, **P**(**x**) and **Q**(**x**), in Equation 8.12 are modified quantitatively, the projection of the unaffected bifurcation net $C(c) = 0$ appears geometrically deformed or scaled

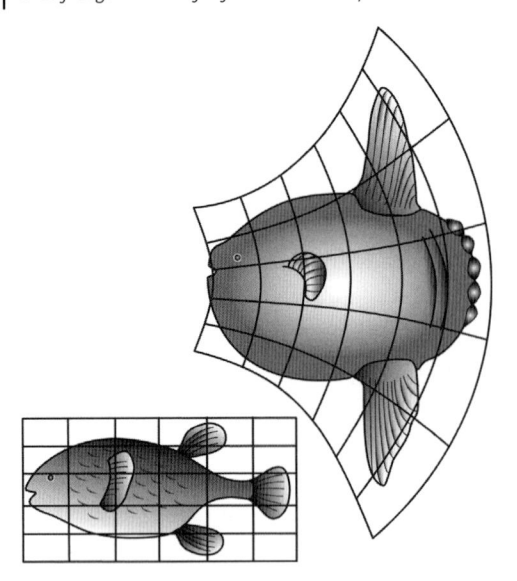

Figure 8.14 Topological invariance under geometrical distortion ("morphing") is frequently observed among biological forms and must be supported in a robust and simple manner by mathematical models for morphogenesis. (adapted from Glaser, 1975).

as $C(\mathbf{c}(\mathbf{r})) = 0$ in $\mathbf{R}_3$. This conservation of morphogenetic topology is an important physical property for individual growth or differentiation between species (Glaser, 1975; Donald, 2008) (see Figure 8.14). In this model, the actual geometrical shape is easily controlled by chemical reaction rates, which is the natural steering knob used by biological systems.

8.9
Neuronal Networks

Recent neurons are extremely specialized cells. In the course of morphogenetic evolution, they must have appeared very early; the freshwater polyp *Hydra* is a primitive metazoon and possesses a nervous system, and parts of that system are the same throughout the animal kingdom (Quach *et al.*, 1991; David and Hager, 1994; Schaller, Hermans-Borgmeyer, and Hoffmeister, 1996; Takahashi *et al.*, 2009a). Marine animals, that is, heterotrophic organisms, must detect and catch their prey in the surrounding water. Olfactory chemoreceptors of parasitic zooplankton may have been the root for the development of neurons used by predators. Recent *Hydra* activates its tentacles when it detects glutathione molecules that are usually produced by living water flea *Daphnia* (Loomis, 1955). Very first sensoric cells probably had the two functions of signal perception and communication combined in a joint form to trigger the activity of an affector cell. It is plausible that at a later stage, the functions could differentiate, that is, that special signal propagation cells developed from the original receptor cell. In fact, several arguments hint on receptor cells of the ectoderm as the phylogenetic origin of recent neurons (Biesold and Matthies, 1977). Likely, olfactory sensation was the evolutionary seed of many complex neural functions of higher organisms, such as of social insects (Farris and Schulmeister, 2010).

The emergence of neurons as a cell type specialized for internal communication was a ritualization transition similar to the schematic of Figure 8.11; it can be understood as just a special case of the general morphogenetic transition to various specialized cell types described in Section 8.8. When a receptor cell is separated from the contact to the external world and can further on recognize only molecules produced by other cells of the same organism, the particular chemical species used for that signaling purpose may gradually and arbitrarily modify as long as transmitter and receptor continue to understand each other by coevolution (Hoyle, 2010). Such transmitter substances responsible for neuron–neuron communication could successively be reduced to symbols of the originally produced and received substances and gained the freedom of coding invariance together with the option for diversification. As a result, the pool of substances actually chosen by recent species reflects the historical pathway of the evolution of the nervous system. The authors are not aware, though, of any attempts made to construct evolutionary trees derived from the biochemical similarity between known transmitter substances.

While the evolution of neurotransmitters and that of morphogenetic substances were apparently not very different processes from the ritualization viewpoint, in comparison to other cell types, the development of neurons may have resulted in the emergence of a second novel, neutrally stable degree of freedom that permitted the independent differentiation of single cells (Feistel, 1990; Ebeling and Feistel, 1994). Neurons are morphogenetically uniform but in their individual shape and functionality much more differentiated than cells of any other tissues (Eccles, 1976; Biesold and Matthies, 1977). The currently commonly accepted "neuron theory" was developed by Ramón y Cajal in 1888 (Eccles, 1976; López-Muñoz, Boya, and Alamo, 2006) and by Waldeyer-Hartz (1891); it assumes that single-nerve cells live their individual lives and develop individual shapes and functionalities, in contrast to cells of other tissues and organs. The specialization for information processing may have liberated neurons from various physical and chemical duties that all other tissues do for keeping the organism alive, such as importing or exporting certain chemicals or performing mechanical work. The development of receptor cells to nerve cells was a ritualization transition which permitted their individual diversification. Neurons that are fed and live in a protected stable environment can redirect large parts of the gene-expression machinery to the task of processing symbolic information, in the form of recognizing certain input substances and responding with other output substances at distant locations. In other words, former use activities, such as to maintain metabolism and energy supply, could be transformed into signal activities with symbolic substances that originate from metabolic substrates and products.

Neurotransmitter substances act as symbolic signals that developed from chemicals involved in metabolic and morphogenetic functions, such as adenosine triphosphate (ATP), the universal power supply of organisms, which appears as a transmitter in basic functions of the autonomic nervous system (Khakh, 2001; Burnstock, 2009;

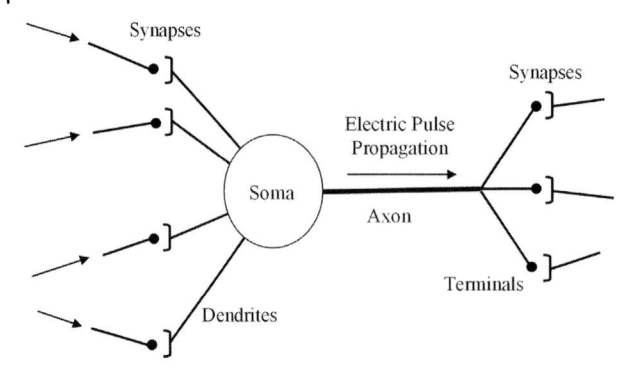

Figure 8.15 Simplified schematic of a neuron. Dendrites receive external transmitter substances that influence the local mRNA transcription. The resulting amplified and filtered chemical signals from the different dendrites are combined in the cell body, the soma, by accumulation and superposition. If a critical threshold is exceeded, an electric excitation pulse for fast, loss-free, long-distance transmission is generated. At the axon terminals, transmitter substances are released at a rate controlled by the pulse frequency.

Teschemacher and Johnson, 2009). As soon as the nature of such substances was no longer dictated by the specifics of metabolic functionality and efficiency, diversification of those substances in coevolution with their genetic expression apparatus could take place (Li, Lee, and Black, 2007). Since mechanical intracellular transport of newly assembled molecules is much too slow over long distances, the synthesis of signal substances had to be decentralized and established at the locations where they are required, in the dendrites and in the axon (Eberwine et al., 2001; Shan et al., 2003; Jung and Holt, 2010) (see the schematic in Figure 8.15). In this picture, electric pulses propagating rapidly along the axon are binary-coded instructions for the synthesis and secretion of neurotransmitters, similar to the pulse-density modulation (PDM) technology. For its good noise resistance, PDM is used today for Super Audio CDs. Like PDM, nerve pulses are probably just a fast and low-noise, long-distance transmission method that has nothing to do with the actual information contents. The question which particular neuron is firing and what transmitter substance it passes to a successor neuron may be more relevant for the functioning principle of the network. Most neurons in the nervous system appear to contain and release more than one chemical acting as a neurotransmitter or neuromodulator (Trudeau and Gutiérrez, 2007; Ko et al., 2011). The set of transmitters released by the axon terminals of a particular neuron is a characteristic and is sometimes regarded as the neuron's individual "transmitter phenotype."

During ontogenetic development and later in adulthood, the neurotransmitter phenotype of neurons can be highly plastic, that is, it can adjust dynamically to changing conditions (Trudeau and Gutiérrez, 2007). The production of peptides from mRNA localized in the axon and in the dendrites seems to control their growth or retreat in dependence on the neuronal activity and the chemical composition of their surrounding. Functioning neuronal gene expression is crucial to learning (Peleg et al., 2010). Thus, the mechanisms that are assumed to form the shape and function of organs in morphogenesis seem to play a similar role for separate branches of single neurons (Tran et al., 2009; Xu et al., 2009).

In his book, Eccles (1976) expressed his conviction that it will still take centuries to reveal the function of the human brain, the most complex structure in the universe we are aware of. Even of much simpler nervous systems such as those of bees or beetles, we know only very little (Sarma et al., 2010; Farris and Schulmeister, 2010). The human brain has estimated numbers of $n = 10^{11}$ neurons and $k = 10^{14}$ synapses (Williams and Herrup, 1988). From the viewpoint of graph theory, the coordination number $c = k/(n-1) \approx 1000$ counts as "strongly coordinated" (Feistel, 1979) (see Section 4.4), even though this value is still far below that of a compact graph with $c = n$. Thus, in a statistical sense, the brain could be modeled in some "thermodynamic limit" $n \to \infty$, $k \to \infty$, taken at constant c. With respect to the applicability of statistical mean-field approaches to complex networks such as nervous systems, the skepticism of neurobiologists is plausible. On the other hand, already the sheer number of neurons and synapses excludes that their blueprints may be precoded in detail in the genetic information and just implemented following that inherited plan, as we believe morphogenesis is functioning. Thus, the brain must develop its complexity from a rather homogenous or rudimentary initial state subject to a few boundary conditions and to relatively general parental genetic influences (Gregg et al., 2010). There must be so-far unknown fundamental construction principles similar to the "fire-and-wire" rule that guide the self-organization of the brain in a stable and robust manner. The amazing plasticity of the nervous system has often been described (Sagan, 1978; Donald, 2008; Lomber, Meredith, and Kral, 2010; Neumann et al., 2011) but is poorly understood. An extreme case is the story of "Sassezki," a Russian student and soldier, who was severely wounded in World War II and was able to report on 3000 pages of his diary the way he gradually recovered from his massive losses of mental functions, including speech and memory (Luria, 1991).

Even though we, similar to the classifying biologists before Darwin, do not know yet the elementary rules how the brain is functioning, learning, and repairing itself, also in this case models for mutation and selection between individual neurons appear as a promising approach to solve the conundrum (Shors et al., 2001). Active neurons may multiply and differentiate, passive ones may die. Ramifications may randomly appear, and disappear if they are not successful. But, how can a specific neuron know whether it is doing fine or bad when we laboriously learn to write, to dance, or to drive a car? The successfully functioning neuronal network is evidently a stable attractor state with respect to "plastic" fluctuations of single neurons, their dendrites, and axons, or even larger parts of the structure. It remains a scientific challenge to discover how the system's integral evaluation as more or less "successful" is continuously measured and fed back in the form of local physico-chemical conditions that either reward or punish each of the nerve cells individually.

There are two complementary formal models for the propagation of excitation in a hard-wired neuronal network. The first considers synapses as elements and neurons as binary irreflexive relations between them. If i is a number associated with a synapse in the network, and j is the number associated with a transmitter substance, then let $s_{ij}(t)$ be the concentration of j at the synapse i at the time t. The balance equation reads

$$\frac{ds_{ij}}{dt} = \sum_k \int_0^\infty f_{ijk}(\mathbf{s}(t-\tau), \tau) \, d\tau - \lambda_{ij} s_{ij} \qquad (8.15)$$

where $f_{ijk}(\mathbf{s}(t-\tau), \tau)$ is the secretion rate after delay τ of the transmitter j by the neuron that connects the synapse k with the synapse i, and λ is some decay rate. Since there is only one connection from the synapse k to the synapse i, the matrix f_{ijk} is sparsely populated similar to a permutation matrix. Each neuron is represented in this "multipole" model by a subset of matrix elements $\{f_{ijk}\}$ that contains all connections from the set of input synapses to the set of output synapses of that particular neuron (Figure 8.15). These matrix elements are coupled by their joint excitation dynamics.

In the more convenient complementary model, neurons represent the elements and synapses represent the binary relations between them. Let the neuron i at the time t has a firing rate of $x_i(t)$. The neuron i receives input from the neuron j by means of a synapse, that is,

$$x_i(t) = \sum_j \int_0^\infty g_{ij}(x_j(t-\tau)) \, d\tau \qquad (8.16)$$

The transmitter released by neuron j and read and evaluated by neuron i is implicitly described by the matrix element g_{ij}. The matrix $\mathbf{g}$ can be considered as an adjacency matrix of a directed graph (see Figure 8.16). Nodes of the graph without predecessors are effector cells (receptors), those without successors are affector cells, such as secretory or muscle cells.

For simplicity, we consider a cycle-free network $\mathbf{g}$. Seen from the input, any internal neuron i is the top of a "pyramid" of predecessors down to a certain subset of effector cells (see Figure 8.16). Through several layers of preprocessing, specific excitation

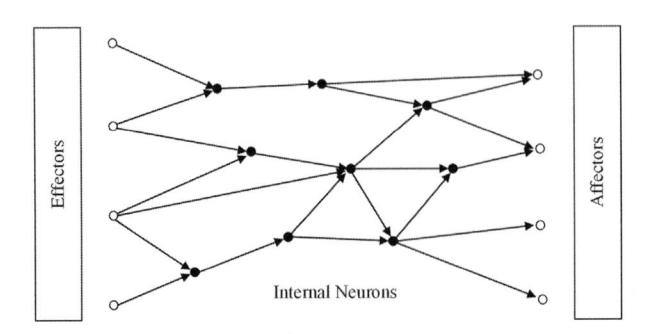

Figure 8.16 Graph representation of a simple cycle-free neuronal network. Nodes represent neurons, directed edges their connections via synapses (Figure 8.15). Nodes of the graph without predecessors are effectors, such as chemical receptors, those without successors are affectors, such as muscle cells; the remaining ones are internal neurons.

patterns of the effectors will excite the particular neuron i. In other words, the excitation state of that neuron is a symbol of a special external situation; it detects a specific structure, object, or process (Jia et al., 2010; Cerf et al., 2010). The longer the distance from the effector layer is, the more effector cells are included in the evaluation of the input pattern, and the more general the situation is that the excited neuron i is symbolizing.

At the same time, seen from the output, the neuron i is also the top of a "pyramid" of successors up to a certain subset of affector cells (see Figure 8.16). Through several layers of postprocessing, specific response patterns will be triggered by an excited neuron i. In other words, the excitation state of that neuron is a symbol of a coordinated response activity, or a subroutine; it starts or stops a certain combined action of muscle cells, or glands. The longer the distance to the affector layer is, the more affector cells are included in the execution of the output pattern, and the more general or "abstract" the activity is that the excited neuron i is symbolizing.

As a result of the structure shown in Figure 8.16, any single effector cell has partial influence on all the affector cells, and vice versa, any single affector cell responds to the combined information provided by all of the receptors. If we think of a simple linear model of this signal transfer (Equation 8.16) and perform a thermodynamic limit to continuous effector and affector fields, $x(\mathbf{r})$ and $y(\mathbf{r})$, respectively, then the mapping of the input to the output pattern takes the form of an integral transform

$$y(\mathbf{r}) = \int G(\mathbf{r}, \mathbf{r}') x(\mathbf{r}') \, d\mathbf{r}' \tag{8.17}$$

where the kernel G represents the neuronal network, or a learning matrix (Kämmerer, 1974). Mathematically, properties of integral transforms are very well known, in particular those of Fourier and Laplace transforms. Physically, a hologram is a Fourier transform of some spatial structure. We believe that this general qualitative property of networks similar to Figure 8.16 is a reasonable explanation for the perplexing similarities found between holograms and nervous systems (Iwanow, 1973; Ostrowski, 1973, 1987).

In the case of a network, the symbols representing certain recognized external situations or controlled response activities do not form a linear sequence as in the case of genetic or language information. The coding invariance of this symbolic information means that there is no physical law that dictates by which means or chemicals a symbol is represented in the network. This leaves room for the diversification of individual neurons. Corresponding to the general rules of ritualization, the physicochemical structure of the neuron, in particular its transmitter phenotype, holds information on the evolutionary history of the network.

We can easily imagine a scenario for the beginning of that history. Prokaryote cells can detect light and can move a flagellum. When several of such cells are aggregated in a colony, the light may need to be detected at the one side and the flagella to work on the other side of the cell cluster. It is of selective advantage when the cells divide their labor into light detection and propulsion, and that the light detector develops a structure to transmit its signal over some distance, fast and loss-free. Next step is that there may be situations when "light on" must be translated into "motor off," that is, a logical NOT gate must be inserted in between effector and affector cell. The gate must be enabled or disabled, depending on a separate input signal or state variable (Gong et al., 2010), similar to the typical function of a transistor or of an allosteric enzyme. That gate cell was probably derived from the receptor cell by differentiation. As soon as the gate cell was reduced to a logical operator between input and output signals, that is, to an internal neuron, the ritualization transition to a symbolic cell was complete, as shown in Figure 8.11.

8.10
Spoken Language

Speech is the default linguistic signaling mode for all human cultures (Fitch, 2010). *Language is what makes us human. Maybe it's the only thing that makes us human*
(Bickerton, 2009).

Our own ease of understanding speech belies its underlying complexity
(Lewicki, 2010).

The evolution of spoken language required the combination of three very different basic things, the ability of speech (talking parrots have it, Pepperberg, 2002; Berger, 2008), the ability to produce and receive symbolic sequences (deaf people use gestures rather than sound) (Fitch, 2010), as well as a selective advantage for those who developed and used those techniques (Bickerton, 2009). In addition, the ability to hear sound in the appropriate frequency range and to decompose it spectrally is also required for communication by speech. It may be assumed that this ability was highly developed already and may have changed only little, if at all, during the evolution of the spoken language.

The first new thing a human baby does after its birth is crying. The English word "stillborn" expresses literally that a silent baby is a dead baby. Usually, the first cry is a signal with high emotional impact for the mother. The baby's crying has a lot of beneficial effects, breathing, clearing mouth and nose, and so on. The fact that it represents a signal activity independent of those effects suggests that crying is a ritualized activity in the classical ethological sense, similar to the sound signals produced by many animals. In the original construction plan of land-living vertebrata, the mouth is used to uptake liquids and solids into the esophagus, and the nose to exchange gases through the trachea. Many animals combined those channels; elephants can drink through the trunk, dogs can pant through the mouth. As a result, feeding tools such as lips, tongue, or jaw could be used to influence the breathing airflow and the noise generated by the turbulence. In a typical ethological process of ritualization, the noise formed as a by-product could be heard by others and developed into actively produced sound, used as signals for various purposes such as the song of birds or the bark of dogs. The crying of human babies was probably the result of a similar development.

When early humanoids began their new way of life in the African savannah, they learned to efficiently walk and run upright on two legs and to use their liberated hands for other purposes than walking. They lost their fur and developed special pigments to protect their naked skin against sunburn (Rogers, Iltis, and Wooding, 2004; Reichholf, 2004; Jablonski, 2006; compare Table 3.4 in Chapter 3). To be carried along by their mothers, some animal babies are riding on the mother's back, others are hanging in the fur under the belly or are sitting in a kangaroo bag. No human baby of a naked, sweaty, upright-walking mother is able to do that; it must be carried in its mother's arms. We can imagine various situations in which a mother needs both her hands to do something else and must put her baby away. If the mother is out of sight, the best way for a hungry baby to get fed is to produce a loud acoustic signal. In the open wilderness, a crying helpless infant is an easy prey; the crying behavior could only develop in a protected social environment. Hence, it seems likely that babies began crying in a long phase of coevolution with upright bipedality, loss of fur and skin pigmentation, and social cooperation. Those processes apparently commenced 5 Myr ago with dramatic changes in the Mediterranean and the Red Sea, and the resulting climate of North Africa (see Table 3.4). In the same context, in contrast to chimpanzees and other mammals, about 1.8 Myr ago humans developed the ability to deliberately control the airflow of breathing, which is crucial for fluent speech (and for singing and laughter, Provine, 2000; Levithin, 2007; Berger, 2008; Ross, Owren, and Zimmermann 2009). Apparently, babies crying for food or caretaking had selective advantages over their silent brothers and sisters.

From the first days after birth on, babies can suckle and drink. The neuronal and mechanical control of lips, tongue, jaw, and throat is well advanced, in contrast to many other skills. By those tools, babies are able to modulate the airflow of crying and to start babbling. Typical sounds produced by periodically closing the lips or pressing the tongue against the palate are "mamama..." and "dadada..."; the reader may easily try that. Parents tend to repeat and to imitate that sound, and after some time, babies start to repeat and to imitate the parent's sound. If those signals are correlated with convenient or unpleasant situations for the baby, it will learn to indicate or to recognize such situations by the signal, that is, to transmit or to receive a symbolic message. The first oral communication happens between babies and their mothers. It is very plausible that the phylogenetic origin of the spoken language was similar; crying babies could also babble, and developed an information exchange with the mother with respect to the baby's needs. The diversification of available acoustic signals was supported by the development of a complicated bifurcation diagram with various kinetic phase transitions between the dynamical regimes of the physical vocal production system, such that small changes in the airflow may generate different and easily distinguishable sounds (Herzel et al., 1994; Fitch, Neubauer, and Herzel, 2002; Tokuda et al., 2007). The ability to generate sound in a controlled manner is not lost when the child grows older; kids may use it also to communicate with other children, and later among adults. Adults in turn can use the oral signal system, once available physiologically, to coordinate important social activities such as defence against predators or enemies, or hunting in a group. Warning sounds are probably among the earliest nuclei of acoustic communication within a group. We instinctively produce them in cases of sudden hurt ("ouch"), disgust ("ugh"), or surprise ("oops"); these "words" are incompletely ritualized and are not part of the regular vocabulary and grammar (Jackendoff, 2002; Berger, 2008). By adult oral communication, a positive feedback loop is closed since it is that kind of social activity that allows the baby to cry without running a deadly risk. The hypothesis that human speech is phylogenetically closely related to feeding is supported by physiological evidence (MacNeilage, 2008; Fitch, 2010). The cries of human infants show melodic similarities with the mother's language and later with that of the child (Wermke, Leising, and Stellzig-Eisenhauer, 2007; Mampe et al., 2009).

This "infant hypothesis" (Jonas and Jonas, 1975; Berger, 2008; Falk, 2009; Fitch, 2010) for the origin of language and communication is consistent with the way we learn to speak during our childhood (Janson, 2006) but in contrast to the common "adult hypothesis" that assumes that collaboration and division of labor among adults was the primary driving force for the development of communication (Tomasello, 2010).

With high probability, the very first word of the human language was "ma" or "mama" or something very similar. Modern infants often develop their individual "baby language" which is only understood by their own parents. This observation suggests that there is some randomness in the origin of human vocabularies, and that they show coding invariance with the association of given meanings with arbitrary sound patterns. From the age of 3 months on, infants can reproduce vowel sounds demonstrated by adults, and can adjust their vocabulary to that of the parents. Many words in human languages are of onomatopoetic origin, as already noticed by Herder in 1772 (Fitch, 2010), and have later undergone a ritualization process in which they lost their close similarity to the sound they emerged from, and were generalized, differentiated, and modified afterward in various ways (Ebeling and Feistel, 1982). We may guess, for example, that the word "babbling" is imitating the sound of an infant, and that "baby" is just shorthand for "the little babbling one." In the course of evolution, words may differentiate and take different meanings in different contexts. Likely, the Rolling Stones had something else in mind when they sang "Have you seen your mother, baby, standing in the shadow?"

While many science-fiction authors have already at hand automatic translation tools for the communication with people from other planets, the reality is that we do not even know much about the spoken language of our close relatives on earth, the dolphins. It was shown that dolphins can acoustically identify each other by their speech and coordinate social activities such as hunting (Herman, Richards, and Wolz, 1984; Díaz López and Shirai, 2009), that is, they are probably using a symbolic information system. Provided this is true, we may assume that those symbols emerged through a ritualization process. It is a plausible hypothesis that the dolphin speech consists of onomatopoetic elements in the very direct sense, namely, of simulated acoustic caricatures of the echo-sounding patterns caused by the objects or processes they "see" (Sagan and Agel, 1973; Ebeling and Feistel, 1994; Kassewitz and Reid, 2008). Whether those symbols are already in a stage of strong fluctuations and neutral drift, regional dialects may be decided experimentally by communication tests between dolphins from, say, the Black Sea and the Pacific Sea, similar to the dialects in bird songs (Derryberry, 2011). The authors are not aware of any such findings.

As Table 3.4 in Chapter 3 shows, the emergence of spoken language is estimated for the period when *Homo erectus* suffered a genetic bottleneck. In retrospect, any superior mutation may appear as such a bottleneck, i.e., as a local genetic variant from which all successful descendants subsequently derived. Thus, such bottlenecks do not necessarily imply a critical extinction risk for the population, as concluded by some authors (Hawks et al., 2000). For instance, our common maternal ancestor "Eve" was not necessarily the only woman of her time, she was just the only woman whose mitochondrial DNA has survived (Wells, 2002).

8.11
Possession

Human possession is a right of use or disposal; as such, it is part of the human culture and does not apply to other than social systems. Possession of an object is an emergent rather than a physical category. If the ownership of something is changing, the particular object itself may not change at all in the moment when the rights on it are transferred from the old to the new owner. If the deal was orally agreed between them, all what physically changed in the transfer act is the memory content in the brains of those two persons.

Possession originates from ethological claims, pretensions, and drives of animals with respect to their environment. From insects to primates, many animals claim territories, sexual dominance, or food resources. Claims are usually asserted by physical force. Animals of the same species reduce the cost of realizing their claims by ritualized fights, or just by symbols. Birds use acoustic symbols for claims, coral fishes use optical ones, and cats and dogs prefer chemical marks (Lorenz, 1963; Fugère, Ortega, and Krahe, 2011; Thinh et al., 2011). The development of those symbols is a classical ritualization process; skin color, breathing noise, body movements, or the smell of excrements are recognized by other members of the species, the traits are gradually enhanced as a result of its signaling effect on others (Lorenz, 1963; Tembrock, 1977). Ultimately, though, if the symbolic signal is ignored by an invader, physical force remains the final means to regain or lose the claim.

Similar to apes, early humanoids likely conquered or defended the things they claimed by fights between tribes or individuals (Mitani, Watts, and Amsler, 2010), including what is termed murder, theft, rape, assault, or cheating in today's habits. Physical force could be reduced to symbols of ownership in the presence of a jointly respected authority such a chieftain, a king, or a government. Today, human property is commonly specified by means of written text in legal documents; those symbols are generally respected for the threat of force that is executed by jurisdiction and police in the case of unlawful acts. The fact that paying deference to symbolic property always requires a functioning social structure, such as a family, a tribe, or a state, becomes obvious in times of social turmoil, such as wars, revolutions, or just devastating floods or earthquakes.

Possession in its ritualized, symbolic form is a key category in social systems (Engels, 1972). Balance equations of gain, loss, and transfer of private property are fundamental for dynamical models for the self-organization of economy (Feistel and Ebeling, 1976; Feistel, 1977b, 1991; Ebeling, 1978b; Ebeling and Feistel, 1982; Schweitzer, 2002) (see Section 6.11).

8.12
Written Language

The alphabet was the most significant of the boons conferred upon mankind by Phoenicia. It is generally considered the greatest invention ever made by man

(Hitti, 1961)

When we learn a foreign language, we do not distinguish between courses, say, for spoken and for written Italian. The vocabulary is the same, the grammar is the same, we express acoustically what we read in a textbook, and we write down what the teacher is telling us. In the evolution of the human species, as well as in infancy and childhood, things are quite different. The emergence of the oral language was a prerequisite for the development of the human culture, while written language was its product. Despite of this initial causality, by feedback, the culture accelerated the development of the spoken language, and the written language had a substantial impact on the culture (Logan, 1986). Spoken language may have developed 1–2 million years ago, in contrast to the earliest evidence of written language from the Sumerian's cuneiform writing which dates about 6000 years back (Haarmann, 2001; Janson, 2006), when early seafarers spread their cultural and genetic seed across the Mediterranean (Lawler, 2010). For more than 99% of its lifetime, volatile audible language was not accompanied by a durable visible counterpart.

Researchers discussed various different reasons for the origin of writing, in particular, for religious and magical purposes. But the most convincing argument is the existence of private property in a developed state, for registering cattle and other commodities, fixing the tax to be paid to the sovereign, taking notes of oral contracts and agreements, safely carrying such information over large distances, and storing it away from one year to the next (Janson, 2006). Very probably, it was not the intention of the inventors to create a one-to-one mapping between spoken sound snippets and visible structures. Rather, objects such as cows or huts had to be specified and counted. Similar to the onomatopoetic origin of many spoken words, many first "written words" were just pictograms of the objects they represented. Such a method is robust, does not require schools or dictionaries, and can easily be understood by people of different tongues which certainly belonged to large ancient empires. Those pictograms were already symbols of real objects, simplified caricatures, but they were not completely arbitrary.

The arbitrariness of written symbols came with the ritualization transition caused by social evolution, the need for new and more abstract words, the flexibility of written

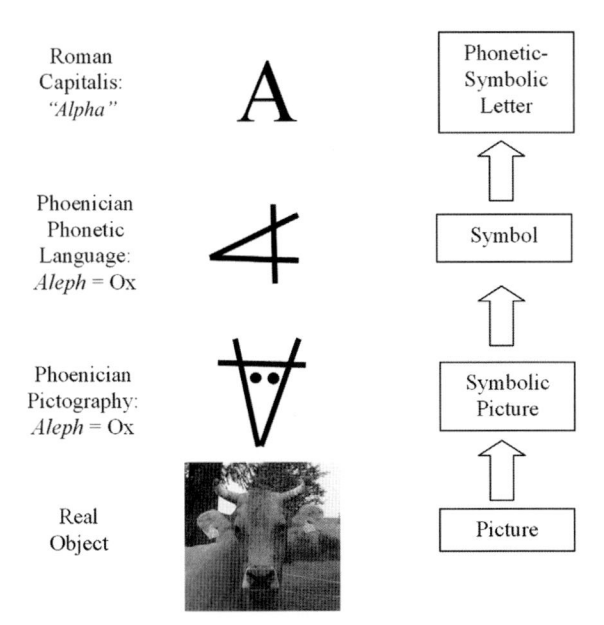

Roman Capitalis: "Alpha"	A		Phonetic-Symbolic Letter
			⇧
Phoenician Phonetic Language: Aleph = Ox	⊿		Symbol
			⇧
Phoenician Pictography: Aleph = Ox	⊽		Symbolic Picture
			⇧
Real Object			Picture

Figure 8.17 Among the most important ritualization transitions in the social evolution of humans was the transformation of real-world pictures into an abstract phonetic alphabet of letters (Klix, 1980; Ebeling and Feistel, 1982; Logan, 1986).

language required to document, say, the personal and family history of an emperor, his glorious victories in wars, or the many titles and crowns carried by him. An instructive example is that of the Phoenician ox (Figure 8.17). It shows how through the use in history the visual representation of an object gradually transformed into an abstract symbol which is now arbitrarily associated with an acoustic pattern. In the text of this book, the original relation between the letter A and an ox has completely disappeared.

Similar to other ritualization transitions, the decoupling of the symbolic picture from the object it represented went along with a neutral drift and diversification of stock of symbols. Hardly any other example shows this process as obviously as the key to the

Table 8.4 First symbols of the Phoenician alphabet (Klix, 1980; Khalaf, 1996)[a].

Sign	Name	Meaning	Greek	Latin
⊿	Aleph	ox	A: alpha	A
⊲	Beth	house	B: beta	B
٦	Gamel	camel	Γ: gamma	C, G
◁	Daleth	door	Δ: delta	D

a) It is obvious that the Latin letters we are using today as well as the names of the related Greek letters preserve traces how they evolved from the objects they originally represented.

secrets of the Mayan writing (Coe, 1999), where graphical and logical flexibility and diversification of symbols turned out to be an expression of stylistic elegance of those writers. The arbitrarily chosen symbols preserve a trace on their evolution history. In Table 8.4, the origin of the first four letters of our modern Latin alphabet is given.

The coding invariance of languages does not only refer to the arbitrariness of letters, it also includes the way those letters are combined to form sequences, that is, the pool of chosen words and the grammar rules for their relative positioning. In addition to natural languages, this does as well apply to their technical descendants such as programming languages and computer codes. Readers may remember that some pocket computers of the first generations required as instruction sequence the input in "reverse Polish notation" to avoid parantheses, just to give an example. We learn that the modern word "alphabet" literally means "ox house" in Phoenician (Table 8.4). The Phoenician alphabet, its symbols, and their standard sequence were conserved (including some neutral drift, extension, and differentiation) through many centuries; the Minoans passed it to the Greeks which in turn brought it to the Etruscans (Haarmann, 2002). Romans learned it from their Etruscan neighbors and distributed it all over Europe. From there, it took another 2000 years for the Latin alphabet to develop into a universal global communication tool. Many humans know Latin letters in addition to their local, say, Chinese, Arabic, or Amharic writing. Cyrillic is a similar descendant of the Greek alphabet. The objects listed in Table 8.4 are of very practical nature, related to the daily life rather than to religious or magic ceremonies. This again is structural information on the evolution history that can be extracted from the physical details of our recent alphabet.

In their diversity, complexity and easy accessibility, natural languages are rather perfect, preserved records on the development of humankind (Feistel, 1990; Ebeling and Feistel, 1992). Comparative language analyses provide insight in the progress of natural sciences, historical events, and ethnographic relations. In particular, in the past two decades, various interesting and excellent reviews and studies were published on the human history derived from the structural information preserved in genetic sequences and in natural languages (Cavalli-Sforza, Menozzi, Piazza, 1994; Schmoeckel, 1999; Cavalli-Sforza, 2001; Janson, 2006; Wells, 2002; Haarmann, 2003; Hamel, 2007; Berger, 2008; Lindsey and Brown, 2009; Reich et al., 2009; Hubbe, Neves, and Harvati, 2010; Currie et al., 2010; Lawler, 2010; Gibbons, 2011; Dunn et al., 2011; Atkinson, 2011). Highly conservative language elements such as geographic names are of special interest for human settlement and migration studies (Urmes, 2004; Hamel, 2007). Similar to other processes of self-organization, new human languages may emerge by dynamic instabilities (Atkinson et al., 2008) and survive by competition (Hull, 2010).

In contrast to the opinion of Hitti (1961) that the alphabet was the greatest invention ever made by human, the modern system of numbers was perhaps an even greater invention. At least, its development took much longer, and the problem was much more demanding. Comparison of numbers does not necessarily require a number system. To compare a herd of sheep with one of cattle, one can drive them pairwise through a gate, one sheep along with one cow. The same can be done with a herd and scratches carved in wood, a scratch for each piece of livestock that passed a gate, one by one. Up to about 30 of such scratches, in groups of 5, were found on 25–30 kyr old wolf spikes (Klix, 1980). This is easy for small numbers but not really handy for larger ones, and practically impossible for, say, counting days in a calendar. In the beginning, simple symbols may help, V for five fingers of a hand, X for two hands, XX for four hands, but at some point, for larger numbers, arbitrary symbols, ciphers or words, are required which no longer immediately resemble the numerical value. For this purpose, Greeks and Romans used certain letters borrowed from the alphabet, such as L, C, D, M, based on pure convention between writers and readers. Mathematical calculations are rather complicated with Roman numbers.

The breakthrough came with the introduction of zero and the position system (Klix, 1980). It permits the representation of fairly large numbers because the length of the sequence grows only logarithmically with the value it represents. It supports simple formal rules for addition and multiplication that are practically very important operations for estimating the harvest expected from a certain area or the equipment required for an army. The decadic position system with zero and "Arabic" numerals is unrivaled in its convenience and effectivity; no serious alternatives have survived in any language. Those symbols reached Europe from India through the Arabs (Ifrah,1985, 1999). Just like the letters in Table 8.4, their sequence, shape, and pronunciation are still similar to the original Sanskrit, sunya, eka, dvi, tri, catur, panca, sas, sapta, asta, and nava, and preserve the trace back to their very roots. Nevertheless, there is still some arbitrariness and diversity preserved. For the two-digit numbers between 10 and 99, different European languages use different traditional irregular number-naming systems, such as eleven or fifteen rather than twenty-one. The related Asian systems are very regular and superior in their use (Miller *et al.*, 1995). In modern computers, numbers represented by the binary position system have completely taken over the job of the letters of an alphabet. In fact, computer bits are used to represent any other symbolic information extracted from the structural information of the real world, such as sounds or pictures, or books, formulas, data, or text in arbitrary languages. This is possible because of the coding invariance; any symbol can be represented by another symbol just by convention without affecting the information it carries.

8.13
Money

Surely there never was so evil a thing as money, which maketh cities into ruinous heaps, and banisheth men from their houses, and turneth their thoughts from good unto evil,
Sophocles warned in Antigone already in –442.

Gold is most excellent, wrote Christopher Columbus to King Ferdinand and Queen Isabella of Spain in 1503 in a letter from Jamaica, *gold is treasure, and he who possesses it does all he wishes to in this world*
(Olson and Bourne, 1906).

Money is the commodity which functions as a measure of value. Therefore, gold is money. The elementary expression of the relative value of a single commodity, such as linen, in terms of the commodity, such as gold, that plays the part of money, is the price form of that commodity

(Marx, 1951a)

The Phoenicians, known for being salesmen *par excellence*, considered money the soul of all things. Already chimpanzees are able to negotiate a fair deal (Melisab, Hareb, and Tomaselloa, 2009). It is not difficult to imagine that early humanoids exchanged objects of mutual interest such as flintstone, salt, dye, or food. This process was certainly intensified with the increasing division of labor, with new production technologies for arms, tools, agriculture, or domestic animals.

Exchange of goods with equal value is not always easily possible, a goat may be less valuable than a cow, and crop may be available at a different season than a winter pelt. Durable, valuable objects such as feathers, fur, pearls, or stones may serve as temporary, intermediate commodities of a barter deal. This is a first ritualization step toward money; in between the actual exchange objects appears an "acceptance bill," a symbol for the value of the thing given away until the desired equivalent is available to make the bargain complete. That symbolic object represents the proper value but may be of little practical benefit except its role as an exchange equivalent (Engels, 1972). The actual nature, size, or shape is irrelevant as long as the parties agree on it, such that in principle a sufficient amount of any substance or bodies may be used for this purpose.

Gold turned out to be the ultimate material of choice. It is rare, it is neither volatile nor fragile, it does not degrade over time, and it can be molten and divided into smaller or larger portions. It has a high value–mass ratio; a fortune can be stored in a small place. For easy handling, pieces of well-defined masses were used. To prevent from faking with copper or silver, such pieces were ornamented with the emperor's face or coat of arms, and for convenience, with a number representing the mass in some arbitrary unit. Those early golden coins carried a symbol of the value they

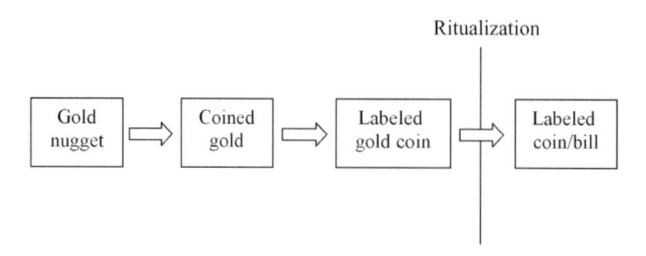

Figure 8.18 Evolution of money. Originally, gold in any form could be used as an exchange equivalent for commodities or services. It was later replaced by gold pieces of certified mass and quality, issued (coined) by an authority. For easier change, different gold coins were labeled with numerical symbols representing their masses, that is, their exchange values. Finally, in the ritualization transition, the use value disappeared from the coin or banknote and was only indicated by a symbolic label; commodity money was replaced by fiat money. The exchange value developed into symbolic information on the value it stands for.

constituted. When the coins are circulating from hand to hand, they are gradually losing mass. This matters much less if the material value of the coin is small and can be replaced from time to time, while it still carries the symbol for the mass of gold it is worth. If there is an authority that grants the equivalence of symbolic money and a certain amount of gold, or a similar valuable commodity, fiat money as a legal tender has a number of advantages over commodity money (Greco, 2001). The money's loss of use value is a ritualization transition (Figure 8.18); the exchange of values is replaced by an exchange of symbols for those values. Those symbols are arbitrary, can be diversified, and are subject to coding invariance, for example, a handful of coins is considered equivalent to a paper note. The shape and structure of symbolic money preserves information on its evolution history. For example, the word "Dollar" printed on US notes derives from "Thaler," such as the money coined in the German towns of Brunswick and Luneburg in 1799. Those coins in turn received their name from the shorthand of "Johannis*thaler* Guldengroschen," a popular currency once emitted in Sankt Johannisthal, Bohemia, in 1520. Hardly any US citizen may be aware that the greenback borrowed its name from a little medieval Bohemian town. The Dutch "Gulden" (Guilder) still refers to the gold it used to be, and the same is true for the Polish "Złoty," which is related to the Russian words "zholty" (yellow, golden) and zoloto (gold).

Money, in particular symbolic (fiat) money, is a special kind of property. As with any other property, it is an emergent rather than a physical quantity, and it relies on social structures that grant the value and the possession by law and if necessary, by force. Similar to usual measurable physical quantities, money has a numerical value with respect to a defined standard unit and obeys a conservation law when it circulates in the society. To say it in the words of Ayres (1994), *money plays a central role in modern economic theory. It is the "universal medium of exchange" required for any market to operate efficiently.* As long as money physically consists of gold, the conservation of money is equivalent to the physical conservation of gold. As soon as money becomes ritualized and turns into a mere symbol for gold, the conservation of money becomes a legal rather than a physical issue, and may then become subject to political decisions and criminal speculations.

9
On the Origin of Life

Life seems to be orderly and lawful behavior of matter, not based exclusively on its tendency to go over from order to disorder.
Erwin Schrödinger: What is Life?

How does newness come into the world?

Of what fusions, translations, conjoinings is it made?

How does it survive, extreme and dangerous as it is?
Salmon Rushdie: The Satanic Verses

Hardly any phenomenon in the world around us is so mysterious, fascinating, sophisticatedly complex, and commonplace at the same time, as life. Despite its individual fragility, mortality, and improbability, life is an extremely robust and practically imperishable state of matter. The conundrum what life is and how it originated is as old as humans are intelligent enough to ask themselves such questions. Life has always inspired natural scientists, philosophers, and theologians to create a wealth of different theories, from more or less reasonable hypotheses to suspicious religious or political dogmas, until the present time. Is the origin of life a problem that can be formulated and treated by means of the language and the tools of theoretical physics? At least since the seminal works of Schrödinger, Oparin, Prigogine, and Eigen, we think the answer should be *yes*. Likely, the development of a "physics of life" may turn out to be a fundamental challenge of similar significance as the origin of the cosmos and the nature of space, time, and elementary particles. We are still at the very beginning of such a theory. What is the difference between a living individual and other physical systems in terms of physical laws? We know already necessary physical conditions for life, such as open systems kept far from thermodynamic equilibrium, but can physical conditions sufficient for life also be formulated? Rigorous answers to questions of this kind are still lacking. Life is an exotic physical state of matter which permanently extracts structural information from the environment, stores it away as symbolic information, and exploits the

Physics of Self-Organization and Evolution, First Edition. Rainer Feistel and Werner Ebeling.
© 2011 Wiley-VCH Verlag GmbH & Co. KGaA. Published 2011 by Wiley-VCH Verlag GmbH & Co. KGaA.

accumulated experience to predict relevant future conditions of the surrounding. Genetic sequences of bacteria as well as this book are two different abstract models of the physical reality, just written in terms of arbitrary languages and codes. Frequencies and correlations of events in the past are recorded, combined, and processed to improve survival chances by predicting probabilities and expected correlations under actually unknown future circumstances. How could such incredibly advanced and complicated capabilities emerge from random accumulations of atoms such as the early solar system?

Thermodynamically, living beings are open systems far from equilibrium which permanently export entropy to maintain their internal structure and functionality; consequently, they need regular import of high-valued energy. This necessary physical boundary condition must reasonably be assumed to hold also during the very first steps on the staircase to the birth of life. In this book, life is understood to consist of (i) *spatially separate individuals*, capable of (ii) *self-reproduction* controlled by (iii) *symbolic information processing*. We consider this triple as sufficient conditions for life; any physical system equipped with those three traits must be classified as being alive, regardless of the particular substances, compounds, solvents, or coding principles it may use. Any plausible hypothesis for the origin of life must explain how the combination of those three properties could possibly emerge by sufficiently probable random fluctuations and dynamical self-organization under conditions similar to those prevailing on the early Earth. In the literature, various rather different problems are addressed under the headline of the origin of life, such as

- the first inorganic synthesis of organic substances,
- the formation and growth of spatial droplets,
- the particular structure of compounds and their detailed reaction chains that are considered to be the most likely ancestors of the present biochemistry,
- the origin of autotrophic energy capture, or
- the appearance of the first cells similar to most primitive recent taxa.

In the approach of this book, proper life began with the ability to accumulate genetic information in symbolic form, with a kinetic transition from physical chemistry to biology (Chapter 8). Certainly, the road to that point was long and winding, with various dead ends and alternative branches, with many open questions and difficult problems; it was a complex process but none the less subject to the "ordinary" laws of physics and chemistry before novel, emergent properties took control over the subsequent evolution. That process may have taken several 100,000,000 years, a time span well beyond all experimental imagination, from the oldest known rocks, 4500 Myr ago (Jackson *et al.*, 2010) to early evidence for marine biology, 3500 Myr ago (Blake, Chang, and Lepland, 2010), see Table 3.1 in Chapter 3.

The hypothesis presented in this chapter is based on ideas developed first in 1978/79 during research visits at the groups of Yuri Klimontovich and Yuri Romanovsky at the Moscow State University, based on the works of Oparin (1924, 1963), Romanovsky, Stepanova, and Chernavsky (1975), and Volkenstein (1978), and presented in preliminary form already in the original German edition of this book (Ebeling and

Feistel, 1982). One of the key theses is the natural combination of the coacervate and hypercycle models (Ebeling and Feistel, 1979; Feistel, Romanovsky, and Vasiliev, 1980). The authors easily remember the winter of that year because it was extreme in Germany (Feistel and Feistel, 2006; Tiesel, 2008), and with enduring temperatures of about -40 °C it was exceptionally cold even for Moscow. When this chapter was written, the exceptional winter days of December 2010 strongly reminded us of the time when the ideas to the German original of this book were developed and collected, and the first texts were typed. In the previous chapters, we have discussed conditions which lead to the formation of spatial structures by self-organization, the kinetics of self-reproduction, the properties of symbolic information, and the physical and chemical environment of the planet Earth. It is in this chapter that we try to paint a picture for the origin of life, consistent with the earlier concepts and scenarios. In contrast to other hypotheses, here we refrain from relying on very specific physical and chemical details as far as possible, and focus on indispensible general functional and structural aspects. The subsequent steps are preferably described here in the form of triples, (i) if known, the need for the particular step in general (physical) terms, (ii) some theoretical model to illustrate the related process more rigorously, and eventually, as far as we are aware of, (iii) observational or experimental evidence for the practical possibility of that step under the conditions of the early Earth. Our aim is primarily to develop a succession of necessary stages and conditions rather than to offer a very specific, virtually sufficient scenario.

We are well aware that the problem of the origin of life belongs to the most difficult questions of modern science, and that beside the famous scenarios of Oparin (1924, 1963) and Eigen (1971) there are various other more recent concepts (Rutten, 1971; Fox and Dose, 1972; Miller and Orgel, 1974; Kaplan, 1978; Körner, 1978; Volkenstein, 1978, 1981, 1984; Eigen et al., 1981; Eigen and Winkler-Oswatitsch, 1981; Posokhov, 1981; Reinbothe and Krauß, 1982; Doolittle, 1984; Gould, 1993; Szostak, Bartel, and Luisi, 2001; Dyson, 2008; Schuster, 2009; Ricardo and Szostak, 2010). All one can seriously suggest at present is a hypothetical staircase of evolutionary stages and events which obey the heuristic principles of plausibility and continuity (Lehninger, 1972; Romanovsky, Stepanova, and Chernavsky, 1975, 1984), and reflect as many experimental facts and theoretical conditions as there are known today (Miller and Orgel, 1974; Küppers, 1979; Eigen et al., 1981). Nevertheless we agree with Dyson (2008) who said: "Now, ..., the time is ripe to ask the questions which Schrödinger avoided. The questions of origin (of life) are now becoming experimentally accessible, just as the questions of structure were becoming experimentally accessible in the nineteen-forties."

9.1
Catalytic Cascades in Underoccupied Networks

The theoretical stage on which the dramatic play "Origin of Life" is performed is the phase space of an extremely underoccupied catalytic network. The players are molecules of various sizes and complexities; the conditions far from thermodynamic equilibrium arrange the actors in permanently changing positions on that scene and randomly introduce new characters.

There are two qualitatively different classes of molecules, *abundant* and *rare* ones. In a chosen reaction volume, perhaps of the size of a cubic millimeter or decimeter, the number of abundant molecules by definition exceeds one, $N > 1$, that is, those molecules are permanently present. Rare molecules have average particle numbers less than one, $N < 1$, that is, those molecules are only occasionally present in a given observation period, or even not at all. Some stable ambient compounds such as N_2, H_2O, CO_2, SO_2, H_2S, SiO_2, or $CaCO_3$ certainly belong to the abundant species of the laboratory "early Earth"; without energy supply they do not decay nor do they form more complex molecules. Typical rare molecules are metastable; they release energy when they decompose, and energy is required for their formation. By thermal fluctuations, such molecules form only occasionally with exponentially decreasing probability,

$$P \propto \exp(-E/kT) \qquad (9.1)$$

as a function of the binding energy, E, and the ambient temperature, T, according to the laws of Maxwell–Boltzmann and Arrhenius. A typical value for the binding energy of organic compounds is $E = m\varepsilon$, where m is the molecular mass and $\varepsilon = 10 \, \text{kcal g}^{-1} = 42 \, \text{MJ kg}^{-1}$ (Cummins and Wuycheck, 1971) is the specific energy of formation, see also Section 3.1. Moreover, the multiplicity of possible spatial arrangements of atoms also increases exponentially with their number. Proteins consisting of a linear chain of λ amino acids selected from a pool of 20 available ones can form 20^λ different configurations. Thus, the chance to find a specific macromolecule with the mass m is an exponentially small fraction of the exponentially small probability P, Equation 9.1. For such molecules, precisely tuned in their structure and count, to be abundant in large quantities at the same place at the same time as required for a primitive organism to immediately start living, the resulting chance for spontaneous abiogenesis is so astronomically small that some researchers considered a divine creation act or a unique giant fluctuation as the only possible "explanation" for life (Monod, 1971; Doolittle, 1984). Scientifically, such kind of explanations are unsatisfactory as they do not explain at all how an observed phenomenon is related to basic natural laws, similar to Landau's famous example of a "gramophone spirit" imprisoned in a box, if suggested as the "physical explanation" for the functioning of a gramophone.

So the first question must be how the abundance of rare molecules could substantially increase beyond their thermodynamic equilibrium concentrations, or beyond their expected frequency of random appearance by thermal collisions, under natural conditions that likely prevailed on Earth about 4000 Myr ago. Obviously, nonequilibrium conditions are required which continuously provide free energy and remove the entropy produced. Chemical catalysis is irrelevant at equilibrium but can strongly influence the properties of nonequilibrium states. Three virtually inexhaustible energy sources were available – solar irradiation, geothermal heat, and the heat produced by impacts of asteroids and comets. Among those, hydrothermal vents are apparently the only known system that provides strong thermal

Figure 9.1 Hydrothermal vent "Geysir Basin" in Yellowstone National Park. Photo taken in September 1999.

gradients rather regularly and over extended periods of time (Früh-Green et al., 2003). They can be found today along mid-ocean ridges at about 3000 m water depth (Simoneit, 1993, 1995; Reed, 2006; Bates et al., 2010) or at about 1000 m depth like the famous Lost City Hydrothermal Field on the Atlantis Massif at 30° 8' N, 42° 8' W (Früh-Green et al., 2003; Boetius, 2005; Bradley, 2008; Proskurowski et al., 2008; Evans, 2010; Brazelton et al., 2010), in shallower waters with geological activity such as the East China Sea (Chen et al., 2005), and in volcanic lakes and springs such as in the African rift valley or in the Yellowstone caldera (Cuhel et al., 2002; Bradley, 2008; Balodis et al., 2008), see Figure 9.1.

The free energy available from hydrothermal vents in the form of strong temperature and concentration gradients is evidently sufficient to feed recent autonomous ecosystems in the deep sea (Grassle, 1986; Lutz and Kennish, 1993; Pledger, Crump, and Baross, 1994; Jeanthon, 2000; van Dover, 2000). Those populations include *Archaea*, the oldest life forms we know in the present biosphere. Regions where glowing magma is in close contact with liquid water, such as deep-sea vents or hot springs, very likely served as strong and perpetual energy sources for various geochemical reaction chains (Wächtershäuser, 1990). The surfaces of cracks and pores in surrounding rocks and precipitated minerals were certainly relevant ingredients for heterogeneous catalysis and may have formed tiny, protected reaction volumes. The vents along the current mid-ocean ridges extend over vast volumes and remain in place over thousands of years. About 3500 Myr ago, when apparently life forms already populated the oceans, water temperatures were at about 70 °C (Robert and Chaussidon, 2006), see also Table 3.1. Present-day deep-ocean vents have temperatures up to 680 °C (Reed, 2006; Sun et al., 2008) and can maintain strong temperature gradients even in hot seawater. Additionally, localized vents provide thermal contrasts in their vicinity, and by fluctuating turbulent currents of the hot plumes also offer more or less regularly changing local physical and chemical conditions. All these arguments make the oceanic vents the most promising candidates for the open systems that formed the cradle for the birth of life.

We mention that the permanent existence of mid-ocean vents may be related to sea-floor spreading, continental drift and the bimodal hypsographic structure of the terrestrial crust, in contrast to, for example, our sister planet Venus with a unimodal distribution (Chapter 3). If those vents were the only place where life could originate with sufficiently high probability, Venus might never have had the same chance, despite very similar initial conditions in the "habitable belt" after the formation of the solar system. Moreover, the "giant impact" that formed the Moon may be the cause for the terrestrial crustal bimodality. Thus, it may turn out that the conditions for the origin of life on Earth were rather specific and were possibly met by our planet only due to exceptional fortunate circumstances.

In the first phase on the route to life, nonequilibrium conditions must have prevailed in spatially and temporally extended chemically active regions. Such conditions are indispensible for expanding the distribution of occupation numbers in the phase space from the region of highly abundant equilibrium species sufficiently far into the less-populated region of rare species. Catalysis in aqueous solutions, either between the solutes or in contact with solid surface structures, must have played a key role. In particular, closed catalytic cycles may have increased local concentrations dramatically. Cycles imply the occupation of their catalytic side chains and of diverse reaction products, that is, of compounds which cannot reasonably be assumed to emerge under equilibrium conditions. Such occasional occupation "waves", randomly invading new regions of the largely unpopulated phase space along catalytic edges, may be termed *catalytic cascades* (Ebeling and Feistel, 1982, 1994), or avalanches. This way, high random abundances of "improbable" molecules, at least locally and temporarily, must have laid the foundation for the next step, just the second one in a fairly long row.

9.2
Formation of Spatial Compartments

Any important catalytic reactions in recent organisms are complex, nonlinear, and include several reactants. This is in particular the case for the molecular genetic expression machinery. As a rule, reactions of this type lead to hyperselection behavior unsuitable for successive improvement by molecular Darwinian evolution (Chapters 5–7). This problem does not occur when complex reactions proceed within separate spatial units which may grow and compete by their growth rates (Ebeling and Feistel, 1979, 1982; Feistel, Romanovsky, and Vasiliev, 1980; Feistel, 1983d, 1983c), see also Section 6.10. More than any other, it is this reason why we consider spatial compartments indispensable already at the earliest stage of molecular evolution. Such "coacervates" or "microspheres" have the additional advantage that they may maintain much higher concentrations of organic compounds than the open solution, even of species that are not soluble at all. They keep rare substances within short distances of each other and support their interaction. They offer a controlled and relatively constant internal physical and chemical milieu. On the other hand, droplets of unspecific mixtures have densities which more or less deviate from that of water;

they will either float on the water surface, or sink to the ground, or stick to solid surfaces, such as to the rocks of hydrothermal vents. While equilibrium droplets can exist without energy supply, nonequilibrium spatial structures can survive only as long as they consume energy and produce entropy.

We imagine that among the various substances that were randomly synthesized in catalytic cascades as described in the previous section, several were hydrophobic or of the form of lipids with hydrophilic and hydrophobic parts. Such molecules tend to spontaneously accumulate to membranes or droplets similar to grease drops on the soup (Cape, Monnard, and Boncella, 2011). For example, experiments of Oparin (1963) and Fox and Dose (1972) demonstrated the self-organization of coacervates and microspheres, respectively, for different conditions. Moreover, the appearance of concentrated droplets may have intensified the catalytic cascades of rare, complex chemicals, thus via positive feedback improving the preconditions for the following evolution phase.

As suggested by the simple dense-pack model developed in Section 6.10, the relative composition, $\mathbf{x}$, of a growing droplet will converge toward an attractor state which we assume here for simplicity to be homogeneous and stationary (otherwise, one may think of suitable averages). Let

$$\frac{dc_i}{dt} = f_i(c_1, c_2, \ldots,) \tag{9.2}$$

be the kinetic equations for the molar concentrations c_i of the constituents of the droplet in the case that the reaction volume V is constant, such as in a dilute solution. In contrast, if the volume is changing as a result of the reactions, Equation 9.2 must correctly be written in terms of particle numbers, $N_i = c_i \times V$, rather than concentrations, in the form

$$\frac{1}{V}\frac{dN_i}{dt} = f_i(c_1, c_2, \ldots) \equiv f_i(\mathbf{c}) \tag{9.3}$$

In a simple dense-pack model, the droplet volume is the sum of the molecular volumes, v_i,

$$V = \sum v_i N_i \tag{9.4}$$

of the particular constituents. The droplet volume grows at the rate

$$\frac{dV}{dt} = \sum v_i \frac{dN_i}{dt} = V \sum v_i f_i \tag{9.5}$$

if we neglect changes of the molecular volumes due to particle interaction. For the molar fractions, $x_i = c_i/c$, we get the dynamic equations

$$\frac{dx_i}{dt} = \frac{d}{dt}\left(\frac{N_i}{\sum_k N_k}\right) = \frac{1}{c}\left\{f_i(\mathbf{x}c) - x_i \sum_k f_k(\mathbf{x}c)\right\} \tag{9.6}$$

where $c = \sum c_i$ is the total concentration which changes according to the equation

$$\frac{dc}{dt} = \frac{1}{V}\sum \frac{dN_i}{dt} - \frac{c}{V}\frac{dV}{dt} = \sum f_i - c\sum v_i f_i \qquad (9.7)$$

When the composition reaches a stationary state, $\mathbf{x} = \mathbf{x}^{(s)}$, Equation 9.6 results in

$$f_i = x_i \sum_k f_k \qquad (9.8)$$

Then, the total concentration, c, in Equation 9.7 is also stationary because of Equations 9.4 and 9.8

$$c\sum_i v_i f_i = c\sum_i v_i x_i \sum_k f_k = \frac{1}{V}\sum_i v_i N_i \sum_k f_k = \sum_k f_k \qquad (9.9)$$

The related constant rate of volume growth, Equation 9.5, takes the form

$$\frac{1}{V}\frac{dV}{dt} = \sum v_i f_i = \frac{1}{c}\sum f_i \qquad (9.10)$$

This important model result tells us that droplets with stationary composition grow or decay always exponentially, similar to biological cells, irrespective of the details of the particular metabolism. The selective value

$$E(\mathbf{x}) = \frac{1}{c}\sum f_i(\mathbf{x}c) \qquad (9.11)$$

is associated with a droplet species of the composition $\mathbf{x}$. Together with the growth rate, this composition follows from the nonlinear "eigenvalue problem," Equation 9.8, in the form

$$\mathbf{f}(\mathbf{c}) = E\mathbf{c} \qquad (9.12)$$

When models of self-reproducing droplets are compared, this simple and convenient equation enables us to compute their particular selective values. Hence, such a value is a functional of the reaction kinetics, $\mathbf{f}$, and of the internal composition, $\mathbf{c}$, that is, of the part of the general catalytic network of all species which is actually occupied by the given system. If Equation 9.12 has multiple non-negative solutions for $\mathbf{c}$, the highest value of E is responsible for the droplet's multiplication rate. This is in particular the case if the internal catalytic network consists of several disconnected competing components, say, of a parent system and a variation of it.

The stage described up to here may be termed the "Oparin stage" as it describes the formation and growth of organic coacervates, droplets or coating layers on rocky surfaces, consisting of random rare molecules accumulated as a result of catalytic acceleration of nonequilibrium chemical reactions. In particular, closed catalytic cycles such as those suggested by Wächtershäuser (1990) are required to produce substantial amounts of rare chemicals in this way. Analysis of Equation 9.12 shows that only cycle members plus their "parasitic" side chains and decay products can form the constituents of steadily growing coacervates, that is, the network has a cycle as its root and a "tail" of secondary compounds, Figure 9.2.

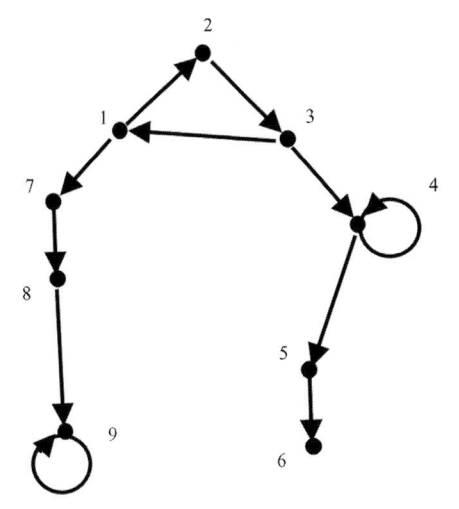

Figure 9.2 Schematic example of a self-reproducing catalytic network. Labeled vertices represent chemical species. An arrow from (a) to (b) means that the production rate of (b) is proportional to the concentration of (a). The cycle (1, 2, 3) is the root with a tail of "parasitic" descendants, species (4, ..., 9), as a solution of Equation 9.12. The reproduction of the loops (4) and (9) must be slower than that of the root cycle, otherwise the faster one would take over the root position and the cycle (1, 2, 3) would get extinct (compare the models discussed in Chapter 6).

This family of species must include those molecules which are required for the physical integrity of the spatial compartment; other, only initially or occasionally present chemicals disappear necessarily after the characteristic replication time, $\tau = 1/E$. In general, such "Oparin systems" are still unable to gradually improve their reproduction rates or to evolve in any systematic manner. But, by populating previously uninhabited regions of the huge underoccupied catalytic network, they are nevertheless crucially important as they may reasonably provide a large pool of otherwise highly improbable, optional candidates for the next, qualitatively new evolution step. In the following, we will use the terms coacervate, compartment, or droplet synonymously, if not specific circumstances demand their distinction.

9.3
Replicating Chain Molecules

Even though experimental or observational evidence is scarce, it seems plausible that under nonequilibrium conditions of early Earth, catalytic cycles could appear in the favorable and relatively stable environment of concentrated spatial compartments such as droplets, coacervates, or rock-surface layers. Only those cycles are of interest here which reproduced the chemical composition of the compartments and maintained their existence against loss and decay. In the vast natural laboratory of millions of liters of water over millions of years time, out of the large pool of arbitrary combinations there may have randomly appeared certain catalytic cycles with special properties. Likely, rather than an act of rigorous necessity this was just an event of reasonable probability in a gigantic lottery of molecules and chemical reactions. Once the novel cycles had appeared, even at small numbers, they may have advanced systematically, leaving behind all the suddenly irrelevant rest of the sticky and slimy chemical community.

Chain molecules usually have properties that almost continuously change with the length of the chain, or by substitution of internal building blocks. Chemistry knows ample examples for this rule, such as the properties of methane, ethane, propane, butane, and so on. The longer the chain, the smoother the change of properties.

In a catalytic network such as the one shown in Figure 9.2, any occasional error copies that appear in the catalytic tail are irrelevant; the permanent production of new copies starting from the root cycle will quickly eliminate any fluctuations of the tail composition. It is therefore necessary that slightly changeable molecules are part of the root cycle to have any effect of modification in the longer term. This requirement implies a severe problem; if an altered member of a cycle appears, it must remain a member of the same or of a slightly modified catalytic cycle which immediately starts reproducing the error copy in a similar way as it formerly reproduced its precursor molecule. The easiest way to imagine this to happen is a binary cycle, with one molecule being the spatial complement of the other. Rather than for globular coils or balls, this may work for molecules that have a nearly planar or linear three-dimensional structure, or both, similar to the technical production of a vinyl disc or a gypsum replica, see Figure 8.3 in Section 8.2. It has been shown experimentally that complementary replication is possible with RNA strands, although only in the presence of other supporting ingredients (Küppers, 1979; Kruger *et al.*, 1982; Lincoln and Joyce, 2009). Thus, for a functioning binary RNA cycle enclosed in a compartment, the cycle must be expected to produce a tail which supports RNA copying and maintains the compartment itself. For simplicity, we represent all the tail members here by just one effective fictitious molecule, regarded as the *replicase*. Similar to suggestions made by Schramm, Grötsch, and Pollmann (1962) and Kaplan (1978), we termed this simplest model of a primary system capable of reproduction and stepwise improvement of its skills an *RNA-replicase cycle* (Ebeling and Feistel, 1979, 1982, 1994; Sonntag, Feistel, and Ebeling, 1981), see Figure 9.3.

The authors are not aware of any detailed experimental investigation of such a system, despite several attempts to achieve the coupled replication of both a peptide and an oligonucleotide (Ghosh and Chmielewski, 2004). The tail's efficiency in catalyzing the RNA replication may be minor, just enough for a positive feedback effect that overcompensates other decay and loss processes. Another necessary condition for the RNA-replicase cycle is that the RNA must offer catalytic properties to produce a tail (Szostak, 2009). It was suggested by Kuhn (1976) that under the

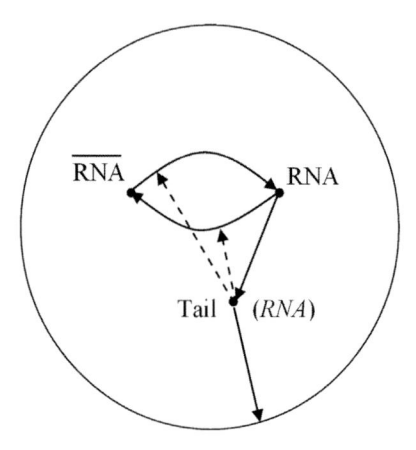

Figure 9.3 Schematic of the RNA-replicase cycle, the simplest model capable of gradual molecular evolution. RNA and $\overline{\text{RNA}}$ are complementary linear strands, *RNA* is the folded version of RNA. Solid arrows indicate catalytic production, dashed ones catalytic support of a reaction. Here, RNA is a representative for a chain molecule that permits (i) gradual property change by addition or substitution of its chain members, (ii) in its linear configuration, acting as a template for a spatially complementary replica ($\overline{\text{RNA}}$), and (iii) in its folded configuration (*RNA*), catalysis of a tail which in turn supports replication (dotted arrows) and produces additional compartment material. For simplicity, the tail is represented here as a single molecule, the replicase, assuming that structural details of the tail network are unimportant for the model at this point. A possibly folded configuration of the complementary strand $\overline{\text{RNA}}$ is omitted for simplicity.

influence of external fluctuations, for instance of the ambient temperature, RNA molecules undergo periodic or occasional configuration changes between a compact structure, suitable as a catalyst for the tail, and a linear strand, better suitable for replication. In our approach, we note that the model of the RNA-replicase cycle does not actually require a bidirectional configuration change of the RNA. It is sufficient for proper functioning that some RNA specimens remain unfolded to permit continual replication, while additional copies may fold and in this form represent a part of the catalytic tail, without the need for stretching those copies again at some later time. This one-way, read-only translation of a molecular sequence to a catalytic function is typical for all later life forms (the "central dogma") and may have been established already at this very first pregenetic stage of evolution.

Here, the RNA chain has the form of a specific sequence of building blocks chosen from a limited pool of available suitable molecules. Although this structure is similar to a "text" consisting of "letters" selected from an "alphabet," here this sequence still represents structural rather than symbolic information (Chapter 8). The criterion for this distinction is that a modified sequence necessarily implies a modified catalytic function; the coding invariance of symbolic information is not yet established. It is the central aim of this chapter to develop a plausible scenario for the implementation of coding invariance which, from our perspective, is equivalent to the beginning of life. Thus, even if a self-replicating system similar to that shown in Figure 9.3 may one day be set up in a laboratory, despite of its relevance, it may not be regarded as the final breakthrough to the origin of life. To arrive at that distant goal, many additional complicated problems remain to be overcome.

It is instructive to derive a mathematical model for the RNA-replicase cycle. If c_0 is the concentration of the replicase, and c_1 and c_2 are those of the RNA and its complement, respectively, we can formulate the related reaction rates by means of formal kinetics, as

$$\begin{aligned} f_0(\mathbf{c}) &= kc_1 - \lambda c_0 \\ f_1(\mathbf{c}) &= \kappa_{12} c_0 c_2 - \lambda c_1 \\ f_2(\mathbf{c}) &= \kappa_{21} c_0 c_1 - \lambda c_2 \end{aligned} \qquad (9.13)$$

For easier calculation, the loss rates, λ, are set here to equal values. We ask now for the stationary compartment composition and the related selective value, Equation 9.12,

$$\mathbf{f}(\mathbf{c}) = E\mathbf{c} \qquad (9.14)$$

and get the equations

$$\begin{aligned} kc_1 &= (E + \lambda)c_0 \\ \kappa_{12} c_0 c_2 &= (E + \lambda)c_1 \\ \kappa_{21} c_0 c_1 &= (E + \lambda)c_2 \end{aligned} \qquad (9.15)$$

By successive elimination of the variables, the solution is found to be

$$c_0 = \frac{E + \lambda}{\sqrt{\kappa_{12}\kappa_{21}}}, \quad c_1 = \frac{(E+\lambda)^2}{k\sqrt{\kappa_{12}\kappa_{21}}}, \quad c_2 = \frac{(E+\lambda)^2}{k\kappa_{12}} \qquad (9.16)$$

The molecular volumes determine the selective value by means of the identity (9.4)

$$\sum v_i c_i = 1 \qquad (9.17)$$

with the result

$$E = -\lambda - \frac{B}{2} \pm \sqrt{\frac{B^2}{4} - C} \qquad (9.18)$$

where the constants are $B \equiv \dfrac{v_0 k}{v_1 + v_2 \sqrt{\kappa_{21}/\kappa_{12}}}$ and $C \equiv -\dfrac{k\kappa_{12}}{v_1 \sqrt{\kappa_{12}/\kappa_{21}} + v_2}$.

Only the plus sign of the root in Equation 9.18 may result in a positive growth rate, E. If we make the plausible assumption that the replication speed is independent of the RNA base sequence, $\kappa_{12} = \kappa_{21} = \kappa$, the formula for the selective value reduces to

$$E = \frac{k}{2} \frac{v_0}{v_1 + v_2} \left\{ \sqrt{1 + 4\frac{v_1 + v_2}{v_0^2} \frac{\kappa}{k}} - 1 \right\} - \lambda \qquad (9.19)$$

It is obvious already at this point that growth rates of compartments depend in a relatively complicated way on the reaction constants and on other properties of the internal metabolism. For example, we see in Equation 9.19 that a long catalytic tail of the RNA-replicase cycle will result in larger values of the total tail volume, v_0, which may turn the value of E down or even negative. Thus, alternative compartments will compete for shorter or less voluminous tails. The catalytic support of the replication rate, κ, must exceed a critical value for a positive replication rate, $E > 0$.

Due to the nonlinearity of the reaction rates, Equation 9.13, we distinguish two qualitatively different competition situations when alternative compositions of the RNA-replicase cycle are compared, as caused by different RNA sequences, namely (i) competition between coacervates and (ii) interaction within a coacervate.

If two coacervates with different compositions, that is, different RNA chains, grow independently, the one with the higher selective value E will survive, even if the initial abundance of that coacervate species is low, with a probability that can be estimated from stochastic competition models (Section 6.4). Coacervate competition overcomes the once-forever hyperselection of nonlinear catalytic networks. This will lead to a mutual fine-tuning of all the properties that influence the selective value by randomly changing the sequence and the length of the RNA. This is an important model result for the evolution of RNA-replicase cycles.

Even more interesting and important is the model's behavior if in a parent coacervate an RNA mutation appears. Under the assumption that the two replicases support RNA replication independent of the actual sequence (Wochner et al., 2011), the related reaction network is shown in Figure 9.4. We may ask for the stability of the parent coacervate with respect to the appearance of an additional RNA chain. We find now six species in the system, with the reaction rates for the parent species (0:

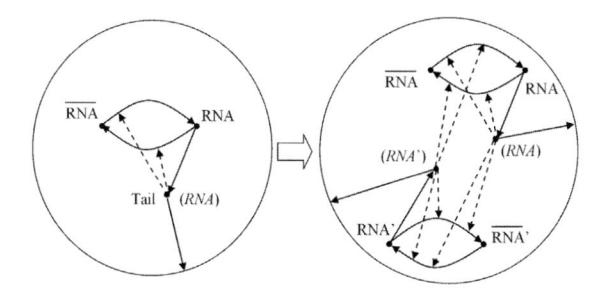

Figure 9.4 Schematic of the RNA-replicase cycle shown on the left, after duplication and appearance of an error copy, RNA′, shown on the right, within the same compartment. The two different strands, RNA and RNA′, form a cycle with their particular complements, $\overline{\text{RNA}}$ and $\overline{\text{RNA}'}$, respectively. The tails of the two cycles are assumed to support RNA replication independent of the particular base sequence, that is, the replication of their own RNA as well as that of the other cycle. Of the double RNA cycle, several links can be inhibited or removed by evolutionary changes without degrading its functionality.

replicase, 1: RNA, 2: $\overline{\text{RNA}}$)

$$f_0(c) = kc_1 - \lambda c_0$$
$$f_1(c) = (\kappa c_0 + \kappa' c_3)c_2 - \lambda c_1 \quad (9.20)$$
$$f_2(c) = (\kappa c_0 + \kappa' c_3)c_1 - \lambda c_2$$

and those for the newcomers (3: replicase, 4: RNA', 5: $\overline{\text{RNA}'}$)

$$f_3(c) = k'c_4 - \lambda c_3$$
$$f_4(c) = (\kappa c_0 + \kappa' c_3)c_5 - \lambda c_4 \quad (9.21)$$
$$f_5(c) = (\kappa c_0 + \kappa' c_3)c_4 - \lambda c_5$$

Here, k' and κ' are catalytic properties of the additional molecules. To keep the formulae simple, all decay rates are taken as equal.

In Section 7.6 it was shown that the problem of stability with respect to new species can be decomposed into one for the parent species alone and one for the newcomers. The Jacobian of the rate functions (9.21), taken at the boundary of the positive cone with the parent properties, Equations 9.16 and 9.19, reads

$$J = \frac{\partial(f_3, f_4, f_5)}{\partial(c_3, c_4, c_5)} = \begin{pmatrix} -\lambda & k' & 0 \\ 0 & -\lambda & E+\lambda \\ 0 & E+\lambda & -\lambda \end{pmatrix} \quad (9.22)$$

The solutions of the secular equation, $\det(J - \mu I) = 0$, are the eigenvalues

$$\mu_1 = E, \mu_2 = -E - 2\lambda, \mu_3 = -\lambda \quad (9.23)$$

The eigenvector of J related to the positive eigenvalue, μ_1, is positive and given by

$$\frac{(E+\lambda)}{k'} c_3 = c_4 = c_5 \quad (9.24)$$

Since the parent system grows at the same rate, $\mu_1 = E$, the newcomers behave neutrally; they are neither amplified nor rejected in linear approximation, independent of their values k' and κ'. To verify this result with the nonlinear equations, we inspect the solutions of the problem (9.14)

$$f(c) = Ec \quad (9.25)$$

with the full set of six functions and variables, Equations 9.20 and 9.21. In explicit form, this system of equations reads

$$(E+\lambda)c_0 = kc_1$$
$$(E+\lambda)c_1 = (\kappa c_0 + \kappa' c_3)c_2$$
$$(E+\lambda)c_2 = (\kappa c_0 + \kappa' c_3)c_1$$
$$(E+\lambda)c_3 = k'c_4 \quad (9.26)$$
$$(E+\lambda)c_4 = (\kappa c_0 + \kappa' c_3)c_5$$
$$(E+\lambda)c_5 = (\kappa c_0 + \kappa' c_3)c_4$$

The solution can be found by successive elimination of the variables as

$$c_0 = \frac{(E+\lambda)k}{k\kappa + k'\kappa'w} \quad c_1 = c_2 = \frac{(E+\lambda)^2}{k\kappa + k'\kappa'w} \tag{9.27}$$

for the parent system, and

$$c_3 = \frac{(E+\lambda)k'}{k\kappa + k'\kappa'w}w \quad c_4 = c_5 = \frac{(E+\lambda)^2}{k\kappa + k'\kappa'w}w \tag{9.28}$$

for the newcomer system. Here, the value of the mixing ratio $w = c_4/c_1$ of the two RNA fractions is an arbitrary non-negative number, that is, the solution of Equation 9.25 is degenerate. This ambiguity of the nonlinear problem confirms the internal neutral stability with respect to the addition of modified RNA chains as long as they can still be copied by the parent replicase. The selective value, E, of the mixed coacervate shown in Figure 9.4 follows from the condition (9.17), $\sum v_i c_i = 1$, as

$$E = -\lambda - \frac{B}{2} \pm \sqrt{\frac{B^2}{4} - C} \tag{9.29}$$

where the constants are $B \equiv \frac{v_0 k + v_3 k'w}{(v_1+v_2)+(v_4+v_5)w}$ and $C \equiv -\frac{k\kappa + k'\kappa'w}{(v_1+v_2)+(v_4+v_5)w}$.

Again, only the positive sign in front of the root in Equation 9.29 may result in a positive growth rate, E. To this selective value of the coacervate, the properties of the old and the altered internal catalytic net contribute jointly in nontrivial combination. In particular it is possible, for instance, that the new cycle does not produce any replicase, $\kappa' = 0$, and relies entirely on the available replication system of the parent system. The related lowering of E may be compensated by other, more beneficial properties of the mutant. An example for such a division of labor is indicated in Figure 9.5 in which the new RNA' has taken over the production of specialized constituents of the coacervate. At the same time the original RNA becomes discharged from this task and may then specialize on more effective support of the replication process.

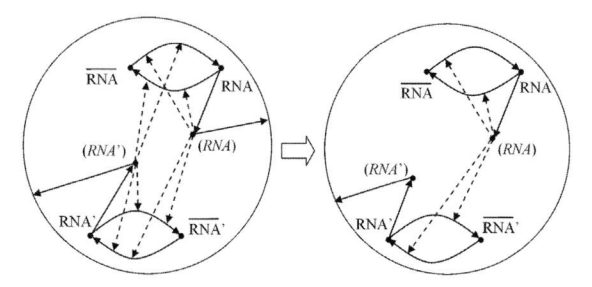

Figure 9.5 Schematic of the double RNA-replicase cycle, on the left, after division of labor between the two catalytic tails, that is, after removal of several arrows, on the right. Symbolically, RNA is now specialized on coding for the replicase and RNA' on constituents of the coacervate.

These results demonstrate that the model of a RNA-replicase cycle embedded in a coacervate is a very promising structure for earliest molecular evolution processes.

Internally, the model behaves neutrally with respect to the appearance of RNA error copies in the coacervate. Similar to symbolic information systems (see Chapter 8), this neutrality is favorable for random modification and diversification of the RNA pool. Each RNA cycle may code for a different tail that supports a specific catalytic or structural function. Neutral system response to specific perturbations ("Goldstone modes," "non-Darwinian evolution") always implies the possibility of macroscopic fluctuations, typically in the vicinity of phase transitions.

Externally, despite the internal neutrality, the different RNA pool members do influence the selective value of the coacervate, similar to Equation 9.29. Separate coacervates compete by their growth rates and select the fittest RNA pools and functions. The direction of the selection pressure on the coacervate properties can be concluded from Fisher's law (Wilson and Bossert, 1973; Feistel and Ebeling, 1982, Section 7.2)

$$\frac{d\mathbf{q}}{dt} = \sigma^2 \frac{\partial E}{\partial \mathbf{q}} \qquad (9.30)$$

where $\mathbf{q}$ is the set of phenotypic properties, such as $\mathbf{q} = \{\lambda, w, k, k', \kappa, \kappa', v_0, \ldots, v_5\}$ in the case of the fitness landscape, Equation 9.29, and σ^2 is the covariance matrix of those properties within the coacervate population. Daughter coacervates may mechanically separate from existing ones in a turbulent fluid, with the parent molecules randomly distributed between the fractions.

At the end of this evolution stage, we imagine a wealth of RNA pieces with different lengths and base sequences, which in their folded form produced catalytic tails of molecules that were specialized for various functions required for faster growth of the self-organized compartments. The scenario resembles primitive cells with primitive genes but still without coding invariance of the RNA sequences and without sequential mounting machinery of more complex catalysts. The evolution of this type of system may be restricted by essentially two problems – limited catalytic capabilities of folded simple RNA strands and the need for catalytic support of $O(n^2)$ interactions when the system permits coding of only $O(n)$ suitable molecules. The two problems have a clever common solution, as we know from today's retrospect, namely the self-organization of the molecular coding language (Section 8.6). This development must have been a very complicated process, leading from plausible beginnings to a plausible end, but with rather hypothetical details in between.

9.4
Molecular Information Processing

From the perspective of mathematical evolution models, the RNA-replicase cycle embedded in a self-organized compartment is the simplest system that is capable of molecular evolution and systematic improvement of its phenotypic properties due to RNA mutation and selection between the coacervates. Here, RNA stands for some chain molecule that (i) can be replicated catalytically, independent of its particular sequence of elements, (ii) possesses a spatially complementary molecule which consists of elements taken from the same pool and by replication reproduces the original template, and (iii) can undergo a spatial configuration change to a catalytically more active three-dimensional structure. Although real RNA molecules are known to fulfill these requirements, the authors are not aware yet of any experimental setup for a functioning RNA-replicase cycle.

To reach the point where life actually began with symbolic information processing, the RNA-replicase coacervate had to evolve to a primitive cell with a linear DNA strand, of which pieces are copied to linear RNA strands (mRNA), of which in turn symbolic fractions (codons) are associated with folded RNA molecules (tRNA) that catalyze the sequential mounting of amino acids to longer chains of different enzymes. The RNA-replicase-cycle model provides in rudimentary form the essential root elements for the development of those advanced functions, as illustrated in Figure 9.6.

To bridge the wide theoretical gap between the distant pillars shown in Figure 9.6 with a reasonable model scenario, it may be helpful to begin at the final side. As in the previous section, we shall focus here on the derivation of a minimum number of apparently necessary conditions for indispensable transition steps, rather than trying to speculate on very chemical details that may form a specific, virtually sufficient series of intermediate steps to smoothly connect the two shores.

The final step to symbolic information processing is the ritualization transition as described in Section 8.6. In this process, a structure is gradually reduced to a symbol ("caricature") of it while its ultimate effect is maintained. In our case, the finally used symbols are RNA codons, and the related structure is the chemically active, folded RNA molecule which provides the link to other members of the catalytic tail. In the original causal chain of the RNA-replicase cycle, Figure 9.6,

$$\overline{RNA} \to RNA \to RNA \to \text{catalytic tail} \qquad (9.31)$$

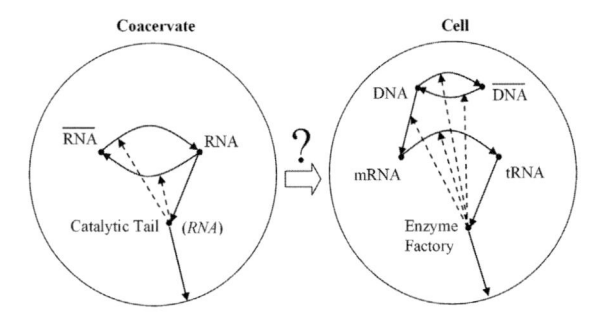

Figure 9.6 Evolution from the RNA-replicase cycle on the left to the first form of life on the right, similar to recent prokaryotic cells. Although all key ingredients are already present in rudimentary form, any details of the successive steps from one to the other are necessarily hypothetical and speculative.

this reduction may perhaps happen in such a way that the two configurations, linear RNA and folded *RNA*, are occasionally concatenated to form two parts of a single, combined pre-tRNA molecule, where the one end complements the structure of the template, $\overline{RNA}$, and the other end controls the catalytic tail,

$$\overline{RNA} \rightarrow [RNA + RNA] \rightarrow \text{catalytic tail} \tag{9.32}$$

If the folded part remains intact and firmly linked to the linear part, the latter may shrink to a shorter identifier that still complements the counterpart section of the $\overline{RNA}$, for instance to speed up the translation process,

$$\overline{RNA} \rightarrow [\text{codon} + RNA] \rightarrow \text{catalytic tail} \tag{9.33}$$

Although the codon is then mechanically fixed to the structure that expresses its chemical meaning, in the end this fixation appears arbitrary and can in principle be replaced by any other codon associated with the same *RNA* structure. This perfects the ritualization transition and gives way to differentiation and diversification of the new symbolic coding system (Chapter 8).

Any successful transition from (9.31) to (9.33) requires two essential preparatory changes in advance. Symbolic information processing implies the existence of symbols along with their spatial arrangement, for example, in the form of sequences. It is essentially the physical possibility of arbitrary configurations of symbols which creates the enormous capacity for information storage, but only if those configurations are stable enough to persist over the time span between writing and reading the information (Chapter 8). In the evolution process indicated in Figure 9.6, the two different but related problems must be solved, symbols must be created, and they must be processed sequentially.

To begin with the problem related to the emergence of symbols, latest at the stage (9.33), the backward reaction part of the replication loop

$$\overline{RNA} \leftrightarrow RNA \tag{9.34}$$

will turn impossible and the cycle will no longer be closed. Without the fundamental RNA cycle, growth and multiplication of the coacervate can no longer proceed. Thus, a backup loop system must be installed that is redundant initially and later takes over the central replication task and simultaneously is liberating the RNA cycle to specialize on the translation task. This can be managed by a trick that is often found in biological evolution; first an existing system is identically duplicated, and later the twins break their symmetry and differentiate for two separate tasks which were previously managed by the one original system alone, although less effectively. In our case, the symmetry breaking is between growth and multiplication activity of the primordial cell. As a model for this process, we can imagine that the early replicase was not very specific and could occasionally replicate also DNA strands at RNA templates, and vice versa.

As a result, four separate cycles appear of which three are redundant, as shown in Figure 9.7. DNA may well take over the linear replication task of RNA but cannot provide similar chemical activity. If the RNA is systematically modified by some

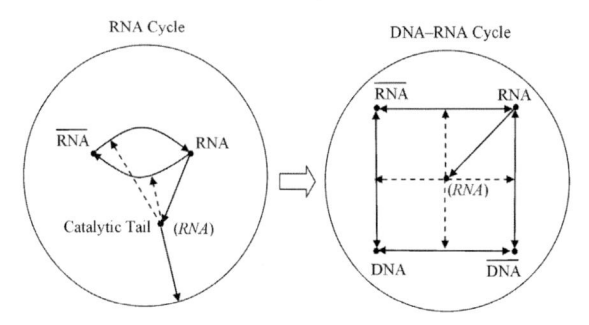

Figure 9.7 Duplication of the RNA-replicase cycle by occasional production of adequate DNA chains supported by an unspecific replicase molecule. To keep the schematic simple, in the right panel the binary cycles are indicated by bidirectional arrows, and the catalytic tail is not drawn except for the replicase function. Of this DNA–RNA cycle, several links can be inhibited or removed by evolutionary changes without degrading its functionality.

post-processing in order to intensify the related catalytic activity, it may no longer be suitable to produce RNA replicas, as shown in Figure 9.8. The DNA–RNA cycle appears as a simple option for a successor of the RNA-replicase cycle which at the same time offers the initial conditions for the ritualization transition (9.31)–(9.33).

These simple models demonstrate that the reduction in RNA chains to codons in the context of the ritualization transition requires the prior installation of a backup system which maintains the task of the RNA loop. We can imagine this process as duplication with subsequent symmetry breaking into the different functions of sequence replication and catalyst production, similar to duplication processes frequently found in the biological evolution of plants and animals (Jiao, 2011). The initially redundant auxiliary DNA loop turns into the backbone of the replication process while RNA becomes liberated to specialize for the translation of the DNA sequence to chemical activity.

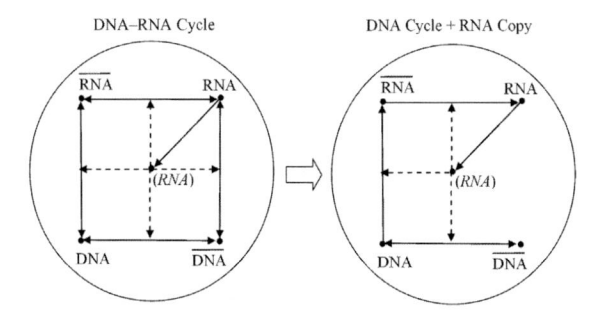

Figure 9.8 RNA postprocessing may delete some replication options from the multiple DNA–RNA cycle, in particular, RNA may no longer function as a template for DNA. The left part corresponds to the right panel of Figure 9.7, and the right part is a schematic of the situation (9.31) which permits subsequent ritualization.

After having considered essential preconditions for the reduction of longer RNA strands to shorter, symbolic identifiers, we must now take care of the second key problem for ritualization, namely the need for the spatial arrangement of symbols in sequences. If we imagine a multi-RNA coacervate similar to Figure 9.5, the reduction in the various RNA pieces to independently floating triplets makes little sense unless those codons queue up in a definite manner to form longer "texts" that can be read by the translation machinery. Very probably, this should have happened already before the reduction process to symbols commenced. Hence, we must think of suitable processes during which independent RNA pieces are subsequently connected to longer chains, and the catalytic tails they are producing will mutually interact and assemble to larger catalytic units.

A reasonable model scenario may start from that end. When a pool of different RNA cycles is coding for different catalytic tails within the same compartment, chemical interactions between the members of those tails are likely. An example may be a situation in which the replicase becomes more effective as soon as some molecule produced within another tail is attached to it. As a rule, bigger globular molecules have more complicated surface structures and more specific catalytic activities than small ones. Self-assembly and aggregation of complex organic molecules has often been observed experimentally (Ghosh and Chmielewski, 2004).

When more effective catalytic tools arise from certain combinations of independent RNA fragments, one may imagine that those fragments become concatenated to ensure their joint presence within each daughter compartment, and to run them in temporal and spatial proximity. A termination marker may be required on the now longer strand to indicate the end of one fragment and the beginning of another one. Such a process may be rather complicated in the chemical detail of its requisite structures and workflows, but we can imagine it as the origin of the recent building-block mounting principle of enzyme synthesis. Such a development can replace a larger number of independent RNA pieces of the multi-RNA-replicase model by fewer but longer assemblies of those. Each of the long, multifragment RNA then effectively codes for a mounting pipeline to fabricate a more complex and more effective catalyst. In this phase of *explicit coding* (Ebeling and Feistel, 1982), the number of possible catalysts obtained from alternative combinations of a few basic "standard" elements, just by permutation of the mounting sequence, could increase significantly. On the other hand, occasionally changing the mounting sequence became more difficult with multifragment RNAs. This unfavorable situation could improve again after the reduction in the full RNA pieces to short symbols. In that case, point mutations of the RNA are equivalent to the substitution of a valid symbol by another valid symbol for a building block taken from the same pool, rather than to an "alien" molecule to be built in.

With the gradual shortening of the RNA to a sequence of symbols which controls the succession of mounting standardized building blocks in order to produce more complicated catalysts, RNA developed to an information carrier similar to the recent mRNA. In the DNA–RNA model, this shortening must have begun in the DNA cycle rather than in the RNA chain because DNA is read-only already at that stage. This distinct role of DNA was permitted by the RNA that was still functioning as a proper messenger between DNA and tRNA, even without the now redundant segments.

Summing up, the transition shown in Figure 9.6, from the primordial coacervates with RNA-replicase cycles to the most primitive cell equipped with a symbolic information-processing system, should have implemented six important qualitative steps (see Ebeling & Feistel, 1982, for a more detailed, stepwise scenario):

1) diversification of RNA cycles and catalytic tails;
2) RNA concatenation, explicit coding, building-block synthesis;
3) RNA–DNA cycles;
4) RNA–DNA symmetry breaking of replication and translation;
5) assembly of linear and folded RNA to a tRNA-like molecule;
6) reduction in DNA to a codon sequence and of explicit RNA to an mRNA-like molecule.

With the appearance of the first self-reproducing microreactors capable of symbolic information storage and processing, a qualitatively new form of matter appeared on Earth: life. Those earliest forms were equipped with proper tools to experience successful Darwinian evolution. In the very first stage, the ritualization transition must advance to a mature state.

9.5
Darwinian Evolution

Symbolic information processing, established as the result of a ritualization transition, is an extremely powerful and flexible tool for problem solving, as discussed in Chapter 8. Digital software control is superior over analog hardware devices, as known from the triumph of talking humans over specialized prey and predator animals, to the market success of computers over mechanical typewriters. The first living cells with primitive genetic code and translation apparatus gained enormous advantage over other, less-advanced self-reproducing competitors and certainly dominated the early scenery very quickly.

From the viewpoint of information, as with any ritualization transition, the liberty gained by releasing the previously inherent structural coupling between the symbol and its meaning, opened the way to a fast diversification of the genetic code and the related pool of standard building blocks available for the series production of catalysts. Perhaps this differentiation began with just a binary purine-pyrimidine code and the two simplest stable amino acids, glycine (CH_2-NH_2COOH) and alanine (CH_3CH-NH_2COOH), see Section 8.6. The incorporation of an additional enzyme requires nothing but an error copy of the template sequence, and by processing that copy the readily available machinery can immediately produce and test the new tool. There was an immense demand for catalytic tools to consolidate and speed up error-free replication and translation, to build membranes and organize the internal cellular structure, to coordinate regular cell division, to explore new sources for energy and material, and so on. As an illustration for the complexity of recent

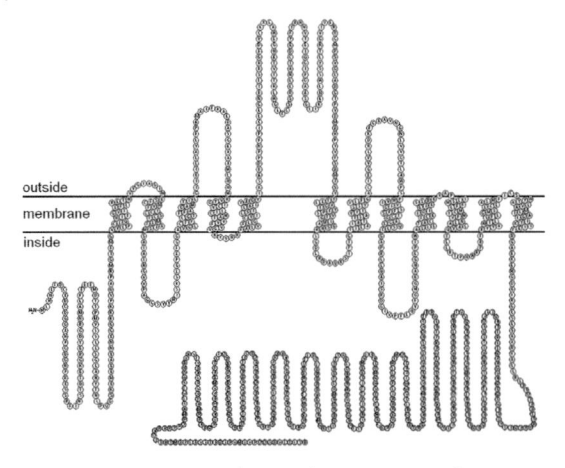

Figure 9.9 Primary and secondary structure of a recent potassium-chloride cotransporter protein KCC2 located in a neuronal membrane (Payne, Stevenson, and Donaldson, 1996; Hartmann, 2010). Schematic courtesy of Anna-Maria Hartmann.

functional proteins, Figure 9.9 shows the primary and secondary structure of a recent KCl cotransporter located in a cell membrane (Payne, Stevenson, and Donaldson, 1996; Hartmann, 2010).

When the rapid expansion phase reached a state of saturation, freezing of the code became more advantageous than adding more symbols and new building blocks. The structure of the recent genetic code, Table 8.1 in Section 8.6 strongly suggests that it was frozen in before any regular, symmetric system was completed, in contrast to what an intelligent designer would have planned in advance before running and testing such a fundamental and universal conversion device in practice. Here, we shall speak of *Darwinian evolution* after the ritualization transition, in contrast to *molecular evolution* before that magnificent crossover from physical chemistry to biology. In our model scenario, the most rudimentary form of the genetic code was invented in the molecular evolution phase but completed and perfected during the subsequent Darwinian evolution phase.

Molecular evolution probably began under conditions that may truly be termed "the hell on Earth," soon after a giant impact that created the Moon, and the formation of a solid crust about 4500 Myr ago, compare Table 3.1, in a hot, anoxic, and perhaps acid ocean. The early Darwinian evolution phase apparently began already before the termination of massive showers of asteroids and comets about 3800 Myr ago, and at about that time resulted in our Universal Common Ancestor (David and Alm, 2011).

With the self-accelerating and expanding population, the consumption of available energy and material resources increased rapidly. So far we had assumed here that the accumulated stock of abiogenically generated, energy-rich molecules was large enough and its replenishment fast enough to maintain the moderate evolution process. The existence of early cells always relied on catalytic activity, that is, they could not survive under equilibrium conditions without continual consumption of free energy and production of entropy. When a population grows exponentially and food is produced at only a constant rate, demand will exceed supply at some point of time. The usual result is a collapse of the existing population, similar to the example shown in Figure 6.3. Competition for fastest growth is suddenly replaced by competition for energetic efficiency of the metabolism, and for exploitation of alternative or spatially remote sources of free energy (Ebeling and Feistel, 1982).

The permanent solar irradiation at the ocean surface provides much more energy than the geothermal vents at the sea floor (Chapter 3). Cells that exploited light energy, for instance by molecules similar to porphine (tetrapyrrole) or to phycocyanin, could develop to populations independent of the original geothermal feeding (Pepper, 1998; Leslie, 2009). As an aside, similar photo-sensitive substances are also used in modern solar cells such as Grätzel cells (Caramori et al., 2010), for which the inventor was recently awarded with the Millenium Technology Prize of 2010. From the geological, genetic, and fossil evidence, see Table 3.1, plausible but speculative evolution scenarios can be painted, beginning 3800 Myr ago (Gould, 1993; Szostak, Bartel, and Luisi, 2001; Ricardo and Szostak, 2010).

Likely, the 3800 Myr old Isua banded-iron formation in Greenland resulted already from biological activity. If so, it is among the oldest traces that earliest life left behind. The mass of oxygen chemically bound globally in those iron oxide ores exceeds several times that in our present atmosphere. These and other problems keep the question open what exactly happened when those rocks were formed, whether biology was involved, and if it was, in which way exactly (Dymek and Klein, 1988; Canfield, Rosing, and Bjerrum, 2006). Available data suggest that cyanobacteria, the ancient producers of oxygen by photosynthesis, did not appear earlier than about 2500 Myr back (Rasmussen et al., 2008; David and Alm, 2011; Schirrmeister, Antonelli, and Bagheri, 2011). Rather, anoxygenic photosynthesis by archaea seems to be the first mass exploitation of light energy without releasing oxygen (Beukes, 2004; Blake, Chang, and Lepland, 2010). Biological origin of the iron formation is supported by carbon isotope data (Mojzsis et al., 1996). Still today, recent archaea populate hydrothermal vents and other anoxic and sulfidic aqueous environments with extreme temperatures and pH values; they probably were the first life form on Earth. Anaerobic processes without free oxygen available in the ocean or in the atmosphere may also have been responsible for banded-iron formations (Kappler et al., 2005; Canfield, Rosing, and Bjerrum, 2006). The most probable phylogenetic tree root consistent with observational and genetic evidence (David and Alm, 2011) is shown in Figure 9.10.

Archaea-type organisms may have been the first which exhaustively populated the ocean from the hydrothermal vents to the surface. Their activity precipitated dissolved iron and changed the chemical environment to conditions more favorable for further evolution steps. The oceans cooled down, and about 2700 Myr ago, apparently due to self-organized, preferable conditions, archaea experienced an unprecedented diversification when the lineages of eukaryotes (2670 Myr) and of cyanobacteria (2500 Myr) diverged from the archaean root, the putative Universal Common Ancestor (Woese, 1998).

Photosynthesis by cyanobacteria must have been strongly superior over that of archaea; the unrivaled sovereigns over 1000 Myr of Earth history were suddenly dethroned by their more effective offsprings. Almost explosively, free oxygen

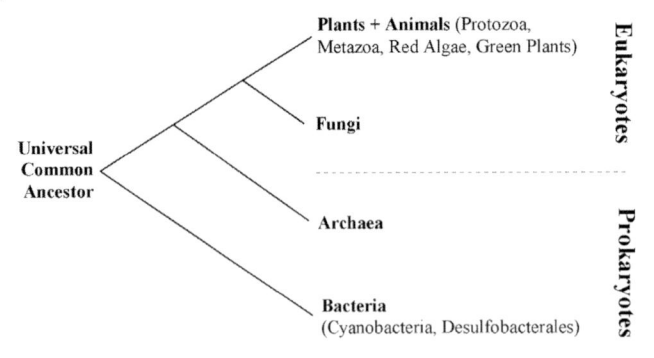

Figure 9.10 Main branches of the earliest evolutionary tree. (Adapted from David and Alm, 2011).

appeared in the ocean and in the atmosphere, culminating in the Great Oxidation Event 2400 Myr ago (Lyons and Reinhard, 2009). Weathered rocks and petrified fossils are undoubted witnesses of that period (Rasmussen *et al.*, 2008; Scott *et al.*, 2008). Up to date, under calm and warm weather conditions cyanobacteria cover the sea surface with dense populations of occasionally toxic species, see Figure 9.11. Their scarcely predictable occurrence is, therefore, carefully monitored in affected seas (Nausch, 2006; Feistel, Nausch, and Wasmund, 2008a). In the light of the dense population visible in Figure 9.11, it is not surprising that fossil evidence of colonial organisms and eukaryotes from 2100 and 2000 Myr ago, respectively, could be discovered (El Albani *et al.*, 2010; Zimmer, 2009b).

The energy available from respiration of free oxygen was a powerful driving force for evolution, such as for the development of very complex eukaryotic cells with their sophisticated genetic apparatus and cell division procedure. Intense, widespread phytosynthesis at the sea surface must have produced vast quantities of dead

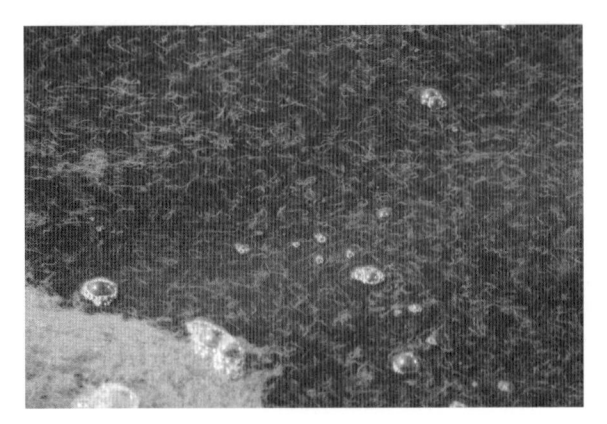

Figure 9.11 Dense population of cyanobacterium *Nodularia spumigena* observed at the Baltic Sea in August 2006 by r/v "Prof. Albrecht Penck" (Nausch, 2006). Photo courtesy of Toralf Heene.

organisms that sank to the sea floor, similar to the situations observed today in productive upwelling areas (Mohrholz et al., 2007; Stramma et al., 2008; Chan et al., 2008; Diaz and Rosenberg, 2008; Savchuk, 2010, see also Section 3.8), or during the plankton blooms in euphotic estuaries such as the Baltic Sea. The available stock of biomass in the deep water is a niche that could quickly be exploited by bacteria or other organisms. As soon as dissolved oxygen is exhausted and not replenished by deep-water ventilation, available sulfate is reduced to hydrogen sulfide by specialized bacteria such as *Desulfobacterales* in the Baltic Sea of our days. Similar processes happened very likely about 1900 Myr back when anoxic and sulfidic conditions extensively prevailed in the ocean over a "boring billion" of years, see Table 3.1. Owing to the widespread distribution of the anoxic conditions in space and in time, it is believed that different sulfate-reducing bacteria developed repeatedly and independently, beginning as early as 3700 Myr ago (Baumgartner et al., 2006).

While in oxidized waters the solubility of phosphate is low, the opposite is true for anoxic conditions (Nausch, Nehring, and Nagel, 2008; Wasmund and Siegel, 2008; Reissmann et al., 2009). When dead organisms sink down from the sea surface and are decomposed in anoxic deep water, ammonium and phosphate are released and become available again as nutrients in the surface layer sooner or later, depending on the vertical transport rate. At low oxygen concentrations near the anoxic/oxygenic transition zone ("redoxcline"), nitrate is partly transformed to gaseous nitrogen and released to the atmosphere, while phosphorus remains in the ocean. As a result of slow but enduring nitrogen loss over long periods, the dominating anoxic ocean of the "cyanobacteria age" from 1900 to 800 Myr ago was probably enriched with phosphates and deficient of inorganic nitrogen compounds. The circumstance that phosphate circulated in the anoxic ocean but dissolved nitrate gradually disappeared may easily explain why cyanobacteria "learned" to capture atmospheric molecular nitrogen, and also the apparent paradox (Ricardo and Szostak, 2010) that phosphorus is so common in organisms even though it is almost unavailable in ordinary oxygenic ocean waters. Hence, also in today's Baltic Sea, cyanobacteria find preferable conditions for their growth when the deep water is anoxic and the N:P ratio in the surface layer is low (Reissmann et al., 2009). They can grow abundantly if just excess phosphorus is available while other phytoplankton is suffering from nitrogen deficiency.

The first fossil evidence for sexual reproduction is for red algae *Bangiomorpha pubescens* from the 1200 Myr old Hunting Formation in northern Canada (Butterfield, 2000). It is assumed that the successful management of complex morphogenetic multicell organization was the main selective advantage of generative recombination, based on the advanced eukaryotic cell structure (Ramakrishnan, 2011). Rhodoplasts, the photosynthetically active units of red-algae cells, are organized similar to cyanobacteria (Yoon et al., 2006). Red algae are autotrophic specialists for subsurface dim-light conditions and exploit the optical blue–green absorption gap of cyanobacteria, see Figure 9.12. Those facts indicate that, perhaps by endosymbiosis, red algae may have developed in an ecological niche at greater ocean depths, beneath the bright surface layer dominated by mats of cyanobacteria, similar to conditions also found for the Ediacaran period in lakes (Gingras et al., 2011).

Figure 9.12 Recent red alga *Delesseria sanguinea* in the Baltic Sea. Photo taken in April 2004 off Boltenhagen, courtesy of Dirk Schories.

It seems as if it took a 1000 Myr period of global marine cyanobacteria regency until the first fungi managed to leave the ocean and to conquer the land, some 900 Myr ago (Lücking *et al.*, 2009), immediately followed by (and likely in combination with) the general establishment of photosynthesis on land (Knauth and Kennedy, 2009). Many of today's land plants live in symbiosis with fungi (Maillet *et al.*, 2011). The suddenly "green Earth" experienced severe climate changes and turned into the cryogenian "snowball Earth." Thermophilic cyanobacteria possibly suffered from that enduring cold, or other reasons may have been responsible for the dramatic transitions that permitted a "second great oxidation event" about 600 Myr ago (Scott *et al.*, 2008), but this time possibly sustained by photosynthetic eukaryotes rather than by cyanobacteria. Similar to the first event, abundantly available oxygen stimulated a rapidly progressing evolution phase which cumulated in the "Cambrian explosion" of eukaryotic, multicellular organisms. After the two long periods of archaean and cyanobacteria dominance, it is this third fundamental phase of Darwinian evolution which still continues today.

Even on the longest time scales of our hypothetical scenario, we recognize the fundamental working principle of evolution processes outlined in this book, the alternating sequence of fluctuations, instabilities, and irreversible self-organization steps. Early Earth turned unstable 4500 Myr ago, with respect to certain fluctuations of the aqueous catalytic network. The subsequent molecular self-organization process resulted eventually in the very first ritualization transition, that is, the origin of life on Earth. The novel living beings initiated the period of archaea that began 3800 Myr back. In the wake of the first dominating ecosystem on the planetary scale, photosynthetic cyanobacteria and sulfate-reducing bacteria prepared the next instability. They triggered a new self-organization phase which made cyanobacteria dominate the world ocean. With the Great Oxidation Event 2400 Myr ago, they established completely new physical, chemical, and climatic conditions all over the globe, including the atmosphere and the land surface. In that lasting period of time, in the "boring billion" of years, under extended anoxic conditions in the deeper world ocean, fungi and metazoans gradually developed until they were able to induce a new global instability – the uninhabited continents suddenly turned green and caused the second Great Oxidation Event. As its result, about 600 Myr ago the Cambrian Explosion changed the world once again completely and gave way for the unprecedented self-organization of plants, animals, and humans.

10
Conclusion and Outlook

Act only according to that maxim whereby you can at the same time will that it should become a universal law.
        Immanuel Kant: Groundwork for the Metaphysics of Morals

10.1
Basic Physical Concepts and Results

Let us summarize some of the basic views which we developed in this book. The first point is connected with the relation between physical law and the properties of evolutionary systems.

There is very little doubt that the canonical equations of classical mechanics

$$\frac{dq_k}{dt} = \frac{\partial H}{\partial p_k}, \frac{dp_k}{dt} = -\frac{\partial H}{\partial q_k} \tag{10.1}$$

or alternatively those of quantum mechanics for, say, microscopic effects in biochemical reactions or the absorption of photons are correct and valid for practically all processes we observe in our daily life, from the motion of stars in the dark sky to driving a car through downtown traffic or listening to an organ concert in a cathedral. Except for special processes at very high velocities, in very strong gravity fields, or in nuclear reactions, no violation of the classical or quantum-mechanical equations could ever be observed, in particular not even in "unphysical" biological or in social systems. So, what is the problem, then?

One problem is that those equations are too detailed and too correct for answering almost any practical question. To compute the next solar eclipse, the "correct" physical way would be to determine the masses, coordinates, and momenta of all atoms of Sun, Earth, and Moon at a given point of time, and integrate the equations of motion for those initial conditions (ignoring for the moment all the rest of the world). This list of atoms would not only include terrestrial rocks and turbulent gases in solar protuberances, this "correct" approach would also require the knowledge of the position and curvature of each hair on each human's head on Earth, the properties of each droplet falling down from the clouds, and of each wing of each buzzing fly. In

Physics of Self-Organization and Evolution, First Edition. Rainer Feistel and Werner Ebeling.
© 2011 Wiley-VCH Verlag GmbH & Co. KGaA. Published 2011 by Wiley-VCH Verlag GmbH & Co. KGaA.

addition, even if we made reasonable approximations for all those initial conditions, how shall we detect a solar eclipse from the vast number of predicted trajectories of moving atoms? Clearly, this is a hopeless enterprise.

The only way out is a rigorous simplification of our mathematical model at the cost of losing accuracy. We need a substantially reduced set of relevant variables to formulate and to solve our problem. In the case of celestial bodies, we know the recipe. We divide the space into three boxes which contain the Sun, the Earth, and the Moon. In each box we take the total mass of all contained particles, their total momentum, and the barycenter. If we formulate dynamical equations for those mean values by summing up over the atoms contained in each box, we obtain in good approximation a set of closed differential equations which relate the few mean values to each other. Those equations can be solved, for example numerically or in some approximation also analytically, and whenever the positions of the "mass points" fulfill certain geometric inequalities, an eclipse can be predicted.

For the solution of practical problems, simplified mathematical descriptions in terms of suitable reduced sets of variables are required. Unfortunately, we do not possess any universally valid rules or procedures of how this can be achieved. Many problems, such as the oscillation frequency of a mass fixed at a spring, can be solved sufficiently accurately in the way described before, namely by taking suitable averages over certain parts of the given system. The atoms of the spring and of the mass are spatially separate, and the timescale on which the mass oscillates is much slower than that of the thermal motion of the atoms. Under such circumstances, we can derive from Equations 10.1 for interacting atoms a simple dynamical equation for the mechanical harmonic oscillator

$$m\frac{d^2x}{dt^2} = -kx \tag{10.2}$$

where the properties and interactions of all the atoms of the spring and the mass are represented by a reduced set of just three variables, the mass m, the spring constant k, and the position x. The recipe to arrive at an equation of the form (10.2) is to separate the original processes into three classes:

1) processes that are much faster and on much smaller scales than those we are looking for,
2) processes that are much slower and on larger scales than those we are looking for, and
3) processes that we are mainly interested in.

In Haken's terminology, processes (1) are represented by *slaved variables*, (2) by *control parameters*, and (3) by *order parameters*. For the slaved variables, their integrals or stable attractor states must be determined, or reasonable assumptions must be made, such as random motion. For the control parameters, given constant values can be assumed. What remains are a few equations between the order parameters which can be solved numerically or studied analytically by the methods we discussed in Chapter 4. In the case of Equation 10.2, there is one order parameter, x, and there are two control parameters, m and k. All the details of microscopic motion and

10.1 Basic Physical Concepts and Results

interaction are either neglected or indirectly, effectively represented by the three remaining parameters, x, m, and k. It must be emphasized that the elimination procedure for the slaved variables involves three important qualitative impacts on the resulting equations:

1) The resulting equations for the order parameters exhibit memory effects and are no longer strictly Markovian, that is, instantaneous.
2) Since integrals of dynamical equations have in general different symmetries than the equations themselves, the resulting equations for the order parameters may exhibit symmetries different than those of the original Equations 10.1.
3) Neglected or unresolved internal and external processes appear as random noise acting on the order parameters.

Looking closer at the oscillator example (10.2), a memory effect (1) results for instance from the fact that the spring and the mass are not exactly point-like or strictly rigid bodies. When they move, elastic sound waves propagate across the mass and along the spring, and overlay the forces between the mass and the spring in a time-dependent way, such as described by the equation

$$m\frac{d^2x}{dt^2} = -k\int_0^\infty R(\tau)x(t-\tau)d\tau \tag{10.3}$$

where R is a response function that expresses some delay between the acting force and the motion of the bodies. As soon as the oscillation period is comparable with the memory effect, the function R can no longer be approximated by a Dirac delta.

Looking closer at the oscillator example (10.2), a symmetry effect (2) is that the spring deformation is never exactly elastic. For instance, propagating elastic sound waves results in local temperature gradients and dissipation of heat so that a damped harmonic oscillator

$$m\frac{d^2x}{dt^2} = -kx - \gamma\frac{dx}{dt} \tag{10.4}$$

is an improved approximation for the order-parameter equation. While Equation 10.2 is reversible, Equation 10.4 is irreversible; the time symmetry of fundamental equations 10.1 is broken at the description level of the order parameters.

Looking closer at the oscillator example (10.2), a noise effect (3) is that the oscillator may be mounted in a laboratory with turbulent air motion which may result in additional random forces, such as described by the Langevin equation

$$m\frac{d^2x}{dt^2} = -kx + \xi(t) \tag{10.5}$$

The random force ξ is unknown in detail but certain statistical properties can reasonably be associated with it. As a result, predictions based on Equation 10.5 can provide results only in terms of probabilities, $P(x, t)$, and related uncertainties.

The modifications (10.3)–(10.5) are standard methods used for dynamical models using order parameters. It must be underlined that there are no universal, naturally given order parameters; their choice is to some extent arbitrary and depends on the particular problem and on our specific interest in scales of space and time that we subjectively consider as the relevant ones. The physical process itself is of course unaffected by the spot at which we focus our magnifying glass. As an example, theoretical chemistry with its wide range of time scales and reaction rates is traditionally using separation techniques between fast and slow reactions (Hahn, 1974). Similarly, processes in oceanography extend from milliseconds and millimeters to millennia and thousands of kilometers and require a careful preparation of equations which govern the processes in a selected spectral window of interest.

Two problems pose significant difficulties in this general order-parameter concept and must be considered as basically unsolved:

1) A clear separation between the temporal and spatial scales of the three classes of parameters may not always be possible, and the resulting system of chosen order parameters may be too large, too complicated, or too crude for practical use.
2) The order parameters may not simply be explicit functions of the microscopic variables and may represent complicated integrals of the microscopic equations.

As an example for the problem (1), we may consider turbulent or climate processes. We do not know what the best order parameters are for the description of those processes, not even their exact number (i.e., the dimension of their attractors), nor do we possess closed and sufficiently accurate dynamical equations in terms of those parameters. Practical approaches in such cases are often based on numerical solutions on space–time grids, that is, by taking averages over certain regions, over subgrid processes. Dynamics of clouds, for instance, is a typical unresolved subgrid process in climate models that is either ignored or very roughly approximated by a procedure termed "parameterization."

Problem (2) is of central interest in the context of this book. Order parameters of this kind are also termed *emergent properties*:

- Emergent properties are certain integrals of the microscopic motion under given initial and boundary conditions.
- Emergent properties cannot be expressed as functions of the microscopic variables.
- Emergent properties represent holistic, irreducible properties of the system.
- Emergent properties may possess different symmetries than the microscopic equations.
- Emergent properties are consistent with the fundamental laws of physics

From the virial theorem of mechanics (Landau and Lifschitz, 1967a), a formal but instructive example for a physical emergent property, $L(t)$, is known as defined by the functional

$$L(t) = \frac{1}{t}\int_0^t dt' \sum_k \left(p_k(t') \frac{\partial H}{\partial p_k} - q_k(t') \frac{\partial H}{\partial q_k} \right) \tag{10.6}$$

where the integral is carried out along the phase–space trajectory $\mathbf{p}(t)$, $\mathbf{q}(t)$. If the system is bound so that neither velocities nor coordinates may take infinite values, such as for the solar system, for a galaxy, or for gas particles in a given volume, $L(t)$ is an irreversible function of time which asymptotically converges to zero,

$$\lim_{t \to \infty} L(t) = 0 \tag{10.7}$$

Nontrivial emergent physical properties are those which describe thermal processes, in particular the entropy. According to the Boltzmann formula for the equilibrium entropy

$$S(E, V) = k_B \ln W \tag{10.8}$$

the phase volume W of the energy shell, $H(\mathbf{p}, \mathbf{q}) = E$, of a system with the spatial volume V can be interpreted as an integral similar to Equation 10.6 carried out along the phase trajectory, that is, over the microscopic attractor.

Emergent properties reflect the complexity of physical processes; most of those properties do not belong to the standard basic toolkit of physics. They represent the dominating, collective, coordinated motion of many atoms in certain dynamical modes on spatial and temporal scales of our interest in our models of reality. They are not reducible to the motion of single atoms. Entropy or life are not some kind of mysterious aethereal medium or divine spirit such as "phlogiston" or "vis vitalis" which could possibly be discovered under a suitable microscope. Magnifying a few atoms of a living cell or a water droplet cannot tell whether those atoms are alive or possess a certain entropy value.

Emergent properties gain increasing importance for the description of more complex macroscopic systems. For the dynamics of biological and social systems, physical constrains imposed by the conservation laws of mass and energy as well as the second law are still highly relevant and cannot be violated, but they do no longer dominate the evolution of the system. The number of microscopic variables is equally high for simple physical or complex biological systems; the important difference between the two is that simple systems need only very few order parameters for their description (such as temperature and volume, two quantities, for an equilibrium state), while complex systems possess much more degrees of freedom in terms of order parameters required for their dynamical models.

Deducing order parameters of complex systems from the fundamental equations of motion is difficult. Rather, order parameters are preferably derived from experience, from observational or experimental evidence. Temperature is a typical example; it controls the heat flow between macroscopic systems when they are in thermal contact. For its high relevance for the survival of organisms, we are able to sense temperature immediately. In contrast, a rigorous theoretical–mechanical proof is still lacking for the fundamental observation that heat is always flowing down the temperature gradient.

Of particular importance for the dynamics of irreversible systems are properties termed *"values."* Entropy can be interpreted as the value of energy and controls the direction of irreversible thermodynamic processes. Selective values determine the evolution of biological species. Economic exchange values on the market rule the competition among commodities and producers. Information values of genetic sequences can be identified with selective values. The value of information exchanged between humans can be related to the exchange value of the commodity "information."

Beyond the fact that all those values are related to one another, they have interesting features in common:

- Values assigned to elements of a system represent holistic properties ("The whole is more than the sum of its parts").
- Value concepts are of central importance for the quantitative and qualitative description of the systems they belong to.
- The irreversible dynamics of the systems is intrinsically connected with extremum principles of the time evolution of values.
- None of those value concepts is thoroughly understood by now.

In contrast to the reversible fundamental equations of physics, values are emergent properties which express the empirically observed irreversible tendencies of energy to dissipate, of biological fitness to improve, and of economic productivity to grow. A related central question of this book is whether those universal *irreversible tendencies* imply the existence of universal *extremum principles*, of ultimate goal states, or unique attractors toward which evolutionary processes gradually proceed. The answer is *yes* for isolated thermodynamic systems, where the tendency of positive entropy production implies the maximization of entropy which eventually results in a unique equilibrium state. In contrast, the general answer is *no*. In the succession of discrete states, the directional tendency is reflected by *asymmetry* of the binary relation between "successor" between subsequent states. For the existence of a global order relation between the states also *transitivity* is necessary. Asymmetry is not sufficient for transitivity. Similarly, in the time development of continuous states, irreversibility can be expressed by local Pfaffian differential forms similar to the Prigogine–Glansdorff evolution criterion. If the Pfaffian form is integrable to a proper function (i.e., if the differential expression is an exact differential), the local tendency implies the existence of a global extremum principle. For the general Prigogine–Glansdorff criterion, this is possible in the limiting case of the immediate vicinity of the equilibrium state in the form of Prigogine's theorem on the minimum of entropy production. Similarly, for several biological and chemical models discussed in this book, this condition is fulfilled only in very special, even though important, cases. As a consequence, a global criterion of *survival of the fittest* may in general not reasonably exist and may better be expressed in the form of a local rule: *survival of the fitter*.

A similar contrast exists between the uniqueness of equilibrium states and the historicity of evolution and of other processes far from equilibrium. The equilibrium state is unique and will be approached by a system starting from any arbitrary initial state. Once the equilibrium is reached, the system's particular history is forgotten and cannot be extracted anymore from its macroscopic properties. For evolving systems, the situation is very different. Typically, very specific events, more precisely, critical fluctuations, such as mutations, inventions, and innovations, but as well accidental random events chose the actual evolution pathway from a multitude of alternatives. We know many examples from the history of our planet, from the history of our society, or of our family where singular natural events or decisions made by single persons caused irreversible consequences visible ever after.

This fundamental property of historicity is strongly evident in the biological and social evolution which is based on the collection and exploitation of symbolic information in the form of genes and of human language. Exploitation of this extremely long memory and the resulting prediction capability is the key to life and survival. The self-organization of symbolic information by a process termed ritualization is similar to kinetic phase transitions known from various dynamical systems. The very first ritualization transition was the origin of life. Model scenarios discuss how this transition may have occurred under the conditions of the early Earth, 3–4 billion years ago. A similar transition occurred when humans separated from their animal ancestors by the use of orally transferred information, perhaps 2 million years back, and especially with the development of written language, only about 6000 years before today.

We conclude this book with some final remarks on the future of evolution.

10.2
Quo Vadis Evolutio?

Most of this book is concerned with the understanding of the evolutionary processes which created the world surrounding us. Let us try now to take a view into the possible future. In particular, we like to discuss the role of the research on self-organization and evolution phenomena from the perspective of global problems in the environment, in the finance, and climate systems, which humankind is currently confronted with. According to our view, the physics of self-organization and evolution may – in collaboration with other branches of science – play a helpful role for the solution of the pending problems. Let us discuss this by means of several examples.

One of the most urgent problems is global warming. We all know about the Kyoto Protocol which was a first international agreement linked to the United Nations Framework Convention on Climate Change. The major feature of the Kyoto Protocol is that it sets binding targets for 37 industrialized countries and the European community for reducing greenhouse gas (GHG) emissions. Those amount to an average of 5% with respect to 1990 levels over the five-year period 2008–2012. In 2007, the Intergovernmental Panel on Climate Change (IPCC) argued that GHG emissions from industrialized countries must fall by 25–40% by 2020 to keep global warming below a maximum of 2 K. These and other experts predict serious consequences for humankind if the increase within this century exceeds 2 K (IPCC, 2007; Bunde, Kropp, and Schellnhuber, 2002; Schellnhuber et al., 2006; Schellnhuber, 2011). Following the warnings of the experts, the representatives of many governments in the United Nations Climate Change Conferences recommended at several meetings (such as at Copenhagen and Cancun in 2010) to limit the increase to at most 2 K. We agree of course completely with this aim, although we would prefer more scientific formulations of the targets and at the same time more flexible conclusions. As we have tried to make clear at several places in this book, and in particular in Chapters 3 and 4, any prediction on the future of evolving systems, and the climate is one of those, is inevitably associated with large unavoidable uncertainties. A century of research in stochastic theory and 50 years of chaos theory have taught us that we can predict only mean values, dispersions, and probability distributions, but never individual trajectories. This is a matter of principle.

Another point we emphasized in this book is the role of atmospheric vapor and clouds. In fact, water vapor (rather than CO_2) is the strongest GHG, as already stated a century ago by Emden (1913). When the atmosphere gets a little warmer, it stores more vapor, which in turn increases the greenhouse effect, and so on. In a simple linear radiation-balance model, this instability amplifies the initial warm or cold temperature fluctuation. The positive feedback results in either the complete evaporation of all oceans on a Venus-like hot planet, or a complete freezing of all oceans on a Mars-like cold planet, that is, the model does not possess a stable stationary state with liquid oceans and moderate temperatures. Hence, simple radiation–evaporation balance models of the Arrhenius type are dynamically unstable for the ocean. Our conclusion is that the climate attractor is chaotic rather than a randomly fluctuating stable steady state, as sometimes suggested by the popular greenhouse argument. The role of water in the atmosphere and in the climate system urgently requires more research work. We agree with the point of view expressed by many authors that we actually need to understand more details of the functioning of that nonlinear complex system (Bunde, Kropp, and Schellnhuber, 2002). Solutions to the complicated problems of evolution were never simple and easy, neither we are able to offer simple recipes here.

Coming back to the problem of global warming, even the best strategy which promises that the mean temperature increase remains within the planned limits cannot guarantee that the attractor's trajectory will actually stay below that boundary. To prevent the temperature curve from exceeding the borderline of 2 K, we may need to fulfill much more stringent limitations, and even then the effect remains uncertain for a number of reasons. The hope that the global mean temperature can actively be steered in a rather linear manner on a timescale of, say, a decade of years, by just a single turning knob, the atmospheric CO_2 level, appears overly optimistic. The reduction of anthropogenically induced changes to a minimum is the least risky option we have, but even in that case climatic change cannot fully be excluded, as a look at the prehuman terrestrial history clearly indicates. Ice ages and warm periods formed the human evolution rather than vice versa. Our geophysical understanding of the causes and feedback mechanisms during ice-age cycles is still only fragmentary. Considering the complexity of the climate system, suspicious countermeasures by "geo-engineering" to reduce global warming by additional, alternative human intervention is very dangerous as long as the climate dynamics is neither sufficiently understood nor predictable. By ocean currents and continental ice caps, the climate system has a memory of many thousands of years. Historicity is inherent to evolution. Actions once taken cannot be withdrawn, with all the consequences they may imply on the long term.

Taking into account the inherent uncertainties, more flexible political and economic strategies are urgently needed for the worst and the best-case scenarios, that is, for larger deviations from the mean, which might be improbable but can – in a stochastic nonlinear system – never be excluded entirely. Beside the stochastic aspects, we put the focus on the thermodynamics balances and stability problems as the entropy balance of the planet Earth, the entropy production, the stability, and sensitivity of the climate system. In particular, we underlined the predominant thermodynamic role of water for the functioning principles of the "steam-engine Earth" (Ebeling and Feistel, 1994), and developed a picture of the dynamical, self-organized structure of our planet as a coupled system from the inner core to the top of the atmosphere.

With respect to the development of the climate, we have pointed out in Chapters 3 and 4 that following the targets posed by the agreements in Copenhagen and Cancun, the temperature curve could end in the year 2100 at $+1$ K or at $+6$ K, according to stochastic theory. This outcome may be unlikely, but it may happen with some low probability even if humankind succeeds to keep the most probable warming below 2 K. What we have to learn from evolution is that biological and social communities have to be extremely flexible and adaptive. In particular, the strategies for overcoming climatic problems have to be quite adaptive, which includes, for example, the ability to move agriculture to other geographic regions, in our case probably to regions in north-eastern Europe. In the recent history after the last ice age, several climatic transitions happened, deserts or swamps turned into fertile soils or vice versa, with large-scale human migration processes as the consequence. In today's world of selfish and sometimes even adverse countries, such a perspective poses substantial political, economic, and technical challenges.

Another lesson which we might learn is that the best strategies are often based on the self-organization of nature, which should be supported by appropriate measures. Natural changes are always going on; human intentions to hinder those by technical means may eventually fail. We should not forget that 2–3 billions of years ago, the atmosphere was reducing and full of CO_2. Through the process of photosynthesis using water, oxygen is produced as a waste product. Consequently, the oxygen concentration in the air increased very rapidly. And, as a result of the permanent photosynthesis in autotrophic organisms such as plants or phytoplankton, using water, CO_2, and light, which were available in unlimited amounts, Earth became rich in oxygen and the present ecosystem was formed. Only the great evolutionary invention of photosynthesis reduced the content of CO_2 and generated free oxygen. So our atmosphere is the result of self-organization and evolution of the ecosystem on Earth.

Still the influence of photosynthesis on the CO_2 concentration in the atmosphere cannot be neglected. As well known, there is a seasonal cycle in the Keeling curve of CO_2 concentration associated primarily with the Northern Hemisphere growing season. Since photosynthesis is the most natural way of CO_2 reduction, we are convinced that besides dreaming about purely technical solutions for the GHG problem, as the erection of hundreds of thousands of new windmills and millions of sun collectors in the deserts, one should check carefully all possibilities for the extension of photosynthesis on land and in the oceans. In fact, as a result of 2000 years of modern civilization, the forests in Europe have been reduced to about one-third.

Just looking at the old civilizations in Greece, in Italy, and in Spain, they all succeeded in reducing the woods to a gloomy rest. This way, since the middle ages, the CO_2-binding capacity of woods was reduced to about one-third. Nevertheless this capacity of the forests is still of the order of 10% of the industrial emissions according to recent estimates (Hofmann, Jenssen, and Anders, 2002; Jenssen, 2009). Instead of stronger support for reforestation, the European Union imports every year a very large amount of timber from tropical countries such as Brazil or Indonesia, where still some rain forest survived. We should never forget that photosynthesis is the natural method which evolution invented to fight against a reducing and CO_2-rich atmosphere. The great advantage of any natural strategy based on photosynthesis is that CO_2 is consumed and radiation energy is condensed to a valuable resource and transformed into useful forms of energy. So at the end, getting energy from wood is CO_2-neutral. We should be aware that a tree is a wonderful machine, the best and most effective collector and transformer of energy from the sun (Blankenship et al., 2011). Instead of tolerating that everywhere in the world the forests are cut down in an uncontrolled way and photosynthesis is decreased every year by a large amount, we should fight for each tree. Wood should be used more carefully, starting with use in construction and building and only at the end of the chain by burning it and generating heat.

We are well aware that photosynthesis alone cannot stop the observed growth of CO_2 in the atmosphere, which is in fact due to the increased need in high-valued forms of energy on our planet. Photosynthesis can reduce atmospheric CO_2 only to the extent at which the produced organic compounds are deposited rather than oxidized again. As of October 2010, CO_2 in Earth's atmosphere is at a level of 388 ppm by volume. Atmospheric concentrations of CO_2 fluctuate slightly in the annual course, driven primarily by seasonal plant growth in the northern hemisphere. Concentrations of CO_2 fall during the northern spring and summer as plants consume the gas, and rise during the northern autumn and winter as plants go dormant, die, and decay. This seasonal process is influenced by higher solubility of CO_2 in the colder ocean surface water during the winter (Takahashi et al., 2009b), and release during, and release during warming up in summer. Taking all this into account, the concentration of CO_2 grew by about 2 ppm in 2009 and has reached twice the preindustrial level. To put the numbers in a prehistoric perspective, see Chapter 3, atmospheric CO_2 levels were at 1500 ppm when the Antarctic glaciation began about 35 million years ago, and went down to 280 ppm about 20 million years later after grasslands and ruminants had conquered the continents. Since the onset of quaternary ice ages, concentrations undulated between 180 ppm in cold periods and 300 ppm in warm interglacials.

Human energy consumption amounts to only a minor fraction (about 0.01%) of the total solar energy transfer through the "photon mill." There is enough energy available for our tiny needs. Using alternatives to fossil carbon is an economic, technical, and political rather than a physical or thermodynamic problem (Schiermeier et al., 2008). As physicists, we are convinced that final solutions for the problem of CO_2 increase can be expected only from mastering the mechanism of Sun's energy generation, namely the fusion of light nuclei as hydrogen and deuterium to more heavy nuclei as helium. However this needs still more time for research. It might well be that we will have power stations working on fusion and imitating the mechanism of the Sun not before the end of this century. In this situation, we need good buffers for CO_2 and the best and most natural buffers are still pools based on our forests. Investigations for the example of the East German countries which have an area of about 100,000 km^2 with about 30% forests are able to bind approximately 4–5 million tons of carbon per year. In spite of the fact that this is only about 10% of the carbon-emissions from industry, households, and traffic, it is a remarkable amount and could be improved by more effective forestry based on ideas of the self-organization of forests (Hofmann, Jenssen, and Anders, 2002; Jenssen, 2009). This buffer could provide us one or two more years which could be used for fusion research. Of course we should do also more research for the growing of more effective and more resistant plants and algae, possibly based on new biotechnological methods. Being physicists by education, we should in principle welcome that solutions based on physics such as photovoltaics are so popular among people, and methods based on biotechnology are often considered as suspicious and potentially dangerous. On the other hand, as physicists we understand well that a photovoltaic system with a power of 1 kW needs a very large amount of energy and material resources to come to work; a new species of tree grown in a forest nursery needs less of a percent of this effort. At present, we seem to have big programs for solving the energy problem, but as we believe, there is not enough support for new solutions to develop a more effective photosynthesis on Earth and in particular in the oceans and in lakes or swamps, to develop new strategies in forestry (Jenssen, 2009).

The famous physicist Freeman Dyson was dreaming of developing trees that produce electricity rather than wood. Such a system requires advanced methods of biotechnology. We believe that this research goal is difficult, but may be not more difficult and even less expensive than the CERN research aiming to the exploration of the conditions of the cosmic evolution in the first microseconds, or the research on nuclear fusion in Garching, Greifswald, and elsewhere. What we mean is that we have to learn from evolution that one needs many adaptive and flexible available strategies (Behringer, 2007; Heithoff, 2011).

Exploring the evolutionary history of life and humankind on our planet which survived so many dramatic and dangerous events, we may get lessons how to meet difficult situations in future. Evolution is an inventive process and we see one of the goals of evolutionary research in developing strategies for solving our present-day problems by studying the strategies of nature. There are already many convincing examples how this can help solving problems of technology and computing (Rechenberg, 1973, 1994; Schwefel, 1977; Voigt *et al.*, 1996). As the present authors are deeply convinced, the evolutionary approach might be also very useful in attacking our global problems.

References

Aad, G., ATLAS Collaboration *et al.* (2010) *Phys. Rev. Lett.*, **105**, 252303.

Aamodt, K., ALICE Collaboration *et al.* (2010) *Phys. Rev. Lett.*, **105**, 252302.

Abbot, C.G. and Fowle, F.E., Jr. (1908) Income and outgo of heat from the Earth, and the dependence of its temperature thereon, *Ann. Astrophys. O.* (Smithsonian Institution, Washington DC) **2**, 159–176.

Abel, D.L. (2009) The capabilities of chaos and complexity. *Int. J. Mol. Sci.*, **10**, 247–291.

ABM (2010) *Australian Climate Change and Variability.* Commonwealth of Australia 2010, Bureau of Meteorology, http://www.bom.gov.au/climate/change/aus_cvac.shtml.

Abramowitz, M. and Stegun, I.A. (1968) *Handbook of Mathematical Functions*, Dover Publications, New York.

Agladse, K.I., Krinsky, V.I., Panfilov, A.V., Linde, H., and Kuhnert, L. (1989) Three-dimensional vortex with a spiral filament in a chemical active medium. *Physica D*, **39**, 38–42.

Albert, R. and Barabási, A.-L. (2002) Statistical mechanics of complex networks. *Rev. Mod. Phys.*, **74**, 47–97.

Albert, R., Jeong, H., and Barabási, A.-L. (1999) Diameter of the world-wide web. *Nature*, **401**, 130–131.

Albert, R., Jeong, H., and Barabási, A.-L. (2000) Error and attack tolerance of complex networks. *Nature*, **406**, 378–382. correction: 409, 542.

Alberti, P.M. and Uhlmann, A. (1981) *Dissipative Motion in State Spaces*, Teubner, Leipzig.

Alberty, R.A. (2001) Use of Legendre transforms in chemical thermodynamics. *Pure Appl. Chem.*, **73**, 1349–1380.

Alboussiere, T., Deguen, R., and Melzani, M. (2010) Melting-induced stratification above the Earth's inner core due to convective translation. *Nature*, **466**, 744–747.

Albrecht, F. (1940) *Untersuchungen über den Wärmehaushalt der Erdoberfläche in verschiedenen Klimagebieten*, Julius Springer, Berlin.

Alekseyev, V.V. (1976) Dinamicheskiye modeli vodnykh biogeotsenosov (Dynamical models of aqueous biogeocenoses), in *Chelovek i Biosfera (Man and Biosphere)* (ed. V.D. Vedorova), MGU, Moscow.

Alekseyev, V.V. and Kostin, I.K. (1974) *Biologicheskiye Sistemy v Semledelii i Lesovodstve (Biological Systems in Agriculture and Forestry)*, Nauka, Moscow.

Allen, P.M. (1975) Darwinian evolution and a predator-prey ecology. *B. Math. Biol.*, **37**, 1–17.

Allen, P.M. (1976) Evolution, population dynamics, and stability. *Proc. Natl. Acad. Sci. U.S.A.*, **73**, 665–668.

Allen, P. and Ebeling, W. (1983) On the stochastic description of a predator-prey ecology. *BioSystems*, **16**, 113–126.

Allen, W.L., Cuthill, I.C., Scott-Samuel, N.E., and Baddeley, R. (2011) Why the leopard got its spots: relating pattern development to ecology in felids. *Proc. R. Soc. B*, **278**, 1373–1380.

Alvarez, L.W., Alvarez, W., Asaro, F., and Michel, H.V. (1980) Extraterrestrial cause for the cretaceous-tertiary extinction. *Science*, **208**, 1095–1108.

Anderson, B.J., Acuña, M.H., Korth, H., Slavin, J.A., Uno, H., Johnson, C.L., Purucker, M.E., Solomon, S.C., Raines, J.M., Zurbuchen, T.H., Gloeckler, G., and McNutt Jr., R.L. (2010) The magnetic field of Mercury. *Space Sci. Rev.*, **152**, 307–339.

Anderson, P.W. (1972) More is different – broken symmetry and the nature of the hierarchical structure of science. *Science*, **177**, 393.

Anderson, P.W. (1978) Local moments and localized states. *Rev. Mod. Phys.*, **50**, 191–201.

Andrieux, D., Gaspard, P., Ciliberto, S., Garnier, N., Joubaud, S., and Petrossyan, A. (2007) Thermodynamic time asymmetry. *Phys. Rev. Lett.*, **98**, 150601.

Andrieux, D., Gaspard, P., Ciliberto, S., Garnier, N., Joubaud, S., and Petrossyan, A. (2008) Thermodynamic time asymmetry in nonequilibrium fluctuations. *J. Stat. Mech. Theor. Exp.*, **2008**, P01002. doi: 10.1088/1742-5468/2008/01/P01002

Andronow, A.A., Chaikin, S.E., and Witt, A.A. (1965) *Theorie der Schwingungen*, Akademie Verlag, Berlin.

Anishchenko, V.S. (1995) *Dynamical Chaos – Models and Experiments*, World Scientific, Singapore.

Anishchenko, V.S., Astakhov, V.V., Neiman, A.B., Vadivasova, T.E., and Schimansky-Geier, L. (2002) *Nonlinear Dynamics of Chaotic and Stochastic Systems*, Springer, Berlin, Heidelberg.

Antonov, J.I., Levitus, S., and Boyer, T.P. (2002) Steric sea level variations during 1957–1994: Importance of salinity. *J. Geophys. Res.*, **107** (C12). doi: 10.1029/2001JC000964

Archer, C.L. and Jacobson, M.Z. (2005) Evaluation of global wind power. *J. Geophys. Res.*, **110**, D12110. doi: 10.1029/2004JD005462

Aris, R. (1965) Prolegomena to the rational analysis of systems of chemical reactions. *Arch. Ration. Mech. An.*, **19**, 81–99.

Aris, R. (1968) Prolegomena to the rational analysis of systems of chemical reactions – II: Some addenda. *Arch. Ration. Mech. An.*, **27**, 356–364.

Armitage, S.J., Jasim, S.A., Marks, A.E., Parker, A.G., Usik, V.I., and Uerpmann, H.-P. (2011) The southern route "Out of Africa": evidence for an early expansion of modern humans into Arabia. *Science*, **331**, 453–456.

Arrhenius, S. (1896) On the influence of carbonic acid in the air upon the temperature of the ground. London, Edinburgh, and Dublin. *Phil. Mag. J. Sci. (fifth series)*, **41**, 237–275.

Atkinson, Q.D., Meade, A., Venditti, C., Greenhill, S.J., and Pagel, M. (2008) Languages evolve in punctuational bursts. *Science*, **319**, 588.

Atkinson, Q.D. (2011) Phonemic diversity supports a serial founder effect model of language expansion from Africa. *Science*, **332**, 346–349.

Avery, J. (2003) *Information Theory and Evolution*, World Scientific, Singapore.

Ayres, R.U. (1994) *Information, Entropy, and Progress – A New Evolutionary Paradigm*, AIP Press, Woodbury.

Bak, P., Tang, C., and Wiesenfeld, K. (1987) Self-organized criticality: an explanation of 1/f noise. *Phys. Rev. Lett.*, **59**, 381–384.

Baldwin, M.P., Gray, L.J., Dunkerton, T.J., Hamilton, K., Haynes, P.H., Randel, W.J., Holton, J.R., Alexander, M.J., Hirota, I., Horinouchi, T., Jones, D.B.A., Kinnersley, J.S., Marquardt, C., Sato, K., and Takahashi, M. (2001) The quasi-biennial oscillation. *Rev. Geophys.*, **39**, 179–229.

Ball, P. (2009) In retrospect: the physics of sand dunes. *Nature*, **457**, 1084–1085.

Balodis, E., Mann, V., Trevani, L., and Tremaine, P. (2008) Ionization Constants, Equation of State Parameters, and Thermal Decomposition Kinetics of Nucleic Acid Bases under Hydrothermal Conditions. ICPWS XV, Berlin, September 8–11, 2008, http://www.icpws15.de/papers/09_Geo-10_balodis.pdf.

Balter, M. (2011) Was North Africa the launch pad for modern human migrations? *Science*, **331**, 20–23.

Bannon, P.R. (2003) Hamiltonian description of idealized binary geophysical fluids. *J. Atmos. Sci.*, **60**, 2809–2819.

Baosen, L. (1989) The latent and sensible heat fluxes over the western tropical pacific and its relationship to ENSO. *Adv. Atmos. Sci.*, **6**, 467–474.

Barker, P.F., Diekmann, B., and Escutia, C. (2007) Onset of cenozoic antarctic glaciation. *Deep-Sea Res. II*, **54**, 2293–2307.

Barnola, J.-M., Raynaud, D., Lorius, C., and Barkov, N.I. (2003) *Historical CO_2 Record from the Vostok ice Core. In Trends: A Compendium of*

Data on Global Change, Carbon Dioxide Information Analysis Center, Oak Ridge National Laboratory, U.S., Department of Energy, Oak Ridge, TN., USA, http://cdiac.ornl.gov/trends/co2/vostok.html.

Barrick, J.E., Yu, D.S., Yoon, S.H., Jeong, H., Oh, T.K., Schneider, D., Lenski, R.E., and Kim, J.F. (2009) Genome evolution and adaptation in a long-term experiment with *Escherichia coli*. *Nature*, **461**, 1243–1247.

Bartholomay, A.F. (1958a) On the linear birth and death processes of biology as Markoff chains. *B. Math. Biophys.*, **20**, 97–118.

Bartholomay, A.F. (1958b) Stochastic models for chemical reactions: I. Theory of the unimolecular reaction process. *B. Math. Biophys.*, **20**, 175–190.

Bartholomay, A.F. (1959) Stochastic models for chemical reactions: II. The unimolecular rate constant. *B. Math. Biophys.*, **21**, 363–373.

Bateman, R.M., Crane, P.R., DiMichele, W.A., Kenrick, P.R., Rowe, N.P., Speck, T., and Stein, W.E. (1998) Early evolution of land plants: phylogeny, physiology, and ecology of the primary terrestrial radiation. *Annu. Rev. Ecol. Syst.*, **29**, 263–292.

Bates, A.E., Lee, R.W., Tunnicliffe, V., and Lamare, M.D. (2010) Deep-sea hydrothermal vent animals seek cool fluids in a highly variable thermal environment. *Nat. Communications*, **1** (14). doi: 10.1038/ncomms1014

Baumgartner, A. and Reichel, E. (1975) *The World Water Balance*, R. Oldenbourg Verlag, München.

Baumgartner, L.K., Reid, R.P., Dupraz, C., Decho, A.W., Buckley, D.H., Spear, J.R., Przekop, K.M., and Visscher, P.T. (2006) Sulfate reducing bacteria in microbial mats: changing paradigms, new discoveries. *Sediment. Geol.*, **185**, 131–145.

Baus, M. and Tejero, C.F. (2008) *Equilibrium Statistical Physics*, Springer, Berlin.

Beal, L.M., De Ruijter, W.P.M., Biastoch, A., and Zahn, R., SCOR/WCRP/IAPSO Working Group 136 (2011) On the role of the Agulhas system in ocean circulation and climate. *Nature*, **472**, 429–436.

Beardmore, R.E., Gudelj, I., Lipson, D.A., and Hurst, L.D. (2011) Metabolic trade-offs and the maintenance of the fittest and the flattest. *Nature*, **472**, 342–346.

Becks, L. and Agrawal, A.F. (2010) Higher rates of sex evolve in spatially heterogeneous environments. *Nature*, **468**, 89–92.

Beer, C., Reichstein, M., Tomelleri, E., Ciais, P., Jung, M., Carvalhais, N., Rödenbeck, C., Arain, M.A., Baldocchi, D., Bonan, G.B., Bondeau, A., Cescatti, A., Lasslop, G., Lindroth, A., Lomas, M., Luyssaert, S., Margolis, H., Oleson, K.W., Roupsard, O., Veenendaal, E., Viovy, N., Williams, C., Woodward, F.I., and Papale, D. (2010) Terrestrial gross carbon dioxide uptake: global distribution and covariation with climate. *Science*, **329**, 834–838.

Behringer, W. (2007) *Kulturgeschichte des Klimas. Von der Eiszeit bis zur globalen Erwärmung*, C.H. Beck, München (English edition 2010: Cultural history of climate Polity Press Cambridge).

Bekenstein, J.D. (1973) Black holes and entropy. *Phys. Rev. D*, **7**, 2333–2346.

Bellott, D.W., Skaletsky, H., Pyntikova, T., Mardis, E.R., Graves, T., Kremitzki, C., Brown, L.G., Rozen, S., Warren, W.C., Wilson, R.K., and Page, D.C. (2010) Convergent evolution of chicken Z and human X chromosomes by expansion and gene acquisition. *Nature*, **466**, 612–616.

Berger, R. (2008) *Warum der Mensch spricht*, Eichborn, Frankfurt am Main.

Berner, R.A. (1999) Atmospheric oxygen over phanerozoic time. *Proc. Natl. Acad. Sci. U.S.A.*, **96**, 10955–10957.

Bestehorn, M. (2006) *Hydrodynamik und Strukturbildung*, Springer, Berlin.

Bethell, T. and Bergin, B. (2009) Formation and survival of water vapor in the terrestrial planet-forming region. *Science*, **326**, 1675–1677.

Beukes, N. (2004) Biogeochemistry: early options in photosynthesis. *Nature*, **431**, 522–523.

Beyer, W., Smith, T.F., Stein, M.L., and Ulam, S.M. (1974) A molecular sequence metric and evolutionary trees. *Math. Biosci.*, **19**, 9–25.

Bianconi, G. and Barabási, A.-L. (2001) Competition and multiscaling in evolving networks. *Europhys. Lett.*, **54**, 436–442.

Bickerton, D. (2009) *Adam's Tongue*, Hill and Wang, New York.

Biesold, D. and Matthies, H. (1977) *Neurobiologie*, Gustav Fischer, Jena.

Biess, G. (1976) *Graphentheorie. Mathematik für Ingenieure, Naturwissenschaftler,*

Ökonomen und Landwirte, Band 21/2, BSB B.G, Teubner Verlagsgesellschaft, Leipzig.

Biggin, A.J., de Wit, M.J., Langereis, Cor, G., Zegers, T.E., Voûte, S., Dekkers, M.J., and Drost, K. (2011) Palaeomagnetism of Archaean rocks of the Onverwacht Group, Barberton Greenstone Belt (southern Africa): Evidence for a stable and potentially reversing geomagnetic field at ca. 3.5 Ga. Earth Planet. Sci. Lett., 302, 314–328.

Bijl, P.K., Houben, A.J.P., Schouten, S., Bohaty, S.M., Sluijs, A., Reichart, G.-J., Sinninghe Damsté, J.S., and Brinkhuis, H. (2010) Transient middle eocene atmospheric CO_2 and temperature variations. Science, 330, 763–764.

Bindschadler, R. (2006) The environment and evolution of the West Antarctic ice sheet: setting the stage. Philos. Trans. R. Soc. Lond. A, 364, 1583–1605.

Biswas, S.N., Gupta, K.C., and Karmakar, B.B. (1977) Brillouin–Wigner perturbation solution of Volterra's prey-predator system. J. Theor. Biol., 64, 253–260.

Bjerrum, N. (1952) Structure and properties of ice. Science, 115, 385–390.

Blake, R.E., Chang, S.J., and Lepland, A. (2010) Phosphate oxygen isotopic evidence for a temperate and biologically active Archaean ocean. Nature, 464, 1029–1032.

Blankenship, R.E., Tiede, D.M., Barber, J., Brudvig, G.W., Fleming, G., Ghirardi, M., Gunner, M.R., Junge, W., Kramer, D.M., Melis, A., Moore, T.A., Moser, C.C., Nocera, D.G., Nozik, A.J., Ort, D.R., Parson, W.W., Prince, R.C., and Sayre, R.T. (2011) Comparing photosynthetic and photovoltaic efficiencies and recognizing the potential for improvement. Science, 332, 805–809.

Bland, P.A., Howard, L.E., Prior, D.J., Wheeler, J., Hough, R.M., and Dyl, K.A. (2011) Earliest rock fabric formed in the Solar System preserved in a chondrule rim. Nature Geoscience, 4, 244–247.

Bo, S., Siegert, M.J., Mudd, S.M., Sugden, D., Fujita, S., Xiangbin, C., Yunyun, J., Xueyuan, T., and Yuansheng, L. (2009) The Gamburtsev mountains and the origin and early evolution of the Antarctic Ice Sheet. Nature, 459, 690–639.

Boetius, A. (2005) Lost City life. Science, 307, 1420–1422.

Böhme, H., Hagemann, R., and Löther, R. (1976) Beiträge zur Genetik und Abstammungslehre, Volk und Wissen, Berlin.

Böhme, M. (2003) The miocene climatic optimum: evidence from ectothermic vertebrates of Central Europe. Palaeogeogr. Palaeocl., 195, 389–401.

Bojadziev, G. (1978) The Krylov-Bogoliubov-Mitropolski method applied to models of population dynamics. B. Math. Biol., 40, 335–345.

Boltzmann, L. (1884a) Über eine von Hrn. Bartoli entdeckte Beziehung der Wärmestrahlung zum zweiten Hauptsatze. Wiedemanns Annalen der Physik und Chemie, 22, 31–39.

Boltzmann, L. (1884b) Ableitung des Stefanschen Gesetzes, betreffend die Abhängigkeit der Wärmestrahlung von der Temperatur aus der elektromagnetischen Lichttheorie. Wiedemanns Annalen der Physik und Chemie, 22, 291–294.

Bonhoeffer, S. and Sadler, P.F. (1993) Error thresholds on correlated fitness landscapes. J. Theor. Biol., 164, 359–372.

Bornholdt, S. and Schuster, H.G. (2003) Handbook of Graphs and Networks: From the Genome to the Internet, Wiley-VCH Verlag, Weinheim.

Bottke, W.F., Walker, R.J., Day, J.M.D., Nesvorny, D., and Elkins-Tanton, L. (2010) Stochastic late accretion to earth, the moon, and mars. Science, 330, 1527–1530.

Bouvier, A. and Wadhwa, M. (2010) The age of the solar system redefined by the oldest Pb–Pb age of a meteoritic inclusion. Nat. Geoscience., 1–5. doi: 10.1038/NGEO941

Boyer, T.P., Levitus, S., Antonov, J.I., Locarnini, R.A., and Garcia, H.E. (2005) Linear trends in salinity for the world ocean, 1955–1998. Geophys. Res. Lett., 32, L01604. doi: 10.1029/2004GL021791

Bradley, A.S. (2008) Organic geochemical biosignatures in alkaline hydrothermal ecosystems. PhD Thesis, MIT, http://dspace.mit.edu/handle/1721.1/42920.

Brazelton, W.J., Ludwig, K.A., Sogin, M.L., Andreishcheva, E.N., Kelley, D.S., Shen, C.-C., Edwards, R.L., and Baross, J.A. (2010) Archaea and bacteria with surprising microdiversity show shifts in dominance over 1,000-year time scales in hydrothermal chimneys. *Proc. Natl. Acad. Sci. U.S.A.*, **107**, 1612–1617.

Brembs, B. (2011) Towards a scientific concept of free will as a biological trait: spontaneous actions and decision-making in invertebrates. *Proc. R. Soc. B*, **278**, 930–939.

Bremermann, H.J. (1970) A method of unconstrained global optimization. *Math. Biosci.*, **9**, 1–15.

Brillouin, L. (1953) Negentropy principle of information. *J. Appl. Phys.*, **24**, 1152–1163.

Bristow, T.F., Kennedy, M.J., Derkowski, A., Droser, M.L., Jiang, G., and Creaser, R.A. (2009) Mineralogical constraints on the paleoenvironments of the Ediacaran Doushantuo formation. *Proc. Natl. Acad. Sci. U.S.A.*, **106**, 13190–13195.

Broecker, W.S. (1987) The biggest chill. *Nat. Hist. Magazine*, **97**, 74–82.

Brown, K.S., Marean, C.W., Herries, A.I.R., Jacobs, Z., Tribolo, C., Braun, D., Roberts, D.L., Meyer, M.C., and Bernatchez, J. (2009) Fire as an engineering tool of early modern humans. *Science*, **325**, 859–862.

Bruckner, E., Ebeling, W., Jiménez-Montaño, M.A., and Scharnhorst, A. (1996) Nonlinear stochastic effects of substitution: an evolutionary approach. *J. Evol. Econ.*, **6**, 1–30.

Bruckner, E., Ebeling, W., and Scharnhorst, A. (1990) The application of evolution models in scientometrics. *Scientometrics*, **18**, 21–41.

Brumfiel, G. (2010) Elemental shift for kilo. *Nature*, **467**, 892.

Bryden, H.L., Longworth, H.R., and Cunningham, S.A. (2005) Slowing of the Atlantic meridional overturning circulation at 25° N., *Nature*, **438**, 655–657.

Bryson, B. (2003) *A Short History Of Nearly Everything*. Broadway Books, New York.

Buchdahl, H.A. (1949) On the principle of Caratheodory. *Am. J. Phys.*, **17**, 41–43.

Budyko, M.I. (1969) The effect of solar radiation variations on the climate of the earth. *Tellus*, **21**, 611–619.

Budyko, M.I. (1978) *Teplovy balans Zemli (Heat Balance of the Earth)*, Gidrometeoizdat, Leningrad.

Büntgen, U., Tegel, W., Nicolussi, K., McCormick, M., Frank, D., Trouet, V., Kaplan, J.O., Herzig, F., Heussner, K.-U., Wanner, H., Luterbacher, J., and Esper, J. (2011) 2500 years of European climate variability and human susceptibility. *Science*, **331**, 578–582.

Bunde, A., Kropp, J., and Schellnhuber, H.J. (2002) *The Science of Disaster, Climate Disruptions, Heart Attacks, and Market Crashes*, Springer, Berlin, Heidelberg.

Buffett, B.A. (2009) Onset and orientation of convection in the inner core. *Geophys. J. Int.*, **179**, 711–719.

Buffett, B.A. (2010) Tidal dissipation and the strength of the earth's internal magnetic field. *Nature*, **468**, 952–954.

Buis, A. (2010) *Chilean Quake May Have Shortened Earth Days*, NASA, 1 March 2010, http://www.nasa.gov/topics/earth/features/earth-20100301.html.

Burnstock, G. (2009) Purinergic cotransmission. *Exp. Physiol.*, **94**, 20–24.

Burrows, W.J. (1992) *Weather Cycles: Real or Imaginary?*, Cambridge University Press, Cambridge.

Busacker, R.G. and Saaty, T.L. (1965) *Finite Graphs and Networks: An Introduction with Applications*, International Series in Pure and Applied Mathematics, McGraw-Hill, New York, London.

Butchvarov, P. (1998) *Skepticism About the External World*. Oxford University Press, New York.

Butenin, N.V., Neimark, Yu.L., and Fufayev, N.A. (1976) *Vvedeniye v Teoriyu Nelineinych Kolebanii (Introduction to the Theory of Nonlinear Oscillations)*, Nauka, Moscow.

Butterfield, N.J. (2000) *Bangiomorpha pubescens* n. gen., n. sp.: implications for the evolution of sex, multicellularity, and the Mesoproterozoic/Neoproterozoic radiation of eukaryotes. *Paleobiology*, **26**, 386–404.

Callen, H.B. (1960) *Thermodynamics*, John Wiley & Sons, New York & London.

Callender, C. (2010) *Introducing Time*, Faber & Faber, London.

Campbell Jr., K.E., Prothero, D.R., Romero-Pittman, L., Hertel, F., and Rivera, N. (2010) Amazonian magnetostratigraphy: dating the first pulse of the Great American faunal interchange. *J. S. Am. Earth Sci.*, **29**, 619–626.

Canfield, D.E., Rosing, M.T., and Bjerrum, C. (2006) Early anaerobic metabolisms. *Philos. Trans. R. Soc. B*, **361**, 1819–1836.

Canup, R.M. (2010) Origin of Saturn's rings and inner moons by mass removal from a lost Titan-sized satellite. *Nature*, **468**, 943–926.

Cape, J.L., Monnard, P.-A., and Boncella, J.M. (2011) Prebiotically relevant mixed fatty acid vesicles support anionic solute encapsulation and photochemically catalyzed trans-membrane charge transport. *Chem. Sc.*, **2**, 661–671.

Caramori, S., Cristino, V., Boaretto, R., Argazzi, R., Bignozzi, C.A., and Di Carlo, A. (2010) New components for dye-sensitized solar cells. *Int. J. Photoenergy*, doi:10.1155/2010/458614

Carathéodory, C. (1909) Untersuchungen über die Grundlagen der Thermodynamik. *Math. Ann.*, **67**, 355–386.

Carrigan, C.R. and Gubbins, D. (1984) Wie entsteht das Magnetfeld der Erde? in *Ozeane und Kontinente*, Spektrum der Wissenschaft, Heidelberg, p. 40–53.

Casti, J.L. (1979) *Connectivity Complexity and Catastrophe in Large-Scale Systems*, vol. 7, International Series on Applied Systems Analysis, John Wiley & Sons, Chichester, New York.

Catling, D.C. and Zahnle, K.J. (2010) Wenn die Atmosphäre ins All entweicht. *Spektrum der Wissenschaft*, Januar, 24–31.

Cavalli-Sforza, L.L., Menozzi, P., Piazza, A. (1994) *The History and Geography of Human Genes*. Princeton University Press, Princeton.

Cavalli-Sforza, L.L. (2001) *Gene, Völker und Sprachen*. dtv, München (Italian original: Geni, populi e lingue, Adelphi Edizioni, Milano, 1996).

Caviedes, C.N. (2001) *El Niño in history. Storming through the ages*, University Press of Florida, Gainesville.

Cerf, M., Thiruvengadam, N., Mormann, F., Kraskov, A., Quian Quiroga, R., Koch, C., and Fried, I. (2010) On-line, voluntary control of human temporal lobe neurons. *Nature*, **467**, 1104–1108.

Chan, F., Barth, J.A., Lubchenco, J., Kirincich, A., Weeks, H., Peterson, W.T., and Menge, B.A. (2008) Emergence of Anoxia in the California Current Large Marine Ecosystem. *Science*, **319**, 920.

Chandrasekhar, S. (1943) Stochastic problems in physics and astronomy. *Rev. Mod. Phys.*, **15**, 1–89.

Chen, Z., Mo, S., Hu, P., Jiang, S., Wang, G., and Cheng, X. (2010) Entropy flow, entropy generation, exergy flux, and optimal absorbing temperature in radiative transfer between parallel plates. *Front. Energy Power Eng. China*, **4**, 301–305.

Chen, C.-T.A., Zeng, Z., Kuo, F.-W., Yang, T.F., Wang, B.-J., and Tu, Y.-Y. (2005) Tide-influenced acidic hydrothermal system offshore NE Taiwan. *Chem. Geol.*, **224**, 69–81.

Cheng, H., Fleitmann, D., Edwards, R.L., Wang, X., Cruz, F.W., Auler, A.S., and Mangini, A., Wang, Y., Kong, X., Burns, S.J., Matter, A. (2009) Timing and structure of the 8.2 kyr B.P. event inferred from $\delta^{18}O$ records of stalagmites from China, Oman, and Brazil. *Geology*, 37, 1007–1010.

Chernavsky, D.S. (2001) *Synergetics and Information: Dynamical Information Theory (in Russian)*, Nauka, Moscow.

Choiniere, J.N., Xu, X., Clark, J.M., Forster, C.A., Guo, Y., and Han, F. (2010) A basal Alvarezsauroid Theropod from the early late Jurassic of Xinjiang, China. *Science*, **327**, 571–574.

Christen, R., Ratto, A., Baroin, A., Perasso, R., Grell, K.G., and Adoutte, A. (1991) An analysis of the origin of metazoans, using comparisons of partial sequences of the 28S RNA, reveals an early emergence of triploblasts. *EMBO J.*, **10**, 499–503.

Clarke, J.A., Ksepka, D.T., Salas-Gismondi, R., Altamirano, A.J., Shawkey, M.D., D'Alba, L., Vinther, J., DeVries, T.J., and Baby, P. (2010) Fossil evidence for evolution of the shape and color of penguin feathers. *Science*, **330**, 954–957.

Clausius, R. (1865) Über verschiedene für die Anwendung bequeme Formen der Hauptgleichungen der mechanischen Wärmetheorie. *Annalen der Physik und Chemie*, **125**, 353–400, http://gallica.bnf.fr/ark:/12148/bpt6k152107/f369.image.pagination.langEN.

Clausius, R. (1876) *Die mechanische Wärmetheorie. Zweite umgearbeitete und vervollständigte Auflage des unter dem Titel "Abhandlungen über die mechanische Wärmetheorie" erschienenen Buches*, Friedrich Vieweg und Sohn, Braunschweig.

Clement, A.C., Burgman, R., and Norris, J.R. (2009) Observational and model evidence for positive low-level cloud feedback. *Science*, **325**, 460–464.

Clementz, M.T. and Sewall, J.O. (2011) Latitudinal gradients in greenhouse seawater $\delta^{18}O$: Evidence from Eocene Sirenian tooth enamel. *Science*, **332**, 455–458.

Closs, H., Giese, P., and Jacobshagen, V. (1984) Alfred Wegeners Kontinentalverschiebung aus heutiger Sicht, in *Ozeane und Kontinente*, Spektrum der Wissenschaft, Heidelberg, p. 40–53.

Clough, S.A., Ianoco, M.J., and Moncet, J.-L. (1992) Line-by-line calculations of atmospheric fluxes and cooling rates: application to water vapor. *J. Geophys. Res.*, **97**, 15,761–15,785.

Cockell, C. (2008) *An Introduction to the Earth-Life System*, Cambridge University Press, Cambridge.

Coe, M.D. (1999) *Breaking the Maya Code*, Thames & Hudson, New York.

Collet, P. and Eckmann, J.P. (1980) *Iterated Maps on the Interval as Dynamical Systems*, Birkhäuser, Basel.

Conrad, M. (1978) Evolution of the adaptive landscape. *Lect. Notes in Biomathematics*, **21**, 147–169.

Conrad, M. (1982) Natural selection and the evolution of neutralism. *Biosystems*, **15**, 83–85.

Conrad, M. (1983) *Adaptability: The Significance of Variability from Molecule to Ecosystems*, Plenum Press, New York.

Conrad, M. and Ebeling, W. (1992) M.V. Volkenstein, evolutionary thinking and the structure of fitness landscapes. *BioSystems*, **27**, 125–130.

Conrad, M. and Rizki, M.M. (1980) Computational illustration of the bootstrap effect. *BioSystems*, **13**, 57–64.

Crick, F.H.C. (1966) Codon-anticodon pairing: the wobble hypothesis. *J. Mol. Biol.*, **19**, 548–555.

Crick, F.H.C. (1968) The origin of the genetic code. *J. Mol. Biol.*, **38**, 367–379.

Cuhel, R.L., Aguilar, C., Anderson, P.D., Maki, J.S., Paddock, R.W., Remsen, C.C., Klump, J.V. and Lovalvo, D. (2002) Underwater domains in Yellowstone Lake hydrothermal vent geochemistry and bacterial chemosynthesis. in Yellowstone Lake: Hotbed of Chaos or Reservoir of Resilience? Proceedings of the 6th Biennial Conference on the Greater Yellowstone Ecosystem. October 8–10, 2001, Mammoth Hot Springs Hotel, Yellowstone National Park. Yellowstone National Park, WY and Hancock, MI: Yellowstone Center for Resources and The George Wright Society (eds R.J. Anderson and D. Harmon), pp. 27–53.

Cummins, K.W. and Wuycheck, J.C. (1971) Caloric equivalents for investigations in ecological energetics. *Mitteilungen der Internationalen Vereinigung für Theoretische und Angewandte Limnologie*, **18**, 158, http://www.humboldt.edu/cuca/cummins/documents/caloric.pdf.

Cunningham, S.A., Kanzow, T., Rayner, D., Baringer, M.O., Johns, W.E., Marotzke, J., Longworth, H.R., Grant, E.M., Hirschi, J.J.-M., Beal, L.M., Meinen, C.S., and Bryden, H.L. (2007) Temporal variability of the Atlantic meridional overturning circulation at 26.5° N. *Science*, **317**, 935–938.

Currie, T.E., Greenhill, S.J., Gray, R.D., Hasegawa, T., and Mace, R. (2010) Rise and fall of political complexity in island Southeast Asia and the Pacific. *Nature*, **467**, 801–804.

Curry, R., Dickson, B., and Yashayev, I. (2003) A change in the freshwater balance of the Atlantic Ocean over the past four decades. *Nature*, **426**, 826–829.

Czajkowski, G. (1975) Kinetic Phase Transitions in Nonlinear Thermodynamics. Thesis, Nicolaus Copernicus University, Toruń.

Dahl, T.W., Hammarlund, E.U., Anbar, A.D., Bond, D.P.G., Gill, B.C., Gordon, G.W., Knoll, A.H., Nielsen, A.T., Schovsbo, N.H., and Canfield, D.E. (2010) Devonian rise in atmospheric oxygen correlated to the radiations of terrestrial plants and large predatory fish. *Proc. Natl. Acad. Sci. U.S.A.*, **107**, 17911–17915.

Dai, A. (2006) Recent climatology, variability, and trends in global surface humidity. *J. Climate*, **19**, 3589–3605.

Dallai, L., and Burgess, R. (2011) A record of Antarctic surface temperature between 25 and 50 m.y. ago. *Geology*, **39**, 423–426.

Darlington, P.J. (1977) The cost of evolution and the imprecision of adaptation. *Proc. Natl. Acad. Sci. U.S.A.*, **74**, 1647–1651.

Darwin, C. (1845) *Journal of researches into the natural history and geology of the countries* visited during the voyage of the H.M.S. Beagle round the world: Under the command of Capt. Fitz Roy. John Murray, London.

Darwin, C. (1911) *The Origin of Species by Means of Natural Selection or the Preservation of Favored Races in the Struggle for Life.* Reprinted from the Sixth London Edition, with Additions and Corrections. Hurst and Company Publishers, New York (first published in 1859).

Dautcourt, G. (1976) *Relativistische Astrophysik*, Akademie-Verlag, Berlin.

David, L.A. and Alm, E.J. (2011) Rapid evolutionary innovation during an Archaean genetic expansion. *Nature*, **469**, 93–96.

David, C.N. and Hager, G. (1994) Formation of a primitive nervous system: nerve cell differentiation in the polyp hydra. *Perspect. Dev. Neurobiol.*, **2**, 135–140.

Davis, T.M. (2010) Verliert das Universum Energie? *Spektrum der Wissenschaft*, November 2010, 22–29.

Dawkins, R. (1976) *The Selfish Gene*, Oxford University Press, New York.

Dawkins, R. (1996a) *The Blind Watchmaker*, W.W. Norton & Co., New York.

Dawkins, R. (1996b) *Climbing Mount Improbable*, W.W. Norton & Co., New York.

Decker, P. (1975) Evolution in biods. *Origins Life Evol. B.*, **6**, 211–218. doi: 10.1007/BF01372407

De Gregorio, B.T., Sharp, T.G., Flynn, G.J., Wirick, S., and Hervig, R.L. (2009) Biogenic origin for Earth's oldest putative microfossils. *Geology*, **37**, 631–634.

De Groot, S.R. and Mazur, P. (1962, 1984) *Non-equilibrium thermodynamics*, North Holland Publishing Company, Dover Publications, Amsterdam, New York.

Demb, J.B. and Brainard, D.H. (2010) Vision: neurons show their true colours. *Nature*, **467**, 670–671.

deMenocal, P.B. (2011) Climate and human evolution. *Science*, **331**, 540–542.

Denton, G.H., Anderson, R.F., Toggweiler, J.R., Edwards, R.L., Schaefer, J.M., and Putnam, A.E. (2010) The last glacial termination. *Science*, **328**, 1652–1656.

Derryberry, E.P. (2011) Male response to historical and geographical variation in bird song. *Biol. Lett.*, **7**, 57–59.

Desmurget, M., Reilly, K.T., Richard, N., Szathmari, A., Mottolese, C., and Sirigu, A. (2009) Movement intention after parietal cortex stimulation in humans. *Science*, **324**, 811–813.

Dessler, A.E. (2010) A determination of the cloud feedback from climate variations over the past decade. *Science*, **330**, 1523–1527.

Dessler, A.E. and Sherwood, S.C. (2009) A matter of humidity. *Science*, **323**, 1020–1021.

Dessler, A.E., Zhang, Z., and Yang, P. (2008) Water-vapor climate feedback inferred from climate fluctuations, 2003–2008. *Geophys. Res. Lett.*, **35**, L20704. doi: 10.1029/2008GL035333

DGFI (2010) Geodetic results of the earthquake in Chile (2010-02-27). Deutsches Geodätisches Forschungsinstitut, http://www.dgfi.badw.de/.

Diaz, H.F. and Markgraf, V. (2000) *El Niño and the Southern Oscillation*, Cambridge University Press, Cambridge.

Diaz, R.J. and Rosenberg, R. (2008) Spreading dead zones and consequences for marine ecosystems. *Science*, **321**, 926–929.

Díaz López, B. and Shirai, J.A.B. (2009) Mediterranean common bottlenose dolphin's repertoire and communication use, in *Dolphins: Anatomy, Behavior and Threats* (eds G.P. Agustin and M.C. Lucia), Nova Science Publishers, Hauppauge, NY, pp. 129–148.

Dixon, R.M.W. (1997) *The Rise and Fall of Languages*, Cambridge University Press, Cambridge.

Domínguez-Villar, D., Fairchild, I.J., Baker, A., Wang, X., Edwards, R.L., and Cheng, H. (2009) Oxygen isotope precipitation anomaly in the North Atlantic region during the 8.2 ka event. *Geology*, **37**, 1095–1098.

Donahue, T.M. and Hodges, Jr., R.R. (1992) Past and present water budget of Venus. *J. Geophys. Res.*, **97**, 6083–6091.

Donald, M. (2008) *Triumph des Bewusstseins*. Klett-Cotta, Stuttgart. American original (2001) A Mind so Rare: The Evolution of Human Consciousness, W.W. Norton & Co., New York.

Donlon, C.J., Minnett, P.J., Gentemann, C., Nightingale, T.J., Barton, I.J., Ward, B., and Murray, M.J. (2002) Toward improved validation of satellite sea surface skin temperature measurements for climate research. *J. Climate*, **15**, 353–369,

http://journals.ametsoc.org/doi/full/
10.1175/1520-0442%282002%29015%
3C0353%3ATIVOSS%3E2.0.CO%3B2.

Doob, J.L. (1953) *Stochastic Processes*, John Wiley and Sons, New York.

Doolittle, R. (1984) The probability and origin of life, in *Scientists Confront Creationism* (ed. L.R. Godfrey), W.W. Norton, New York, pp. 85–97.

Dorogovtsev, S.N., Mendes, J.F.F., and Samukhin, A.N. (2003) Principles of statistical mechanics of uncorrelated random networks. *Nucl. Phys. B*, **666**, 396–416.

Dosi, G., Freeman, C., Nelson, R., and Suete, L. (1988) *Technical change and Economic Theory*, Pinter, London.

Dunaeva, A.N., Antsyshkin, D.V., and Kuskov, O.L. (2010) Fazovaya diagramma H_2O: termodinamicheskiye funktsii fazovykh perekhodov ldov vysokogo davleniya (Phase diagram of H_2O: Thermodynamic functions of phase transitions of high-pressure ices). *Astronomicheskii Vestnik*, **44**, 222–243.

Durack, P.J. and Wijffels, S.E. (2010) Fifty-year trends in global ocean salinities and their relationship to broad-scale warming. *J. Climate*, **23**, 4342–4362.

Dutkiewicz, A., Volk, H., George, S.C., Ridley, J., and Buick, R. (2006) Biomarkers from Huronian oil-bearing fluid inclusions: an uncontaminated record of life before the Great Oxidation Event. *Geology*, **34**, 437–440.

Dutton, J.A. (1973) The global thermodynamics of atmospheric motion. *Tellus*, **25**, 89–110.

Dunn, M., Greenhill, S.J., Levinson, S.C., and Gray, R.D. (2011) Evolved structure of language shows lineage-specific trends in word-order universals. *Nature*, **473**, 79–82.

Dymek, R.F. and Klein, C. (1988) Chemistry, petrology and origin of banded iron-formation lithologies from the 3800 MA Isua supracrustal belt, West Greenland. *Precambrian Res.*, **39**, 247–302.

Dyson, F. (1959) Search for artificial stellar sources of infrared radiation. *Science*, **131**, 1667–1668.

Dyson, F. (2008) Origins of life. *Lect. Notes Phys.*, **764**, 71–97.

East, E.M. (1918) The role of reproduction in evolution. *Am. Nat.*, **52**, 273–289.

Ebeling, W. (1965) Statistisch-mechanische Ableitung von verallgemeinerten Diffusionsgleichungen. *Ann. Phys.-Berlin*, **471**, 147–159. doi: 10.1002/andp.19654710306

Ebeling, W. (1974) Statistical derivation of the mass-action law. *Physica*, **73**, 573–584.

Ebeling, W. (1976a) *Strukturbildung bei irreversiblen Prozessen*, Teubner, Leipzig (1st revised Russian edition 1979: Mir Moscow 2nd revised Russian edition 2004: R&C Dynamics Moskva-Ishevsk).

Ebeling, W. (1976b) Physikalische Aspekte der qualitativen Sprünge bei Entwicklungsprozessen. *Wissenschaftliche Zeitschrift der Humboldt-Universität Berlin*, GSR 25, 20–24.

Ebeling, W. (1978a) Physikalisch-chemische Grundlagen der Strukturbildung bei Entwicklungsprozessen, in *Struktur und Prozeß* (ed. K.F. Wessel), Deutscher Verlag der Wissenschaften, Berlin.

Ebeling, W. (1978b) Ein stochastisches Modell für das Wirken des Marx'schen Wertgesetzes im Kapitalismus der freien Konkurrenz. *Rostocker Physikalische Manuskripte*, 3/I, 47–54.

Ebeling, W. (1981) Structural stability of stochastic systems. *Springer Ser. Synergetics*, **11**, 188–198.

Ebeling, W. (1989) *Chaos, Ordnung und Information*, Urania-Verlag Leipzig, Verlag H. Deutsch, Frankfurt am Main.

Ebeling, W. (1992) On the relation between various entropy concepts and the valoric interpretation. *Physica A*, **182**, 108–115.

Ebeling, W. (1993) Entropy and information in processes of self-organisation: uncertainty and predictability. *Physica A*, **194**, 563–575.

Ebeling, W. (1994) Entropie, Vorhersagbarkeit und nichtlineare Dynamik. *Sitzungsberichte der Leibniz-Sozietät Heft*, 1/2, 33–50.

Ebeling, W. (2006) Value in physics and self-organization. *Nat. Society and Thought*, **19**, 133–143.

Ebeling, W. (2008) Max Planck on entropy and irreversibility, in *Max Planck: Annalen Papers* (ed. D. Hoffmann), Wiley-VCH, Weinheim.

Ebeling, W. (2010) Physik und Interdisziplinarität an der Humboldt-Universität zu Berlin 1979 bis 2010. In: Girnus, W., Meier, K. (eds.): *Die Humboldt-Universität unter den Linden 1945 bis 1990, Zeitzeugen – Einblicke – Analysen*. Leipziger Universitätsverlag, Leipzig.

Ebeling, W., Engel, A., Esser, B., and Feistel, R. (1984) Diffusion and reaction in random media and models of evolutionary processes. *J. Stat. Phys.*, **37**, 369–384.

Ebeling, W. and Engel-Herbert, H. (1989) Entropy lowering and attractors in phase space. *Acta Phys. Hung.*, **66**, 339–348.

Ebeling, W., Engel, A., and Feistel, R. (1990) *Physik der Evolutionsprozesse*, Akademie-Verlag, Berlin.

Ebeling, W., Engel, H., and Herzel, H.-P. (1990) *Selbstorganisation in der Zeit*, Akademie-Verlag, Berlin.

Ebeling, W. and Feistel, R. (1974) Zur Kinetik molekularer Replikationsprozesse mit Selektionscharakter. *Stud. Biophys.*, **46**, 183–195.

Ebeling, W. and Feistel, R. (1975) Stochastic theory of molecular replication processes with selection character. Proceedings of the IUPAP Conference on Statistical Physics, Budapest, 1975.

Ebeling, W. and Feistel, R. (1976a) Zur nichtlinearen Kinetik von homogenen Konkurrenzreaktionen. *Z. Phys. Chem. (Leipzig)*, **257**, 705–720.

Ebeling, W. and Feistel, R. (1976b) Physikalische Aspekte der qualitativen Sprünge bei Entwicklungsprozessen. II. Modellsysteme. *Wissenschaftliche Zeitschrift der Humboldt-Universtät Berlin*, GSR 25, 25–28.

Ebeling, W. and Feistel, R. (1977) Stochastic theory of molecular replication processes with selection character. *Ann. Phys-Leipzig*, **34**, 81–160.

Ebeling, W. and Feistel, R. (1979) On the evolution of biological macromolecules. I. physico-chemical self-organization. *Stud. Biophys.*, **75**, 131–146.

Ebeling, W. and Feistel, R. (1982) *Physik der Selbstorganisation und Evolution*, Akademie-Verlag, Berlin.

Ebeling, W. and Feistel, R. (1983) Thermodynamik irreversibler Prozesse und spontane Strukturbildung mit Beispielen aus Physik und Geophysik. Sitzungsberichte der Akademie der Wissenschaften der DDR 10 N, 22–29.

Ebeling, W. and Feistel, R. (1992) Theory of Selforganization: The Role of Entropy, Information and Value. *J. Nonequilibrium Thermodyn.*, **17**, 303–332.

Ebeling, W. and Feistel, R. (1994) *Chaos und Kosmos: Prinzipien der Evolution*, Spektrum Akademischer Verlag, Heidelberg, Berlin, Oxford.

Ebeling, W. and Feistel, R. (2005) *Khaos i Kosmos: Sinergetika Evolyutsii (Chaos and Cosmos: Synergetics of Evolution)*, R&C Dynamics, Moskva-Izhevsk.

Ebeling, W. and Feistel, R. (2008) Überlegungen zur Evolution des Klimas, http://www2.hu-berlin.de/leibniz-sozietaet/debatte/meinungen_1/Ebeling_Feistel%20zu%20Lanius.pdf.

Ebeling, W., Feistel, R., Hartmann-Sonntag, I., Schimansky-Geier, L., and Scharnhorst, A. (2006) New species in evolving networks – stochastic theory of sensitive networks and applications on the metaphorical level. *BioSystems*, **85**, 65–71.

Ebeling, W., Feistel, R., and Jiménez-Montaño, M.A. (1977) On the theory of stochastic replication and evolution of molecular sequences. *Rostocker Physikalische Manuskripte MNR*, **2**, 105–127.

Ebeling, W., Freund, J., and Schweitzer, F. (1998) *Komplexe Strukturen: Entropie und Information*, B.G. Teubner-Verlag, Stuttgart, Leipzig.

Ebeling, W., Ivanov, Chr., and Schimansky-Geier, L. (1977) Stochastic theory of nucleation in bistable reaction systems. *Rostocker Physikalische Manuskripte MNR*, **2**, 93–103.

Ebeling, W. and Jiménez-Montaño, M.A. (1980) On grammars, complexity and information measures of biological macromolecules. *Math. Biosci.* **52**, 53–71.

Ebeling, W., Karmeshu, and Scharnhorst, A. (2001) Dynamics of economic and technological search processes in complex adaptive landscapes. *Ad. Complex Syst.*, **40**, 71–88.

Ebeling, W. and Klimontovich, Yu.L. (1984) *Self-Organization and Turbulence in Liquids*, Teubner, Leipzig.

Ebeling, W., Kraeft, W.D., and Kremp, D. (1976) Theory of bound states and ionization equilibrium in plasmas and solids, in *Ergebnisse der Plasmaphysik und der Gaselektronik* (eds R. Rompe and M. Steenbeck), Bd. 5, Akademie-Verlag, Berlin.

Ebeling W., Molgedey, L., Kurths, J., and Schwarz, U. (2002) Entropy, complexity, predictability and data analysis of time series and letter sequences, in *The Science of Disaster: Climate Disruptions, Heart Attacks, and Market Crashes* (eds A. Bunde, J. Kropp, and H.-J. Schellnhuber), Springer, Berlin, Heidelberg.

Ebeling, W., Molgedey, L., and Reimann, A. (2000) Stochastic urn models of innovation and search dynamics. *Physica A*, **287**, 599–612.

Ebeling, W. and Muschik, W. (1993) *Statistical Physics and Thermodynamics of Nonlinear Nonequilibrium Processes*, World Scientific, Singapore.

Ebeling, W. and Reimann, A. (2002) Ensemble-based control of search dynamics with applications to string optimization. *Z. Phys. Chem.*, **216**, 65–75.

Ebeling, W. and Peschel, M. (1986) *Lotka-Volterra Approach to Cooperation and Competition in Dynamic Systems*, Akademie-Verlag, Berlin.

Ebeling, W. and Scharnhorst, A. (1986) Self-organization models for field mobility of physicists. *Czechoslovak J. Phy. B*, **36**, 43–46.

Ebeling, W. and Scharnhorst, A. (2000) Evolutionary models of innovation dynamics, in *Traffic and Granular Flow 99: Social, Traffic, and Granular Dynamics* (eds D. Helbing, H.J. Herrmann, and M. Schreckenberg), Springer, Berlin, Heidelberg, New York.

Ebeling, W., Scharnhorst, A., Jiménez-Montaño, M.A., and Karmeshu, A. (1999) Evolution and innovation dynamics as search processes in complex adaptive landscapes, in *Komplexe Systeme und Nichlineare Dynamik in Natur und Gesellschaft* (ed. K. Mainzer), Springer, Berlin, Heidelberg, New York, pp. 446–473.

Ebeling, W. and Schmelzer, J. (1980) Koexistenz von Sorten in nichtlinearen autokatalytischen Parallelreaktionen. *Z. Phys. Chem. (Leipzig)*, **261**, 677–696.

Ebeling, W. and Sokolov, I.M. (2005) *Statistical Thermodynamics and Stochastic Theory of Nonequilibrium Systems*, World Scientific, Singapore.

Ebeling, W. and Sonntag, I. (1986) A stochastic description of evolutionary processes in underoccupied systems. *BioSystems*, **19**, 91–100.

Ebeling, W., Sonntag, I., and Schimansky-Geier, L. (1981) On the evolution of biological macromolecules II: catalytic networks. *Stud. Biophys.*, **84**, 87–88.

Ebeling, W. and Volkenstein, M.V. (1990) Entropy and the evolution of information. *Physica A*, **163**, 398–402.

Eberwine, J., Miyashiro, K., Kacharmina, J.E., and Job, C. (2001) Local translation of classes of mRNAs that are targeted to neuronal dendrites. *Proc. Natl. Acad. Sci. U.S.A.*, **98**, 7080–7085.

Eccles, J.C. (1976) *Das Gehirn des Menschen*. Piper, München, Zürich. American original (1973): *The Understanding of the Brain*, McGraw-Hill, New York.

Eckmann, J.P. and Ruelle, D. (1985) Ergodic theory of chaos and strange attractors. *Rev. Mod. Phys.*, **57**, 617–656.

Edler, F. and Engert, J. (2007) Rauschthermometrie bei tiefen und hohen Temperaturen. *PTB-Mitteilungen*, **117**, 41–45.

Ehrenfest, P. and Ehrenfest, T. (1907) Über zwei bekannte Einwände gegen das Boltzmannsche H-Theorem. *Physikalische Zeitschrift*, **8**, 311–314. Reprinted in Klein, M.J. ed. (1959): Paul Ehrenfest: Collected Scientific Papers. North-Holland Publishing, Amsterdam.

Ehrmann, W. (2000) Smectite content and crystallinity in sediments from CRP-2/2A, Victoria Land Basin, Antarctica. *Terra Antartica*, **7**, 1–6.

Eigen, M. (1971) The self-organisation of matter and the evolution of biological macromolecules. *Naturwissenschaften*, **58**, 465–523.

Eigen, M. (1976) Wie entsteht Information? Prinzipien der Selbstorganisation in der Biologie. *Berichte der Bunsengesellschaft für physikalische Chemie*, **80**, 1059–1081.

Eigen, M. (1994) The origin of genetic information. *Origins Life Evol. Biospheres*, **24**, 241–262.

Eigen, M., Gardiner, W., Schuster, P., and Winkler-Oswatitsch, R. (1981) The origin of genetic information. *Sci. Am.*, **244**, 88–118.

Eigen, M., McCaskill, J., and Schuster, P. (1989) The molecular quasi-species. *Adv. Chem. Phys.*, **75**, 149–263.

Eigen, M. and Schuster, P. (1977) The hypercycle: a principle of natural self-organization. Part A: emergence of the hypercycle. *Naturwissenschaften*, **64**, 541–565.

Eigen, M. and Schuster, P. (1978a) The hypercycle: a principle of natural self-organization. Part B: the abstract hypercycle. *Naturwissenschaften*, **65**, 7–41.

Eigen, M. and Schuster, P. (1978a) The hypercycle: a principle of natural self-organization. Part C: the realistic hypercycle. *Naturwissenschaften*, **65**, 341–369.

Eigen, M. and Schuster, P. (1979b) *The Hypercycle: A Principle of Natural Self-Organization*, Springer, New York.

Eigen, M. and Winkler, R. (1975) *Das Spiel. Naturgesetze steuern den Zufall*, Piper Verlag, München.

Eigen, M. and Winkler-Oswatitsch, R. (1981) Transfer-RNA, an early gene? *Naturwissenschaften*, **68**, 282–292.

Eiraku, M., Takata, N., Ishibashi, H., Kawada, M., Sakakura, E., Okuda, S., Sekiguchi, K., Adachi, T., and Sasai, Y. (2011) Self-organizing optic-cup morphogenesis in three-dimensional culture. *Nature*, **472**, 51–56.

El Albani, A., Bengtson, S., Canfield, D.E., Bekker, A., Macchiarelli, R., Mazurier, A., Hammarlund, E.U., Boulvais, P., Dupuy, J.-J., Fontaine, C., Fürsich, F.T., Gauthier-Lafaye, F., Janvier, P., Javaux, E., Ossa Ossa, F., Pierson-Wickmann, A.-C., Riboulleau, A., Sardini, P., Vachard, D., Whitehouse, M., and Meunier, A. (2010) Large colonial organisms with coordinated growth in oxygenated environments 2.1 Gyr ago. *Nature*, **466**, 100–104.

Elliot, J.L., Person, M.J., Zuluaga, C.A., Bosh, A.S., Adams, E.R., Brothers, T.C., DuPré, K., Pasachoff, J.M., Souza, S.P., Rosing, W., Secrest, N., Bright, L., Dunham, E.W., Sheppard, S.S., Kakkala, M., Tilleman, T., Berger, B., Briggs, J.W., Jacobson, G., Valleli, P. *et al.* (2010) Size and albedo of Kuiper belt object 55636 from a stellar occultation. *Nature*, **465**, 897–900.

Emanuel, K.A. (1994) *Atmospheric convection*, University Press, New York, Oxford.

Emden, R. (1913) Über Strahlungsgleichgewicht und atmosphärische Strahlung. Ein Beitrag zur Theorie der oberen Inversion. *Sitzungsberichte der mathematisch-physikalischen Klasse der Königlich Bayerischen Akademie der Wissenschaften zu München*. Jahrgang, **1913**, 55–142.

Emery, W.J., Talley, L.D., and Pickard, G.L. (2006) *Descriptive Physical Oceanography*, Elsevier, Amsterdam.

Engel, A. and Ebeling, W. (1987a) Interaction of moving interfaces with obstacles. *Phys. Lett. A*, **122**, 20–24.

Engel, A. and Ebeling, W. (1987b) Comment on diffusion in a random potential, hopping as a dynamical consequence of localization. *Phys. Rev. Lett.*, **39**, 1979.

Engel, A. (1983) Selektion und Diffusion in stochastischen Feldern. Diploma Thesis, Humboldt University, Berlin.

Engel, A. (1985) *Reaktions-Diffusionsprozesse in stochastischen Medien*. Dissertation A, Humboldt University, Berlin.

Engel, M.S. and Grimaldi, D.A. (2004) New light shed on the oldest insects. *Nature*, **427**, 627–630.

Engel, A.E.J., Nagy, B., Nagy, L.A., Engel, C.G., Kremp, G.O.W., and Drew, C.M. (1968) Alga-like forms in Onverwacht series, South Africa: oldest recognized lifelike forms on Earth. *Science*, **161**, 1005–1008.

Engels, F. (1972) *The Origin of the Family, Private Property and the State*, Pathfinder Press, New York (originally published in 1884 in Hottingen-Zurich).

England, P.C. and Katz, R.F. (2010) Melting above the anhydrous solidus controls the location of volcanic arcs. *Nature*, **467**, 700–703.

Erdős, P. and Rényi, A. (1960) The evolution of random graphs. *Magyar Tudományos Akadémia Matematikai Kutató Intézetének közleményei*, **5**, 17–61.

Essex, C. (1984) Radiation and the violation of bilinearity in the thermodynamics of irreversible processes. *Planet. Space Sci.*, **32**, 1035–1043.

Evans, B.W. (2010) Lizardite versus antigorite serpentinite: magnetite, hydrogen, and life(?). *Geology*, **38**, 959–960.

Evans, D.J., Cohen, E.G.D., and Morriss, G.P. (1993) Probability of second law violations in shearing steady flows. *Phys. Rev. Lett.*, **71**, 2401–2404.

Evans, D.J., Searles, D.J., and Mittag, E. (2001) Fluctuation theorem for Hamiltonian systems – Le Chatelier's principle. *Phys. Rev. E*, **63** (4), 051105.

Faber, M.M. and Proops, J. (1990) *Evolution, Time, Ppoduction and the Environment*, Springer, Berlin.

Fairall, C.W., Bradley, E.F., Hare, J.E., Grachev, A.A., and Edson, J.B. (2003) Bulk parameterization of air–sea fluxes: updates and verification for the COARE algorithm. *J. Climate*, **16**, 571–591.

Falk, D. (2009) *Finding our Tongues. Mothers, Infants and the Origins of Language*, Basic Books, New York.

Falkenhagen, H. and Ebeling, W. (1971) *Theorie der Elektrolyte*, S. Hirzel, Leipzig.

Farías, M., Vargas, G., Tassara, A., Carretier, S., Baize, S., Melnick, D., and Bataille, K. (2010) Land-level changes produced by the M_w 8.8 2010 Chilean earthquake. *Science*, **329**, 916.

Farris, S.M. and Schulmeister, S. (2010) Parasitoidism, not sociality, is associated with the evolution of elaborate mushroom bodies in the brains of hymenopteran insects. *Proceedings of the Royal Society B*, **278**, 940–951.

Feingold, G., Koren, I., Wang, H., Xue, H., and Brewer, W.A. (2010) Precipitation-generated oscillations in open cellular cloud fields. *Nature*, **466**, 849–852.

Feistel, R. (1976a) *Ein dynamisches Modell für die differenzierende und stimulierende Funktion des Wertgesetzes im Kapitalismus der freien Konkurrenz*, Wilhelm-Pieck-Universität Rostock, Belegarbeit.

Feistel, R. (1976b) *Anwendungen der Theorie stochastischer Systeme auf lineare und nichtlineare Problem der Flüssigkeitsphysik*. Dissertation A, Wilhelm-Pieck-Universität Rostock. Germany.

Feistel, R. (1977a) Betrachtung der Realisierung stochastischer Prozesse aus automatentheoretischer Sicht. *Wissenschaftliche Zeitschrift der Wilhelm-Pieck-Universität Rostock, Mathematisch-Naturwissenschaftliche Reihe*, **26**, 671–678.

Feistel, R. (1977b) Ein dynamisches Modell zur differenzierenden und stimulierenden Funktion des Wertgesetzes. *Rostocker Physikalische Manuskripte*, **1**, 48–65.

Feistel, R. (1977c) Zur Beschreibung von Evolutionsprozessen bei diploider Sequenzreplikation. *Wissenschaftliche Zeitschrift der Wilhelm-Pieck-Universität Rostock, Mathematisch-Naturwissenschaftliche Reihe*, **26**, 655–662.

Feistel, R. (1979) Selektion und nichtlineare Oszillationen in chemischen Modellreaktionen. Dissertation sc. nat., Wilhelm-Pieck-Universität, Rostock.

Feistel, R. (1982) Selbsterregte Schwingungen. Presented at the 4th Conference "Probleme der Theoretischen Physik", Leipzig, 11–13 February 1981. Physikalische Gesellschaft der DDR, Berlin.

Feistel, R. (1983a) Conservation quantities in selection processes. *Stud. Biophys.*, **95**, 107–113.

Feistel, R. (1983b) Extremum principles in selection processes. *Stud. Biophys.*, **96**, 133–139.

Feistel, R. (1983c) On the evolution of biological macromolecules. III. Precellular organization. *Stud. Biophys.*, **93**, 113–120.

Feistel, R. (1983d) On the evolution of biological macromolecules. IV. Holobiotic competition. *Stud. Biophys.*, **93**, 121–128.

Feistel, R. (1985) Stochastic theory of evolution of biochemical networks. *Biomed. Biochim. Acta*, **44**, 953–957.

Feistel, R. (1986) Physik und Evolution. *Wissenschaftliche Zeitschrift der Humboldt-Universität Berlin, Mathematisch-Naturwissenschaftliche Reihe*, **35**, 412.

Feistel, R. (1990) Ritualisation und die Selbstorganisation der Information, in *Selbstorganisation und Determination* (eds U. Niedersen and L. Pohlmann), Duncker & Humblot, Berlin, pp. 83–98.

Feistel, R. (1991) On the value concept in economy, in *Models of Selforganization in Complex Systems MOSES* (eds W. Ebeling, M. Peschel, and W. Weidlich), Akademie-Verlag, Berlin, pp. 37–44.

Feistel F R. (1993) Equilibrium thermodynamics of seawater revisited. *Prog. Oceanogr.*, **31**, 101–179.

Feistel, R. (2003) A new extended Gibbs thermodynamic potential of seawater. *Prog. Oceanogr.*, **58**, 43–115.

Feistel, R. (2008a) A Gibbs function for seawater thermodynamics for −6 to 80 °C and salinity up to 120 g/kg. *Deep-Sea Res. I*, **55**, 1639–1671.

Feistel, R. (2008b) Physics and information: structures, symbols, and self-organisation. Presented at the 401st Heraeus Seminar,

Physikzentrum Bad Honnef, 21–23 January 2008.

Feistel, R. (2011a) Entropy flux and entropy production of stationary black-body radiation. *J. Non-Equil. Thermody.*, in press.

Feistel, R. (2011b) TEOS-10: a new international oceanographic standard for seawater, ice, fluid water and humid air. *Int. J. Thermophys.*, in press. doi: 10.1007/s10765-010-0901-y

Feistel, R. (2011c) Stochastic ensembles of thermodynamic potentials. *Accredit. Qual. Assur.*, **16**, 225–235.

Feistel, R. (2011d) Radiative entropy balance and vertical stability of a gray atmosphere. *European J. Phys. B*, in press.

Feistel, R. and Ebeling, W. (1976) Dynamische Modelle zum Selektionsverhalten offener Systeme. *Wissenschaftliche Zeitschrift der Wilhelm-Pieck-Universität Rostock, MNR*, **25**, 507–513.

Feistel, R. and Ebeling, W. (1977) Stochastic investigation of limit cycle systems: generalized Lotka-Selkov reaction. *Rostocker Physikalische Manuskripte*, **2**, 55–70.

Feistel, R. and Ebeling, W. (1978a) Deterministic and stochastic theory of sustained oscillations in autocatalytic reaction systems. *Physica A*, **93**, 114–137.

Feistel, R. and Ebeling, W. (1978b) On the Eigen-Schuster concept of quasi-species in the theory of natural selection. *Stud. Biophys.*, **71**, 139 Microfiche 1/44-53.

Feistel, R. and Ebeling, W. (1981) On the thermodynamics of irreversible processes in ecosystems. *Stud. Biophys.*, **86**, 237–244.

Feistel, R. and Ebeling, W. (1982) Models of Darwinian processes and evolutionary principles. *BioSystems*, **15**, 291–299.

Feistel, R. and Ebeling, W. (1984) Stochastic models of evolutionary processes, in *Thermodynamics and Regulation of Biological Processes* (eds I. Lamprecht and A.I. Zotin), Walter de Gruyter, Berlin.

Feistel, R. and Ebeling, W. (1989) *Evolution of Complex Systems: Selforganisation, Entropy and Development*, Deutscher Verlag der Wissenschaften, Kluwer Academic Publishers, Berlin, Dordrecht, Boston, London.

Feistel, R. and Ebeling, W. (2005) Statistical theory of electrolytic skin effects, in *Nucleation Theory and Applications* (eds J.W.P. Schmelzer, G. Röpke, and V.B. Priezzhev), Joint Institute for Nuclear Research, Dubna, pp. 401–419.

Feistel, R. and Feistel, S. (2006) Die Ostsee als thermodynamisches System, in *Physik Irreversibler Prozesse und Selbstorganisation* (eds L. Schimansky-Geier, H. Malchow, and T. Pöschel), Logos-Verlag, Berlin, pp. 81–98.

Feistel, R. and Hagen, E. (1993) Climate and the physics of evolution. Presentation on the Chapman/IAPSO Conference on Fractals, Chaos and Predictability in Oceanography and Meteorology, September 20–22, 1993, Galway, Ireland.

Feistel, R. and Hagen, E. (1994) Thermodynamic quantities in oceanography, in *The Oceans: Physical-Chemical Dynamics and Human Impact* (eds S.K. Majumdar, E.W. Miller, G.S. Forbes, R.F. Schmalz, and A.A. Panah), The Pennsylvania Academy of Science, Easton, pp. 1–16.

Feistel, R., Hagen, E., and Grant, K. (2003) Climate changes in the subtropical Southeast Atlantic: The St. Helena Island Climate Index (1893–1999). *Prog. Oceanogr.*, **59**, 321–337. Time series available from http://www.io-warnemuende.de/hix-st-helena-island-climate-index.html.

Feistel, R. and Mahnke, R. (1977) Konkurrenzprozesse in einfachen Räuber-Beute-Systemen. *Wissenschaftliche Zeitschrift der Wilhelm-Pieck-Universität Rostock, MNR*, **26**, 507–513.

Feistel, R., Nausch, G. and Wasmund, H. (2008a) *State and Evolution of the Baltic Sea, 1952–2005. A Detailed 50-Year Survey of Meteorology and Climate, Physics, Chemistry, Biology, and Marine Environment*, John Wiley & Sons, Hoboken.

Feistel, R., Romanovsky, Yu.M., and Vasiliev, V.A. (1980) Evolyutsiya gipertsiklov Eigena, protekayushchikh v koatservatakh (Evolution of Eigen's hypercycles existing in coacervates). *Biofizika*, **25**, 882–887.

Feistel, R. and Sändig, R. (1977) Zur Analyse hierarchischer Strukturen mit Methoden der Graphen- und Matrizentheorie. *Wissenschaftliche Zeitschrift der Wilhelm-Pieck-Universität Rostock, Mathematisch-Naturwissenschaftliche Reihe*, **26**, 625–634.

Feistel, R. and Wagner, W. (2005) High-pressure thermodynamic Gibbs functions of ice and sea ice. *J. Mar. Res.*, **63**, 95–139.

Feistel, R. and Wagner, W. (2006) A new equation of state for H$_2$O ice Ih. *J. Phys. Chem. Ref. Data*, **35**, 1021–1047.

Feistel, R. and Wagner, W. (2007) Sublimation pressure and sublimation enthalpy of H$_2$O ice Ih between 0 and 273.16 K. *Geochim. Cosmochim. Acta*, **71**, 36–45.

Feistel, R., Wright, D.G., Jackett, D.R., Miyagawa, K., Reissmann, J.H., Wagner, W., Overhoff, U., Guder, C., Feistel, A., and Marion, G.M. (2010a) Numerical implementation and oceanographic application of the thermodynamic potentials of liquid water, water vapour, ice, seawater and humid air – Part 1: background and equations. *Ocean Sci.*, **6**, 633–677, http://www.ocean-sci.net/6/633/2010/.

Feistel, R., Wright, D.G., Kretzschmar, H.-J., Hagen, E., Herrmann, S., and Span, R. (2010b) Thermodynamic properties of sea air. *Ocean Sci.*, **6**, 91–141 http://www.ocean-sci.net/6/91/2010/.

Feistel, R., Wright, D.G., Miyagawa, K., Harvey, A.H., Hruby, J., Jackett, D.R., McDougall, T.J., and Wagner, W. (2008b) Mutually consistent thermodynamic potentials for fluid water, ice and seawater: a new standard for oceanography. *Ocean Sci.*, **4**, 275–291, http://www.ocean-sci.net/4/275/2008/.

Fekry, M.I., Tipton, P.A., and Gates, K.S. (2011) Kinetic consequences of replacing the internucleotide phosphorus atoms in DNA with arsenic. *ACS Chem. Biol.*, **6**, 127–130.

Fellmuth, B. (2003) Redefinition of the temperature unit? *PTB-News 2003*. 2, Physikalisch-Technische Bundesanstalt (PTB), Braunschweig and Berlin, Germany, http://www.ptb.de/en/publikationen/news/html/news032/artikel/03205.htm.

Fellmuth, B., Buck, W., Fischer, J., Gaiser, C., and Seidel, J. (2007) Neudefinition der Basiseinheit Kelvin. *PTB-Mitteilungen*, **117**, 67–73.

Feng, X., Li, Y., Gu, J., Zhuo, Y., and Yang, H. (2007) Error thresholds for quasispecies on single peak Gaussian-distributed fitness landscapes. *J. Theor. Biol.*, **246**, 28–32.

Field, C.B., Behrenfeld, M.J., Randerson, J.T., and Falkowski, P. (1998) Primary production of the biosphere: integrating terrestrial and oceanic components. *Science*, **281**, 237–240.

Finnegan, S., Bergmann, K., Eiler, J.M., Jones, D.S., Fike, D.A., Eisenman, I., Hughes, N.C., Tripati, A.K., and Fischer, W.W. (2011) The magnitude and duration of late ordovician – early silurian glaciation. *Science*, **331**, 903–906.

Fiquet, G., Auzende, A.L., Siebert, J., Corgne, A., Bureau, H., Ozawa, H., and Garbarino, G. (2010) Melting of peridotite to 140 gigapascals. *Science*, **329**, 1516–1518.

Fischer, J., de Podesta, M., Hill, K., Moldover, M., Pitre, L., Rusby, R., Steur, P., Tamura, O., White, R., and Wolber, L. (2011) Present estimates of the differences between thermodynamic temperatures and the ITS-90. *Int. J. Thermophys.*, **32**, 12–25.

Fisher, R.A. (1930) *The Genetical Theory of Natural Selection*, Clarendon Press, Oxford.

Fitch, W.T. (2005) The evolution of language: a comparative review. *Biol. Philos.*, **20**, 193–230.

Fitch, W.T. (2010) *The Evolution of Language*, Cambridge University Press, Cambridge.

Fitch, W.T., Neubauer, J., and Herzel, H. (2002) Calls out of chaos: the adaptive significance of nonlinear phenomena in mammalian vocal production. *Anim. Behav.*, **63**, 407–418.

Fortak, H.G. (1979) Entropy and climate, in *Developments in Atmospheric Science*, vol. **10**, Man's Impact on Climate (eds W. Bach, J. Pankrath, and W. Kellogg), Elsevier Science., New York, pp. 1–14.

Försterling, H.D. and Kuhn, H. (1971) *Physikalische Chemie in Experimenten - Ein Praktikum*, Verlag Chemie, Weinheim.

Försterling, H.D., Kuhn, H., and Tews, H. (1972) Computermodell zur Bildung selbstorganisierter Systeme. *Angew. Chem.*, **84**, 862–865.

Foster, S.S. (2004) Reconstruction of solar irradiance variations, for use in studies of global climate change: application of recent SoHO observations with historic data from the Greenwich observations. Ph.D. Thesis, University of Southampton.

Fox, S.W. and Dose, K. (1972) *Molecular Evolution and the Origin of Life*, Freeman, San Francisco.

Frankignoul, C. (1985) Sea surface temperature anomalies, planetary waves, and air-sea feedback in the middle latitudes. *Rev. Geophys.*, **23**, 357–390.

Frankignoul, C. and Hasselmann, K. (1977) Stochastic climate models, Part II.

Application to sea-surface temperature anomalies and thermocline variability. *Tellus*, **29**, 298–305.

Freiwald, W.A. and Tsao, D.Y. (2010) Functional compartmentalization and viewpoint generalization within the macaque face-processing system. *Science*, **330**, 845–851.

Fröhlich, C. (2009) Observational evidence of a long-term trend in total solar irradiance. *Astron. Astrophys.*, **501**, 27–30. doi: 10.1051/0004-6361/200912318

Früh-Green, G.L., Kelley, D.S., Bernasconi, S.M., Karson, J.A., Ludwig, K.A., Butterfield, D.A., Boschi, C., and Proskurowski, G. (2003) 30,000 years of hydrothermal activity at the Lost City vent field. *Science*, **301**, 495–498.

Fugère, V., Ortega, H., and Krahe, R. (2011) Electrical signalling of dominance in a wild population of electric fish. *Biology Lett.*, **7**, 197–200.

Gadjimuradov, I. and Schmoeckel, R. (2005) *Das Geheimnis des Anatolischen Meeres*. Van-See-Gesellschaft, Bonn.

Gallavotti, G. and Cohen, E.G.D. (1995a) Dynamical ensembles in nonequilibrium statistical mechanics. *Phys. Rev. Lett.*, **74**, 2694–2697.

Gallavotti, G. and Cohen, E.G.D. (1995b) Dynamical ensembles in stationary states. *J. Stat. Phys.*, **80**, 931–970.

Galluccio, S. (1997) Exact solution of the quasispecies model in a sharply peaked fitness landscape. *Phys. Rev. E*, **56**, 4526–4539.

Gantmacher, F.R. (1971) *Matrizenrechnung II*, Deutscher Verlag der Wissenschaften, Berlin, Russian original (1954): Teoriya matrits, chast II. Gosudarstvennoye izdatelstvo tekhniko-teoreticheskoy literatury, Moscow.

Garcia-Castellanos, D., Estrada, F., Jiménez-Munt, I., Gorini, C., Fernàndez, M., Vergés, J., and De Vicente, R. (2009) Catastrophic flood of the Mediterranean after the Messinian salinity crisis. *Nature*, **462**, 778–781.

Gardner, R.M. and Ahsby, W.R. (1970) Connectance of large dynamic (Cybernetic) systems: critical values for stability. *Nature (London)*, **228**, 784.

Gardiner, C.W. and Chaturvedi, S. (1977) The Poisson representation I. A new technique for chemical master equations. *J. Stat. Phys.*, **17**, 429–468.

Gaspard, P. (1998) *Chaos, Scattering and Statistical Mechanics*, Cambridge University Press, Cambridge.

Gaspard, P. (2007) Temporal ordering of nonequilibrium fluctuations as a corollary of the second law of thermodynamics. *C. Rendus Physique*, **8**, 598–608.

Gaspard, P. and Nicolis, G. (1983) What can we learn from homoclinic orbits in chaotic dynamics? *J. Stat. Phys.*, **31**, 499–518.

Gaunt, M.W. and Miles, M.A. (2002) An insect molecular clock dates the origin of the insects and accords with palaeontological and biogeographic landmarks. *Mol. Biol. Evol.*, **19**, 748–761.

Gause, G.F. and Witt, A.A. (1935) Behavior of mixed populations and the problem of natural selection. *Am. Nat.*, **69**, 596–609.

Gavalas, G.R. (1968) *Nonlinear differential equations of chemically reacting systems*, Springer, Berlin, New York.

Gavrilets, S. (2004) *Fitness landscapes and the Origin of Species*, Princeton University Press, Princeton.

Gerschgorin, S. (1931) Über die Abgrenzung der Eigenwerte einer Matrix. *Izvestiya Akademii Nauk SSSR Otdeleniye Fizicheskoy-Matematicheskoy Nauk*, **7**, 749–754.

Ghosh, I. and Chmielewski, J. (2004) Peptide self-assembly as a model of proteins in the pre-genomic world. *Curr. Opin. Chem. Biol.*, **8**, 640–644.

Giauque, W.F. and Stout, J.W. (1936) The entropy of water and the third law of thermodynamics. The heat capacity of ice from 15 to 273 K. *J. Am. Chem. Soc.*, **58**, 1144–1150.

Gibbons, A. (2009) A new kind of ancestor: *ardipithecus* unveiled. *Science*, **326**, 36–40.

Gibbons, A. (2011) A new view of the birth of homo sapiens. *Science*, **331**, 392–394.

Gibbs, J.W. (1873) Graphical methods in the thermodynamics of fluids. *Trans. Connecticut Acad. Arts Sci.*, **2**, 309–342.

Gibbs, J.W. (1902) *Elementary Principles of Statistical Mechanics*, Charles Scribner's Sons, Edward Arnold, New York, London.

Gierer, A. and Meinhardt, H. (1972) A theory of biological pattern formation. *Kybernetik*, **12**, 30–39.

Gill, A.E. (1982) *Atmosphere Ocean Dynamics*, Academic Press, San Diego.

Gill, B.C., Lyons, T.W., Young, S.A., Kump, L.R., Knoll, A.H., and Saltzman, M.R. (2011) Geochemical evidence for widespread euxinia in the later cambrian ocean. *Nature*, **469**, 80–83.

Gillespie, D.T. (1977) Exact stochastic simulation of coupled chemical reactions. *J. Phys. Chem.*, **81**, 2340–2361.

Gingras, M., Hagadorn, J.D., Seilacher, A., Lalonde, S.V., Pecoits, E., Petrash, D., and Konhauser, K.O. (2011) Possible evolution of mobile animals in association with microbial mats. *Nature Geoscience*, published online, doi: 10.1038/ngeo1142

Glansdorff, P. and Prigogine, I. (1971) *Thermodynamic Theory of Structure, Stability and Fluctuations*, Wiley-Interscience, New York.

Glaser, R. (1975) *Biologie einmal anders*, Aulis Verlag Deubner, Köln.

Goel, N.S., Maitra, S.C., and Montroll, E.W. (1971) On the volterra and other nonlinear models of interacting populations. *Rev. Mod. Phys.*, **43**, 231–276.

Gong, Z., Liu, J., Gou, C., Zhou, Y., Teng, Y., and Liu, L. (2010) Two pairs of neurons in the central brain control drosophila innate light preference. *Science*, **330**, 499–502.

Gopnik, A., Meltzoff, A., and Kuhl, P. (2000) *Forschergeist in Windeln*. Ariston, München. American original (1999): *The Scientist in the Crib*, William Morris, New York.

Görke, L. (1970) *Mengen – Relationen – Funktionen*, Volk und Wissen, Berlin.

Götzinger, G. (1909) Studien über das Eis des Lunzer Unter- und Obersees. *Internationale Revue der gesamten Hydrobiologie und Hydrographie*, **2**, 386–396.

Gould, S.J. (1993) *The Book of Life*, Ebory Hutchinson/Random House, London.

Gould, S.J. and Eldredge, N. (1977) Punctuated equilibria: the tempo and mode of evolution reconsidered. *Paleobiology*, **3**, 115–151.

Gradstein, F., Ogg, J., and Smith, A. (2005) *A Geologic timescale*, Cambridge University Press, Cambrdige.

Graham, R. (1973) Statistical theory of instabilities in stationary non-equilibrium systems with applications to lasers and nonlinear optics. *Springer Tracts Mod. Phys.*, **66**, 111.

Graham, R. (1981) Models of stochastic behaviour, in *Scattering Techniques* (eds S.H. Chen, B., Chu and R. Nossal), Plenum Press, New York, John Wiley & Sons, New York, London, Sydney.

Graham F D. (2010) Geochemistry: relict mantle from Earth's birth. *Nature*, **466**, 822–823.

Grasby, S.E., Sanei, H., and Beauchamp, B. (2011) Catastrophic dispersion of coal fly ash into oceans during the latest Permian extinction. *Nat. Geoscience*, **4**, 104–107.

Grassl, H. (1976) The dependence of the measured cool skin of the ocean on wind stress and total heat flux. *Bound-Lay. Meteorol.*, **10**, 465–474. doi: 10.1007/BF00225865

Grassl, H. (1981) The climate at maximum entropy production by meridional atmospheric and oceanic heat fluxes. *Q. J. Roy. Meteor. Soc.*, **107**, 153–166.

Grassle, J.F. (1986) *The Ecology of Deep-Sea Hydrothermal Vent Communities*, Academic Press, London.

Gray, R.D. and Atkinson, Q.D. (2003) Language-tree divergence times support the Anatolian theory of Indo-European origin. *Nature*, **426**, 435–439.

Greco, T.H. (2001) *Money: Understanding and Creating Alternatives to Legal Tender*, Chelsea Green, White River Junction, VT.

Greene, B. (2004) *The Fabric of the Cosmos*, Alfred A. Knopf, New York.

Green, D.H., Hibberson, W.O., Kovác, I., and Rosenthal, A. (2010) Water and its influence on the lithosphere–asthenosphere boundary. *Nature*, **467**, 448–451.

Green III, H.W., Chen, W.-P., and Brudzinski, M.R. (2010) Seismic evidence of negligible water carried below 400-km depth in subducting lithosphere. *Nature*, **467**, 828–831.

Greenberg, R., Wacker, J.F., Hartmann, W.K., and Chapman, C.R. (1978) Planetesimals to planets: numerical simulation of collisional evolution. *Icarus*, **35**, 1–26.

Gregg, C., Zhang, J., Weissbourd, B., Luo, S., Schroth, G.P., Haig, D., and Dulac, C. (2010) High-resolution analysis of parent-of-origin allelic expression in the mouse brain. *Science*, **329**, 643–648.

Gregory, S.G., Barlow, K.F., McLay, K.E., Kaul, R., Swarbreck, D., Dunham, A., Scott, C.E., Howe, K.L., Woodfine, K., Spencer, C.C.A., Jones, M.C., Gillson, C., Searle, S., Zhou, Y.,

Kokocinski, F., McDonald, L., Evans, R., Phillips, K., Atkinson, A., Cooper, R., Jones, C., Hall, R.E., Andrews, T.D., Lloyd, C., Ainscough, R., Almeida, J.P., Ambrose, K.D., Anderson, F., Andrew, R.W., Ashwell, R.I.S., Aubin, K., Babbage, A.K., Bagguley, C.L., Bailey, J., Beasley, H., Bethel, G., Bird, C.P., Bray-Allen, S., Brown, J.Y., Brown, A.J., Buckley, D., Burton, J., Bye, J., Carder, C., Chapman, J.C., Clark, S.Y., Clarke, G., Clee, C., Cobley, V., Collier, R.E., Corby, N., Coville, G.J., Davies, J., Deadman, R., Dunn, M., Earthrowl, M., Ellington, A.G., Errington, H., Frankish, A., Frankland, J., French, L., Garner, P., Garnett, J., Gay, L., Ghori, M.R.J., Gibson, R., Gilby, L.M., Gillett, W., Glithero, R.J., Grafham, D.V., Griffiths, C., Griffiths-Jones, S., Grocock, R., Hammond, S., Harrison, E.S.I., Hart, E., Haugen, E., Heath, P.D., Holmes, S., Holt, K., Howden, P.J., Hunt, A.R., Hunt, S.E., Hunter, G., Isherwood, J., James, R., Johnson, C., Johnson, D., Joy, A., Kay, M., Kershaw, J.K., Kibukawa, M., Kimberley, A.M., King, A., Knights, A.J., Lad, H., Laird, G., Lawlor, S., Leongamornlert, D.A., Lloyd, D.M., Loveland, J., Lovell, J., Lush, M.J., Lyne, R., Martin, S., Mashreghi-Mohammadi, M., Matthews, L., Matthews, N.S.W., McLaren, S., Milne, S., Mistry, S., Moore, M.J.F., Nickerson, T., O'Dell, C.N., Oliver, K., Palmeiri, A., Palmer, S.A., Parker, A., Patel, D., Pearce, A.V., Peck, A.I., Pelan, S., Phelps, K., Phillimore, B.J., Plumb, R., Rajan, J., Raymond, C., Rouse, G., Saenphimmachak, C., Sehra, H.K., Sheridan, E., Shownkeen, R., Sims, S., Skuce, C.D., Smith, M., Steward, C., Subramanian, S., Sycamore, N., Tracey, A., Tromans, A., Van Helmond, Z., Wall, M., Wallis, J.M., White, S., Whitehead, S.L., Wilkinson, J.E., Willey, D.L., Williams, H., Wilming, L., Wray, P.W., Wu, Z., Coulson, A., Vaudin, M., Sulston, J.E., Durbin, R., Hubbard, T. Wooster, R., Dunham, I., Carter, N.P., McVean, G., Ross, M.T., Harrow, J., Olson, M.V., Beck, S., Rogers, J., and Bentley, D.R. (2006) The DNA sequence and biological annotation of human chromosome 1. *Nature*, **441** 315–321.

Greif, S. and Siemers, B.M. (2010) Innate recognition of water bodies in echolocating bats. *Nat. Communications*, **1**, 107. online, doi: 10.1038/ncomms1110

Grena, R. (2008) An algorithm for the computation of the solar position. *Sol. Energy*, **82**, 462–470.

Grmela, M. and Öttinger, H.C. (1997) Dynamics and thermodynamics of complex fluids. *Phys. Rev. E*, **56**, 6620–6632 6633–6655.

Gubbins, D. Sreenivasan, B., Mound, J., and Rost, S. (2011) Melting of the Earth's inner core. *Nature*, **473**, 361–363.

Guggenheim, E.A. (1949) *Thermodynamics*, North-Holland Publishing Company, Amsterdam.

Günther, E. (1978) *Grundriß der Genetik*, Gustav Fischer, Jena.

Gutierrez, M.C., Brisse, S., Brosch, R., Fabre, M., Omaïs, B., Marmiesse, M., Supply, P., and Vincent, V. (2005) Ancient origin and gene mosaicism of the progenitor of mycobacterium tuberculosis. *PLoS Pathogens*, **1**, e5. doi: 10.1371/journal.ppat.0010005

Gutzow, I.S. and Schmelzer, J.W.P. (2011) Glasses and the third law of thermodynamics, in *Glasses and the Glass Transition* (eds J.W.P. Schmelzer and I.S. Gutzow), Wiley-VCH, Weinheim, pp. 357–378.

Haarmann, H. (2001) *Kleines Lexikon der Sprachen*, C.H. Beck, München.

Haarmann, H. (2002) *Lexikon der untergegangenen Sprachen*, C.H. Beck, München.

Haarmann, H. (2003) *Geschichte der Sintflut*, C.H. Beck, München.

Haase, R. (1963) *Thermodynamik der irreversiblen Prozesse*, Steinkopff, Darmstadt.

Haeckel, E. (1874) Die Gastrea-Theorie, die phylogenetische Classification des Tierreichs und die Homologie der Keimblätter. *Jenaische Zeitschrift fur Naturwissenschaft*, **8**, 1–55.

Hagemann, R. (1976) *Beiträge zur Genetik und Abstammungslehre*, Volk und Wissen, Berlin.

Hagen, E. (2009) Atlantic exploration and climate. in: *Selected Contributions on Results of Climate Research in East Germany* (the former GDR), (eds P. Hupfer and K. Dethloff), Reports on Polar and Marine Research 588, 80–95.

Hagen, E., Agenbag, J.J., and Feistel, R. (2005) The winter St. Helena climate index and extreme Benguela upwelling. *J. Marine Syst.*, **57**, 219–230.

Hagen, E. and Feistel, R. (2008) Baltic climate change, in *State and Evolution of the Baltic Sea, 1952–2005. A Detailed 50-Year Survey of Meteorology and Climate, Physics, Chemistry, Biology, and Marine Environment* (eds R. Feistel, G. Nausch, and N. Wasmund), John Wiley & Sons, Hoboken, NJ, pp. 93–120.

Hagen F E., Feistel, R., Agenbag, J.J., and Ohde, T. (2001) Seasonal and interannual changes in Intense Benguela Upwelling (1982–1999). *Oceanol. Acta*, **24**, 1–12.

Hagen, E., Mittelstaedt, E., Feistel, R., and Klein, H. (1994) *Hydrographische Untersuchungen im Ostrandstromsystem vor Portugal und Marokko 1991–1992*, Berichte des BSH, Hamburg, Nr.2.

Hagen, E., Schemainda, R., Michelchen, N., Postel, L., Schulz, S., and Below, M. (1981) Zur küstensenkrechten Struktur des Kaltwasserauftriebs vor der Küste Namibias. *Geodätische und Geophysikalische Veröffentlichungen*, **36**, 1–99.

Hagen, E., Zülicke, Ch., and Feistel, R. (1996) Near surface structures in the Cape Ghir filament off Morocco. *Oceanol. Acta*, **19**, 577–597.

Hager, B.H. and Gurnis, M. (1987) Mantle convection and the state of the Earth's interior. *Rev. Geophys.*, **25**, 1277–1285.

Hahn, H. (1974) Geometrical aspects of the pseudo steady state hypothesis in enzyme reactions, in *Physics and Mathematics of the Nervous System. Lecture Notes in Biomathematics 4* (eds M. Conrad, W. Güttinger, and M. Dal Cin), Springer, Berlin, Heidelberg, New York.

Haigh, J.D. (2010) Climate sensitivity to solar variability. WMO-BIPM workshop on measurement challenges for global observation systems for climate change monitoring: Traceability, stability and uncertainty: 30 March to 1 April 2010, World Meteorological Organisation, Geneva, IOM-Report 105, WMO/TD 1557, http://www.bipm.org/ws/BIPM/WMO-BIPM/Allowed/Session_C/C.06-Haigh.pdf.

Haigh, J.D., Winning, A.R., Toumi, R., and Harder, J.W. (2010) An influence of solar spectral variations on radiative forcing of climate. *Nature*, **467**, 696–699.

Haile-Selassie, Y., Latimer, B.M., Alene, M., Deino, A.L., Gibert, L., Melillo, S.M., Saylor, B.Z., Scott, G.R., and Lovejoy, C.O. (2010) An early Australopithecus afarensis postcranium from Woranso-Mille, Ethiopia. *Proc. Natl. Acad. Sci. U.S.A.*, **107**, 12121–12126.

Hain, M.P., Sigman, D.M., and Haug, G.H. (2011) Shortcomings of the isolated abyssal reservoir model for deglacial radiocarbon changes in the mid-depth Indo-Pacific Ocean. *Geophysical Research Letters*, **38**, L04604, doi:10.1029/2010GL046158

Haken, H. (1970) *Laser Theory*, Springer, Berlin/Heidelberg/New York.

Haken, H. (1978) *Synergetics: An Introduction*, Springer, Berlin, Heidelberg, New York.

Haken, H. (1981) *Erfolgsgeheimnisse der Natur*. Deutsche Verlagsanstalt, Stuttgart.

Haken, H. (1988) *Information and Selforganization: A Macroscopic Approach to Complex Systems*, Springer, Berlin, Heidelberg, New York.

Haken, H. and Graham, R. (1971) Synergetik. Die Lehre vom Zusammenwirken. *Umschau*, **6**, 191–195.

Halliday, A.N. and Wood, B.J. (2009) How did earth accrete? *Science*, **325**, 44–45.

Hamel, E. (2007) *Das Werden der Völker in Europa*, Tenea, Bristol, Berlin.

Hancock, G., Pankhurst, R., and Willetts, D. (1983) *Under Ethiopian Skies*, H&L Communications, London.

Hanks, T.C. and Kanamori, H. (1979) Moment magnitude scale. *J. Geophys. Res.*, **84**, 2348–2350.

Hansen, K. (2010) Unstable Antarctica: What's Driving Ice Loss? NASA's Earth Science News Team, published online 15 December. 2010, http://www.nasa.gov/topics/earth/features/unstable-antarctica.html.

Hausman, R.E. and Cooper, G.M. (2004) *The Cell: A Molecular Approach*, ASM Press, Washington DC.

Hawking, S.W. (1974) Black hole explosions? *Nature*, **248**, 30–31.

Harary, F., Norman, R.Z., and Cartwright, D. (1965) *Structural Models. An Introduction to the Theory of Directed Graphs*, Wiley-Interscience Publishers, New York.

Hardy, G.H. (1908) Mendelian proportions in a mixed population. *Science*, **28**, 49–50.

Harmer, R., Danos, V., Feret, J., Krivine, J., and Fontana, W. (2010) Instrinsic information carriers in combinatorial dynamical systems. *Chaos*, **20**, 037108–1.

Hartmann, A.-M. (2010) Untersuchungen zur Regulation und Struktur von Kationen-Chlorid-Kotransportern. Dissertation, Carl-von-Ossietzky University of Oldenburg.

Hartmann, D.L., Holton, J.R., and Fu, Q. (2001) The heat balance of the tropical tropopause, cirrus, and stratospheric dehydration. *Geophys. Res. Lett.*, **28**, 1969–1972.

Hawks, J., Hunley, K., Lee, S.H., and Wolpoff, M. (2000) Population bottlenecks and pleistocene human evolution. *Mol. Biol. Evol.*, **17**, 2–22.

Hartmann-Sonntag, I., Scharnhorst, A., and Ebeling, W. (2009) Sensitive networks – modelling self-organization and innovation processes in networks, in *Innovative Networks* (eds A. Pyka and A. Scharnhorst), Springer, Dordrecht, Heidelberg, London, New York.

Hassold, N.J.C., Rea, D.K., van der Pluijm, B.A., and Parés, J.M. (2009) A physical record of the Antarctic Circumpolar Current: Late Miocene to recent slowing of abyssal circulation. *Palaeogeogr. Palaeocl.*, **275**, 28–36.

Hazen, R.M. (2010) Die Evolution der Minerale. *Spektrum der Wissenschaft*, August 2010, 80–87.

Haszprunar, G. (2009) *Evolution und Schöpfung. Versuch einer Synthese*, EOS Verlag, Sankt Ottilien.

Head, J.W., III, Fassett, C.I., Kadish, S.J., Smith, D.E., Zuber, M.T., Neumann, G.A., and Mazarico, E. (2010) Global distribution of large lunar craters: implications for resurfacing and impactor populations. *Science*, **329**, 1504–1507.

Hegerl, G.C. and Solomon, S. (2009) Risks of climate engineering. *Science*, **325**, 955–956.

Heinrich, H. (1988) Origin and consequences of cyclic ice rafting in the Northeast Atlantic Ocean during the past 130,000 years. *Quaternary Res.*, **29**, 142–152.

Heinrich, R. and Sonntag, I. (1981) Analysis of the selection equations for a multivariable population model: deterministic and stochastic solutions and discussion of the approach for populations of self-reproducing biochemical networks. *J. Theor. Biol.*, **93**, 325–361.

Heinrich, R. and Schuster, S. (1998) The modelling of metabolic systems. Structure, control and optimality. *BioSystems*, **47**, 61–77.

Heithoff, J. (2011) Sichere Energieversorgung ohne Geowissenschaften nicht denkbar. *Deutsche Geolische Gesellschaft*, **1**, 4–14.

Helbing, D., Herrmann, H.J., and Schreckenberg, M. (2000) *Traffic and Granular Flow 99: Social, Traffic, and Granular Dynamics*, Springer, Berlin, Heidelberg, New York.

Held, R., Ostrovsky, Y., deGelder, B., Gandhi, T., Ganesh, S., Mathur, U., and Sinha, P. (2011) The newly sighted fail to match seen with felt. *Nat. Neurosci.*, **14**, 551–553.

Hemmer, P.C., Holden, H., and Kjelstrup Ratkje, S. (1996) *The Collected Works of Lars Onsager (with commentary)*, World Scientific, Singapore.

Herman, L.M., Richards, D.G., and Wolz, J.P. (1984) Comprehension of sentences by bottlenosed dolphins. *Cognition*, **16**, 129–219.

Herzel, H., Berry, D., Titze, I.R., and Saleh, M. (1994) Analysis of vocal disorders with methods from nonlinear dynamics. *J. Speech Hear. Res.*, **37**, 1008–1019.

Hewitt, G. (2000) The genetic legacy of the Quaternary ice ages. *Nature*, **405**, 907–913.

Hill, P.J. and Exon, N.F. (2004) Tectonics and Basin development of the offshore Tasmanian area incorporating results from deep ocean drilling. *Geophys. Monogr.*, **151**, 19–42.

Hitti, P.K. (1961) *The Near East in History: A 5000 Year Story*, Van Nostrand, Princeton, NJ.

Hirose, K. (2010) Deep mantle properties. *Science*, **327**, 151–152.

Hodych, J.P. and Dunning, G.R. (1992) Did the Manicouagan impact trigger end-of-Triassic mass extinction? *Geology*, **20**, 51–54.

Hoffmann, D. (2008) *Max Planck: Annalen Papers*, Wiley-VCH, Weinheim.

Hofmann, G., Jenssen, M., and Anders, S. (2002) Kohlenstoffpotentiale mitteleuropäischer Wälder. *AFZ Der Wald*, **57**, 605–607.

Hohl, R. (1977) *Unsere Erde. Eine moderne Geologie*, Urania-Verlag, Leipzig, Jena, Berlin.

Hohl, R. (1981) *Die Entwicklungsgeschichte der Erde*, Brockhaus, Leipzig.

Holland, D.M. (2000) Transient sea-ice polynya forced by oceanic flow variability. *Prog. Oceanogr.*, **48**, 403–460.

Hollandt, J., Hartmann, J., Gutschwager, B., and Struß, O. (2007) Strahlungsthermometrie – Temperaturen berührungslos messen. *PTB-Mitteilungen*, **117**, 52–61.

Holliday, N.P., Hughes, S.L., Borenäs, K., Feistel, R., Gaillard, F., Lavín, A., Loeng, H., Mork, K.-A., Nolan, G., Quante, M., and Somavilla, R. (2011) Long-term physical variability in the North Atlantic Ocean, in *ICES Position Paper on Climate Change* (eds L. Valdés and J. Alheit), in press.

Hoorn F C., Wesselingh, F.P., ter Steege, H., Bermudez, M.A., Mora, A., Sevink, J., Sanmartín, I., Sanchez-Meseguer, A., Anderson, C.L., Figueiredo, J.P., Jaramillo, C., Riff, D., Negri, F.R., Hooghiemstra, H., Lundberg, J., Stadler, T., Särkinen, T., and Antonelli, A. (2010) Amazonia through time: Andean Uplift, climate change, landscape evolution, and biodiversity. *Science*, **330**, 927–931.

Hoover, W.G. (2001) *Time Reversibility, Computer Simulation, and Chaos*, World Scientific, Singapore.

Horowitz, A. (2001) *The Jordan Rift Valley*, Swets & Zeitlinger B.V., Lisse.

Horst, S., Yelle, R.V., Buch, A., Carrasco, N., Cernogora, G., Dutuit, O., Quirico, E., Sciamma-O'Brien, E., Smith, M.A., Somogyi, A., Szopa, C., Thissen, R., and Vuitton, V. (2010) Formation of Amino Acids and Nucleotide Bases in a Titan Atmosphere Simulation Experiment. Presentation at the 42nd Annual Meeting of the Division for Planetary Sciences of the American Astronomical Society, October 3–8, 2010, Pasadena, CA, http://dps.aas.org/meetings/2010/.

Hoyle, C.H.V. (2010) Evolution of neuronal signalling: transmitters and receptors. *Auton. Neurosci.* doi: 10.1016/j.autneu.2010.05.007

Hoyng, P. (2006) *Relativistic Astrophysics and Cosmology*. Springer, Dordrecht.

Hren, M.T., Tice, M.M., and Chamberlain, C.P. (2009) Oxygen and hydrogen isotope evidence for a temperate climate 3.42 billion years ago. *Nature*, **462**, 205–208.

Hsü, K.J. (2000) *Klima macht Geschichte. Menschheitsgeschichte als Abbild der Klimaentwicklung*, orell füssli, Zürich.

Hubbe, M., Neves, W.A., and Harvati, K. (2010) Testing evolutionary and dispersion scenarios for the settlement of the new world. *PLoS ONE*, **5**, e11105. doi: 10.1371/journal.pone.0011105

Huberman, B.A. (2001) *The Laws of the Web: Patterns in the Ecology of Information*, The MIT Press, Cambridge, MA.

Hull, D.L. (2010) Science and language, in *For a Philosophy of Biology* (eds I. Jahn and A. Wessel), Kleine, München, pp. 35–36.

Hunt, D.M., Dulai, K.S., Cowing, J.A., Julliot, C., Mollon, J.D., Bowmaker, J.K., Li, W.-H., and Hewett-Emmett, D. (1998) Molecular evolution of trichromacy in primates. *Vision. Res.*, **38**, 3299–3306.

Hurrell, J.W. (1995) Decadal trends in the North Atlantic Oscillation: regional temperatures and precipitation. *Science*, **269**, 676–679.

Huxley, J.S. (1914) The courtship-habits of the great crested grebe (*Podiceps cristatus*); with an addition to the theory of sexual selection. *Proc. Zool. Soc. London*, **1914**, 491–562.

Huybrechts, P. (2009) West-side story of Antarctic ice. *Nature*, **458**, 295–296.

IAPWS (2008) *Release on the IAPWS Formulation 2008 for the Thermodynamic Properties of Seawater*, The International Association for the Properties of Water and Steam, Berlin, Germany, September 2008, http://www.iapws.org.

IAPWS (2009a) *Revised Release on the IAPWS Formulation 1995 for the Thermodynamic Properties of Ordinary Water Substance for General and Scientific Use*, The International Association for the Properties of Water and Steam, Doorwerth, The Netherlands, September 2009, http://www.iapws.org.

IAPWS (2009b) *Supplementary Release on a Computationally Efficient Thermodynamic Formulation for Liquid Water for Oceanographic Use*, The International Association for the Properties of Water and Steam, Doorwerth, The Netherlands, September 2009, http://www.iapws.org.

IAPWS (2009c) *Revised Release on the Equation of State 2006 for H_2O Ice Ih*, The International Association for the Properties of Water and Steam, Doorwerth, The Netherlands, September 2009, http://www.iapws.org.

IAPWS (2010) *Guideline on an Equation of State for Humid Air in Contact with Seawater and*

Ice, Consistent with the IAPWS Formulation 2008 for the Thermodynamic Properties of Seawater, The International Association for the Properties of Water and Steam, Niagara Falls, Canada, July 2010, http://www.iapws.org.

Ibe, O.C. (2011) *Fundamentals of Stochastic Networks*. Wiley-Blackwell, Hoboken.

Ifrah, G. (1985) *From One to Zero. A Universal History of Numbers*, Viking Penguin, New York.

Ifrah, G. (1999) *The Universal History of Numbers: From Prehistory to the Invention of the Computer*, John Wiley & Sons, New York.

IOC, SCOR, IAPSO (2010) The international thermodynamic equation of seawater - 2010: Calculation and use of thermodynamic properties. Intergovernmental Oceanographic Commission, Manuals and Guides No. 56, UNESCO (English), 196 pp., Paris, http://www.TEOS-10.org.

IPCC (2007) IPCC Fourth Assessment Report: Climate Change 2007 (AR4). Intergovernmental Panel on Climate Change, http://www.ipcc.ch/publications_and_data/publications_and_data_reports.shtml#1.

Iqbal, N., Khan, M.S., and Masood, T. (2011) Entropy changes in the clustering of galaxies in an expanding universe. *Nat. Sci.*, **3**, 65–68.

Iwanow, S. (1973) *Der Abdruck des Siegelrings*, Mir, Moscow.

Jablonski, N.G. (2006) *Skin: A Natural History*, University of California Press, Berkeley.

Jackendoff, R. (2002) *Foundations of Language (Brain, Meaning, Grammar, Evolution)*, Oxford University Press, Oxford.

Jackson, M.G., Carlson, R.W., Kurz, M.D., Kempton, P.D., Francis, D., and Blusztajn, J. (2010) Evidence for the survival of the oldest terrestrial mantle reservoir. *Nature*, **466**, 853–856.

Jacobs, G.H. (2009) Evolution of colour vision in mammals. *Philos. Trans. R. Soc. Lond. B*, **364**, 2957–2967.

Jacobson, M.Z. (2005) *Fundamentals of Atmospheric Modeling*, Cambridge University Press, Cambridge.

Jaglom, A.M. and Jaglom, I.M. (1984) *Wahrscheinlichkeit und Information*, Deutscher Verlag der Wissenschaften, Berlin, 1984.

Jain, S. and Krishna, S. (2003) Graph theory and the evolution of autocatalytic networks, in *Handbook of Graphs and Networks* (eds S. Bornholdt and H.G. Schuster), John Wiley & Sons, Weinheim.

Janson, T. (2006) *Eine kurze Geschichte der Sprachen*, Elsevier, München, English original (2002): A Short History of Languages. Oxford University Press, Oxford.

Jaramillo, C., Ochoa, D., Contreras, L., Pagani, M., Carvajal-Ortiz, H., Pratt, L.M., Krishnan, S., Cardona, A., Romero, M., Quiroz, L., Rodriguez, G., Rueda, M.J., de la Parra, F., Morón, S., Green, W., German Bayona, G., Montes, C., Quintero, O., Ramirez, R., Mora, G., Schouten, S., Bermudez, H., Navarrete, R., Parra, F., Alvarán, M., Osorno, J., Crowley, J.L., Valencia, V., and Jeff Vervoort, J. (2010) Effects of rapid global warming at the paleocene-eocene boundary on neotropical vegetation. *Science*, **330**, 957–961.

Jarzynski, C. (1996) Nonequilibrium equality for free energy differences. *Phys. Rev. Lett.*, **78**, 2690–2693; Phys. Rev. E, 56, 5018-5035.

Javaux, E.J., Marshall, C.P., and Bekker, A. (2010) Organic-walled microfossils in 3.2-billion-year-old shallow-marine siliciclastic deposits. *Nature*, **463**, 934–938.

Jeans, J.H. (1902) The stability of a spherical nebula. *Philos. Trans. R. Soc. Lond. A*, **199**, 1–53.

Jeanthon, C. (2000) Molecular ecology of hydrothermal vent microbial communities. *Antonie van Leeuwenhoek*, **77**, 117–133.

Jensen, J.L.W.V. (1906) Sur les fonctions convexes et les inégalités entre les valeurs moyennes. *Acta Math-Djursholm.*, **30**, 175–193.

Jenssen, M. (2009) Relating plant diversity in forests with the spatial scale of ecosystem processes. *Int. J. Ecology*. doi: 10.1155/2009/683061

Jeong, H., Tombor, B., Albert, R., Oltvai, Z.N., and Barabási, A.-L. (2000) The large-scale organization of metabolic networks. *Nature*, **407**, 651–654.

Jia, H., Rochefort, N.L., Chen, X., and Konnerth, A. (2010) Dendritic organization of sensory input to cortical neurons *in vivo*. *Nature*, **464**, 1307–1312.

Jiao, Y., Wickett, N.J., Ayyampalayam, S., Chanderbali, A.S., Landherr, L., Ralph, P.E., Tomsho, L.P. Hu, Y., Liang, H., Soltis, P.S., Soltis, D.E., Clifton, S.W., Schlarbaum, S.E., Schuster, S.C., Ma, H., Leebens-Mack, J., and dePamphilis, C.W. (2011) Ancestral polyploidy in seed plants and angiosperms. *Nature*, **473**, 97–100.

Jiang, S.-Y., Pi, D.H., Heubeck, C., Frimmel, H., Liu, Y.-P., Deng, H.-L., Ling, H.-F., and Yang, J.-H. (2009) Early Cambrian ocean anoxia in South China. *Nature*, **459**, E5–E6. doi: 10.1038/nature08048

Jiménez-Montaño, M.A. (1975) Information Thermodynamics of Chemical systems. PhD Thesis, Nicolaus Copernicus University, Toruń.

Jiménez-Montaño, M.A., de la Mora-Basanez, C.R., and Pöschel, T. (1994) On the hypercube structure of the genetic code. in Proc. Bioinformatics and Genome Research (eds H.A., Lim and C.A. Cantor), p. 445 (World Scientific, 1994), http://arxiv.org/abs/cond-mat/0204044.

Jiménez-Montaño, M.A., de la Mora-Basanez, C.R., and Pöschel, T. (1996) The hypercube structure of the genetic code explains conservative and non-conservative aminoacid substitutions *in vivo* and *in vitro*. *BioSystems*, **39**, 117–125.

Jiménez-Montaño, M.A. and Ebeling, W. (1980) A stochastic evolutionary model of technological change. *Collective Phenomena*, **3**, 107–114.

Jiménez-Montaño, M.A., Feistel, R., and Diez-Martínez, O. (2004) On the information hidden in signals and macromolecules. *Nonlinear Dyn. Psychol., Life Sci.*, **8**, 445–478.

Jiménez-Montaño, M.A. and He, M. (2009) Irreplaceable amino acids and reduced alphabets in short-term and directed protein evolution, in *Bioinformatics Research and Applications* (eds I. Mandoiu, G., Narasimhan and Y. Zhang), Springer, Berlin, Heidelberg, pp. 297–309.

Jin, Z., Charlock, T.P., Smith, W.L., Jr., and Rutledge, K. (2004) A parameterization of ocean surface albedo. *Geophys. Res. Lett.*, **31**, L22301. doi: 10.1029/2004GL021180

Johanson, D.C. and Edey, M. (1990) *Lucy: The Beginnings of Humankind*, Simon & Schuster, New York.

Johari, G.P. and Jones, S.J. (1975) Study of the low-temperature "transition" in ice Ih by thermally stimulated depolarization measurements. *J. Chem. Phys.*, **62**, 4213–4223.

Jonas, D. and Jonas, D. (1975) Gender differences in mental functions: a clue to the origin of language. *Curr. Anthropol.*, **16**, 626–630.

Jones, B.L. (1977a) A solvable selfreproductive hypercycle model for the selection of biological molecules. *J. Math. Biol.*, **4**, 187–193.

Jones, B.L. (1977b) Analysis of Eigen's equations for selection of biological molecules with fluctuating mutation rates. *B. Math. Biol.*, **39**, 311–316.

Jones, B.L., Enns, R.H., and Rangnekar, S.S. (1976) On the theory of selection of coupled macromolecular systems. *B. Math. Biol.*, **38**, 15–28.

Jones, N. (2009) The new & improved Kelvin. *Nature*, **459**, 902–904.

Jones, P.D., Ogilvie, A.E.J., Davies, T.D., and Briffa, K.R. (2001) *History and Climate: Memories of the Future?*, Kluwer Academic/Plenum Publishers, New York.

Jørgensen, S.E. and Svireshev, Yu.M. (2004) *Towards a Thermodynamic Theory for Ecological Systems*, Elsevier Ltd., Oxford.

Jou, D., Casas-Vazques, J., and Lebon, G. (1993) *Extended Irreversible Thermodynamics*, Springer, Berlin.

Joughin, I., Smith, B.E., and Holland, D.M. (2010) Sensitivity of 21st century sea level to ocean-induced thinning of Pine Island Glacier, Antarctica. *Geophys. Res. Lett.*, **37**, L20502. doi: 10.1029/2010GL044819

Jouzel, J., Barkov, N.I., Barnola, J.M., Bender, M., Chappellaz, J., Genthon, C., Kotlyakov, V.M., Lipenkov, V., Lorius, C., Petit, J.R., Raynaud, D., Raisbeck, G., Ritz, C., Sowers, T., Stievenard, M., Yiou, F., and Yiou, P. (1993) Extending the Vostok ice-core record of palaeoclimate to the penultimate glacial period. *Nature*, **364**, 407–412.

Jouzel, J., Lorius, C., Petit, J.R., Genthon, C., Barkov, N.I., Kotlyakov, V.M., and Petrov, V.M. (1987) Vostok ice core: a continuous isotope temperature record over the last climatic cycle (160,000 years). *Nature*, **329**, 403–408.

Jouzel, J., Waelbroeck, C., Malaize, B., Bender, M., Petit, J.R., Stievenard, M., Barkov, N.I., Barnola, J.M., King, T., Kotlyakov, V.M., Lipenkov, V., Lorius, C., Raynaud, D., Ritz, C., and Sowers, T. (1996) Climatic interpretation of the recently extended

Vostok ice records. *Clim. Dynam.*, **12**, 513–521.

Jung, H. and Holt, C.E. (2010) *Local Translation of mRNAs in Neural Development*, Wiley Interdisciplinary Reviews, New York http://wires.wiley.com/WileyCDA/WiresArticle/wisId-WRNA53.html. doi: 10.1002/wrna.53

Jung, M., Reichstein, M., Ciais, P., Seneviratne, S.I., Sheffield, J., Goulden, M.L., Bonan, G., Cescatti, A., Chen, J., de Jeu, R., Dolman, A.J., Eugster, W., Gerten, D., Gianelle, D., Gobron, N., Heinke, J., Kimball, J., Law, B.E., Montagnani, L., Mu, Q., Mueller, B., Oleson, K., Papale, D., Richardson, A.D., Roupsard, O., Running, S., Tomelleri, E., Viovy, N., Weber, U., Williams, C., Wood, E., Zaehle, S., and Zhang, K. (2010) Recent decline in the global land evapotranspiration trend due to limited moisture supply. *Nature*, **467**, 951–954.

Kabelac, S. (1994) *Thermodynamik der Strahlung*, Vieweg, Braunschweig, Wiesbaden.

Kac, M. (1947) Random walk and the theory of Brownian motion. *Am. Math. Mon.*, **54**, 369–391.

Kaldasch, J. (2011) Evolutionary model of an anonymous consumer durable market. *Physica A*, **390**, 2692–2715.

Kämmerer, W. (1974) *Einführung in mathematische Methoden der Kybernetik*, Akademie-Verlag, Berlin.

Kamshilov, M.M. (1977) *Das Leben auf der Erde*, Mir, Moskau, Urania-Verlag, Leipzig/Jena/Berlin, Translation from Russian: Evolyutsia biosfery, Nauka, Moskva 1974.

Kantz, H. and Schreiber, T. (1997, 2003) *Nonlinear Time Series Analysis, Cambridge* Nonlinear Science Series 7, Cambridge University Press, Cambridge.

Kaplan, M.R., Schaefer, J.M., Denton, G.H., Barrell, D.J.A., Chinn, T.J.H., Putnam, A.E., Andersen, B.G., Finkel, R.C., Schwartz, R., and Doughty, A.M. (2010) Glacier retreat in New Zealand during the Younger Dryas stadial. *Nature*, **467**, 194–197.

Kaplan, R.W. (1978) *Der Ursprung des Lebens*, Georg Thieme, Stuttgart.

Kappler, A., Pasquero, C., Konhauser, K.O., and Newman, D.K. (2005) Deposition of banded iron formations by anoxygenic phototrophic Fe(II)-oxidizing bacteria. *Geology*, **33**, 865–868.

Kassewitz, J. and Reid, J.S. (2008) Songs from the sea: deciphering dolphin language with picture words. *Cymascope Press Realease*, http://www.cymascope.com/cyma_research/oceanography.html

Kauffman, S. (1995) *At Home in the Universe: The Search for Laws of Self-Organization and Complexity*, Oxford University Press, New York.

Kaufman, D.S., Schneider, D.P., McKay, N.P., Ammann, C.M., Bradley, R.S., Briff, K.R., Miller, G.H., Otto-Bliesner, B.L., Overpeck, J.T., and Vinther, B.M. (2009) Arctic Lakes 2k Project Members. Recent warming reverses long-term arctic cooling. *Science*, **325**, 1236–1239.

Keeling, C.D. and Whorf, T.P. (1998) *Atmospheric CO_2 Concentrations (ppmv) Derived from in situ Air Samples Collected at Mauna Loa Observatory*, Scripps Institute of Oceanography, Hawaii, August 1998. See also http://cdiac.esd.ornl.gov/trends/co2/contents.htm.

Keller, J.U. (1977) *Thermodynamik der irreversiblen Prozesse*, De Gruyter, Berlin, New York.

Kennett, J.P. and Exon, N.F. (2004) Paleoceanographic evolution of the Tasmanian Seaway and its climatic implications. *Geophys. Monogr.*, **151**, 345–367.

Kenrick, P. and Crane, P.R. (2000) The origin and early diversification of plants on land, in *Shaking the Tree* (ed. H. Gee), Nature/McMillan Magazines, New York, pp. 217–250.

Kerner, E.H. (1957) A statistical mechanics of interacting biological species. *B. Math. Biophys.*, **19**, 121–146.

Kerr, R.A. (2005) Earth's inner core is running a tad faster than the rest of the planet. *Science*, **309**, 1313.

Kerr, R.A. (2009) Clouds appear to be big, bad player in global warming. *Science*, **325**, 376.

Khakh, B.S. (2001) Molecular physiology of P2X receptors and ATP signalling at synapses. *Nat. Rev. Neurosci.*, **2**, 165–174.

Khalaf, S.G. (1996) *Table of the Phoenician Alphabet*, Phoenician International Research Center (PIRC), Lebanon http://phoenicia.org/tblalpha.html.

Khinchin, A.I. (1943) *Matematicheskiye Osnovaniya Statisticheskoy Mekhaniki*

(*Mathematical Foundations of Statistical Mechanics*), State Publishing House, Moscow, English translation (1949) Dover, New York.

Kimball, J.C. (1978) Localisation and spectra in solid state systems. *J. Phys. C*, **11**, 4347–4354.

Kimura, M. (1968) Evolutionary rate at the molecular level. *Nature*, **217**, 624–626.

Kimura, M. (1983) *The Neutral Theory of Molecular Evolution*, Cambridge University Press, Cambridge.

King, J.L. and Jukes, T.H. (1969) Non-Darwinian evolution. *Science*, **164**, 788–789.

King, E.M., Stellmach, S., Noir, J., Hansen, U., and Arnou, J.M. (2009) Boundary layer control of rotating convection systems. *Nature*, **457**, 301–304.

Kirman, A. (2003) Economic networks, in *Handbook of Graphs and Networks: From the Genome to the Internet* (eds S. Bornholdt and H.G. Schuster), Wiley-VCH Verlag, Weinheim, pp. 273–294.

Kittel, C. (1969) *Thermal Physics*, John Wiley & Sons, New York.

Klaus, G. (1968) *Wörterbuch der Kybernetik*, Dietz Verlag, Berlin.

Klemm, K., Flamm, C., and Stadler, P.F. (2008) Funnels in energy landscapes. *European J. Phys. B*, **63**, 387–391.

Kleidon, A., Malhi, Y., and Cox, P.M. (2010) Maximum entropy production in environmental and ecological systems. *Philos. Trans. R. Soc. Lond. B Biol. Sci.*, **365**, 1297–1302.

Kleeman, R. and Power, S.B. (1995) A simple atmospheric model of surface heat flux for use in ocean modeling studies. *J. Phys. Oceanography*, **25**, 92–105.

Klimontovich, Yu.L. (1982) *Statisticheskaya fizika (Statistical Physics)*, Nauka, Moscow, English Translation (1986): Harwood Academic Publishers.

Klimontovich, Yu.L. (1991) *Turbulent Motion, the Structure of Chaos*, Kluwer Academic Publishers, Dordrecht, Boston, London.

Klimontovich, Yu.L. (1995) *Statistical Theory of Open Systems. Vol. 1: A Unified Approach to Kinetic Description of Processes in Active Systems*, Kluwer Academic Publishers, Dordrecht, Boston, London.

Klix, F. (1980) *Erwachendes Denken. Eine Entwicklungsgeschichte der menschlichen Intelligenz*, Deutscher Verlag der Wissenschaften, Berlin.

Klix, F. (1992) *Die Natur des Verstandes*, Hogrefe, Göttingen.

Klix, F. and Lanius, K. (1999) *Wege und Irrwege der Menschenartigen. Wie wir wurden, wer wir sind*, Verlag W. Kohlhammer, Stuttgart, Berlin, Köln.

Kluge, G. and Neugebauer, G. (1976) *Grundlagen der Thermodynamik*, VEB Deutscher Verlag der Wissenschaften, Berlin.

Knauth, L.P. and Kennedy, M.J. (2009) The late precambrian greening of the earth. *Nature*, **460**, 728–732.

Knox, R.S. and Douglass, D.H. (2010) Recent energy balance of earth. *Int. J. Geosc.*, **2010**, 99–101.

Kohlrausch, K.W.F. and Schrödinger, E. (1926) Das ehrenfestsche Modell der H-Kurve. *Physikalische Zeitschrift*, **27**, 306–313. Reprinted in: Schrödinger, E. (1984) Collected Papers, Vol. 1, Vieweg, Braunschweig, Wiesbaden, pp. 349–357.

Ko, H., Hofer, S.B., Pichler, B., Buchanan, K. A., Sjöström, P.J., and Mrsic-Flogel, T.D. (2011) Functional specificity of local synaptic connections in neocortical networks. *Nature*, **473**, 87–91.

Kolmogorov, A.N. (1936) Sulla teoria di Volterra della lotta per l'esistenza. *Giornale del Instituto Italiano degli Attnari*, **7**, 74–80.

Körner, U. (1978) *Probleme der Biogenese*, Gustav Fischer, Jena.

Kramer, B. (1988) *The Art of Measurement. Metrology in Fundamental and Applied Physics*, VCH, Weinheim.

Kröber, G. (1967) Strukturgesetz und Gesetzesstruktur. *Deutsche Zeitschrift für Philosophie*, **15**, 202–216.

Krug, H.-J. and Pohlmann, L. (1997) *Selbstorganisation – Jahrbuch für Komplexität*, Duncker & Humblot, Berlin.

Kruger, K., Grabowski, P.J., Zaug, A.J., Sands, J., Gottschling, D.E., and Cech, T.R. (1982) Self-splicing RNA: autoexcision and autocyclization of the ribosomal RNA intervening sequence of Tetrahymena. *Cell*, **31**, 147–57.

Ksanfomaliti, L.V. (1985) *Planeten*, Mir, Urania-Verlag, Moscow, Leipzig, Jena, Berlin.

Kuhn, H. (1976) Model consideration for the origin of life. Environmental structure as stimulus for the evolution of chemical systems. *Naturwissenschaften*, **63**, 68–80.

Kullback, S. (1951) *Information Theory and Statistics*, John Wiley & Sons, New York.

Kullback, S. and Leibler, R.A. (1951) On information and sufficiency. *Ann. Math. Stat.*, **22**, 79–86.

Küppers, B. (1975) The general principles of selection and evolution at the molecular level. *Prog. Biophys. Mol. Biol.*, **30**, 1–22.

Küppers, B. (1979) Towards an experimental analysis of molecular self-organization and precellular Darwinian evolution. *Naturwissenschaften*, **66**, 228–235.

Küppers, B.-O. (1983) *Molecular Theory of Evolution — Outline of a Physico-Chemical Theory of the Origin of Life*, Springer, Berlin, Heidelberg, New York.

Kurosch, A.G. (1970, 1972) *Gruppentheorie*, vol. 1, 2, Akademie-Verlag, Berlin.

Kürschner, W.M., Kvaček, Z., and Dilcher, D.L. (2008) The impact of Miocene atmospheric carbon dioxide fluctuations on climate and the evolution of terrestrial ecosystems. *Proc. Natl. Acad. Sci. U.S.A.*, **105**, 449–453.

Küster, H. (2002) *Die Ostsee. Eine Natur- und Kulturgeschichte*. C.H. Beck, München.

Kump, L.R. (2010) Earth's second wind. *Science*, **330**, 1490–1491.

Kurochkin, E.N., Dyke, G.J., and Karhu, A.A. (2002) A new presbyornithid bird (aves, anseriformes) from the Late Cretaceous of southern Mongolia. *Am. Mus. Novit.*, **3386**, 1–11.

Kwok, R., Cunningham, G.F., Wensnahan, M., Rigor, I., Zwally, H.J., and Yi, D. (2009) Thinning and volume loss of the Arctic Ocean sea ice cover: 2003–2008. *J. Geophys. Res.*, **114**, C07005. doi: 10.1029/2009JC005312

Lacis, A.A., Schmidt, G.A., Rind, D., and Ruedy, R.A. (2010) Atmospheric CO_2: principal control knob governing earth's temperature. *Science*, **330**, 356–359.

Laepple, T., Werner, M., and Lohmann, G. (2011) Synchronicity of Antarctic temperatures and local solar insolation on orbital timescales. *Nature*, **471**, 91–94.

Lammer, H. and Bauer, S.J. (2003) Isotopic fractionation by gravitational escape. *Space Sci. Rev.*, **106**, 281–291.

Lancaster, P. (1969) *Theory of Matrices*, Academic Press, New York, London.

Landa, P. (2001) *Regular and Chaotic Oscillations*, Springer, Berlin, Heidelberg.

Landau, L.D. and Lifschitz, E.M. (1966) *Statistische Physik*, Akademie-Verlag, Berlin (English edition 1980: Statistical Physics Reed Educational and Professional Publishing Oxford).

Landau, L.D. and Lifschitz, E.M. (1967a) *Mechanik*, Akademie-Verlag, Berlin.

Landau, L.D. and Lifschitz, E.M. (1967b) *Klassische Feldtheorie*, Akademie-Verlag, Berlin.

Landau, L.D. and Lifschitz, E.M. (1974) *Hydrodynamik*, Akademie-Verlag, Berlin.

Landau, L.D. and Lifschitz, E.M. (1988) *Quantenmechanik*, Akademie-Verlag, Berlin.

Landauer, R. (1973) Entropy changes for steady-state fluctuations. *J. Stat. Phys.*, **9**, 351–371.

Landauer, R. (1976) Fundamental limitations in the computational process. *Berichte der Bunsengesellschaft für physikalische Chemie*, **80**, 1048.

Langseth, M.G., Keihm, S.J., and Peters, K. (1976) Revised lunar heat-flow values. Proceedings of the 7th Lunar Science Conference, Houston, p. 3143–3171.

Lanius, K. (1994a) *Natur im Wandel*, Spektrum Akademischer Verlag, Heidelberg, Berlin, Oxford.

Lanius, K. (1994b) *Die Erde im Wandel, Grenzen des Vorhersagbaren*, Spektrum Akademischer Verlag, Heidelberg, Berlin, Oxford.

Lanius, K. (2009) *Klima, Umwelt, Mensch. Sozial-ökonomische Systeme und ihre Überlebens(un)fähigkeit*, Pahl-Rugenstein, Bonn.

Lass, H.U. and Mohrholz, V. (2008) On the interaction between the subtropical gyre and the Subtropical Cell on the shelf of the SE Atlantic. *J. Marine Syst.*, **74**, 1–49.

Laubscher, H.P. (1984) Der Bau der Alpen, in *Ozeane und Kontinente*, Spektrum der Wissenschaft, Heidelberg, p. 144–157.

Laue, R. (1970) *Elemente der Graphentheorie und ihre Anwendung in den biologischen Wissenschaften*, Geest & Portig, Leipzig.

Lawler, A. (2010) Tracking the Med's Stone Age Sailors. *Science*, **330**, 1472–1473.

Lee, T. and McPhaden, M.J. (2010) Increasing intensity of El Niño in the central-equatorial

Pacific. *Geophys. Res. Lett.*, **37**, L14603. 5 pp., doi: 10.1029/2010GL044007

Lee, P.A. and Ramakrishnan, T.V. (1985) Disordered electronic systems. *Rev. Mod. Phy.*, **57**, 287–337.

Lehninger, A.L. (1972) *Biochemistry*, Worth Pubishers, New York.

Leipe, T., Harff, J., Meyer, M., Hille, S., Pollehne, F., Schneider, R., Kowalski, N., and Brügmann, L. (2008) Sedimentary records of environmental changes and anthropogenic impacts during the past decades, in *State and Evolution of the Baltic Sea, 1952–2005. A Detailed 50-Year Survey of Meteorology and Climate, Physics, Chemistry, Biology, and Marine Environment* (eds R. Feistel, G. Nausch, and N. Wasmund), John Wiley & Sons, Hoboken, NJ, pp. 395–440.

Lemmon, E.W. and Span, R. (2010) Multi-parameter Equations of State for Pure Fluids and Mixtures, in *Applied Thermodynamics of Fluids*. (eds A.R.H. Goodwin, J.V. Sengers, and C.J. Peters), RSC Publishing, Cambridge.

Lemmon, E.W., Jacobsen, R.T., Penoncello, S.G., and Friend, D.G. (2000) Thermodynamic properties of air and mixtures of nitrogen, argon and oxygen from 60 to 2000 K at pressures to 2000 MPa. *J. Phys. Chem. Ref. Data*, **29**, 331–362.

Leslie, M. (2009) On the origin of photosynthesis. *Science*, **323**, 1286–1287.

Levithin, D.J. (2007) *This is Your Brain on Music*, PLUME, New York.

Lewicki, M.S. (2010) A signal take on speech. *Nature*, **466**, 821–822.

Li, Q., Lee, J.-A., and Black, D.L. (2007) Neuronal regulation of alternative pre-mRNA splicing. *Nat. Rev.*, **8**, 819–831.

Lieb, E.H. and Yngvason, J. (1999) The physics and mathematics of the second law of thermodynamics. *Phys. Rep.*, **310**, 1–96.

Lifshitz, I.M., Gredeskul, S.A., and Pastur, L.A. (1982) *Vvedeniye v Teoriyu Besporyadochnykh Sistem (Introduction to the Theory of Disordered Systems)*, Nauka, Moscow.

Lincoln, T.A. and Joyce, G.F. (2009) Self-sustained replication of an RNA enzyme. *Science*, **323**, 1229–1232.

Linde, H. (1978) Letter to W. Ebeling dated 18.09.1978 with permission for use.

Linde, H. and Eckert, K. (2010) Letter to W. Ebeling dated Sept. 2010 with permission to use here.

Linde, H. and Engel, H. (1991) Autowave propagation in heterogeneous active media. *Physica D*, **49**, 13–20.

Linde, H. and Schwarz, E. (1963) Freie Oberflächenkonvektion beim Stoffübergang an fluiden Grenzen. *Z. Phys. Chem-Leipzig*, **224**, 331–335.

Linde, H., Schwarz, E., and Gröger, J. (1967) Zum Auftreten des oszillatorischen Regimes der Marangoniinstabilität beim Stoffübergang. *Chem. Eng. Sci.*, **22**, 823–836.

Linde, H. and Winkler, K. (1964) Hydrodynamische Stabilität einer fluiden Phasengrenze. *Z. Phys. Chem-Leipzig*, **225**, 223–230.

Linde, H. and Zirkel, Ch. (1991) Immobilized catalyst of the BZR in gelatinous matrix. *Z. Phys. Chem.*, **174**, 145–161.

Linde, H., Velarde, M.G., Waldhelm, W., Loeschcke, K., and Wierschem, A. (2005) Interfacial wave motions due to marangoni instability. *Ind. Eng. Chem. Res.*, **44**, 1396–1412.

Linde, H., Velarde, M.G., Waldhelm, W., and Wierschem, A. (2001) Interfacial wave motions due to Marangoni instability. *J. Colloid Interf. Sci.*, **236**, 214–224.

Lindsey, D.T. and Brown, A.M. (2008) World color survey color naming reveals universal motifs and their within-language diversity. *Proc. Natl. Acad. Sci. U.S.A.*, **106**, 19785–19790.

Lippmann-Pipke, J., Sherwood Lollar, B., Niedermann, S., Stroncik, N.A., Naumann, R., van Heerden, E., and Onstott, T.C. (2011) Neon identifies two billion year old fluid component in Kaapvaal Craton. *Chemical Geology*, doi: 10.1016/j.chemgeo.2011.01.028

Liu, J. and Curry, J.A. (2010) Accelerated warming of the Southern Ocean and its impacts on the hydrological cycle and sea ice. *Proc. Natl. Acad. Sci. U.S.A.*, **107**, 14987–14992.

Liu, Y.-Y., Slotine, J.-J., and Barabási, A.-L. (2011) Controllability of complex networks. *Nature*, **473**, 167–173.

Logan, R.K. (1986) *The Alphabet Effect*, William Morrow and Company, New York.

Lohmann, J.U. and Bosch, T.G.C. (2000) The novel peptide HEADY specifies apical fate in a simple radially symmetric metazoan. *Gene. Dev.*, **14**, 2771–2777.

Lomber, S.G., Meredith, A.A., and Kral, A. (2010) Cross-modal plasticity in specific auditory cortices underlies visual compensations in the deaf. *Nat. Neurosci.*, **13**, 1421–1427.

Loomis, W.F. (1955) Glutathione control of the specific feeding reactions of hydra. *Ann. N.Y. Acad. Sci.*, **62**, 211–227.

López-Muñoz, F., Boya, J., and Alamo, C. (2006) Neuron theory, the cornerstone of neuroscience, on the centenary of the Nobel Prize award to Santiago Ramón y Cajal. *Brain Res. Bull.*, **70**, 391–405.

Lorenz, K. (1963) *Das sogenannte Böse*, Borotha-Schoeler, Wien.

Lorenz, R.D. (2002) Planets, life and the production of entropy. *Int. J. Astrobiol.*, **1**, 3–13.

Lorito, S., Romano, F., Atzori, S., Tong, X., Avallone, A., McCloskey, J., Cocco, M., Boschi, E., and Piatanesi, A. (2011) Limited overlap between the seismic gap and coseismic slip of the great 2010 Chile earthquake. *Nat. Geoscience.*, **4**, 173–177.

Lotka, A.J. (1910) Zur Theorie der periodischen Reaktionen. *Z. Phys. Chem.*, **72**, 508–511.

Lovell-Smith, J.W. and Pearson, H. (2006) On the concept of relative humidity. *Metrologia*, **43**, 129–134.

Love, G.D., Grosjean, E., Stalvies, C., Fike, D.A., Grotzinger, J.P., Bradley, A.S., Kelly, A.E., Bhatia, M., Meredith, W., Snape, C.E., Bowring, S.A., Condon, D.J., and Summons, R.E. (2009) Fossil steroids record the appearance of Demospongiae during the Cryogenian period. *Nature*, **457**, 718–721.

Lovelock, J. (2000) *Gaia: A New Look at Life on Earth*, Oxford University Press, Oxford.

Lozán, J.L., Graßl, H., and Hupfer, P. (1998) *Warnsignal Klima*, GEO, Hamburg.

Lozier, M.S. (2010) Deconstructing the conveyor belt. *Science*, **328**, 1507–1511.

LSBV (2010) *Alleen in Mecklenburg-Vorpommern*, Landesamt für Straßenbau und Verkehr Mecklenburg-Vorpommern, http://strassenbauverwaltung.mvnet.de/cms2/LSBV_prod/LSBV/de/vi/Sehenswerte_Alleen/index.jsp.

Lu, X., McElroy, M.B., and Kiviluoma, J. (2009) Global potential for wind-generated electricity. *Proc. Natl. Acad. Sci. U.S.A.*, **106**, 10933–10938, http://www.pnas.org/content/106/27/10933.

Lübke, H., Schmölck, U., and Tauber, F. (2011) Mesolithic hunter-fishers in a changing world: a case study of submerged sites on the Jäckelberg, Wismar Bay, northeastern Germany, in *Submerged Prehistory* (eds J. Benjamin, C. Bonsall, C. Pickard, and A. Fischer), Oxbow Books, Oxford, p. 3–37.

Lücking F R., Huhndorf, S., Pfister, D.H., Plata., E.R., and Lumbsch, H.T. (2009) Fungi evolved right on track. *Mycologia*, **101**, 810–822.

Lukeš, J., Leander, B.S., and Keeling, P.J. (2009) Cascades of convergent evolution: the corresponding evolutionary histories of euglenozoans and dinoflagellates. *Proc. Natl. Acad. Sci. U.S.A.*, **106**, 9963–9970.

Lunine, J.I. and Lorenz, R.D. (2009) Rivers, lakes, dunes, and rain: crustal processes in Titan's methane cycle. *Ann. Rev. Earth Platetary Sci.*, **37**, 299–320.

Lunt, D.J., Foster, G.L., Haywood, A.M., and Stone, E.J. (2008) Late Pliocene Greenland glaciation controlled by a decline in atmospheric CO_2 levels. *Nature*, **454**, 1102–1105.

Luria, A.R. (1991) *Der Mann, dessen Welt in Scherben ging. Zwei neurologische Geschichten*, Rowohlt, Reinbek.

Lurie, D. and Wagensberg, J. (1979) Entropy balance in biologivcal development and heat dissipation in embryogenesis. *J. Non-Equil. Thermody.*, **4**, 127–132.

Lutz, R.A. and Kennish, M.J. (1993) Ecology of deep-sea hydrothermal vent communities: a review. *Rev. Geophys.*, **31**, 211–242.

Luzzi, R., Vasconcellos, A.R., and Ramos, J.G. (2000) *Statistical Foundations of Irreversible Thermodynamics*, Teubner, Stuttgart, Leipzig, Wiesbaden.

Lyman, J.M., Good, S.A., Gouretski, V.V., Ishii, M., Johnson, G.C., Palmer, M.D., Smith, D.M., and Willis, J.K. (2010) Robust warming of the global upper ocean. *Nature*, **465**, 334–337.

Lyons, T.W. and Reinhard, C.T. (2009) Oxygen for heavy-metal fans. *Nature*, **461**, 179–181.

Machlup, S. and Onsager, L. (1953) Fluctuations and irreversible processes. II. Systems with kinetic energy. *Phys. Rev.*, **91**, 1512–1515.

Mackas, D.M. (2011) Does blending of chlorophyll data bias temporal trend? *Nature*, **472**, E4–E5.

Mackey, M.C. (1992) *Time's Arrow: The Origins of Thermodynamic Behavior*, Springer, Berlin.

MacNeilage, P.F. (2008) *The Origin of Speech*, Oxford University Press, Oxford.

Mahnke, R. and Feistel, R. (1985) On the kinetics of Ostwald ripening as a competitive growth in a self-organizing system. *Rostocker Physikalische Manuskripte*, **8**, 54–64.

Maillet, F., Poinsot, V., André, O., Puech-Pagès, V., Haouy, A., Gueunier, M., Cromer, L., Giraudet, D., Formey, D., Niebel, A., Andres Martinez, E., Driguez, H., Bécard, G., and Dénarié, J. (2011) Fungal lipochitooligosaccharide symbiotic signals in arbuscular mycorrhiza. *Nature*, **469**, 58–63.

Makarieva, A.M. and Gorshkov, V.G. (2009) Condensation-induced kinematics and dynamics of cyclones, hurricanes and tornadoes. *Phys. Lett. A*, **373**, 4201–4205.

Malchow, H., Ebeling, W., and Feistel, R. (1983a) Polarity, symmetry, and gradient in reaction–diffusion systems (II). *Stud. Biophys.*, **97**, 231–236.

Malchow, H., Ebeling, W., Feistel, R., and Schimansky-Geier, L. (1983b) Stochastic bifurcations in a bistable reaction–diffusion system with Neumann boundary conditions. *Ann. Phys.-Leipzig*, **40**, 151–160.

Malchow, H. and Feistel, R. (1982) Polarity, symmetry, and gradient in reaction–diffusion systems (I). *Stud. Biophys.*, **88**, 125–130.

Mampe, B., Friederici, A.D., Christophe, A., and Wermke, K. (2009) Newborns' cry melody is shaped by their native language. *Curr. Biol.*, **19**, 1994–1997.

Manasrah, R., Badran, M., Lass, H.U., and Fennel, W. (2004) Circulation and winter deep-water formation in the northern Red Sea. *Oceanologia*, **46**, 5–23.

Margenau, H. and Murphy, G.M. (1943) *The Mathematics of Physics and Chemistry*, D. van Nostrand, New York.

Marijuan, P.C. (2003) Foundations of information science. Selected papers from FIS 2002. *Entropy*, **5**, 214–219.

Marion, G.M., Kargel, J.S., Catling, D.C., and Jakubowski, S.D. (2005) Effects of pressure on aqueous chemical equilibria at subzero temperatures with applications to Europa. *Geochim. Cosmochim. Acta*, **69**, 259–274.

Marris, E. (2008) Language: the language barrier. *Nature*, **453**, 446–448.

Marshall, J. and Schott, F. (1999) Open-ocean convection: observations, theory, and models. *Rev. Geophys.*, **37**, 1–64.

Martinez, R.N., Sereno, P.C., Alcober, O.A., Colombi, C.E., Renne, P.R., Montañez, I.P., and Currie, B.S. (2011) A basal dinosaur from the dawn of the dinosaur era in southwestern Pangaea. *Science*, **331**, 206–210.

Marx, K. (1951a) *Das Kapital, Erster Band*, Dietz, Berlin (Original: London, 1867).

Marx, K. (1951b) *Das Kapital, Dritter Band*, Dietz, Berlin (Original: London, 1894).

Mathieson, I.K. (1990) The agricultural climate of St. Helena (with reference to Ascension). Overseas development administration. Eland House/Department of Agriculture and Forestry St. Helena, WS Atkins, London/Cambridge.

Mauersberger, P. (1981) Entropie und freie Enthalpie im aquatischen Ökosystem. *Acta Hydrophysica*, **26**, 67–90.

Maule, C.F., Purucker, M.E., Olsen, N., and Mosegaard, K. (2005) Heat flux anomalies in Antarctica revealed by satellite magnetic data. *Science*, **309**, 464–467.

Maurer, S.M. and Huberman, B.A. (2000) Competitive dynamics of web sites. *J. Econ. Dyn. Control*, **27**, 2195–2206.

May, R.M. (1973) *Stability and Complexity in Model Ecosystems*, Princeton University Press, New Jersey.

McDougall, T.J. (1987) Neutral surfaces. *J. Phys. Oceanogr.*, **17**, 1950–1964.

McDougall, T.J. and Feistel, R. (2003) What causes the adiabatic lapse rate? *Deep-Sea Res. I*, **50**, 1523–1535.

McDougall, T.J. and Jackett, D.R. (1988) On the helical nature of neutral trajectories in the ocean. *Prog. Oceanogr.*, **20**, 153–183.

McDougall, T.J. and Jackett, D.R. (2007) The thinness of the ocean in space and the implications for mean diapycnal advection. *J. Phys. Oceanogr.*, **37**, 1714–1732.

McDougall, T.J., Jackett, D.R., and Millero, F.J. (2009) An algorithm for estimating Absolute Salinity in the global ocean. *Ocean Sci.*

Discuss., **6**, 215–242, http://www.ocean-sci-discuss.net/6/215/2009/.

McFadden, G.I. and van Dooren, G.G. (2004) Evolution: red algal genome affirms a dispatch common origin of all plastids. *Curr. Biol.*, **14**, R514–R516.

McFadden, K.A., Huang, J., Chu, X., Jiang, G.J., Kaufman, A.J., Zhou, C., Yuan, X., and Xiao, S. (2008) Pulsed oxidation and biological evolution in the Ediacaran Doushantuo Formation. *Proc. Natl. Acad. Sci. U.S.A.*, **105**, 3197–3202.

McPhaden, M.J., Zebiak, S.E., and Glantz, M.H. (2006) ENSO as an integrating concept in Earth science. *Science*, **314**, 1740–1745.

McPherron, S.P., Alemseged, Z., Marean, C.W., Wynn, J.G., Reed, D., Geraads, D., Bobe, R., and Béarat, H.A. (2010) Evidence for stone-tool-assisted consumption of animal tissues before 3.39 million years ago at Dikika, Ethiopia. *Nature*, **466**, 857–860.

Meehl, G.A., Arblaster, J.M., Matthes, K., Sassi, F., and van Loon, H. (2009) Amplifying the Pacific climate system response to a small 11-year solar cycle forcing. *Science*, **325**, 1114–1118.

Meijer, P.T. and Krijgsman, W. (2005) A quantitative analysis of the desiccation and re-filling of the Mediterranean during the Messinian Salinity Crisis. *Earth Planet. Sci. Lett.*, **240**, 510–520.

Meinhardt, H. and Klingler, M. (1987) A model for pattern formation on the shells of molluscs. *J. Theor. Biol.*, **126**, 63–89.

Melisab, A.P., Hareb, B., and Tomaselloa, M. (2009) Chimpanzees coordinate in a negotiation game. *Evol. Hum. Behav.*, **30**, 381–392.

Meyer, K.M., Yu, M., Jost, A.B., Kelley, B.M., and Payne, J.L. (2011) $\delta^{13}C$ evidence that high primary productivity delayed recovery from end-Permian mass extinction. *Earth Planet. Sci. Lett.*, **302**, 378–384.

Miller, K.G. and Fairbanks, R.G. (1983) Evidence for Oligocene–Middle Miocene abyssal circulation changes in the western North Atlantic. *Nature*, **306**, 250–253.

Miller, K.G., Fairbanks, R.G., and Mountain, G.S. (1987) Tertiary oxygen isotope synthesis, sea level history, and continental margin erosion. *Paleoceanography*, **2**, 1–19.

Miller, K.F., Smith, C.M., Zhu, J., and Zhang, H. (1995) Preschool origins of cross-national differences in mathematical competence: the role of number-naming systems. *Psychol. Sci.*, **6**, 56–60.

Miller, S.L. (1955) Production of some organic compounds under possible primitive Earth conditions. *J. Am. Chem. Soc.*, **77**, 2351–2361.

Miller, S.L. and Orgel, L.E. (1974) *The Origins of Life on Earth*, Prentice Hall, Englewood Cliffs, NJ.

Millero, F.J. (2010) History of the equation of state of seawater. *Oceanography*, **23**, 18–33.

Millero, F.J., Huang, F., Woosley, R.J., Letscher, R.T., and Hansell, D.A. (2010) Effect of dissolved organic carbon and alkalinity on the density of Arctic Ocean waters. *Aquat. Geochem.*, 1–16. doi: 10.1007/s10498-010-9111-2

Milton, J. (2010) Plants set stage for evolutionary drama. *Nature News*. doi: 10.1038/news.2010.497

Min, S.-K., Zhang, X., Zwiers, F.W., and Hegerl, G.C. (2011) Human contribution to more-intense precipitation extremes. *Nature*, **470**, 378–381.

Minschwaner, K. and Dessler, A.E. (2003) Water vapor feedback in the tropical upper troposphere: model results and observations. *J. Climate*, **17**, 1272–1282.

Mitani, J.C., Watts, D.P., and Amsler, S.J. (2010) Lethal intergroup aggression leads to territorial expansion in wild chimpanzees. *Curr. Biol.*, **20**, R507–R508.

Mitchell, J.F.B. (1989) The "Greenhouse" effect and climate change. *Reviews of Geophysics*, **27**, 115–139.

Mlawer, E.J., Taubman, S.J., Brown, P.D., Iacono, M.J., and Claugh, S.A. (1997) Radiative transfer for inhomogeneous atmospheres: RRTM, a validated correlated-k model for the longwave. *J. Geophys. Res.*, **102**, 16663–16682.

Mo, K.C. and White, G.H. (1985) Teleconnections in the southern hemisphere. *Mon. Weather Rev.*, **113**, 22–37.

Mohrholz, V., Bartholomae, C.H., van der Plas, A.K., and Lass, H.U. (2007) The seasonal variability of the northern Benguela undercurrent and its relation to the oxygen budget on the shelf. *Cont. Shelf Res.*, **28**, 424–441.

Mojzsis, S.J., Arrhenius, G., McKeegan, K.D., Harrison, T.M., Nutman, A.P., and Friend,

C.R.L. (1996) Evidence for life on Earth before 3,800 million years ago. *Nature*, **384**, 55–59.

Monin, A.S. (1982) *Vvedeniye v Teoriyu Klimata (Introduction to the Theory of Climate)*, Gidrometeoizdat, Leningrad.

Monin, A.S. and Shishkov, Yu.A. (1979) *Istoriya Klimata (History of Climate)*, Gidrometeoizdat, Leningrad.

Monod, J. (1971) *Chance and Necessity. An Essay on the Natural Philosophy of Modern Biology*, Alfred A. Knopf, New York.

Morawitz, W.M.L., Sutton, P.J., Worcester, P.F., Cornuelle, B.D., Lynch, J.F., and Pawlowicz, R. (1996) Three-dimensional observations of a deep convective chimney in the Greenland Sea during winter 1988/89. *J. Phys. Oceanogr.*, **26**, 2316–2343.

Moreno, M., Rosenau, M., and Oncken, O. (2010) 2010 Maule earthquake slip correlates with pre-seismic locking of Andean subduction zone. *Nature*, **467**, 198–202.

Morran, L.T., Parmenter, M.D., and Phillips, P.C. (2009) Mutation load and rapid adaptation favour outcrossing over self-fertilization. *Nature*, **462**, 350–352.

Moser, J. (1978) Is the solar system stable? *Math. Intell.*, **1**, 65–71.

Mott, N.F. (1967) Electrons in disordered systems. *Adv. Phys.*, **16**, 49–144.

Mott, N.F. and Davis, E.A. (1971) *Electronic Processes in Non-Crystalline Materials*, Clarendon Press, Oxford.

Müller, R.D., Gaina, C., and Clark, S. (2000) Seafloor spreading around Australia, in *Billion-Year Earth History of Australia and Neighbours in Gondwanaland* (ed. J.J. Veevers), Gemoc Press, Sydney, pp. 18–28.

Muschik, W. and Berezovski, A. (2004) Thermodynamic interaction between two discrete systems in non-equilibrium. *J. Non-Equil. Thermody.*, **29**, 237–255.

Muschik, W. (2008) Survey of some branches of thermodynamics. *J. Non-Equil. Thermody.*, **33**, 165–198.

Muschik, W. (2009) Contact quantities and non-equilibrium entropy of discrete systems. *J. Non-Equil. Thermody.*, **34**, 75–92.

Muschik, W. and Borzeszkowski, H.-H.v. (2008) Entropy identity and equilibrium conditions in relativistic thermodynamics. *Gen. Relat. Gravit.*, **41**, 1285–1304.

Nagle, J.F. (1966) Lattice statistics of hydrogen-bonded crystals. I. The residual entropy of ice. *J. Math. Phys.*, **7**, 1484–1491.

Narbonne, G.M. (2010) Ocean chemistry and early animals. *Science*, **328**, 53–54.

NASA (2001) Unmasking the Face on Mars, http://science.nasa.gov/science-news/science-at-nasa/2001/ast24may_1/.

Nathans, J., Thomas, D., and Hogness, D.S. (1986) Molecular genetics of human color vision: the genes encoding blue, green, and red pigments. *Science*, **232**, 193–202.

Nausch, G. (2006) IOW Cruise Report 40/06/20, 22–24 August 2006. Leibniz Institute for Baltic Sea Research, Warnemünde.

Nausch, G., Nehring, D., and Nagel, K. (2008) Nutrient concentrations, trends and their relation to eutrophication, in *State and Evolution of the Baltic Sea 1952–2005* (eds R. Feistel, G., Nausch and N. Wasmund), John Wiley & Sons, Inc., Hoboken, NJ, pp. 337–366.

NCDC (2010) *Global Surface Temperature Anomalies*, National Climatic Data Center, http://www.ncdc.noaa.gov/cmb-faq/anomalies.html#mean.

Negre F C., Zahn, R., Thomas, A.L., Masqué, P., Henderson, G.M., Martínez-Méndez, G., Hall, I.R., and Mas, J.L. (2010) Reversed flow of Atlantic deep water during the last glacial maximum. *Nature*, **468**, 84–88.

Neimark, Yu.L. (1972) *The Method of Point Maps in the Theory of Nonlinear Oscillations (in Russian)*, Nauka, Moscow.

Neimark, Yu.L. and Landa, P.S. (1987) *Stochastic and Chaotic Oscillations (in Russian)*, Nauka, Moscow.

Nepomnyashchy, A.A., Velarde, M.G., and Colinet, P. (2002) *Interfacial Phenomena and Convection*, Chapman & Hall, London.

Nernst, W. (1918) *Die Theoretischen und Experimentellen Grundlagen des neuen Wärmesatzes*, Verlag W. Knapp, Halle.

Neugebauer, G. (1977) Entropy and gravitation. *Int. J. Theor. Phys.*, **16**, 241–247.

Neugebauer, G. (1980) *Relativistische Thermodynamik*, Akademie-Verlag, Berlin.

Neugebauer, G. (1998) Black hole thermodynamics. *Lect. Notes Phys.*, **514**, 319–382.

Neumann, B., Nguyen, K.C., Hall, D.H., Ben-Yakar, A., and Hilliard, M.A. (2011) Axonal regeneration proceeds through specific axonal fusion in transected C. elegans neurons. *Dev. Dyn.*, **240**, 1365–1372.

Nicholas, J.V., Dransfield, T.D., and White, D.R. (1996) Isotopic composition of water used for triple point of water cells. *Metrologia*, **33**, 265–267.

Nicolis, C. (1999) Entropy production and dynamical complexity in a low-order atmospheric model. *Q. J. Roy. Meteor. Soc.*, **125**, 1859–1878.

Nicolis, C. (2010) Stability, complexity and the maximum dissipation conjecture. *Q. J. Roy. Meteor. Soc.*, **136**, 1161–1169.

Nicolis, G. and Nicolis, C. (1980) On the entropy balance of the earth-atmosphere system. *Q. J. Roy. Meteor. Soc.*, **106**, 691–706.

Nicolis, G. and Portnow, J. (1973) Chemical oscillations. *Chem. Rev.*, **73**, 365–384.

Nicolis, G. and Prigogine, I. (1977) *Self-Organization in Non-Equilibrium Systems: From Dissipative Structures to Order Through Fluctuations*, John Wiley & Sons, New York.

Nicolis, G. and Prigogine, I. (1987) *Die Erforschung des Komplexen*, Piper-Verlag, München-Zürich (English original: Exploring Complexity: An Introduction. W.H. Freeman & Company, New York 1989).

Niedźwiedzki, G., Szrek, P., Narkiewicz, K., Narkiewicz, M., and Ahlberg, P.E. (2010) Tetrapod trackways from the early middle Devonian period of Poland. *Nature*, **463**, 43–48.

Nield, T. (2007) *Supercontinent. Ten Billion Years in the Life of Our Planet*, Harvard University Press, Cambridge, MA.

NOAA (2010) Contiguous U.S. Temperature January-September 1895–2010. National Oceanic and Atmospheric Administration, National Climatic Data Center, http://www.ncdc.noaa.gov/temp-and-precip/time-series/.

Nomura, R., Ozawa, H., Tateno, S., Hirose, K., Hernlund, J., Muto, S., Ishii, H., and Hiraoka, N. (2011) Spin crossover and iron-rich silicate melt in the Earth's deep mantle. *Nature*, **473**, 199–202.

Nowacki, A., Wookey, J., and Kendall, J.-M. (2010) Deformation of the lowermost mantle from seismic anisotropy. *Nature*, **467**, 1091–1094.

Nowak, M.A. (1992) What is a Quasi-species? *Trends Ecol. Evol.*, **7**, 118–121.

Nyquist, H. (1928) Thermal agitation of electric charge in conductors. *Phys. Rev.*, **32**, 110–113.

O'Brien, D.M. and Stephens, G.I. (1995) Entropy and climate. II: simple models. *Q. J. Roy. Meteor. Soc.*, **121**, 1773–1796.

O'Connell, J.F. and Allen, J. (2004) Dating the colonization of Sahul (Pleistocene Australia-New Guinea). *J. Archaeol. Sci.*, **31**, 835–853.

Okazaki, Y., Timmermann, A., Menviel, L., Harada, N., Abe-Ouchi, A., Chikamoto, M.O., Mouchet, A., and Asahi, H. (2010) Deepwater formation in the North Pacific during the last glacial termination. *Science*, **329**, 200–204.

Olson, J.E. and Bourne, E.G. (1906) *The Northmen, Columbus and Cabot, 985–1503: The Voyages of the Northmen, The Voyages of Columbus and of John Cabot*, Charles Scribner's Sons, New York.

O'Nions, R.K., Hamilton, P.J., and Evensen, N.M. (1984) Die chemische Entwicklung des Erdmantels, in *Ozeane und Kontinente*, Spektrum der Wissenschaft Heidelberg, pp. 66–78.

Onsager, L. and Machlup, S. (1953) Fluctuations and irreversible processes. *Phys. Rev.*, **91**, 1505–1512.

Oparin, A.I. (1924) *Proizkhozhdeniye zhizni (Origin of Life)*, Moskovski Rabochi, Moscow.

Oparin, A.I. (1963) *Das Leben – seine Natur, Herkunft und Entwicklung*, Gustav Fischer, Jena.

Orlob, G.T. (1983) *Mathematical Modeling of Water Quality: Streams, Lakes, and Reservoirs*, John Wiley & Sons, Chichester, New York, Brisbane, Toronto, Singapore.

Oró, J. (1965) Prebiological organic systems, in *The Origin of Prebiological Systems* (ed. S.W. Fox), Academic Press, New York.

Ostrowski, Ju.I. (1973) *Dreidimensionale Bilder durch Holographie*, Teubner, Leipzig.

Ostrowski, Ju.I. (1987) *Holografie – Grundlagen, Experimente und Anwendungen*, Teubner, Leipzig.

Ott, E. (1993) *Chaos in Dynamical Systems*, Cambridge University Press, Cambridge.

Öttinger, H.C. (2005) *Beyond Equilibrium Thermodynamics*, Wiley Interscience, London.

Ozawa, H., Ohmura, A., Lorenz, R.D., and Pujol, T. (2003) The second law of thermodynamics and the global climate system: a review of the maximum entropy production principle. *Rev. Geophys.*, **41** (1018), 24.

Pagani, M., Pedentchouk, N., Huber, M., Sluijs, A., Schouten, S., Brinkhuis, H., Damste, J.S., Dickens, G.D., and The Expedition 302 Scientists (2006) Arctic hydrology during global warming at the Palaeocene/Eocene thermal maximum. *Nature*, **442**, 671–675.

Pagani, M., Zachos, J.C., Freeman, K.H., Tipple, B., and Bohaty, S. (2005) Marked decline in atmospheric carbon dioxide concentrations during the paleogene. *Science*, **309**, 600–603.

Palle, E., Goode, P.R., Montañes-Rodriguez, P., and Koonin, S.E. (2004) Changes in Earth's reflectance over the past two decades. *Science*, **304**, 1299–1301.

Paltridge, G.W. (1975) Global dynamics and climate – a system of minimum entropy exchange. *Q. J. Roy. Meteor. Soc.*, **101**, 475–484.

Paltridge, G.W. (2001) A physical basis for a maximum of thermodynamic dissipation of the climate system. *Q. J. Roy. Meteor. Soc.*, **127**, 305–313.

Panin, G.N. and Brezgunov, V.S. (2007) Influence of the salinity of water on its evaporation. *Izv. Atmos. Oceanic Phys.*, **43**, 663–665.

Papentin, F. (1973) A Darwinian evolutionary system. I. Definition and basic properties. *J. Theor. Biol.*, **39**, 397–415.

Parnell, J., Boyce, A.J., Mark, D., Bowden, S., and Spinks, S. (2010) Early oxygenation of the terrestrial environment during the Mesoproterozoic. *Nature*, **468**, 290–293.

Parker, E.T., Cleaves, H.J., Dworkin, J.P., Glavin, D.P., Callahan, M., Aubrey, A., Lazcano, A., and Bada, J.L. (2011) Primordial synthesis of amines and amino acids in a 1958 Miller H_2S-rich spark discharge experiment. *Proceedings of the National Academy of Sciences of the United States of America*, March 21, 2011, published online, doi: 10.1073/pnas.1019191108, www.pnas.org/cgi/doi/10.1073/pnas.1019191108

Paterson, S., Vogwill, T., Buckling, A., Benmayor, R., Spiers, A.J., Thomson, N.R., Quail, M., Smith, F., Walker, D., Libberton, B., Fenton, A., Hall, N., and Brockhurst, M.A. (2009) Antagonistic coevolution accelerates molecular evolution. *Nature*, **464**, 275–278.

Pauen, M. and Roth, G. (2008) *Freiheit, Schuld und Verantwortung. Grundzüge Einer Naturalistischen Theorie der Willensfreiheit*, Suhrkamp, Frankfurt am Main.

Pauling, L. (1935) The structure and entropy of ice and of other crystals with some randomness of atomic arrangement. *J. Am. Chem. Soc.*, **57**, 2680–2684.

Pauling, L. and Pauling, P. (1975) *Chemistry*, W.H. Freeman & Co., San Francisco.

Payne, J.A., Stevenson, T.J., and Donaldson, L.F. (1996) Molecular characterization of a putative K-Cl cotransporter in rat brain. *J. Biol. Chem.*, **271**, 16245–16252.

Pedlosky, J. (1987) *Geophysical Fluid Dynamics*, Springer, New York.

Peixoto, J.P. and Oort, A.H. (1992) *Physics of Climate*, American Institute of Physics, New York.

Peixoto, J.P., Oort, A.H., De Almeida, M., and Tomé, A. (1991) Entropy budget of the atmosphere. *J. Geophys. Res.*, **96**, 10981–10988. doi: 10.1029/91JD00721

Peleg, S., Sananbenesi, F., Zovoilis, A., Burkhardt, S., Bahari-Javan, S., Agis-Balboa, R.C., Cota, P., Wittnam, J.L., Andreas Gogol-Doering, A., Opitz, L., Salinas-Riester, G., Dettenhofer, M., Kang, H., Farinelli, L., Chen, W., and Fischer, A. (2010) Altered histone acetylation is associated with age-dependent memory impairment in mice. *Science*, **328**, 753–756.

Pelkowski, J. (1994) Towards an accurate estimate of the entropy production due to radiative processes: results with a gray atmosphere model. *Meteorol. Atmos. Phys.*, **53**, 1–17.

Pelkowski, J. (1995) *Entropieerzeugung eines strahlenden Planeten: Studien zu ihrer Rolle in der Klimatheorie*. Verlag Harry Deutsch, Thun, Frankfurt am Main.

Pelkowski, J. (1997) Entropy production in a radiating layer near equilibrium: Assaying its variational properties. *J. Non-Equil. Thermody.*, **22**, 48–74.

Penn, M. and Livingston, W. (2010) Long-term evolution of sunspot magnetic fields. To appear in IAU Symposium No. 273, http://arxiv.org/abs/1009.0784v1.

Pennisi, E. (2010) *Volvox* genome shows it doesn't take much to be multicellular. *Science*, **329**, 128–129.

Penny, A.H. (1948) A theoretical determination of the elastic constants of ice.

Proc. Cambrian Philosophical Society, **44**, 423–439.

Pepper, A.E. (1998) Molecular evolution: old branches on the phytochrome family tree. *Curr. Biol.*, **8**, R117–R120.

Pepperberg, I.M. (2002) In search of king Solomon's ring: cognitive and communicative studies of Grey parrots (*Psittacus erithacus*). *Brain, Behav. Evol.*, **59**, 54–67.

Peschel, M. and Mende, W. (1986) *The Predator-Prey Model: Do we Live in a Volterra world?*, Akademie-Verlag, Berlin.

Peters, S.E., Carlson, A.E., Kelly, D.C., and Gingerich, P.D. (2010) Large-scale glaciation and deglaciation of Antarctica during the Late Eocene. *Geology*, **38**, 723–726.

Petersen, K.D., Nielsen, S.B., Clausen, O.R., Stephenson, R., and Gerya, T. (2010) Small-scale mantle convection produces stratigraphic sequences in sedimentary basins. *Science*, **329**, 827–830.

Petit, J.R., Jouzel, J., Raynaud, D., Barkov, N.I., Barnola, J.-M., Basile, I., Bender, M., Chappellaz, J., Davis, M., Delayque, G., Delmotte, M., Kotlyakov, V.M., Legrand, M., Lipenkov, V.Y., Lorius, C., Pepin, L., Ritz, C., Saltzman, E., and Stievenard, M. (1999) Climate and atmospheric history of the past 420,000 years from the Vostok ice core, Antarctica. *Nature*, **399**, 429–436, http://www.ncdc.noaa.gov/paleo/icecore/antarctica/vostok/vostok_co2.html.

Petit, J.R., Raynaud, D., Lorius, C., Jouzel, J., Delaygue, G., Barkov, N.I., and Kotlyakov, V.M. (2000) Historical isotopic temperature record from the Vostok ice core. In Trends: A Compendium of Data on Global Change. Carbon Dioxide Information Analysis Center, Oak Ridge National Laboratory, U.S. Department of Energy, Oak Ridge, Tenn., U.S.A. doi: 10.3334/CDIAC/cli.006 http://cdiac.esd.ornl.gov/trends/temp/vostok/jouz_tem.htm

Petty, G.W. (2006) *A First Course in Atmospheric Radiation*. Sundog Publishing, Madison, Wisconsin.

Piechura, J. and Walczowski, W. (2009) Warming of the West Spitsbergen Current and sea ice north of Svalbard. *Oceanologia*, **51**, 147–164, http://www.iopan.gda.pl/oceanologia/51_2.html#A1

Pigliucci, M. (2008) Sewall Wright's adaptive landscapes: 1932 vs. 1988. *Biology and Philosophy*, **23**, 591–603.

Planavsky, N.J., Rouxel, O.J., Bekker, A., Lalonde, S.V., Konhauser, K.O., Reinhard, C.T., and Lyons, T.W. (2010) The evolution of the marine phosphate reservoir. *Nature*, **467**, 1088–1090.

Planck, M. (1906) *Theorie der Wärmestrahlung*, Johann Ambrosius Barth, Leipzig.

Planck, M. (1935) Bemerkungen über Quantitätsparameter, Intensitätsparameter und stabiles Gleichgewicht. *Physica II*, **2**, 1029–1032. Reprinted in Ebeling, W., Hoffmann, D. (eds.), 2008, Über Thermodynamische Gleichgewichte, von Max Planck. Verlag Harry Deutsch, Frankfurt am Main, pp. 236.

Planck, M. (1954) *Vorlesungen über Thermodynamik*, Walter de Gruyter, Berlin.

Pleasance, E.D., Cheetham, R.K., Stephens, P.J., McBride, D.J., Humphray, S.J., Greenman, C.D., Varela, I., Lin, M.-L., Ordóñez, G.R., Bignell, G.R., Ye, K., Alipaz, J., Bauer, M.J., Beare, D., Butler, A., Carter, R.J., Chen, L., Cox, A.J., Edkins, S., Kokko-Gonzales, P.I., Gormley, N.A., Grocock, R.J., Haudenschild, C.D., Hims, M.M., James, T., Jia, M., Kingsbury, Z., Leroy, C., Marshall, J., Menzies, A., Mudie, L.J., Ning, Z., Royce, T., Schulz-Trieglaff, O.B., Spiridou, A., Stebbings, L.A., Szajkowski, L., Teague, J., Williamson, D., Chin, L., Ross, M.T., Campbell, P.J., Bentley, D.R., Futreal, P.A., and Stratton, M.R. (2010) A comprehensive catalogue of somatic mutations from a human cancer genome. *Nature*, **463**, 191–196.

Pledger, R.J., Crump, B.C., and Baross, J.A. (1994) A barophilic response by two hyperthermophilic, hydrothermal vent Archaea: an upward shift in the optimal temperature and acceleration of growth rate at supra-optimal temperatures by elevated pressure. *FEMS Microbiol. Ecol.*, **14**, 233–241.

Pogliani, L. and Berberan-Santos, M.N. (2000) Constantin Carathéodory and the axiomatic thermodynamics. *J. Math. Chem.*, **28**, 313–324.

Pollack, H.N., Hurter, S.J., and Johnson, J.R. (1993) Heat flow from the Earth's interior: analysis of the global data set. *Rev. Geophys.*, **30**, 267–280.

Posokhov, E.V. (1981) *Khimicheskaya Evolyutsiaya Gidrosfery (Chemical Evolution of the Hydrosphere)*, Gidrometeoizdat, Leningrad.

Poston, T. and Stewart, I. (1978) *Catastrophe Theory and its Applications*, Pitman, London, San Francisco, Melbourne.

Povolotskaya, I.S. and Kondrashov, F.A. (2010) Sequence space and the ongoing expansion of the protein universe. *Nature*, **465**, 922–926.

Prescott, J.A. and Collins, J.A. (1951) The lag of temperature behind solar radiation. *Q. J. Roy. Meteor. Soc.*, **77**, 121–126

Prigogine, I. (1947) *Etude thermodynamique des phénomènes irreversibles*, Desoer, Liege, Paris.

Prigogine, I. (1955, 1967) *Introduction to Thermodynamics of Irreversible Processes*, Interscience Publishers, Wiley Interscience, New York.

Prigogine, I. (1980) *From Being To Becoming*. Freeman, San Francisco.

Prigogine, I., Mayne, F., George, C., and de Haan, M. (1977) Microscopic theory of irreversible processes. *Proc. Natl. Acad. Sci. U.S.A.*, **74**, 4152–4156.

Prigogine, I., Nicolis, G., and Babloyantz, A. (1972) Thermodynamics of evolution. *Phys. Today*, **25**, 97–99.

Prigogine, I. and Stengers, I. (1993) *Time, Chaos and the Quantum: Towards the Resolution of the Time Paradox*, Harmony Books, New York.

Pritchard, H.D., Arthern, R.J., Vaughan, D.G., and Edwards, L.A. (2009) Extensive dynamic thinning on the margins of the Greenland and Antarctic ice sheets. *Nature*, **461**, 971–975.

Proskurowski, G., Lilley, M.D., Seewald, J.S., Früh-Green, G.L., Olson, E.J., Lupton, J.E., Sylva, S.P., and Kelley, D.S. (2008) Abiogenic hydrocarbon production at Lost City hydrothermal field. *Science*, **319**, 604–607.

Provine, R. (2000) *Laughter: A Scientific Investigation*, Penguin Books, New York.

Pujol, T., and Fort, J. (2002) States of maximum entropy production in a onedimensional vertical model with convective adjustment. *Tellus*, **54A**, 363–369.

Pujol, T., Llebot, J.E., and Fort, J. (1999) Greenhouse gases and climatic states of minimum entropy production. *J. Geophys. Res.*, **104**, 24257–24263.

Pyka, A. and Küppers, G. (2002) *Innovation Networks: Theory and Practice*, New Horizons in the Economics of Innovation Series, Edward Elgar Publishing, Cheltenham, Northampton, MA.

Pyka, A. and Scharnhorst, A. (2009) *Innovation Networks. New Approaches in Modeling and Analyzing*, Springer, Dordrecht, Heidelberg, London, New York.

Quach, T.T., Duchemin, A.-M., Oliver, A.P., Schrier, B.K., and Wyatt, R.J. (1991) Hydra head activator peptide has trophic activity for eukaryotic neurons. *Developmental. Brain Res.*, **68**, 97–102.

Quinn, T. (2007) A short history of temperature scales. *PTB-Mitteilungen*, **117**, 23–30.

Raethjen, P. (1961) Gleichgewichtsstörungen im nichtstationären Jet Stream, in *Hamburger Geophysikalische Einzelschriften*, Heft 4, Cram, de Gruyter & Co., Hamburg.

Rahmstorf, S. (2000) The thermohaline ocean circulation – a system with dangerous thresholds? *Climate Change*, **46**, 247–256.

Ramakrishnan, V. (2011) The Eukaryotic ribosome. *Science*, **331**, 681–682.

Rasmussen, B., Fletcher, I.R., Brocks, J.J., and Kilburn, M.R. (2008) Reassessing the first appearance of eukaryotes and cyanobacteria. *Nature*, **455**, 1101–1104.

Rathmann, J. (2008) Klima- und Zirkulationsvariabilität im südhemisphärischen Afrika seit Beginn des 20. Jahrhunderts. Dissertation, Augsburg University.

Rechenberg, I. (1973) *Evolutionsstrategie – Optimierung technischer Systeme nach Prinzipien der Biologischen Information*, Friedrich Frommann Verlag (Günter Holzboog KG), Stuttgart-Bad Cannstadt.

Rechenberg, I. (1994) *Evolutionsstrategie 94*, Fromman-Holzboog-Verlag, Stuttgart-Bad Cannstadt.

Redko, V.G. (1986) Behavior of sysers in coacervates (in Russian). *Biofizika*, **31**, 701–703.

Reed, C. (2006) Marine science: boiling points. *Nature*, **439**, 905–907.

Reich, D., Green, R.E., Kircher, M., Krause, J., Patterson, N., Durand, E.Y., Viola, B., Briggs, A.W., Stenzel, U., Johnson, P.L.F., Maricic,

T., Good, J.M., Marques-Bonet, T., Alkan, C., Fu, Q., Mallick, S., Li, H., Meyer, M., Eichler, E.E., Stoneking, M., Richards, M., Talamo, S., Shunkov, M.V., Derevianko, A.P., and Hublin, J.-J. (2010) *et al.* Genetic history of an archaic hominin group from Denisova Cave in Siberia. *Nature*, **468** 1053–1060.

Reich, D., Thangaraj, K., Patterson, N., Price, A.L., and Singh, L. (2009) Reconstructing Indian population history. *Nature*, **461**, 489–494.

Reichholf, J.H. (2004) *Das Rätsel der Menschwerdung*, dtv, München.

Reinbothe, H. and Krauß, G.-J. (1982) *Entstehung und molekulare Evolution des Lebens*, Gustav Fischer, Jena.

Reinhard, C.T., Raiswell, R., Scott, C., Anbar, A.D., and Lyons, T.W. (2009) A late archean sulfidic sea stimulated by early oxidative weathering of the continents. *Science*, **326**, 713–716.

Reissmann, J.H., Burchard, H., Feistel, R., Hagen, E., Lass, H.U., Mohrholz, V., Nausch, G., Umlauf, L., and Wieczorek, G. (2009) Vertical mixing in the Baltic Sea and consequences for eutrophication – A review. *Prog. Oceanogr.*, **82**, 47–80.

Resulaj, A., Kiani, R., Wolpert, D.M., and Shadlen, M.N. (2009) Changes of mind in decision-making. *Nature*, **461**, 263–266.

Reynard, B., Nakajima, J., and Kawakatsu, H. (2010) Earthquakes and plastic deformation of anhydrous slab mantle in double Wadati-Benioff zones. *Geophys. Res. Lett.*, **37**, L24309. doi: 10.1029/2010GL045494

Ricardo, A.R. and Szostak, J.W. (2010) *Der Ursprung des irdischen Lebens*, Spektrum der Wissenschaft, März, pp. 44–51.

Richardson, M.K. and Jeffrey, J.E. (2002) Editorial: Haeckel and modern biology. *Theor. Biosci.*, **121**, 247–251.

Richardson, S.H., Gurney, J.J., Erlank, A.J., and Harris, J.W. (1984) Origin of diamonds in old enriched mantle. *Nature*, **10**, 198–202.

Ridgway, W.L., Harshvardan, and Arking, A. (1991) Computation of atmospheric cooling rates by exact and approximate methods. *J. Geophys. Res*, **96**, 8969–8984.

Risken, H. (1984) *The Fokker-Planck Equation. Methods of Solution and Applications*, Springer, Berlin, Heidelberg, New York, Tokyo.

Robert, F. and Chaussidon, M. (2006) A palaeotemperature curve for the Precambrian oceans based on silicon isotopes in cherts. *Nature*, **443**, 969–972.

Robert, C., Diester-Haass, L., and Paturel, J. (2009) Clay mineral assemblages, siliciclastic input and paleoproductivity at ODP Site 1085 off Southwest Africa: A late Miocene–early Pliocene history of Orange river discharges and Benguela current activity, and their relation to global sea level change. *Mar. Geol.*, **216**, 221–238.

Roebroeks, W. and Villa, P. (2011) On the earliest evidence for habitual use of fire in Europe. *Proceedings of the National Academy of Sciences of the United States of America*, published online March 14, 2011, doi: 10.1073/pnas.1018116108

Rogers, A.R., Iltis, D., and Wooding, S. (2004) Genetic variation at the MC1R locus and the time since loss of human body hair. *Curr. Anthropol.*, **45**, 105–108.

Romanovsky, Yu.M., Stepanova, M.V., and Chernavsky, D.S. (1975) *Matematicheskoye Modelirovaniye v Biofizike (Mathematical Modelling in Biophysics)*, Nauka, Moscow, Revised English edition (2004) Moscow-Ishevsk.

Romanovsky, Yu.M., Stepanova, N.V., and Chernavsky, D.S. (1984) *Matematicheskaya Biofizika (Mathematical Biophysics)*, Nauka, Moscow.

Rose, H., Ebeling, W., and Asselmeyer, T. (1996) The density of states – a measure of the difficulty of optimisation problems in H.-M. Voigt *et al.*, eds., loc. cit.

Rosing, M.T., Bird, D.K., Sleep, N.H., and Bjerrum, C.J. (2010) No climate paradox under the faint early sun. *Nature*, **464**, 744–747.

Ross, M.D., Owren, M.J., and Zimmermann, E. (2009) Reconstructing the evolution of laughter in great apes and humans. *Curr. Biol.*, **19**, 1106–1111.

Rozwadowska, A. (1991) A model of solar energy input into the Baltic Sea. Studia i materialy oceanologiczne. *Marine Physics*, **59**, 223–242.

Rozwadowska, A. and Isemer, H.J. (1998) Solar fluxes at the surface of the Baltic Proper. Part 1. Mean annual cycle and influencing factors. *Oceanologia*, **40**, 307–330.

Rubenstein, D.R. and Lovette, I.J. (2009) Reproductive skew and selection on female ornamentation in social species. *Nature*, **462**, 786–789.

Rubie, D.C., Frost, D.J., Mann, U., Asahara, Y., Nimmo, F., Tsuno, K., Kegler, P., Holzheid, A., and Palme, H. (2011) Heterogeneous accretion, composition and core–mantle differentiation of the Earth. *Earth Planet. Sci. Lett.*, **301**, 31–42.

Rubinstein, C.V., Gerrienne, P., De La Puente, G.S., Astini, R.A., and Steemans, P. (2010) Early middle ordovician evidence for land plants in Argentina (eastern Gondwana). *New Phytol.*, **188**, 365–369.

Rudtsch, S. and Fischer, J. (2008) Temperature measurements according to the International Temperature Scale of 1990 and its associated uncertainties. *Accredit. Qual. Assur.*, **13**, 607–609.

Ruegg, J.C., Rudloff, A., Vigny, C., Madariaga, R., de Chabalier, J.B., Campos, J., Kausel, E., Barrientos, S., and Dimitrov, D. (2009) Interseismic strain accumulation measured by GPS in the seismic gap between Constitución and Concepción in Chile. *Phys. Earth Planet. In.*, **175**, 78–85.

Ruelle, D. and Takens, F. (1971) On the nature of turbulence. *Commun. Math. Phys.*, **20**, 167–192.

Ruelle, D. (1993) *Zufall und Chaos*, Springer, Berlin.

Rutten, M.G. (1971) *The Origin of Life by Natural Causes*, Elsevier, Amsterdam.

Rust, J., Singh, H., Rana, R.S., McCann, T., Singh, L., Anderson, K., Sarkar, N., Nascimbene, P.C., Stebner, F., Thomas, J.C., Solórzano Kraemer, M., Williams, C.J., Engel, M.S., Sahni, A., and Grimaldi, D. (2010) Biogeographic and evolutionary implications of a diverse paleobiota in amber from the early Eocene of India. *Proc. Natl. Acad. Sci. U.S.A.*, **107**, 18360–18365.

Ryan, W.B. and Pitman, W.C. (2000) *Noah's Flood: The New Scientific Discoveries About the Event that Changed History*, Simon & Schuster, New York.

Saaty, T.L. (1966) Operationsanalyse. Methoden und Struktur von Operationen, in *Die Mathematik für Physik und Chemie Band II*. Teubner, Leipzig. American original (1964): *Operational Analysis: Methods and Structure of Operations. The Mathematics of Physics and Chemistry*, vol. **2** (eds H. Margenau and G.M. Murphy), D. van Nostrand Company, New York.

Sagan, C. (1978) *Die Drachen von Eden*. Droemer Knaur, München, Zürich (American original: The Dragons of Eden. Random House, New York, 1977).

Sallan, L.C., Kammer, T.W., Ausich, W.I., and Cook, L.A. (2011) Persistent predator–prey dynamics revealed by mass extinction. *Proc. Natl. Acad. Sci. USA*, **108**, 8335–8338.

Sagan, C. and Agel, J. (1973) *Cosmic Connection: An Extraterrestrial Perspective*, Anchor Press/Doubleday, Garden City, NY.

Santer, B.D., Mears, C., Wentz, F.J., Taylor, K.E., Gleckler, P.J., Wigley, T.M.L., Barnett, T.P., Boyle, J.S., Brüggemann, W., Gillett, N.P., Klein, S.A., Meehl, G.A., Nozawa, T., Pierce, D.W., Stott, P.A., Washington, W.M., and Wehner, M.F. (2007) Identification of human-induced changes in atmospheric moisture content. *Proc. Natl. Acad. Sci. U.S.A.*, **104**, 15248–15253.

Santer, B.D., Wehner, M.F., Wigley, T.M.L., Sausen, R., Meehl, G.A., Taylor, K.E., Ammann, C., Arblaster, J., Washington, W.M., Boyle, J.S., and Brüggemann, W. (2003) Contributions of anthropogenic and natural forcing to recent tropopause height changes. *Science*, **301**, 479–483.

Sarma, M.S., Rodriguez-Zas, S.L., Gernat, T., Nguyen, T., Newman, T. and Robinson, G.E. (2010) Distance-responsive genes found in dancing honey bees. *Genes, Brain Behav.*, **9**, 825–830.

Savchuk, O.P. (2010) Large-scale dynamics of hypoxia in the Baltic Sea, in *Chemical Structure of Pelagic Redox Interfaces: Observation and Modeling. The Handbook of Environmental Chemistry* (ed. E.V. Yakushev), Springer, Berlin, Heidelberg, pp. 1–24. doi: 10.1007/698_2010_53

Saviotti, P.P. (1996) *Technological Evolution, Variety and the Economy*, Edward Elgar Publishing, Cheltenham, Northampton, MA.

Saviotti, P.P. and Mani, G.S. (1995) Competition, variety and technological evolution: a replicator dynamics model. *J. Evol. Econ.*, **5**, 369–392.

Saviotti, P.P. and Nooteboom, B. (2000) *Technology and Knowledge: From the Firm to Innovation Systems*, Edward Elgar Publishing, Cheltenham, Northampton, MA.

Schaller, H.C., Hermans-Borgmeyer, I., and Hoffmeister, S.A. (1996) Neuronal control of development in hydra. *Int. J. Dev. Biol.*, **40**, 339–344.

Scharnhorst, A. (1998) Citations - networks, science landscapes and evolutionary strategies. *Scientometrics*, **43**, 95–106.

Scharnhorst, A. (2001) Constructing knowledge landscapes within the framework of geometrically oriented evolutionary theories, in *Integrative Systems Approaches to Natural and Social Sciences – Systems Science 2000* (eds M. Matthies, H., Malchow and J. Kriz), Springer, Berlin, pp. 505–515.

Scharnhorst, A. and Ebeling, W. (2005) Evolutionary search agents in complex landscapes – a new model for the role of competence and meta-competence (EVOLINO and other simulation tools), arXiv:physics/0511232v1.

Schell, I.I. (1968) On the relation between the winds off Southwest Africa and the Benguela Current and Agulhas Current penetration in the South Atlantic. *Deutsche Hydrographische Zeitschrift*, **21**, 109–117.

Schellnhuber, H.J. (2011) Globaler Klimaschutz – eine unlösbare Aufgabe? Sitzungsberichte Leibniz – Sozietät der Wissenschaften, http://www.leibniz-sozietaet.de/, Lecture on 13 January 2011, in press.

Schellnhuber, H.J., Cramer, W., Nakicenovic, N., Wigley, T., and Yohe, G. (2006) *Avoiding Dangerous Climate Change*, Cambridge University Press, Cambridge.

Scher, H.D. and Martin, E.E. (2006) Timing and climatic consequences of the opening of Drake Passage. *Science*, **312**, 428–430.

Schiermeier, Q. (2008) North Atlantic cold-water sink returns to life. Nature News, doi: 10.1038/news.2008.1262

Schimansky-Geier, L. (1981) *Stochastische Theorie der Nichtgleichgewichtsübergänge in bistabilen chemischen Reaktionssystemen*. Dissertation A, Humboldt University, Berlin.

Schirrmeister, B.E., Antonelli, A., and Bagheri, H.C. (2011) The origin of multicellularity in cyanobacteria. *BMC Evol. Biol.* 2011, 11:45, doi: 10.1186/1471-2148-11-45, http://www.biomedcentral.com/1471-2148/11/45

Schiermeier, Q. (2010) The real holes in climate science. *Nature*, **463**, 284–287.

Schiermeier, Q. (2011) Race against time for raiders of the lost lake. *Nature*, **469**, 275.

Schiermeier, Q., Tollefson, J., Scully, T., Witze, A., and Morton, O. (2008) Energy alternatives: electricity without carbon. *Nature*, **454**, 816–823.

Schimansky-Geier, L., Zülicke, Ch., and Schöll, E. (1991) Domain formation due to Ostwald ripening in bistable systems far from equilibrium. *Z. Phys. B: Condens. Matter*, **84**, 433–441.

Schlögl, F. (1972) Chemical reaction models for non-equilibrium phase transitions. *Z. Phys. A*, **253**, 147–161.

Schlüssel, P., Emery, W.J., Grassl, H., and Mammen, T. (1990) On the depth skin temperature difference and its impact on satellite remote sensing of sea surface temperature. *J. Geophys. Res.*, **95**, 13341–13356.

Schmelzer, J. (1979) *Zur Koexistenz von Sorten in der nichtlinearen Kinetik Homogener Konkurrenzreaktionen*. Dissertation A, Wilhelm-Pieck-Universität Rostock, Rostock

Schmelzer, J. (1985) Zur Kinetik des Keimwachstums in Lösungen. *Z. Phys. Chem.*, **266**, 1057–1070.

Schmelzer, J. and Gutzow, I. (1986) Ostwald ripening in viscoelastic media, in *Selforganization by Non-Linear Irreversible Processes*, Springer Series in Synergetics, vol. 33 (eds W. Ebeling and H. Ulbricht), Springer, Berlin, pp. 144–148.

Schmetz, J. and Raschke, E. (1990) Bewölkung und Strahlungshaushalt der Erde, in *Atmosphäre, Klima, Umwelt*, Spektrum der Wissenschaft, Heidelberg, pp. 26–36.

Schmoeckel, R. (1999) *Die Indoeuropäer. Aufbruch aus der Vorgeschichte*, Bastei Lübbe, Germany, p. 1999.

Schmutz, W. (2010) Total solar irradiance: Challenges for the future. WMO-BIPM Workshop on Measurement Challenges for Global Observation Systems for Climate Change Monitoring: Traceability, Stability and Uncertainty: 30 March to 1 April 2010, Geneva, IOM-Report 105, WMO/TD 1557, http://www.bipm.org/ws/BIPM/WMO-

BIPM/Allowed/Session_C/C.02-Schmutz.pdf.

Schmutz, W., Fehlmann, A., Hüsen, G., Meindl, P., Winkler, R., Thuillier, G., Blattner, P., Buisson, F., Egorova, T., Finsterle, W., Fox, N., Gröbner, J., Hochedez, J.-F., Koller, S., Meftah, M., Meisonnier, M., Nyeki, S., Pfiffner, D., Roth, H., Rozanov, E., Spescha, M., Wehrli, C., Werner, L., and Wyss, J.U. (2009) The PREMOS/PICARD instrument calibration. *Metrologia*, **46** S202–S206.

Schneider, S.H. (1990) Veränderungen des Klimas, in *Atmosphäre, Klima, Umwelt*, Spektrum der Wissenschaft, Heidelberg, pp. 188–205.

Schöll, M., Steinhilber, F., Beer, J., Haberreiter, M., and Schmutz, W. (2007) Long-term reconstruction of the total solar irradiance based on neutron monitor and sunspot data. *Adv. Space Res.*, **40**, 996–999.

Schönbächler, M., Carlson, R.W., Horan, M.F., Mock, T.D., and Hauri, E.H. (2010) Heterogeneous accretion and the moderately volatile element budget of earth. *Science*, **328**, 884–887.

Schöpf, H.G. (1978) *Von Kirchhoff bis Planck*, Akademie-Verlag, Berlin.

Schöpf, H.G. (1984) Rudolf Clausius – ein Versuch, ihn zu verstehen. *Ann. Phys-Leipzig*, **41**, 185–195.

Schoppe, C.M. and Schoppe, S.G. (2004) *Atlantis und die Sintflut*, Books on Demand, Norderstedt.

Schramm, G., Grötsch, H., and Pollmann, W. (1962) Nicht-enzymatische Synthese von Polysacchariden, Nucleosiden und Nucleinsäuren und die Entstehung selbstvermehrungsfähiger Systeme. *Ang. Chem.*, **74**, 53–59.

Schrödinger, E. (1944) *What is Life - the Physical Aspect of the Living Cell*, Cambridge University Press, Cambridge.

Schulson, E.M. (1999) The structure and mechanical behavior of ice. *Jom-J. Min. Met. Mat. S.*, **51**, 21–27.

Schulte, P., Alegret, L., Arenillas, I., Arz, J.A., Barton, P.J., Bown, P.R., Bralower, T.J., Christeson, G.L., Claeys, P., Cockell, C.S., Collins, G.S., Deutsch, A., Goldin, T.J., Goto, K., Grajales-Nishimura, J.M., Grieve, R.A.F., Gulick, S.P.S., Johnson, K.R., Kiessling, W., Koeberl, C., Kring, D.A., MacLeod, K.G., Matsui, T., Melosh, J., Montanari, A., Morgan, J.V., Neal, C.R., Nichols, D.J., Norris, R.D., Pierazzo, E., Ravizza, G., Rebolledo-Vieyra, M., Reimold, W.U., Robin, E., Salge, T., Speijer, R.P., Sweet, A.R., Urrutia-Fucugauchi, J., Vajda, V., Whalen, M.T., and Willumsen, P.S. (2010) The Chicxulub asteroid impact and mass extinction at the Cretaceous-Paleogene boundary. *Science*, **327**, 1214–1218.

Schulz, M. (2002) On the 1470-year pacing of Dansgaard-Oeschger warm events. *Paleoceanography*, **17**, 1014.

Schuster, H.G. (1984, 1995) *Deterministic Chaos*, Physik-Verlag, Weinheim.

Schuster, P. (1972) Vom Makromolekül zur primitiven Zelle. *Chem. Unserer Zeit*, **6**, 1–16.

Schuster, P. (2009) Genotypes and phenotypes in the evolution of molecules. *Eur. Rev.*, **17**, 282–319.

Schwartz, S.E. (2007) Heat capacity, time constant, and sensitivity of Earth's climate system. *J. Geophys. Res.*, **112**, D24S05, doi: 10.1029/2007JD008746

Schwefel, H.-P. (1977) *Numerische Optimierung von Computermodellen mittels der Evolutionsstrategie*, Birkhäuser Verlag, Basel/Stuttgart.

Schweitzer, F. (1997) *Self-Organization of Complex Structures: From Individual to Collective Dynamics*, Gordon and Breach Science Publishers, Amsterdam.

Schweitzer, F. (2002) *Modelling complexity in Economic and Social Systems*, World Scientific Publishing, Singapore.

Schweitzer, F., Fagiolo, G., Sornette, D., Vega-Redondo, F., Vespignani, A., and White, D.R. (2009) Economic networks: the new challenges. *Science*, **325**, 422–425.

Sclater, J.H. and Tapscott, C. (1984) Die Geschichte des Atlantik, in *Ozeane und Kontinente*, Spektrum der Wissenschaft, Heidelberg, pp. 118–130.

Scott, G.R. and Gibert, L. (2009) The oldest hand-axes in Europe. *Nature*, **461**, 82–85.

Scott, A.C. and Glasspool, I.J. (2006) The diversification of Paleozoic fire systems and fluctuations in atmospheric oxygen concentration. *Proc. Natl. Acad. Sci. U.S.A.*, **103**, 10861–10865.

Scott, C., Lyons, T.W., Bekker, A., Shen, Y., Poulton, S.W., Chu, X., and Anbar, A.D.

(2008) Tracing the stepwise oxygenation of the Proterozoic ocean. *Nature*, **452**, 456–459.

Screen, J.A. and Simmonds, I. (2010) The central role of diminishing sea ice in recent Arctic temperature amplification. *Nature*, **464**, 1334–1337.

Secord, R., Gingerich, P.D., Lohmann, K.C., and MacLeod, K.G. (2010) Continental warming preceding the Palaeocene–Eocene thermal maximum. *Nature*, **467**, 955–958.

Seidel, J., Engert, J., Fellmuth, B., Fischer, J., Hartmann, J., Hollandt, J., and Tegeler, E. (2007) Die internationalen Temperaturskalen: ITS-90 und PLTS-2000. *PTB-Mitteilungen*, **117**, 16–22.

Seifert, U. (2005) Entropy production along a stochastic trajectory and an integral fluctuation theorem. *Phys. Rev. Lett.*, **95**, 040602.

Sell, A., Bryant, G.A., Cosmides, L., Tooby, J., Sznyce, D., von Rueden, C., Krauss, A., and Gurven, M. (2010) Adaptations in humans for assessing physical strength from the voice. *Proc. R. Soc. B*, **277**, 3509–3518.

Sellers, W.D. (1969) A global climate model based on the energy balance of the earth–atmosphere system. *J. Appl. Meteorol.*, **8**, 392–400.

Seo, H. and Lee, D. (2009) Persistent feedback. *Nature*, **461**, 50–51.

Semaw, S., Rogers, M.J., Quade, J., Renne, P.R., Butler, R.F., Domínguez-Rodrigo, M., Stout, D., Hart, W.S., Pickering, T., and Simpson, S.W. (2003) 2.6-Million-year-old stone tools and associated bones from OGS-6 and OGS-7, Gona, Afar, Ethiopia. *J. Hum. Evol.*, **45**, 169–177.

Sempf, M., Dethloff, K., Handorf, D., and Kurgansky, M.V. (2007) Circulation regimes due to attractor merging in atmospheric models. *J. Atmos. Sci.*, **64**, 2029–2044.

Šesták, J., Hubík, P., and Mareš, J.J. (2010) Thermal analysis scheme aimed at better understanding of the Earth's climate changes due to the alternating irradiation. *J. Therm. Anal. Calorim.*, **101**, 567–575.

Shan, J., Munro, T.P., Barbarese, E., Carson, J.H., and Smith, R. (2003) A molecular mechanism for mRNA trafficking in neuronal dendrites. *J. Neurosci.*, **23**, 8859–8866.

Shannon, C.E. (1948) A mathematical theory of communication. *AT&T Tech. J.*, **27**, 379–423. 623–656.

Shannon, C.E. (1951) Prediction and entropy of printed English. *AT&T Tech. J.*, **30**, 50–64.

Shannon, L.V., Boyd, A.J., Brundrit, G.B., and Taunton-Clark, J. (1986) On the existence of an El Niño-type phenomenon in the Benguela system. *J. Mar. Res.*, **44**, 495–520.

Shapiro, N.M.S. and Ritzwoller, M.H. (2004) Inferring surface heat flux distributions guided by a global seismic model: particular application to Antarctica. *Earth Planet. Sci. Lett.*, **232**, 213–224.

Sharpton, V.L. and Head, J.W. III, (1985) Analysis of regional slope characteristics on Venus and Earth. *J. Geophys. Res.*, **90**, 3733–3740.

Sherwood, O.A., Lehmann, M.F., Schubert, C.J., Scott, D.B., and McCarthy, M.D. (2011) Nutrient regime shift in the western North Atlantic indicated by compound-specific $\delta^{15}N$ of deep-sea gorgonian corals. *Proc. Natl. Acad. Sci. U.S.A.*, **108**, 1011–1015.

Shors, T.J., Miesegaes, G., Beylin, A., Zhao, M., Rydel, T., and Gould, E. (2001) Neurogenesis in the adult is involved in the formation of trace memories. *Nature*, **410**, 372–376.

Siegert, M.J., Ellis-Evans, J.C., Tranter, M., Mayer, C., Petits, J.-R., Salamantin, A., and Priscu, J.C. (2001) Physical, chemical and biological processes in Lake Vostok and other Antarctic subglacial lakes. *Nature*, **414**, 603–609.

Simon, H. (1962) The architecture of complexity. *P. Am. Philos. Soc.*, **106**, 467.

Simon, L., Bousquet, J., Lévesque, R.C., and Lalonde, M. (1993) Origin and diversification of endomycorrhizal fungi and coincidence with vascular land plants. *Nature*, **363**, 67–69.

Simoneit, B.R. (1993) Aqueous high-temperature and high-pressure organic geochemistry of hydrothermal vent systems. *Geochim. Cosmochim. Acta*, **57**, 3231–3243.

Simoneit, B.R. (1995) Evidence for organic synthesis in high temperature aqueous media-facts and prognosis. *Origins Life Evol. B.*, **25**, 119–140.

Singer, S.J., Kuo, J.-L., Hirsch, T.K., Knight, C., Ojamäe, L., and Klein, M.L. (2005) Hydrogen-bond topology and the ice VII/VIII and Ih/XI proton ordering phase transitions. *Phys. Rev. Lett.*, **94**, 135701 [4 pages]. doi: 10.1103/PhysRevLett.94.135701

Sinnott, E.W. (1946) Substance or system: the riddle of morphogenesis. *Am. Nat.*, **80**, 497–505.

Sirocko, F. (2010) Wetter, Klima, Menschheitsentwicklung von der Eiszeit bis ins 21. Jahrhundert. Wissenschaftliche Buchgesellschaft, Darmstadt.

Slimak, L., Svendsen, J.I., Mangerud, J., Plisson, H., Heggen, H.P., Brugère, A., and Pavlov, P.Y. (2011) Late Mousterian persistence near the Arctic Circle. *Science*, **332**, 841–845.

Smirnow, W.I. (1968) Lehrgang der Höheren Mathematik, Teil IV. Verlag der Wissenschaften, Berlin. Russian original (1953) Kurs vyschey matematiki, tom chetverty. Gosudarvstvennoye Izdatelstvo Tekhniko-Teoreticheskoy Literatury, Moscow.

Smith, F.A., Boyer, A.G., Brown, J.H., Costa, D.P., Dayan, T., Ernest, S.K.M., Evans, A.R., Fortelius, M., Gittleman, J.L., Hamilton, M.J., Harding, L.E., Lintulaakso, K., Lyons, S.K., McCain, C., Okie, J.G., Saarinen, J.J., Sibly, R.M., Stephens, P.R., Theodor, J., and Uhen, M.D. (2010) The evolution of maximum body size of terrestrial mammals. *Science*, **330**, 1216–1219.

Smith, J.M. (1978) *The Evolution of Sex*, Cambridge University Press, Cambridge.

Smith, R.B., Jordan, M., Steinberger, B., Puskas, C.M., Farrell, J., Waite, G.P., Husen, S., Chang, W.-L., and O'Connell, R. (2009) Geodynamics of the Yellowstone hotspot and mantle plume: Seismic and GPS imaging, kinematics, and mantle flow. *J. Volcanol. Geoth. Res.*, **188**, 26–56.

Smoot, G.F. (2006) Cosmic Microwave Background Radiation Anisotropies: Their Discovery and Utilization. Nobel Lecture 2006, Stockholm, http://nobelprize.org/nobel_prizes/physics/laureates/2006/smoot-lecture.html.

Soares, P.C. (2010) Warming power of CO_2 and H_2O: correlations with temperature changes. *Int. J. Geosci.*, **1**, 102–112.

Solé, R.V. and Montoya, J.M. (2001) Complexity and fragility in ecological networks. *Proc. R. Soc. Lond. B Biol. Sci.*, **268**, 2039–2045.

Solomatov, V.S. (2007) Magma oceans and primordial mantle differentiation, in *Treatise on Geophysics*, vol. **9** (ed. G. Schubert), Elsevier, The Netherlands, pp. 91–120.

Solomon F S., Rosenlof, K.H., Portmann, W.R., Daniel, J.S., Davis, S.M., Sanford, T.J., and Plattner, G.-K. (2010) Contributions of stratospheric water vapor to decadal changes in the rate of global warming. *Science*, **327**, 1219–1222.

Sonntag, D. (1990) Important new values of the physical constants of 1986, vapour pressure formulations based on the ITS-90, and psychrometer formulae. *Z. Meteorol.*, **40**, 34–344.

Sonntag, I., Feistel, R., and Ebeling, W. (1981) Random networks of catalytic biochemical reactions. *Biometrical J.*, **23**, 501–515.

Sonntag, I. (1984) Random application of the percolation theory to random networks of biochemical reactions. *Biometrical J.*, **26**, 799–807.

Specht, H.P., Nölleke, C., Reiserer, A., Uphoff, M., Figueroa, E., Ritter S., and Rempe, G. (2011) A single-atom quantum memory. *Nature*, **473**, 190–193.

Speijer, D. (2010) Constructive neutral evolution cannot explain current kinetoplastid panediting patterns. *Proc. Natl. Acad. Sci. U.S.A.*, **107**, E25.

Spencer, J.W. (1971) Fourier series representation of the position of the sun. *Search*, **2**, 172, http://www.mail-archive.com/sundial@uni-koeln.de/msg01050.html.

Spielhagen, R.F., Werner, K., Aagaard Sørensen, S., Zamelczyk, K., Kandiano, E., Budeus, G., Husum, K., Marchitto, T.M., and Hald, M. (2011) Enhanced modern heat transfer to the Arctic by warm Atlantic water. *Science*, **331**, 450–453.

Springer, M. (2009) *Machen uns die Quanten frei?*, Spektrum der Wissenschaft, Juli, p. 26.

Sprott, J.C. (1991) *Numerical Recipes: Routines and Examples in BASIC*, Cambridge University Press, Cambridge.

Sprott, J.C., Bolliger, J., and Mladenoff, D.J. (2002) Self-organized criticality in forest-landscape evolution. *Phys. Lett. A*, **297**, 267–271.

Stadler, P.F. (1996) Landscapes and their correlation functions. *J. Math. Chem.*, **20**, 1–45.

Stadler, G., Gurnis, M., Burstedde, C., Wilcox, L.C., Alisic, L., and Ghattas, O. (2010) The dynamics of plate tectonics and mantle flow:

from local to global scales. *Science*, **329**, 1033–1038.

Stanley, E.H. (1971) *Introduction to Phase Transitions and Critical Phenomena*, Clarendon Press, Oxford.

Steig, E.J., Schneider, D.P., Rutherford, S.D., Mann, M.E., Comiso, J.C., and Shindell, D.T. (2008) Warming of the Antarctic ice-sheet surface since the 1957 International Geophysical Year. *Nature*, **457**, 459–462.

Steinbuch, K. (1961) *Automat und Mensch*, Springer, Heidelberg.

Steinhilber, F., Beer, J., and Fröhlich, C. (2009) Total solar irradiance during the Holocene. *Geophys. Res. Lett.*, **36**, L19704. doi: 10.1029/2009GL040142

Stepanow, W.W. (1982) *Lehrbuch der Differentialgleichungen*. Verlag der Wissenschaften, Berlin. Russian original (1953) *Kurs differentsialnykh uravneniy*. Gosudarvstvennoye izdatelstvo tekhniko-teoreticheskoy literatury, Moscow.

Stephens, G.L. and O'Brien, D.M. (1993) Entropy and climate. I: ERBE observations of the entropy production of the earth. *Q. J. Roy. Meteor. Soc.*, **119**, 121–152.

Steuer, R., Ebeling, W., Russell, D.F., Bahar, S., Neiman, A., and Moss, F. (2001a) Entropy and local uncertainty of data from sensory neurons. *Physical Review E*, **64**, 061911.

Steuer, R., Molgedey, L., Ebeling, W., Jimenez-Montano, M.A. (2001b) Entropy and optimal partition for data analysis. *European Journal of Physics B*, **19**, 265–269.

Stine, A.R., Huybers, P., and Fung, I.Y. (2009) Changes in the phase of the annual cycle of surface temperature. *Nature*, **457**, 435–441.

Stoltzfus, A. (1999) On the possibility of constructive neutral evolution. *J. Mol. Evol.*, **49**, 169–181.

Stommel, H.M. (1961) Thermohaline convection with two stable regimes of flow. *Tellus*, **13**, 224–230.

Stott, R.A., Sutton, R.T., and Smith, D.M. (2008) Detection and attribution of Atlantic salinity changes. *Geophys. Res. Lett.*, **35**, L21702.

Stratonovich, R.L. (1961, 1963, 1967) *Selected Topics in the Theory of Random Noise*, Gordon & Breach, New York.

Stratonovich, R.L. (1965) On information value (in Russian). Izvestnik AN SSSR. *Seriya Tekhnicheskaya Kibernetika*, **5**, 3–12.

Stratonovich, R.L. (1975) *Teoriya Informatsii (Information Theory)*, Sovietskoye Radio, Moscow.

Stratonovich, R.L. (1985) On the problem of the valuability of information, in *Thermodynamics and Regulation of Biological Processes* (eds I. Lamprecht and A.I. Zotin), De Gruyter, Berlin.

Stratonovich, R.L. (1994) *Nonlinear Nonequilibrium Thermodynamics*, Springer, Berlin.

Stramma, L., Gregory C. Johnson, G.C., Sprintall, J., and Mohrholz, V. (2008) Expanding oxygen-minimum zones in the tropical oceans. *Science*, **320**, 655–658.

Strehlow, P. and Seidel, J. (2007) Definition der Temperatur und ihre Grenzen. *PTB-Mitteilungen*, **117**, 7–15.

Strobeck, C. (1973) N-species competition. *Ecology*, **54**, 650–654.

Strother, P.K., Battison, L., Brasier, M.D., and Wellman, C.H. (2011) Earth's earliest non-marine eukaryotes. *Nature*, published online, doi: 10.1038/nature09943

Stuessy, T. (2010) The rise of sunflowers. *Science*, **329**, 1605–1606.

Stugren, B. (1978) *Grundlagen der Allgemeinen Ökologie*, Gustav Fischer, Jena.

Sun, G., Dilcher, D.L., Wang, H., and Chen, Z. (2011) A eudicot from the Early Cretaceous of China. *Nature*, **471**, 625–628.

Sun, H., Feistel, R., Koch, M., and Markoe, A. (2008) New equations for density, entropy, heat capacity, and potential temperature of a saline thermal fluid. *Deep-Sea Res. I*, **55**, 1304–1310.

Surridge, A.K., Osorio, D., and Mundy, N.I. (2003) Evolution and selection of trichromatic vision in primates. *Trends Ecol. Evol.*, **18**, 198–205.

Svatkov, N.M. (1974) *Osnovy Planetarnoy Geograficheskoy Prognozy (Fundamentals of the Planetary Geographic Prognosis)*, Mysl, Moscow.

Svireshev, Yu.M. and Logofet, D.O. (1978) *Ustoichivost Biologicheskych Soobshchestv (Stability of Biological Communities)*, Nauka, Moskva.

Swinney, H.L. and Gollub, J.P. (1981) *Hydrodynamic Instabilities and the Transition to Turbulence*, Springer, New York.

Szilard, L. (1929) Über die Entropieverminderung in einem

thermodynamischen System bei Eingriffen intelligenter Wesen. *Z. Phys.*, **53**, 840–856.

Szostak, J.W. (2009) Systems chemistry on early earth. *Nature*, **459**, 171.

Szostak, J.W., Bartel, D.P., and Luisi., P.L. (2001) Synthesizing life. *Nature*, **409**, 387–390.

Tailleux, R. (2009) Understanding mixing efficiency in the oceans: do the nonlinearities of the equation of state for seawater matter? *Ocean Sci.*, **5**, 271–283, www.ocean-sci.net/5/271/2009/.

Takahashi, T., Koizumi, O., Hayakawa, E., Minobe, S., Suetsugu, R., Kobayakawa, Y., Bosch, T.C.G., David, C.N., and Fujisawa, T. (2009a) Further characterization of the PW peptide family that inhibits neuron differentiation in Hydra. *Dev. Genes Evol.*, **219**, 119–129.

Takahashi, T., Sutherland, S.C., Wanninkhof, R., Sweeney, C., Feely, R.A., Chipman, D.W., Hales, B., Friederich, G., Chavez, F., Sabine, C., Watson, A., Bakker, D.C.E., Schuster, U., Metzl, N., Yoshikawa-Inoue, H., Ishii, M., Midorikawa, T., Nojiri, Y., Körtzinger, A., Steinhoff, T., Hoppema, M., Olafsson, J., Arnarson, T.S., Tilbrook, B., Johannessen, T., Olsen, A., Bellerby, R., Wong, C.S., Delille, B., Bates, N.R., de Baar, H.J.W. (2009b) Climatological mean and decadal change in surface ocean pCO_2, and net sea–air CO_2 flux over the global oceans. *Deep Sea Res. Part II Top. Stud. Oceanogr.*, **56**, 554–577.

Tans, P. (2010) NOAA ESRL DATA ftp://ftp.cmdl.noaa.gov/ccg/co2/trends/co2_mm_mlo.txt.

Tateno, S., Hirose, K., Ohishi, Y., and Tatsumi, Y. (2010) The Structure of iron in earth's inner core. *Science*, **330**, 359–361.

Tattersall, I. (2010) The rise of modern humans. *Evo. Edu. Outreach*, **3**, 399–402.

Tauber, F. (2007) Seafloor exploration with sidescan sonar for geo-archaeological investigations. *Bericht der Römisch-Germanischen Kommission*, **88**, 67–79.

Tauber, F. (2011) Search for palaeo-landscapes in the south-western Baltic Sea with sidescan sonar. *Bericht der Römisch-Germanischen Kommission*, submitted.

Tauber, F. and Kührt, E. (1987) Thermal stress in cometary nuclei. *Icarus*, **69**, 83–90.

Tembrock, G. (1977) *Grundlagen des Tierverhaltens*, Akademie-Verlag, Berlin.

Teolis, B.D., Jones, G.H., Miles, P.F., Tokar, R.L., Magee, B.A., Waite, J.H., Roussos, E., Young, D.T., Crary, F.J., Coates, A.J., Johnson, R.E., Tseng, W.-L., and Baragiola, R.A. (2010) Cassini finds an oxygen–carbon dioxide atmosphere at Saturn's icy moon Rhea. *Science*, **330**, 1813–1815.

Teschemacher, A.G. and Johnson, C.D. (2009) Cotransmission in the autonomic nervous system. *Exp. Physiol.*, **94**, 18–19.

Thess, A. (2007) *Das Entropieprinzip – Thermodynamik für Unzufriedene*, Oldenbourg-Verlag, München, Wien.

Thinh, V.N., Hallam, C., Roos, C., and Hammerschmidt, K. (2011) Concordance between vocal and genetic diversity in crested gibbons. *BMC Evol. Biol.*, **11**, 36. doi: 10.1186/1471-2148-11-36

Thom, R. (1975) *Structural Stability and Morphogenesis*, Benjamin, Reading, MA.

Thoma, M., Smith, A.M., Grosfeld, K., and Woodward, J. (2010) The "tipping" temperature within Subglacial Lake Ellsworth, West Antarctica and its implications for lake access. *The Cryosphere Discussion* 5, 1003-1020, www.the-cryosphere-discuss.net/5/1003/2011/

Thomas, E.R., Wolff, E.W., Mulvaney, R., Steffensen, J.P., Johnsen, S.J., Arrowsmith, C., White, J.W.C., Vaughn, B., and Popp, T. (2007) The 8.2 kyr event from Greenland ice cores. *Quaternary Sci. Rev.*, **26**, 70–81.

Thompson, C.J. and McBride, J.L. (1974) On Eigen's theory of selforganisation of molecules and the evolution of biological macromolecules. *Math. Biosci.*, **21**, 127–142.

Thomson, S.N., Brandon, M.T., Tomkin, J.H., Reiners, P.W., Vásquez, C., and Wilson, N.J. (2010) Glaciation as a destructive and constructive control on mountain building. *Nature*, **467**, 313–317.

Thouless, D.J. (1974) Electrons in disordered systems and the theory of localization. *Phys. Rep.*, **13**, 93–142.

Tiesel, R. (2008a) Weather of the Baltic Sea, in *State and Evolution of the Baltic Sea, 1952–2005. A Detailed 50-Year Survey of Meteorology and Climate, Physics, Chemistry, Biology, and Marine Environment* (eds R. Feistel, G. Nausch, and N. Wasmund), John Wiley & Sons, Hoboken, NJ, pp. 65–91.

Tiesel, R. (2008b) Schneekatastrophen-Winter 1978/1979, http://www.tiesel.de/winter%201978-1979.html.

Tikhonov, A.N. (1952) Systems of differential equations containing small parameters in the derivatives (in Russian). *Matematichesky Sbornik*, **31**, 575–586.

Tillner-Roth, R. (1998) *Fundamental Equations of State*, Shaker Verlag, Aachen.

Timofeeff-Ressovsky, N.V., Voroncov, N.N., and Jablokov, A.N. (1975) *Kurzer Grundriss der Evolutionstheorie. Gustav Fischer, Jena. Russian Original (1969)*, Nauka, Moscow.

Tlusty, T. (2010) A colorful origin for the genetic code: information theory, statistical mechanics and the emergence of molecular codes. *Phys. Life Rev.*, **7**, 362–376.

Tobiska, W. and Nusinov, A. (2006) ISO 21348 – Process for determining solar irradiances. in 36th COSPAR Scientific Assembly, vol. 36 of COSPAR, Plenary Meeting, pp. 2621–2462.

Tokuda, I.T., Horáček, J., Švec, J.G., and Herzel, H. (2007) Comparison of biomechanical modeling of register transitions and voice instabilities with excised larynx experiments. *J. Acoust. Soc. Am.*, **122**, 519–531.

Tomasello, M. (2010) *Origins of Human Communication*, The MIT Press, Cambridge, London.

Torsvik, T.H., Burke, K., Steinberger, B., Webb, S.J., and Ashwal, L.D. (2010) Diamonds sampled by plumes from the core–mantle boundary. *Nature*, **466**, 352–355.

Touboul, M., Kleine, T., Bourdon, B., Palme, H., and Wieler, R. (2007) Late formation and prolonged differentiation of the Moon inferred from W isotopes in lunar metals. *Nature*, **450**, 1206–1209.

Toups, M.A., Kitchen, A., Light, J.E., and Reed, D.L. (2011) Origin of clothing lice indicates early clothing use by anatomically modern humans in Africa. *Mol. Biol. Evol.*, **28**, 29–32.

Tran, T.S., Rubio, M.E., Clem, R.L., Johnson, D., Case, L., Tessier-Lavigne, M., Huganir, R.L., Ginty, D.D., and Kolodkin, A.L. (2009) Secreted semaphorins control spine distribution and morphogenesis in the postnatal CNS. *Nature*, **462**, 1065–1069.

Trenberth, K.E. (2010) The ocean is warming, isn't it? *Nature*, **465**, 304.

Trenberth, K.E., Fasullo, J.T., and Kiehl, J. (2009) Earth's global energy budget. *B. Am. Meteorol. Soc.*, 311–323. doi: 10.1175/2008BAMS2634.1

Trenberth, K.E., Jones, P.D., Ambenje, P., Bojariu, R., Easterling, D., Klein Tank, A., Parker, D., Rahimzadeh, F., Renwick, J.A., Rusticucci, M., Soden, B., and Zhai, P. (2007) Observations: surface and atmospheric climate change, in *Climate Change 2007: The Physical Science Basis. Contribution of Working Group I to the Fourth Assessment Report of the Intergovernmental Panel on Climate Change* (eds S. Solomon, D. Qin, M. Manning, Z. Chen, M. Marquis, K.B. Averyt, M. Tignor, and H.L. Miller), Cambridge University Press, Cambridge, UK, New York, NY, USA, http://www.ipcc.ch/pdf/assessment-report/ar4/wg1/ar4-wg1-chapter3-supp-material.pdf.

Trenberth, K.E. and Solomon, A. (1994) The global heat balance: heat transports in the atmosphere and ocean. *Clim. Dynam.*, **10**, 107–134.

Tripati, A.K., Roberts, C.D., and Eagle, R.A. (2009) Coupling of CO_2 and ice sheet stability over major climate transitions of the last 20 million years. *Science*, **326**, 1394–1397.

Trudeau, L.E. and Gutiérrez, R. (2007) On cotransmitter & neurotransmitter phenotype plasticity. *Mol. Interv.*, **7**, 138–146.

Tsunogai, S., Kusakabe, M., Iizumi, H., Koike, I., and Hattori, A. (1979) Hydrographic features of the deep water of the Bering Sea – the sea of Silica. *Deep Sea Res. Part A*, **26**, 641–659.

Turing, A. (1937) On computable numbers, with an application to the entscheidungsproblem. *P. Lond. Math. Soc.*, **s2-42**, 230–265.

Turing, A.M. (1952) The chemical basis of morphogenesis. *Philos. Trans. R. Soc. Lond. B Biol. Sci.*, **s2-42**, 230–265.

Uhlenbeck, G.E. (1959) The Boltzmann Equation, in *Probability and Related Topics in Physical Sciences* (ed. M. Kac), Interscience Publishers, Ltd., London, pp. 183–203.

Uhlmann, A. (1977) Markov master equation and the behaviour of some entropylike quantities. *Rostocker Physikalische Manuskripte*, **2**, 45.

Urban, M.A., Nelson, D.M., Jiménez-Moreno, G., Châteauneuf, J.-J., Pearson, A., and Hu, F.S. (2010) Isotopic evidence of C_4 grasses in southwestern Europe during the Early

Oligocene–Middle Miocene. *Geology*, **38**, 1091–1094.

Urmes, D. (2004) *Handbuch der geographischen Namen: Ihre Herkunft, Entwicklung und Bedeutung*. Marix Verlag, Wiesbaden.

van Doorn, G.S., Edelaar, P., and Weissing, F.J. (2009) On the origin of species by natural and sexual selection. *Science*, **326**, 1704–1707.

Van Dover, C.L. (2000) *The Ecology of Deep-Sea Hydrothermal Vents*, Princeton University Press, Princeton.

Van Dokkum, P. (2011) Astrophysics: Era of compact disks. *Nature*, **473**, 160–161.

Van Kampen, N. (1981, 1992) *Stochastic Processes in Physics and Chemistry*, North Holland, Amsterdam.

Van Roy, P., Orr, P.J., Botting, J.P., Muir, L.A., Vinther, J., Lefebvre, B., el Hariri, K., and Briggs, D.E.G. (2010) Ordovician faunas of Burgess Shale type. *Nature*, **465**, 215–218.

Van Soest, P.J. (1994) *Nutritional Ecology of the Ruminant*, Cornell University Press, Ithaca.

Van Valen, L. (1973) A new evolutionary law. *Evolutionary Theory*, **1**, 1–30.

van Ypersele, J.P. (1993) Sea-ice interactions in polar regions, in *Energy and Water Cycles in the Climate System* (eds E. Raschke and D. Jacob), Springer, Berlin, pp. 295–322.

Varga, R.S. (1962) *Matrix Iterative Analysis*, Prentice-Hall, Englewood Cliffs, NJ.

Vasilenko, Yu.K. (1978) *Biologicheskaya Khimiya (Biochemistry)*, Vysshaya Shkola, Moscow.

Veevers, J.J. (2000) *Billion-year Earth History of Australia and Neighbours in Gondwanaland*, Gemoc Press, Sydney.

Vekua, A., Lordkipanidze, D., Rightmire, G.P., Agusti, J., Ferring, R., Maisuradze, G., Mouskhelishvili, A., Nioradze, M., Ponce de Leon, M., Tappen, M., Tvalchrelidze, M., and Zollikofer, C. (2002) A new skull of early *Homo* from Dmanisi, Georgia. *Science*, **297**, 85–89.

Venditti, C., Meade, A., and Pagel, M. (2010) Phylogenies reveal new interpretation of speciation and the Red Queen. *Nature*, **463**, 349–352.

Vogelstein, B., Lane, D., and Levine, A.J. (2000) Surfing the p53 network. *Nature*, **408**, 307–310.

Voigt, H.-M., Ebeling, W., Rechenberg, I., and Schwefel, H.-P. (1996) Parallel problem solving from nature – PPSN IV. Proceedings of the 4th International Conference on Parallel Problem Solving from Nature, Berlin, Germany, September 1996. Springer, Berlin, Heidelberg, New York.

Volkenstein, M.V. (1978) *Obshchaya Biofizika (General Biophysics)*, Nauka, Moscow.

Volkenstein, M.V. (1979) Mutations and the value of information. *J. Theor. Biol.*, **80**, 155–165.

Volkenstein, M.V. (1981) *Biofizika (Biophysics)*, Nauka, Moscow, English translation (1983) Nauka, Moscow.

Volkenstein, M.V. (1984) The essence of biological evolution. *Soviet Physics Uspekhi*, **27**, 515. Uspekhi Fizicheskoy Nauk 143, 429.

Volkenstein, M.V. (1986) *Entropiya i informatsiya (Entropy and information)*, Nauka, Moscow, English Translation (2009) Birkhäuser, Basel.

Volkenstein, M.V. (1994) *Physical Approaches to Biological Evolution*, Springer, Berlin.

Volkenstein, M.V. and Chernavsky, D.S. (1979) Fizicheskiye aspekty primeneniya teorii informatsii v biologii (Physical aspects of the application of information theory in biology). *Izvestiya AN SSSR Seriya Biologii*, **4**, 531.

Volterra, V. (1931) *Le$_¿$8cons sur la theorie mathématique de la lutte pour la vie*, Gauthier-Villars et Cie, Paris, (Russian translation by Nauka, Moskva 1976).

Von Neumann, J. (1966) *Theory of Self-Reproducing Automata*, University of Illinois Press, Urbana.

Wächtershäuser, G. (1990) Evolution of the first metabolic cycles. *Proc. Natl. Acad. Sci. U.S.A.*, **87**, 200–204.

Wadhams, P., Holfort, J., Hansen, E., and Wilkinson, J.P. (2002) A deep convective chimney in the winter Greenland Sea. *Geophys. Res. Lett.*, **29**, 1434. doi: 10.1029/2001GL014306

Wagner, W. and Kretzschmar, H.-J. (2008) *International Steam Tables*, 2nd edn, Springer Verlag, Berlin, Heidelberg.

Wagner, W. and Pruß, A. (2002) The IAPWS formulation 1995 for the thermodynamic properties of ordinary water substance for general and scientific use. *J. Phys. Chem. Ref. Data*, **31**, 387–535.

von Waldeyer-Hartz, H.W.G. (1891) Über einige neuere Forschungen im Gebiete der Anatomie des Centralnervensystems.

Deutsche Medicinische Wochenschrift, Berlin, **17**, 1213–1218, 1244–1246, 1287–1289, 1331–1332, 1350–1356.

Ward, P. (2009) *The Medea Hypothesis: Is Life on Earth Ultimately Self-Destructive?* Princeton University Press, Princeton.

Wardlaw, C.W. (1953) A commentary on Turing's diffusion–reaction theory of morphogenesis. *New Phytol.*, **52**, 40–47.

Wasmund, N. and Siegel, H. (2008) Phytoplankton, in *State and Evolution of the Baltic Sea 1952–2005* (eds R. Feistel, G. Nausch, and N. Wasmund), John Wiley & Sons, Inc., Hoboken, NJ, pp. 441–481.

Watts, D.J. and Strogatz, S.H. (1998) Collective dynamics of 'small world' networks. *Nature (London)*, **393**, 440–442.

Weare, B.C., Strub, P.T., and Samuel, M.D. (1981) Annual mean surface heat fluxes in the tropical Pacific Ocean. *J. Phys. Oceanogr.*, **2**, 705–717.

Weber, B.H., Depew, D.J., and Smith, J.D. (1988) *Entropy, Information, and Evolution*, MIT Press, Cambridge.

Weber, R.C., Lin, P.-Y., Garnero, E.J., Williams, Q., and Lognonne, P. (2011) Seismic detection of the lunar core. *Science*, **331**, 309–312.

Wegener, A. (1915) *Die Entstehung der Kontinente und Ozeane*, Vieweg, Braunschweig.

Weidlich, W. (2000) *Sociodynamics: A Systematic Approach to Mathematical Modelling in the Social Sciences*, Harwood Academic Publishers, Amsterdam.

Weinberg, W. (1908) Über den Nachweis der Vererbung beim Menschen. *Jahreshefte des Vereins für vaterländische Naturkunde in Württemberg*, **64**, 368–382.

Weinberger, E.D. (1991) Spatial stability analysis of Eigen's quasispecies model and the less than five membered hypercycle under global population regulation. *B. Math. Biol.*, **53**, 623–638.

Weiss, W. (1996) The balance of entropy on earth. *Continuum. Mech. Therm.*, **8**, 37–51.

Weissman, P.R. (1980) Physical loss of long-period comets. *Astron. Astrophys.*, **85**, 191–196.

Wellnhofer, P. (2008) *Archaeopteryx: Der Urvogel von Solnhofen*, Friedrich Pfeil, München.

Wells, A. and Goff, J. (2007) Coastal dunes in Westland, New Zealand, provide a record of paleoseismic activity on the Alpine fault. *Geology*, **35**, 731–734.

Wells, N.C. and King-Hele, S. (1990) Parametrization of tropical ocean heat flux. *Q. J. Roy. Meteor. Soc.*, **116**, 1213–1224.

Wells, S. (2002) *The Journey of Man: A Genetic Odyssey*. Allen Lane/The Penguin Press, London.

Wermke, K., Leising, D., and Stellzig-Eisenhauer, A. (2007) Relation of melody complexity in infants' cries to language outcome in the second year of life: A longitudinal study. *Clin. Linguist. Phonet.*, **21**, 961–973.

Westerhold, T., Bickert, T., and Röhl, U. (2005) Middle to late Miocene oxygen isotope stratigraphy of ODP site 1085 (SE Atlantic) new constrains on Miocene climate variability and sea-level fluctuations. *Palaeogeogr. Palaeocl.*, **217**, 205–222.

Westman, P., Wastegård, S., Schoning, K., Gustafsson, B., and Omstedt, A. (1999) Salinity change in the Baltic Sea during the last 8,500 years: evidence, causes and models. Technical Report TR-99-38. Svensk Kärnbränslehantering AB, Stockholm, Sweden.

Whipple, F.L. and Huebner, W.F. (1976) Physical processes in comets. *Annu. Rev. Astron. Astr.*, **14**, 143–172.

White, D.H. (1980) A theory for the origin of a self-replicating chemical system. I: natural selection of the autogen from short, random oligomers. *J. Mol. Evol.*, **16**, 121–147.

White, T.D., Asfaw, B., DeGusta, D., Gilbert, H., Richards, G.D., Suwa, G., and Clark Howell, C. (2003b) Pleistocene Homo sapiens from Middle Awash, Ethiopia. *Nature*, **423**, 742–747.

White, D.R., Dransfield, T.D., Strouse, G.F., Tew, W.L., Rusby, R.L., and Gray, J. (2003a) Effects of heavy hydrogen and oxygen on the triple-point temperature of water. American Institute of Physics; CP684, Temperature: Its Measurement and Control in Science and Industry, 7, pp. 221–226.

Whittaker, J.M., Müller, R.D., Leitchenkov, G., Stagg, H., Sdrolias, M., Gaina, C., and Goncharov, A. (2007) Major Australian-Antarctic plate reorganization at

Hawaiian-Emperor bend time. *Science*, **318**, 83–86.

Wielgosz, R. and Calpini, B. (2010) Report on the WMO-BIPM workshop on measurement challenges for global observation systems for climate change monitoring: Traceability, stability and uncertainty: 30 March to 1 April 2010, World Meteorological Organisation, Geneva, IOM-Report 105, WMO/TD 1557, Rapport BIPM-2010/08, http://www.bipm.org/utils/common/pdf/rapportBIPM/2010/08.pdf.

Wielicki, B.A., Wong, T., Loeb, N., Minnis, P., Priestley, K., and Kandel, R. (2005) Changes in Earth's albedo measured by satellite. *Science*, **308**, 825.

Wiener, N. (1952) *Mensch und Menschmaschine. Kybernetik und Gesellschaft*, Alfred Metzner, Frankfurt am Main.

Wignall, P.B. and Twitchett., R.J. (1996) Oceanic anoxia and the End Permian mass extinction. *Science*, **272**, 1155–1158.

Williams, G.D. and Bindoff, N.L. (2003) Wintertime oceanography of the Adélie Depression. *Deep Sea Res. Part II Top. Stud. Oceanogr.*, **50**, 1373–1392.

Williams, W.J., Carmack, E.C., and Ingram, R.G. (2007) *Physical Oceanography of Polynyas, Elsevier Oceanography Series*, vol. **74** SPRINGER, pp. 55–85.

Wiliams, R.W. and Herrup, K. (1988) The control of neuron number. *Annu. Rev. Neurosci.*, **11**, 423–453.

Willis, J.K. (2010) Can in situ floats and satellite altimeters detect long-term changes in Atlantic Ocean overturning? *Geophys. Res. Lett.*, **37**, L06602. doi: 10.1029/2010GL042372

Wilson, E.O. and Bossert, W.H. (1973) *Einführung in die Populationsbiologie*, Springer, Berlin, Heidelberg, New York, American original (1971) A Primer of Population Biology, Sinauer Associates, Stamford, CT.

WMO (2008) Guide to meteorological instruments and methods of observation. World Meteorological Organisation, Geneva, http://www.wmo.int/pages/prog/gcos/documents/gruanmanuals/CIMO/CIMO_Guide-7th_Edition-2008.pdf.

Woese, C. (1998) The universal ancestor. *Proc. Natl. Acad. Sci. U.S.A.*, **95**, 6854–6859.

Woodcock, A.H. (1965) Melt patterns in ice over shallow waters. *Limnol. Oceanogr.*, **10**, 290–297.

Wochner, A., Attwater, J., Coulson, A., and Holliger, P. (2011) Ribozyme-Catalyzed Transcription of an Active Ribozyme. *Science* **332**, 209–212.

Wolff, E.W., Chappellaz, J., Blunier, T., Rasmussen, S.O., and Svensson, A. (2010) Millennial-scale variability during the last glacial: The ice core record. *Quaternary Science Reviews* 29, 2828–2838.

Worley, S.J., Woodruff, S.D., Reynolds, R.W., Lubker, S.J., and Lott, N. (2005) ICOADS release 2.1 data and products. *Int. J. Climatol.* **25**, 823–842.

Wright, A., Siegert, M.J. (2010) The identification and physiographical setting of Antarctic subglacial lakes: an update based on recent geophysical data. In: Siegert, M. J., Kennicutt, C., Bindschadler, B. (eds.) *Subglacial Antarctic Aquatic Environments*. AGU Monograph, Washington D.C.

Wright, S. (1931) Evolution in Mendelian populations. *Genetics*, **16**, 97–159.

Wright, S. (1932) The roles of mutation, imbreeding, crossbreeding and selection in evolution. in Proceedings of the Sixth International Congress of Genetics (ed. D.F. Jones) Vol. I, pp. 356–366, http://www.blackwellpublishing.com/ridley/classictexts/wright.asp.

Wright, S. (2007) Comparative analysis of the entropy of radiative heat transfer and heat conduction. *Int. J. Thermophys*, **10**, 27–35.

Wright, D.G., Feistel, R., Reissmann, J.H., Miyagawa, K., Jackett, D.R., Wagner, W., Overhoff, U., Guder, C., Feistel, A., and Marion, G.M. (2010) Numerical implementation and oceanographic application of the thermodynamic potentials of liquid water, water vapour, ice, seawater and humid air – Part 2: The library routines. *Ocean Sci.*, **6**, 695–718, http://www.ocean-sci.net/6/695/2010/.

Wright, S.E., Scott, D.S., Haddow, J.B., and Rosen, M.A. (2001) On the entropy of radiative heat transfer in engineering thermodynamics. *Int. J. Eng. Sci.*, **39**, 1691–1706.

Wulf, O.R. and Davis, L. Jr., (1952) On the efficiency of the engine driving the atmospheric circulation. *J. Meteorology*, **9**, 80–82.

Wynands, R. and Göbel, E.O. (2010) Die Zukunft von Kilogramm und Co. *Spektrum der Wissenschaft*. March 2010, pp. 34–41.

Xiong, Y., Chen, X., Chen, Z., Wang, X., Shi, S., Wang, X., Zhang, J., and He, X. (2010) RNA sequencing shows no dosage compensation of the active X-chromosome. *Nat. Genet.*, **42**, 1043–1047.

Xu, T., Yu, X., Perlik, A.J., Tobin, W.F., Zweig, J.A., Tennant, K., Jones, T., and Zuo, Y. (2009) Rapid formation and selective stabilization of synapses for enduring motor memories. *Nature*, **462**, 915–919.

Xu, X., Zheng, X., and You, H. (2010) Exceptional dinosaur fossils show ontogenetic development of early feathers. *Nature*, **464**, 1338–1341.

Yan, Y., Gan, Z., and Qi, Y. (2004) Entropy budget of the ocean system. *Geophys. Res. Lett.*, **31** (L14311), 1–4. doi: 10.1029/2004GL019921

Yano, J.-I., Fraedrich, K., and Blender, R. (2001) Tropical convective variability as 1/f noise. *J. Climate*, **14**, 3608–3616.

Yockey, H.P. (2005) *Information Theory, Evolution and the Origin of Life*, Cambridge University Press, Cambridge.

Yoon, H.S., Hackett, J.D., Ciniglia, C., Pinto, G., and Bhattacharya, D. (2004) A molecular timeline for the origin of photosynthetic eukaryotes. *Mol. Biol. Evol.*, **21**, 809–818.

Yoon, H.S., Müller, K.M., Sheath, R.G., Ott, F.D., and Bhattacharya, D. (2006) Defining the major lineages of red algae (rhodophyta). *J. Phycol.*, **42**, 482–492.

Zachos, J., Pagani, M., Sloan, L., Thomas, E., and Billups, K. (2001) Trends, rhythms, and aberrations in global climate 65 Ma to present. *Science*, **292**, 686–693.

Zahn, M. and Storch, H.v. (2010) Decreased frequency of North Atlantic polar lows associated with future climate warming. *Nature*, **467**, 309–312.

Zeldovich, Ya.B. (1983) Disturbances leading to the disconnected domains. *Sov. Phys. Doklady*, **280**, 490.

Zhang, F., Zhou, Z., Xu, X., Wang, X., and Sullivan, C. (2008) A bizarre jurassic maniraptoran from China with elongate ribbon-like feathers. *Nature*, **455**, 1105–1108.

Zhao, M. and Running, S.W. (2010) Drought-induced reduction in global terrestrial net primary production from 2000 through 2009. *Science*, **329**, 940–943.

Zhisheng, A., Kutzbach, J.E., Prell, W.L., and Porter, S.C. (2001) Evolution of Asian monsoons and phased uplift of the Himalaya–Tibetan plateau since Late Miocene times. *Nature*, **411**, 62–66.

Zhong, W. and Haigh, J.D. (1995) Improved broadband emissivity parameterization for water vapor cooling rate calculations. *J. Atmospheric Sci.*, **52**, 124–138.

Ziman, J. (2000) *Technological Innovation as an Evolutionary Process*, Cambridge University Press, Cambridge, New York.

Zimmer, C. (2009a) On the origin of sexual reproduction. *Science*, **324**, 1254–1256.

Zimmer, C. (2009b) On the origin of eukaryotes. *Science*, **325**, 666–668.

Zurek, W.H. (1990) *Complexity, Entropy, and the Physics of Information*, Addison Wesley, Reading, MA.

Index

a

absorber state 223, 227, 228, 230, 232, 256, 261, 264, 299
absorption 93, 94, 110, 119, 120, 131, 227, 440, 443
abundance 7, 10, 11, 258, 269, 287, 288, 293, 303, 338–340, 343, 349, 351, 390, 418, 420, 427
abyss 138
accretion 151
active cliff 87
adenine 214
adenosine triphosphate (ATP) 33, 400
adiabatic 7, 8, 12, 36, 104, 105, 111, 130, 131, 136, 137, 167
adjacency matrix 187, 188, 268, 269, 271, 274–277, 292, 402
adult 300, 400, 405, 406
Afar 159
affector 399, 402–404
affinity 65, 66, 83, 251, 387
Africa 134, 143, 145, 147, 149, 158, 159, 405, 419
Agulhas Current 127
air dryer 119
air-sea flux 111
air-sea interface 111, 114
Akinetes 139
alanine (Ala) 389–391, 435
albedo 89, 96, 101, 110, 119, 123
allele 300–303
allele drift 302
allosteric enzyme 404
alphabet 317, 407–411, 425
Alps 143, 157, 158
aluminium 149
America 143, 159
Amharic 410

amino acid 28, 387, 389–392, 418, 431, 435
ammonium 439
AMOR 112, 133
amphibian 70
Andes 143, 146, 157
Andronov, Aleksandr Aleksandrovich 25, 163, 170, 173
Angola 127
anisotropic 66, 68, 75, 76, 91
annihilation 10
anoxic 138, 139, 147, 155, 371, 436, 437, 439, 441
anoxygenic 139, 437
Antarctic 102, 118, 120, 121, 125, 137, 140, 142, 143, 147, 149, 157–159, 185, 186, 452
Antarctic Bottom Water (AABW) 140
Antarctic Circumpolar Current (ACC) 102, 157–159, 389–391
Antarctic Peninsula 142, 149
anticyclone 125, 126, 135
antisymmetric 66, 166, 167, 292, 295
ants 377
anvil head 131
ape 375, 393, 407
Apex Chert formation 139, 154
Apollo 125
Arabia 158
Arabic 410, 411
archaea 139, 437, 438, 440
archaean 139, 438, 440
Archaeopteryx 147, 372
Archimedes 40, 130
Arctic 118, 120, 138
Arrhenius, Svante 85, 86, 91, 97, 103, 108, 109, 124, 418, 450
Asia 411
Asimov, Isaac 3

Physics of Self-Organization and Evolution, First Edition. Rainer Feistel and Werner Ebeling.
© 2011 Wiley-VCH Verlag GmbH & Co. KGaA. Published 2011 by Wiley-VCH Verlag GmbH & Co. KGaA.

asteroid 151, 418, 437
asymmetry 44, 166, 167, 315, 325, 359, 362, 377, 448
Atlantic 96, 112, 127, 138, 143, 157, 158
Atlantis Massif 419
atmosphere 3, 13, 53, 54, 85, 86, 88–93, 103–112, 117–121, 123, 124, 130–132, 134, 136, 137, 139, 140, 144, 149, 152, 153, 437–439, 441, 450–452
atmospheric pressure 38, 133
atoll 140
attack 201, 206
attractor 32, 54, 77, 86–88, 123, 134, 141, 155, 169, 170, 172, 173, 176, 214, 216, 217, 220, 230, 233, 234, 243, 254, 264, 305, 335, 347, 348, 351–354, 356, 358–361, 382, 394, 395, 401, 421, 444, 446–448, 450
Australia 122, 143, 149, 159
autocatalytic 213–214, 216–218, 238, 246, 253, 266, 267, 271–273, 275, 287, 289, 304, 306, 307
automata 3, 4
autotrophic 154, 155, 291, 292, 296, 416, 440, 451
avalanche 87, 210, 420
Avatar 385
Avogadro, Amedeo 39, 378
Avogadro constant 39
Avogadro number 39
Awash 159
axon 400, 401
Azores High 126

b

babbling 366, 405, 406
baby 366, 367, 404–406
background radiation 9, 94, 177
bacteria 139, 387, 416, 439, 441
bacterium 188, 201, 213, 315
balance equation 47, 63, 64, 106, 111, 151, 179, 197, 310, 402
Baltic Sea 87, 91, 132, 134, 136, 138, 145, 159, 161, 297, 371, 438–440
Bangiomorpha pubescens 439
Barkhausen, Heinrich 25, 163
baroclinic instability 86
barotropic 105
basalt 139, 141, 142, 153
beauty 392, 393
beech tree 375
Belousov-Zhabotinsky reaction 23, 24
Benard cell 23, 68, 125, 140
Benguela 126–129, 157
Benguela Niño 126, 127, 129

Benguela upwelling 127, 129, 157
Berlin 2, 3, 18, 112, 372
beryllium 90
Bible 160
bicycle 206
bifurcation 68, 69, 130, 145, 314, 320, 381, 394, 396, 397, 405
– diagram 314, 381, 395, 405
Big Bang 8
bilinearity 47, 79
bimodality 86, 140, 141, 143, 145, 153, 420
binary 43, 151, 165, 166, 168, 188, 225, 291, 292, 315, 343, 344, 357, 359, 361, 385, 390, 400–402, 411, 424, 433, 435, 448
biological species 200, 223, 305, 329, 448
biomass 288, 289, 291, 305, 439
bipedality 159, 405
bird 147, 160, 300, 375, 376, 392, 404, 406
– songs 406
birth-and-death process 195, 199, 202, 206, 221, 222
bistability 138, 213, 214, 217–220, 223, 231, 232, 242, 289, 377
black-body 7, 11, 12, 75, 76, 85, 91–94
– radiation 42, 71, 103
black hole 53
Black Sea 138, 159, 161, 406
block matrix 270, 283
blossoms 147
Bogolyubov, Nikolay Nikolaevich 173
boiling temperature 37
Boltenhagen 440
Boltzmann constant 21, 42, 76
Boltzmann equation 302
Boltzmann, Ludwig 1, 6, 21, 32, 42, 51, 55, 60, 61, 71, 76, 77, 163, 234, 238, 302, 378, 379, 418, 447
Boltzmann-Planck entropy 234, 238
Boltzmann-Planck principle 77
Boole, George 167, 213, 263, 316, 343, 344, 351
boring billion 155, 160, 439, 441
Bose gas 201
bottleneck 111, 159
boundary layer 114, 119
brain 239, 367, 401
Braun-Le Chatelier principle 118
British Columbia 159
Buck Reef Chert 139
Budyko, Mikhail Ivanovich 88, 89, 91, 106
buoyancy 130, 131, 137, 141, 142
Burgess Shale 147
butterfly effect 86

C

calcium 149
calibration 43
California 138
Cambrian 147, 155, 156, 160, 440, 441
camera 384–386
Canada 439
cancer 239, 397
Cancun 449, 451
cannibalism 393
canonical-dissipative 172, 173, 295
canonical equation 169, 443
Cape Horn 149
Capek, Karel 3
Capetown 102
Carathéodory, Constantin 54, 67, 167, 342
carbon 85, 90, 121, 138, 153, 390, 437, 452
carbon dioxide 85, 86, 106, 119–124, 138, 153, 155, 157, 418, 450–452
caricature 384, 392, 406, 408, 431
carnivores 292
Carnot cycle 90
Carnot, Sadi 12, 18, 43, 90, 148
carrier invariance 369
Carroll, Lewis 313
Cassini-Huygens space probe 155
catalyst 378, 388, 425, 430, 434, 435
catalytic 29, 154, 198, 201, 203, 205, 263, 266–271, 274, 275, 277, 278, 287, 307, 386–388, 391, 392, 417, 420–425, 427–437, 440, 442
catalytic cascades 420, 421
catastrophe theory 394
causality 16, 26, 91, 117, 121, 122, 127, 164, 165, 167, 186, 368, 383, 408, 432
Cenozoic 157, 160
central dogma 425
CERN 7, 9, 453
chalk cliff 87
Chapman-Kolmogorov equation 181, 182
charcoal 147
chemical picture 271
chemical potential 38, 46, 53, 68, 111, 113–116, 291
chemical reactions 65, 187, 221, 246, 250, 255, 304, 306, 307, 342, 422, 424
Chicxulub impact 147, 149, 160
Chile 145, 146, 149
Chiloe Island 146
chimney 135, 137, 138, 149
Chinese 410
Chlamydomonas reinhardtii 394
chromodynamic 10
chromosome 233

circulation 119, 120, 125, 128, 134–136, 138, 143, 145, 158
circulation model 120
cladogenesis 335
Clausius-Clapeyron equation 117
Clausius, Rudolph 1, 6, 18–20, 32, 35, 47, 52, 54, 60, 63, 117, 313, 314, 379
cliff
– active 87
– chalk 87
climate 13, 71, 85, 86, 88, 90–92, 95, 103, 106, 107, 109–111, 119, 120, 122–125, 136, 138, 155–158, 184, 186, 405, 440, 441, 446, 449–451
– attractor 86, 124, 450
climatology 39, 91, 102, 103, 105, 110, 113, 114, 122, 134, 249
clock 381, 384
clouds 86, 103, 119, 120, 122, 123, 131, 150, 152, 153, 269, 328, 374, 376, 443, 446, 450
cluster 8, 11, 12, 53, 189, 201, 205, 206, 270–274, 276–280, 307, 362, 395, 396, 403
coacervate 29, 303, 305–307, 367, 417, 420–424, 427, 429–432, 434, 435
coagulation 151
COBE 9
cod 297
code 29, 30, 116, 334, 366–369, 373, 382–392, 399, 403, 406, 409, 411, 412, 416, 425, 429, 430, 432, 434–436
coding invariance 29, 369, 382–384, 387, 399, 403, 406, 409, 411, 412, 425, 430
coexistence 227, 232, 266, 271, 282–284, 286, 296
collective properties 241
colony 394, 395
Colorado river 123
color receptors 375
colour vision 147
Columbus, Christopher 411
combinatorial explosion 30
comet 150–152, 418, 437
commodity 16, 308, 309, 318, 408, 411, 412, 448
communication 367–369, 380, 388, 393, 396, 399, 404–406, 410
community matrix 253, 291, 292, 294, 295, 336, 337, 339, 341
comparability 39, 308
complexity 1, 4, 14, 15, 30, 31, 111, 167, 168, 209, 211, 306, 308, 315, 347, 369, 373, 376, 384, 390, 392, 401, 404, 410, 436, 447, 450
computer code 334, 365, 410
concatenation 392, 432, 434, 435

concave   325, 328
Concepción   145, 148
connectivity   190, 192–194
Conrad, Michael   241, 321–323, 327, 333
conscious activity   313
conservation of energy   19, 61
conservation of mass   64
contact temperature   76, 94
continent   91, 110, 117–119, 125, 128, 136, 140–145, 148, 149, 152–155, 157, 158, 303, 420, 441, 450, 452
continental plate   125, 141, 142
control parameter   37, 170, 245, 381, 382, 394, 444
convection   23, 68, 72, 88, 125, 126, 130, 131, 137, 138, 140, 143–145, 148, 149, 152, 153, 156, 158, 314
convex   325
conveyor belt   138
cooling, radiative   105, 152
cooling rate   104–106
Copenhagen   449, 451
Copernicus, Nicolaus   25, 164, 376
copper   412
coral fishes   407
coral reef   140
Coriolis force   66, 125, 127, 135, 136
correlation   8, 49, 52, 117, 119, 121, 123, 129, 156, 187, 200, 205, 206, 211, 323, 325, 333–335, 366, 374–376, 385, 388, 416
– function   187
– length   323, 333–335
cosmic background   8, 12, 13, 88–90, 92–95, 101, 103, 106, 148
Coulomb friction   148
Cracow   5
crater   224
craton   141
Cretaceous   146–147, 151, 157
critical fluctuation   86, 344, 387, 448
critical point   58, 59, 87, 216, 388, 395
cross-catalytic   266, 267, 273, 275, 277, 280, 287, 300, 304, 307, 331, 345
crust   13, 42, 86, 88, 124, 140, 141, 143–145, 148, 149, 152–155, 420, 436
crying   366, 404, 405
Cryogenian   139, 155, 160, 440
crystal   51, 60, 106, 119, 150, 377–380
culture   4, 365, 386, 407, 408
cuneiform   408
Curie, Pierre   66, 371
Curie point   371
Curie-Prigogine principle   67
cyanobacteria   90, 139, 437–441

cybernetics   3, 4
cycle   19, 97, 98, 100, 101, 103, 105, 117, 123, 126, 127, 129, 188, 189, 193, 194, 269, 287, 292, 296, 402, 422–427, 429–433, 435, 451
cyclone   86, 125, 126, 135, 137
Cydonia   375
Cyrillic   410
cytosine   214

d

Dampfloch   138, 140
Dansgaard-Oeschger events   126
Daphnia   399
Darwin, Charles   1, 5, 29, 31, 140, 145, 148, 161, 167, 239, 240, 289, 296, 300, 308, 313, 314, 361, 362, 365, 367, 373, 377, 392, 393, 401, 420, 430, 435–437, 440, 442
Darwin principle   240
Debye ocean   374
deception   374
decision-making   376
deep convection   137, 138, 153
deepwater formation   137, 157
degradation   63, 90, 395
Delesseria sanguinea   440
dendrite   400, 401
denudation   141
dependent variable   48
desert   90, 110, 121, 127, 160, 451
desiccation   158
desoxyribonucleic acid (DNA)   28, 30, 214, 233, 342, 387, 392, 431, 433–435, 442
Desulfobacterales   439
deuterium   134, 452
devaluation of energy   61
Devonian   147
diagonal-dominant matrix   267, 269
diamond   141, 149
diffusion   16, 42, 53, 113, 165, 179, 183, 195, 221, 328, 332, 337, 394, 395
diffusion matrix   328
Dikika   159
dinosaurs   147, 151, 157, 160, 300, 349
diode effect   121, 137
Dirac delta   182, 326, 445
discontinuity   88, 124
disorder   21, 22, 32, 51, 60, 61, 83, 323, 331, 332, 415
dissipative structure   3, 23, 86, 88, 90, 95, 124, 125, 134, 313, 314, 380
diversification   29, 157, 383, 397, 399, 400, 403, 405, 408, 430, 432, 435, 438
divine beings   363, 372, 376
Dmasini   159

Dollar 412
dolphins 406
Doushantuo 147
Drake Passage 149, 157, 158
Drake plate 149
dry season 122
Duffing, Georg 25, 163
duplication 427, 434
dust devil 86, 126, 134
dyadic product 328
dynamical map 26, 164
Dyson sphere 91

e

earthquake 145, 146, 148, 407
East China Sea 419
eclipse 443, 444
ecological picture 273, 276, 279
ecology 4, 14–17, 31–33, 90, 165, 169, 195, 199, 221, 232, 234, 239, 241, 246, 261, 266, 271–273, 276, 279, 281, 282, 285, 291, 293, 295, 297, 339, 341, 344, 356, 440
economy 15, 16, 31, 32, 91, 165, 194–196, 199, 206, 212, 221, 239, 241, 290, 308, 310, 380, 407, 413, 448, 450–452
ectoderm 399
Ediacaran 147
effector 402–404
Ehrenfest, Paul and Tatjana 195, 196, 221, 224, 225
eigenfunction 182, 330, 332–334, 336
Eigen-Schuster concept 29
Eigen-Schuster theory 255
eigenvalue 130, 174, 175, 182, 247, 249, 254, 268, 270, 271, 273, 275, 277–279, 284, 307, 308, 330–334, 336, 337, 350, 422, 428
eigenvector 268–270, 275, 276, 278, 279, 307, 331, 428
Einstein, Albert 1, 6, 7, 46, 51, 52, 77, 78, 167, 183
electric power 49, 90
electroweak symmetry 9
El Niño 126, 129, 156, 158
emergent property 5, 28, 31, 39, 240, 241, 271, 309, 318, 320, 369, 370, 416, 446, 448
emergent value 4
emission 71, 76, 86, 92–94, 97, 112, 122
empirical orthogonal function (EOF) 156
endosymbiosis 393, 440
energy 19, 53, 214, 293
– conservation 19, 61
– devaluation of 61, 62, 81
– free 35, 57, 58, 61, 62, 81, 82, 148, 154, 292, 418, 419, 437

– kinetic 44, 53
– potential 148, 330
– shell 61, 77, 447
ensemble 62, 82, 178, 180, 241, 299, 334, 379, 382
enthalpy 62, 81, 94, 96, 100, 104, 113
– free 58, 59, 62, 81, 82, 96, 292
entropy
– export 23, 64, 89, 91, 104–106, 125, 148
– flux 21, 71–76
– lowering 32, 54, 56, 60, 61, 64, 313, 370
– production 23, 44, 53, 56, 57, 59, 60, 63–73, 75, 76, 79, 80, 82, 83, 85, 88–91, 94, 95, 115, 125, 148, 246, 251, 292, 313, 314, 437, 448, 451
– residual 51, 377, 379, 380
– specific 51, 53, 104, 105, 113, 378
enzyme 387, 431, 434, 436
Eocene 157
epicycles 164, 376
equation of state 37–39, 48, 96, 130
equator 125, 126, 131, 135, 158
equipartition theorem 53
equivalence relation 167, 189
ERBE satellite 91
erosion 131, 140, 157
error catastrophe 145
escape velocity 54, 132
Escherichia coli 201
ester 213
eternal truth 377
Ethiopia 157–159
ethyl ethanoate 213
ETOPO5 141, 144
Etruscan 410
Euclidian norm 346
Euclidian space 244, 346
eukaryote 139, 155, 367, 389, 438–440
Euler equation 135
euphotic 439
Europa 155
Europe 3, 138, 143, 147, 157, 159, 410, 411, 449, 451
Evans-Galavotti-Cohen theorem 82
evaporation 43, 54, 86, 110, 111, 113–115, 117–119, 134, 137, 450
evolution history 28, 155, 376, 384, 391, 408, 410, 412
evolution principle 68, 167, 245, 246, 248, 250, 253, 271, 279, 280, 284, 286, 290, 295–297, 306, 309, 314, 329, 341, 342, 364
evolution strategy 241
exchange value 31, 33, 63, 308, 310, 318, 412, 448

exergy   61
experience   1, 5, 40, 51, 85, 123, 130, 135, 372, 374, 376, 388, 393, 416, 435, 447
extensive variable   38, 40, 45, 46, 48, 50, 51, 56, 57, 60
external variable   37, 39
extinction   138, 147, 151, 157, 159, 197, 223, 224, 232, 241, 242, 256, 257, 259, 260, 266, 270, 295, 298, 341, 377
extremum principle   32, 95, 241–243, 245, 248, 249, 252, 260, 261, 313–315, 323, 341, 344, 361, 448
– complete   244
– dynamical   243, 244, 249, 251
– static   244, 253

f

face on Mars   375
face recognition   375
fashion   393
feathers   147, 300, 412
feeding   246, 404, 406, 437
Ferrel cell   125
fertility   335
fiat money   412
fire-and-wire rule   401
first law of thermodynamics   19, 20, 45, 53, 61, 64
Fisher-Eigen equation   250, 253, 303, 306, 307, 313, 315, 323, 329, 331, 332, 335–337, 339–341, 344, 364
Fisher, Ronald Aylmer   32, 163, 250, 253, 297, 303, 306, 307, 313, 315, 323, 325, 326, 328, 329, 331, 332, 335–337, 339–342, 344, 363, 364, 382, 430
fitness   5, 31, 211, 235, 240, 241, 297, 302, 303, 308, 315, 316, 324, 325, 328–332, 335–337, 341, 361, 362, 393, 430, 448
fitness landscape   31, 241, 329, 331, 332, 337, 430
flintstone   411
Floquet coefficient   175
Floquet exponent   175
Florida   159
fluctuation catastrophe   223, 256, 257
fluctuation-dissipation theorem   80
flux temperature   76, 94
Fokker-Planck equation   179, 183
food   30, 32, 33, 90, 154, 201, 205, 235–237, 249, 292, 295, 296, 345, 356, 396, 405, 407, 411, 437
food web   201, 205, 292, 295, 296, 345, 356
forest   145, 159, 161, 375, 451, 453
fossil   17, 120, 138–139, 146, 156, 206, 370–372, 386, 387, 437–439, 452
fossil fuel   120
Fourier, Joseph   100, 403
Fourier transform   403
Fowler, Ralph Howard   43
fox   236, 237
fractal dimension   54
fractionation   150
Fréchet derivative   156, 329
Fredholm integral equation   182
free will   376
freezing point   38, 49, 118, 137
freezing temperature   136, 137
friction   136, 143, 148, 168, 172
Friedman, Alexander Alexandrovich   1, 6, 7
Frobenius theorem   269
front   24, 76, 86, 123, 125, 149, 182, 234, 291, 429
frozen   8, 123, 370, 380, 384, 387, 436
– history   370
frozen-in   9
fruits   147, 160, 375
fungi   139, 147, 160, 440, 441

g

Gaia world   314
galaxy   6, 8, 53, 447
Galerkin, Boris Grigoryevich   170
Galilei, Galileo   25, 40, 167
Gamov, George   1, 6, 7, 214
Garching   453
gas, ideal   49, 54, 58, 104, 114, 116, 149, 150, 250, 297, 380
Gaspard theorem   83
gas thermometry   40, 42, 43
Gause principle   296
Gaussian distribution   78, 325, 326, 330, 338
Gauss theorem   179, 328
generating function   199, 224, 257, 262, 298
GENERIC   81
genetic distance   323
genome   233, 239
genotype   240, 302, 303
geo-engineering   92, 122, 450
Georgia   159
geostrophy   128, 135, 136
geothermal   124, 143, 144, 148, 153, 418, 437
German   4, 15, 57, 87, 158, 381, 412, 416, 453
Germany   140, 160, 375, 417
Gershgorin, Semyon Aranovich   350

giant impact   139, 153, 154, 420, 436
Giauque, William Francis   51, 378
Gibbs energy   58, 59, 62, 81, 82, 96, 291, 292
Gibbs function   59, 96, 104, 112, 113
Gibbs fundamental equation   48, 49, 52, 53, 56, 68
Gibbs, Josiah Willard   35, 37, 45, 47–49, 52, 53, 55–61, 68, 76, 77, 84, 96, 104, 112, 113, 235, 291, 292, 294, 299
Gibbs' phase rule   37
Gierer-Meinhardt model   394
Gilbert, Grove Karl   140
Gillespie algorithm   198
glaciation   121, 155, 157, 158, 371, 452
glacier   118, 125, 159, 371
Glansdorff-Prigogine criterion   67, 68, 251
global warming   85, 90, 92, 111, 118, 121, 123–125, 138, 185, 449, 450
glutathione   398
glycine, Gly   389–391, 435
Gobi desert   110, 122
gold   9, 411–413
Goldstone modes   380, 430
grammar   405, 408, 409
gramophone spirit   418
granite   141, 142, 155
graph   15–17, 163, 165, 168, 187–194, 200–204, 206, 209, 210, 245, 256, 268–271, 275–277, 282, 292, 357, 358, 360–362, 401, 402
gravitational escape   134, 150, 153
gravity   8, 11, 23, 25, 39, 53, 54, 104, 117, 124, 125, 130, 131, 134, 150–153, 371, 443
great oxidation event (GOE)   139, 154, 155, 160, 440
Greece   452
Greek   7, 17, 18, 36, 40, 47, 409, 410
green algae   139, 394
greenhouse   85, 86, 91, 92, 98, 103, 106–109, 111, 119, 120, 122, 123, 134, 145, 152, 153, 162, 185, 186, 449
greenhouse effect   85, 86, 91, 98, 106–109, 111, 120, 122, 123, 152, 153, 162, 185, 186, 450
Greenland   137, 138, 158, 159, 437
Greenland Sea   137, 138
Greifswald   453
ground state   51, 330, 332, 377, 380
group theory   167
guanine   214
Guilder   413
Gulden   413
Gulf of Aqaba   137
Gulf of Lions   137
Gulf Stream   106, 118, 136, 138, 158

h

habitable belt   420
Hadley cell   125
Haldane, John Burdon Sanderson   163, 240
half-life   124
Hamiltonian   77, 169, 171, 172, 174
Hamilton, William Rowan   25, 169, 213
Hamming distance   318, 361
Hardy-Weinberg law   302
hare   236, 356
harmonic oscillator   330, 444, 445
harmonic sum   257
Hausdorff dimension   176
Hawaii   149
heat bath   36, 43, 57, 58, 72, 91, 92
heat capacity   49, 71, 96–98, 103, 104, 110, 115, 116, 137
heat conduction   63, 71, 72, 124, 144
heat exchange   47, 49, 51, 71, 138
heat flux   125
heat, sensible   110, 114, 119
heat transfer   19, 21, 46, 72, 92, 119
Hegel, Georg Wilhelm Friedrich   1, 2
Heinrich events   126
helicity   54
helium   7, 11, 124, 452
Helmholtz energy   35, 57, 58, 61, 62, 81, 82, 148, 154, 292, 418, 419, 437
Helmholtz function   58, 112, 113
Helmholtz, Herrmann   1, 6, 18, 25, 32, 57, 58, 60–62, 112, 113
Hephaistos   7
herbivores   292, 296
Hermite polynomials   330
Hesiod   7
heterotrophic   291, 398
hierarchy   15, 180, 210, 211, 271, 272, 296
Higgs field   8
Higgs ocean   374
Hilbert-Schmidt expansion   336, 337
Himalayas   143, 157
historicity   86, 88, 151, 154, 156, 158, 318, 320, 448, 449
holistic   15, 39, 211, 446, 448
Hollywood   210
hologram   403
Homer   7
Homo erectus   159
Homo heidelbergensis   159
homologous substitution   327
homomorphous   168

Homo sapiens 159
honey bees 377
Hopf bifurcation 381, 382
Hopf, Eberhard 25, 163, 381, 382
hopping 203, 205, 209, 320, 331, 335, 336
horse 206
hot spot 149
Hubble, Edwin 6, 7
human 6, 28, 33, 43, 61, 85, 88, 90, 121, 123, 125, 156, 158–161, 184, 200, 233, 239, 300, 308, 312, 313, 326, 365, 367, 368, 372, 374–376, 381, 384–387, 393, 401, 404–410, 415, 435, 441, 443, 448–451
humanoids 404, 407, 411
human society 90, 125, 368, 385
Humboldt Current 129
humid air 43, 52, 111–114, 116, 130, 131, 133
humidity 59, 86, 92, 105, 111, 114, 117, 119, 120, 123, 131, 133, 242
– relative 109, 111, 114–117, 119, 123, 134, 152, 153
– specific 111, 113, 116, 117, 119, 123, 130, 131, 134
Hunting formation 439
hurricane 86, 126, 134, 136
Huxley, Julian 28, 366, 393
hydrogen 7, 11, 39, 41, 124, 132, 134, 138, 150, 155, 213, 271, 332, 377, 378, 439, 452
hydrogen sulfide
– H_2S 138, 155, 418, 439
hydrological cycle 118
hydrosphere 88, 124, 136, 140
hydrostatic 104, 105, 130, 136
hydrothermal vent 154, 418, 419, 421, 437
hyperbolic growth 219, 220, 230, 242
hypercycle 16, 29, 219, 255, 287, 300, 308, 367, 417
hypercyclic growth 219
hyperselection 232, 242, 266, 290, 300, 303, 315, 347, 420, 427
hypsography 86, 140, 153, 158, 420

i

ice 38, 43, 51, 90, 106, 112, 118, 119, 121, 126, 136–138, 140–142, 145, 150, 153, 155–159, 186, 377–380, 450–452
ice age 138, 159, 451, 452
iceberg 141, 142
ice core 90, 121, 186
ice cover 43, 90, 118, 136, 140, 156
ice I 377–379
ice Ih 377, 379
Icelandic Low 126
ice XI 378

ideal gas 49, 54, 58, 104, 114, 116, 149, 150, 250, 297, 380
Illinoian ice age 159
illusion 374
Immanuel Kant's yellow glasses 374
immigration 343
imprimitive matrix 270
inclination 99
independent variable 20, 37, 58, 59, 81, 346
India 129, 143, 157, 158, 411
Indian Ocean 129, 158
Indus river 120
infant hypothesis 406
inflation 8, 310
information processing 28, 29, 63, 234, 365, 366, 370, 384, 386, 392, 399, 432, 435
information, structural 42, 160, 367, 370–373, 377, 383–385, 389, 392, 393, 396, 410, 411, 415
information, symbolic 28, 214, 365, 367, 369–371, 373, 377, 380, 383–388, 392, 393, 397, 399, 403, 406, 411, 412, 415–417, 425, 430, 431, 435, 449
information value 31, 33, 235
information, value of 31, 33, 235, 237, 365, 448
infrared 86, 91, 92, 106, 107, 112, 119, 120, 122, 136
inheritance 359–362
inner core 85, 124, 143, 145, 148, 451
innovation 3, 195, 196, 199, 202, 203, 205–209, 211, 221, 311, 314, 318, 448
insects 147, 160, 407
insolation 98
Institute for Remote Sensing Applications 112
integrability 48, 251, 342
intensive variable 37, 38, 40, 42, 43, 53
interface 24, 27, 68, 69, 88, 94, 111–114, 125, 131, 137, 145, 148, 367, 374, 383, 385, 393, 395, 397
internal variable 39
International Association for the Properties of Water and Steam (IAPWS) 38, 52, 96, 112, 114, 117, 133
International Thermodynamic Equation of Seawater - 2010 (TEOS-10) 112, 116, 133, 138
interpretation, valoric 60, 63
interstellar gas 150
intertropical convergence zone (ITCZ) 125
invention 5, 206, 224, 314, 318, 367, 372, 407, 410, 448, 451
IOC 38, 54, 71, 96, 110–112, 115, 136–138

IPCC-report   184, 185, 449
IPTS-68, 41
irradiation   92, 94, 98, 102, 103, 108, 111, 123, 131
irreducible   4, 31, 189, 240, 268–271, 273, 274, 276, 278, 279, 307, 331, 344, 350, 355, 358, 362, 446
– matrix   268, 269, 273
– property   4, 31, 240, 446
irreflexive   167, 315, 359, 362, 401
irreversible process   1, 21–23, 35, 44, 47–49, 55, 59, 61, 63, 65, 67, 78, 81, 115
isentropic   47, 104–106, 130, 131
isolated   16, 19, 21, 32, 36, 45–47, 49, 53, 55–57, 59, 61, 77, 88, 137, 165, 194, 195, 221, 243, 354, 355, 370, 448
isomorphous   168
isotope ratio   41, 90, 372
isotopic fractionation   134
isotropic   66, 67, 71, 73, 93, 149
Ispra   112
Isthmus of Panama   158, 159
Isua formation   139, 437
Italy   112, 452
ITS-90   41

j

Jacobian   54, 174, 175, 247, 283, 350, 428
Jamaica   411
Janus, faces of   60
Japanese   145
Jasmund National Park   375
Jeans instability   54
jet stream   86, 125
Jordan river   123
Joule, James Prescott   18
Juan Fernandez Island   146, 148
jungle   90
Jupiter   155
Jurassic   143, 147, 385
Jurassic Park   385
jurisdiction   407

k

Kadet channel   161
Kalahari desert   127
Kant, Immanuel   1, 443
Keeling, Charles David   85, 120, 451
Kelvin, Lord   41, 42
Kepler, Johannes   25
kernel   151, 156, 182, 336, 337, 341, 395, 403
Khinchin, Aleksandr Yakovlevich   297
kilogram   39, 43
kimberlite   149

kinetic energy   44, 53
Klimontovich, Yuri Lvovich   2, 21, 22, 57, 60, 125, 370, 416
knowledge   7, 17, 28, 52, 59, 96, 143, 154, 165, 170, 171, 180, 181, 187, 235, 297, 372, 376, 443
Kohlrausch, Friedrich   195, 221
Kolmogorov, Andrey Nikolaevich   55, 60, 176, 180, 183, 213, 369
Kolmogorov entropy   55, 176
Kolmogorov-Solomonov complexity   369
Kotelyanski, D.M.   350
Kramers-Moyal expansion   183, 395
K-selection   281
Kullback-Leibler entropy   235, 314
kurtosis   325
Kyoto   449

l

Lake Aral   123
Lake Ellsworth   125
Lake Vostok   157
Landau-Ginzburg functional   329
Landau, Lev Davidovich   49, 53, 54, 71, 104, 132, 134, 149, 167, 250, 293, 329, 332, 336, 382, 418, 446
landscape   31, 159, 160, 241, 321–323, 325–337, 339–343, 362
– stochastic   322, 333
landslide   87
Langevin approximation   298
Langevin, Paul   178, 298, 445, 454
language   3, 4, 16, 18, 29, 30, 140, 159, 165, 240, 317, 335, 365, 367, 369, 373, 384, 386, 387, 393, 403, 404, 406, 408–411, 415, 430, 449
La Niña   129
lapse rate   104, 105, 111, 131, 134, 136, 137
Large Hadron Collider (LHC)   7, 9
latent heat   86, 107, 109–111, 113–116, 118, 131, 133, 136
Latin   36, 317, 408–410
learning matrix   377, 403
Legendre, Adrien-Marie   58, 59
Legendre transform   58, 59
Lem, Stanislaw   3, 5, 58, 363
leucine, Leu   389
light   8, 11, 31, 53, 91, 119, 120, 160, 211, 242, 335, 363, 366, 373, 374, 384, 385, 403, 437, 438, 440, 451, 452
limestone   372
limit cycle   171–173, 175, 176, 243, 347, 351, 381
linear irreversible thermodynamics   66, 79

Liouville equation   179, 181
lithosphere   88, 140, 148
Litorina transgression   161
Little Ice Age (LIA)   90, 371
localisation length   333
localised state   333, 334
logistic growth   280, 293
long-wave radiation   108, 110
Lorentz, Hendrik Antoon   167
Lorenz, Konrad   95, 155, 176, 366, 407
Loschmidt, Josef   39
Loschmidt number. *see* Avogadro constant
Lost City Hydrothermal Field   419
Lotka, Alfred   25, 163, 233, 234, 291, 293, 295–298, 323, 336, 339–342, 344, 381
Lotka-Volterra equation   233, 234, 291, 293, 295–298, 323, 336, 339–342, 344, 381
Lucy   159
lunar quakes   125
Lyapunov, Aleksandr Mikhailovich   25, 45, 55, 57–61, 84, 96, 124, 127, 163, 173, 175–177, 213, 243, 247, 263, 286, 294, 349, 380, 381
Lyapunov coefficient   96, 124, 127, 380, 381
Lyapunov function   45, 55, 57–61, 173, 243, 286, 294

m

macaques   375
Mach cone   329
magma   141–143, 148, 153, 155, 419
magma mill   148
magnesium   88
magnetic field   66, 90, 139, 144
magnetic pole   144
malonic acid   23
mammal   151, 157, 375, 405
Manicouagan impact   147
Mann, Thomas   213
mantle   124, 142–145, 148, 149, 153, 158
Marangoni effects   24
Marangoni instability   13, 24, 68, 69
margin of subsistence   281–283
marine ecosystem   155, 292
market   16, 165, 200, 202, 206, 208, 221, 308–311, 318, 385, 413, 435, 448
Markov, Andrey Andreyevich   90, 179–181, 183, 196, 310
Markov chain   180–183, 194, 195, 445
Markov process   180, 181, 183, 196, 310
Mars   86, 134, 153, 450
Marx, Karl   1, 31, 308, 411

master equation   156, 182, 183, 189, 197, 199, 206, 211, 222, 224, 226, 227, 230, 232, 256, 261, 262, 263, 297, 298, 302, 310, 316, 326
matrix
– adjacency   187, 188, 268, 269, 271, 274–277, 292, 402
– diagonal-dominant   267, 269
– imprimitive   270
– irreducible   268, 269, 273
– monotonous   267
– non-negative   254, 268, 277, 279, 292, 326
– reducible   270, 273, 275, 277, 350
Maunder minimum   90
Maxwell-Boltzmann distribution   132
Maxwell, James Clerk   132, 167, 418
Mayan writing   408
Mayer, Robert   6
measurement   36, 39–41, 43, 48, 49, 52, 56, 57, 62, 70, 82, 88, 89, 103, 112, 114, 125, 138, 164, 178, 186, 202, 206, 376
Mecklenburg-Vorpommern   158, 160
Mediterranean   137, 157–159, 384, 405, 408
melting   38, 43, 88, 118, 120, 125, 137, 143, 144, 152
melting curve   38
membrane   30, 303, 386, 421, 436
memory   30, 181, 214, 234, 376, 380, 385, 387, 401, 407, 445, 449, 450
Mendel matrix   301
meridional overturning circulation (MOC)   138
messenger RNA (mRNA)   392, 400, 431, 435
Messinian Salinity Crisis   158
metabolic   30, 201, 306, 308, 373, 381, 394, 396, 399
metabolism   201, 273, 305, 396, 399, 422, 427, 437
metagalaxis   6
metastable   42, 59, 150, 290, 418
metazoans   139, 147, 397, 441
meteorite   149, 152
meteorology   39, 59, 111, 122, 131, 140
metrology   39
Miami   159
Michaelis-Menton law   286
microfossils   139
microsphere   29, 303, 420, 421
mid-ocean ridge   144, 419
Minoans   410
Miocene   157
mobility edge   332
moisture   111, 117, 123, 125, 134
molecular evolution   3, 154, 253, 257, 308, 323, 387, 420, 425, 430, 431, 436

money 33, 235, 236, 310, 372, 386, 411–413
monomer 30, 214
monostability 218
monotonous matrix 267
Monsoon 120
Moon 88, 125, 134, 139, 153, 154, 158, 420, 436, 443, 444
morphing 398
morphogene 396
Moscow 3, 173, 416
mountain 6, 90, 125, 148, 321
multistability 213, 214, 234
music 300, 380
mutation 3, 145, 198, 201, 203, 205, 224, 267, 269, 300, 301, 303, 311, 314, 316–320, 326–329, 331–333, 335–337, 343, 349, 359, 360, 381, 387, 401, 427, 431, 434, 448
– catastrophe 318
– rate 145, 301, 317–319

n

Namibia 127, 128, 138
Neanderthals 159
negentropy 370
Nernst, Walther 51
network 15–17, 165, 178, 189, 190, 194, 200–207, 209–211, 266, 268, 270, 271, 275, 296, 387, 388, 391, 392, 400–403, 417, 422–425, 427, 440
neuron 30, 239, 386, 398–405, 436
neurotransmitter 399, 400
neutral evolution 316, 327
neutral mutation 327, 328
neutral selection 263
neutral stability 29, 130, 131, 224, 380, 382, 387, 392, 399, 429
neutral surface 54
neutrinos 10
Newton, Isaac 25, 26, 45, 130, 164, 167
New World 375
New Zealand 148
niche 232, 266, 281, 282, 339, 341, 356, 359–361, 393, 439, 440
nitrate 439
nitrogen 390, 439
Nodularia spumigena 438
noise
– thermal 42
– thermometry 42
non-negative matrix 254, 268, 277, 279, 292, 326
North America 147
North Atlantic 112, 126, 133, 137, 138
North Atlantic Oscillation (NAO) 126
Northern Territory 122
North Pacific 138
North Pacific Deep Water (NPDW) 138
nuclear fission 124
nuclear fusion 11, 12, 124, 453
nucleation 150, 152, 187, 287
nucleotide 214, 233, 389–391
nucleus 3, 8, 10, 11, 14, 30, 44, 150, 151, 240, 290, 405, 452
nutrients 129, 373, 394, 396, 439

o

ocean 6, 42, 86, 90, 97, 117, 119, 120, 123, 125, 131, 134, 139, 141, 143, 145, 155, 158, 419, 437, 450, 451, 453
ocean-atmosphere 108, 109, 111, 123, 129, 155
oceanography 39, 59, 111, 113, 137, 138, 446
Ohm, Georg Simon 65
Old World 375
olfactory 155
oligonucleotide 424
Olympus 7
onomatopoetic 406, 408
Onsager coefficient 67, 79, 80, 95
Onsager forces 95, 113, 162
Onsager, Lars 26, 66, 67, 79, 80, 95, 113, 162, 165, 314
Onsager relation 66
ontogenetic 400
Onverwacht formation 139
Oparin, Alexander Ivanovich 29, 303, 367, 415–417, 421–423
Orange River 157
order parameter 5, 16, 17, 37, 123, 187, 246, 314, 381, 382, 394, 444–447
order relation 165, 167, 240, 282, 315, 316, 341, 344, 361, 448
organism 13, 30, 33, 70, 139, 155, 201, 213, 214, 239, 272, 291, 292, 300, 301, 304, 313, 327, 342, 366, 367, 370, 373, 374, 376, 388, 393, 394, 397–400, 418, 420, 437, 439, 440, 447, 451
origin of life 5, 29, 30, 33, 88, 153, 154, 205, 214, 366, 367, 386, 388, 415–417, 420, 426, 440, 449
oscillator 60, 172, 330, 382, 445
Ostwald ripening 150
Ostwald, Wilhelm 32, 60, 150
overfishing 223
Ovid (Publius Ovidius Naso) 85
oxygen 39, 41, 129, 138, 139, 147, 153–155, 378, 390, 437, 439, 440, 451
– dissolved 439

– free 132, 154, 437–439, 451
oxygenic 439
ozone 131, 134

p

Pacific 129, 143–145, 158, 406
painting 375
Pakistan 120, 123
Paleocene 157
paleoclimatic 137
paleorecord 121
Panama, Isthmus of 158, 159
parasite 154, 283, 307, 398, 422, 423
Paris 39, 43
parrot 404
Pathfinder 102
pathogen 393
Pauli equation 182
Pauling, Linus Carl 51, 377, 379, 389
Pauli principle 4
penguins 157
Penzias, Arno 7, 11
peptide 397, 400, 424
percolation 193, 206, 333
Permean 147
perpetuum mobile 371, 377
Perron-Frobenius theorems 254, 307, 350
Persian Gulf 159
Pesin entropy 176
Pfaffian differential equation 20, 48, 67, 246, 251, 342, 448
Phanerozoic 147, 160
phase boundary 53, 303, 395
phase diffusion 381, 384
phase rule 37
phase transition 3, 5, 8, 52, 58, 59, 95, 130, 145, 150, 156, 193, 216, 218, 290, 367, 381–383, 386, 388, 405, 430, 449
phenotype 200, 240–242, 248, 258, 300–303, 318, 321, 323, 325–333, 335–341, 351, 387, 393, 400, 403, 430, 431
phenotype space 302, 323, 326–329, 332, 333, 335–337, 340
phenotypic diversity 325, 326, 328, 330, 338
phenotypic property 248, 258, 301, 323, 328, 336, 430, 431
phenotypic trait 303, 387
phenylalanine (Phe) 389
phlogiston 447
Phoenicia 407–410
phonon gas 60
phosphate 439
phosphorus 139, 439
photography 384, 385

photon gas 11, 71–76, 89, 91, 93
photon mill 12, 13, 91, 92, 94, 101, 104, 106, 107, 110, 111, 115, 123, 452
photopigment 147
photosynthesis 30, 90, 139, 147, 154, 437, 439–441, 451–453
phycocyanin 437
phyletic gradualism 335
phylogenetic 300, 393, 399, 405, 437
physical picture 3, 271
phytoplankton 110, 292, 439, 451
pictogram 408
picture
– chemical 271
– ecological 273, 276, 279
– physical 3, 271
pigment 158, 404
Pioneer Venus probe 134
Planck era 8
Planck, Max 8, 21, 39, 42, 51, 52, 55, 60, 61, 70, 71, 76, 89, 91, 120, 179, 183
Planck time 8
plankton 129, 439
plate tectonics 143
Pliocene 159
Pluto 152
Poincaré, Henri 25, 45, 55, 163, 167, 173
Poisson distribution 191, 210, 211, 299
polar 43, 99, 106, 125, 134, 140, 143, 389
polarity 391
pole 85, 125, 126, 131, 134, 143
polynucleotide 29, 213, 214, 233
polynya 140
polypeptide 29, 391
population biology 325
porphine 437
position system 410
positive cone 197, 209, 232, 243, 244, 248, 250, 253, 254, 256, 260, 264, 279, 294, 295, 299, 343, 346–349, 428
positive subspace 244, 253, 254, 284, 295
potassium 23, 436
potential
– chemical 38, 46, 53, 68, 111, 113–116, 291
– energy 148, 330
power-law spectrum 211
Precambrian 139, 153
precipitation 52, 117–119, 129, 137
predator 168, 291, 292, 295–297, 341, 356, 381, 398, 405, 435
prediction 7, 9, 10, 120, 161, 177, 178, 184, 185, 187, 202, 203, 206, 209, 216, 217, 340, 376, 377, 445, 449, 450
prediction capability 120, 161, 374, 376, 449

pressure 5, 19, 30, 37–40, 52, 53, 58, 59, 66, 86, 88, 96, 104, 105, 110, 113, 117, 121, 125–127, 129, 130, 135–137, 144, 150, 152, 158, 203, 205, 333, 374, 376, 377, 379, 382, 388, 394, 430
pressure, atmospheric 38, 133
prey 168, 291, 295–297, 341, 356, 381, 398, 405, 435
price 16, 165, 308–310, 411
Prigogine, Ilya 2–4, 6, 16, 22, 25, 35, 45, 50, 67, 69, 80, 95, 96, 113, 163, 164, 167, 211, 246, 313, 314, 415, 448
Prigogine theorem 67
primary production 90
primates 367, 370, 375, 407
primordial soup 201, 349
process
– irreversible 1, 21–23, 35, 44, 47–49, 55, 59, 61, 63, 65, 67, 78, 81, 115
– reversible 20, 44, 45, 47, 48, 167
prokaryote 327, 389, 403
prophecy 372
propulsion 403
protein 28, 30, 201, 205, 287, 387, 392, 436
protozoa 394
Ptolemaios 376
pulse-density modulation (PDM) 400
punctuated equilibrium theory 335
purine 390, 391, 435
pycnocline 131, 137
pyrimidine 390, 391, 435

q

quake 145, 146, 148, 407
quantum tunneling 331
quark annihilation 9
quark-gluon plasma 9
quasi-biennial oscillation (QBO) 126
quasi species 269, 331, 340
quasi-steady-state hypothesis 273
quaternary 159

r

radiation 7, 8, 11–13, 42, 70–76, 79, 85, 86, 88, 91–94, 97, 98, 100, 101, 103, 106, 108–110, 112, 122, 124, 131, 132, 134, 136, 314, 450, 452
– long-wave 108, 110
– short-wave 111
– thermometry 42, 112
– transfer 70, 111
radiative cooling 105, 152
radioactive decay 88, 144, 152, 242
rainy season 103, 123, 127

random graph 190, 210
random number 317
Rashevsky, Nicolas 25, 163
Rayleigh, Lord 25, 68, 69, 172, 173, 213, 314, 337
Rayleigh number 68, 69, 314
reaction rate 65, 66, 215, 223, 257, 271, 272, 282, 304, 306, 307, 395, 398, 426–428, 446
receptor 366, 373–376, 398, 399, 402–404
red algae 139, 147, 439, 440
redoxcline 439
Red-Queen hypothesis 297, 342, 363
Red Sea 137, 158, 159, 405
reducible 189, 241, 270–277, 309, 344, 350, 355, 358, 362, 447
reducible matrix 270, 273, 275, 277, 350
redundancy 369, 375, 389
reference state 36, 52, 104, 115
reflexive 166, 167, 190, 357
refrigerator 313
regeneration 397
regular fluctuations 349
relative humidity 109, 111, 114–117, 119, 123, 134, 152, 153
relativity, theory of 36, 53, 161, 167
relaxation time 76, 86, 96–98, 101–103, 109, 111, 126, 227, 273, 288, 369, 371, 373
remineralisation 138
remote sensing 111, 120
replica 372, 424, 425
replicase 29, 424–435
reproducibility 39
reptile 375
residence time 117, 197
residual entropy 51, 377, 379, 380
resource 239, 308, 339, 396, 407, 437, 452, 453
rest mass 46, 52, 379
reversible process 20, 44, 45, 47, 48, 167
revolution 99, 100
Rhea 153
ribonucleic acid (RNA) 29, 30, 214, 233, 380, 387–389, 392, 424–435
Rift Valley 419
Riss ice age 159
ritualization 28, 30, 366–368, 381, 383, 385–388, 391–393, 396, 397, 399, 403, 404, 406–409, 412, 414, 431–436, 440, 449
river 120, 123, 159, 161
roboter 3
robustness 200, 201, 205, 206
Rocky Mountains 143
Rodinia 139
Rolling Stones 406

Romans   410
r-selection   281
Rügen island   87, 375
rugged landscape   334
ruminants   157, 452
Runge-Kutta algorithm   252, 316
Russia   55, 120, 123, 140, 170, 173, 186, 235, 339, 401, 413
r/v A. v. Humboldt   112, 128, 133
r/v Glomar Challenger   143
r/v Maria S. Merian   132
r/v Prof. Albrecht Penck   438

S

Sackur-Tetrode equation   54
saddle   130, 171, 174, 187, 234, 322
Sagan, Carl   85, 158, 401, 406
Sahara   122, 159
salinity   53, 59, 96, 97, 110, 113, 115, 117, 118, 136, 137, 145
sand pile   145
Sankt Johannisthal   413
Sanskrit   411
Santa Maria Island   146
Sassezki   401
satellite   9, 89, 103, 120, 372
– ERBE   91
saturated air   114, 132
Saturn   150, 153, 155
savannah   134, 158, 404
scale-free networks   200, 201, 206, 210, 211
scaling invariance   39
scientific evolution   196
sculptures   375
seafarers   408
sea floor   141, 143, 149, 153, 154, 371, 420, 437, 439
sea ice   118, 120, 137
sea level   105, 111, 117, 120, 127, 129, 140, 157, 158, 336
seamount   140
season
– dry   122
– rainy   103, 123, 127
seasonal   89, 97, 98, 100–103, 105, 123, 137, 249, 451, 452
sea surface   93, 97, 102, 109–111, 113, 117–119, 149, 153, 154, 438, 439
sea-surface temperature   97, 101–103, 111, 152
seawater   43, 52, 54, 92, 96, 110–115, 118, 136, 137, 139, 142, 419
second law of thermodynamics   18, 20–22, 32, 35, 44, 47, 50–54, 56, 57, 60–62, 64, 66, 67, 71, 73–75, 78, 79, 81, 88, 91, 313, 314, 370, 372, 373, 385, 447
sediment   138, 147, 149, 371
seesaw pattern   121, 137
segregation principle   245, 294
selection-mutation equation   328, 330, 335
selection principle   19, 294
selective advantage   228, 232, 240, 258, 262, 263, 295, 297, 315, 316, 320, 329, 335, 370, 373, 381, 388, 391, 396, 403–405, 439
selective value   240, 242, 248, 259, 262, 274, 275, 277, 284, 290, 300, 303, 306–308, 315–320, 322, 323, 325–330, 332–335, 362, 393, 422, 426, 427, 429–430, 448
self-organised criticality (SOC)   87, 88, 124, 145, 155, 318
self-reproduction   30, 198, 203, 205, 213–215, 222, 304, 326, 391, 416, 417
semiconductor   323
semi-groups   45
semi-order   167, 189, 315, 344, 359, 361, 362
sensation   366, 374, 375
sensible heat   110, 114, 119
sensitive networks   200, 202, 205, 206, 224
sensitivity   85, 126, 155, 158, 205, 211, 373, 396, 451
separatrix   171, 187, 220, 230, 231, 255, 264, 347, 351, 394
serine, Ser   389–391
settlement   159, 161
sexual recombination   327
sexual reproduction   139, 367, 388, 439
sexual selection   300, 302, 312, 393
Shannon, Claude Elwood   21, 55, 60, 126, 235, 365, 367–369, 380
shoreline   146, 336, 337, 341
short-wave radiation   111
shot noise   177, 255
Siberia   159
silent mutation   327
silicate   138
silicium   88
Silurian   147
silver   412
singular fluctuations   349, 350
SI unit   19, 20, 39–41, 46, 47
skewness   325
skin effect   111, 114, 118
skin pigmentation   159, 405
slaved variable   3, 108, 181, 288, 444
smell   366, 376, 407
Smith, Adam   31, 117, 157, 393
Smoluchowski-Fokker-Planck equation   179, 183

smoothness   322, 323, 333, 391
– postulate   322, 323, 333
snowball Earth   123, 134, 139, 155, 440
society   1, 3, 14, 52, 53, 168, 184, 209, 241, 413, 449
socio-economic   14–17, 31, 165, 195, 196, 200, 202, 208, 212
solar   53, 71, 85, 88–92, 94, 97, 99–102, 106–108, 110, 119, 123, 124, 131, 132, 134, 136, 137, 139, 140, 149, 151–154, 416, 418, 420, 437, 443, 447, 452
– constant   89, 92, 100, 108
– cycle   89
– intensity   89
– irradiation   85, 89–91, 93, 97, 99–102, 106, 108, 110, 119, 123, 134, 136, 140, 152, 153, 418, 437
– radiation   88, 89, 91
Solaris   5
Solnhofen   372
solstice   100, 111
Sophocles   411
sound speed   49, 124, 130
South Africa   127
South America   129, 147, 149
South Atlantic   97, 102, 103, 105, 110, 126, 127, 129, 157, 158
South Atlantic Anticyclone (SAA)   110, 126, 127
Southern Ocean   121
South Pole   147
South Sandwich Trench   149
Spain   411, 452
specific heat   35, 97, 113
specific humidity   111, 113, 116, 117, 119, 123, 130, 131, 134
speech   300, 401, 404–406
speed of light   39, 52, 76
spin glass   380
spoken language   159, 366, 367, 381, 404, 406, 408
spontaneous abiogenesis   418
sprat   297
spring hole   138, 140
stability   25, 27, 39, 85, 86, 124, 126, 130, 131, 136, 137, 150, 154, 173–175, 190, 247, 260, 263, 282–284, 294, 295, 337, 348–350, 427, 428, 451
standard   7, 32, 33, 39, 56, 66, 72, 78, 112, 138, 170, 174, 179, 214, 246, 258, 291, 311, 385, 392, 393, 410, 434, 435, 445, 447
state variable   20, 37, 47–49, 81, 111, 273, 317, 404
statistical ensemble   53, 63, 206

statistical mechanics   36, 59–61, 83, 181, 206, 271, 297, 379
statistical physics   36, 59–61, 83, 181, 206, 271, 297, 379
steam engine   85, 91, 119, 451
steam vent   138, 140
Stefan-Boltzmann constant   73, 92, 108
Stefan-Boltzmann law   71, 85, 110
St. Helena Climate Index (HIX)   129
St. Helena Island   97, 103, 105, 126, 127, 129
Stirling formula   299
stochastic event   255, 317
stochastic landscape   322, 333
stoichiometric coefficient   65
stoichiometry   64, 65
stone axe   206
strange attractor   176, 243, 381
stratification   53, 130, 131, 136, 145
Stratonovich, Ruslan Leontevich   81, 177, 234–237, 365
stratosphere   124, 131, 133
stroboscopic   26, 164
structural information   42, 160, 367, 370–373, 377, 383–385, 389, 392, 393, 396, 410, 411, 415
structural stability   347, 348, 396
subduction   144, 148, 149
subglacial lake   125
sublimation   38, 43, 150
sublimation curve   38
subsaturation   54, 115
subtropical   92, 96–99, 101–103, 105, 110, 112, 118, 119, 121, 125, 127, 133, 135
sulfate   139, 439, 441
sulfide   139, 147, 155, 437, 439
sulfur   390
sulphuric acid   23
Sumerian   408
Sun   6, 70, 88, 89, 92–95, 99, 101, 103, 111, 124, 150, 419, 443, 444
sunrise   99
sunset   99
sun-spot cycle   89
sun spots   89
supernova   149
surface salinity   117
surface temperature   12, 85, 89, 95–98, 103, 104, 106, 108, 125, 153, 155, 156
surface tension   68, 303, 329
survival   33, 202, 206, 228, 232, 240–242, 258–266, 281–284, 313, 314, 342, 343, 351, 374, 375, 377, 387, 393, 416, 447, 449
survival probability   228, 232, 259, 262–264, 343

symbolic information   28, 214, 365, 367, 369–371, 373, 377, 380, 383–388, 392, 393, 397, 399, 403, 406, 411, 412, 415–417, 425, 430, 431, 435, 449
symmetry   3–5, 8, 9, 11, 29, 66, 68, 69, 73, 82, 83, 100, 101, 135, 145, 151, 166, 167, 174, 250, 262, 292, 301, 314, 315, 327, 328, 336, 361, 369, 379, 381–384, 388, 391, 392–394, 432, 434–436, 445, 446
synapse   401, 402
synergetics   3, 25, 163, 212
system
– adiabatic   36
– closed   36, 250, 315, 370
– isolated   19, 21, 32, 36, 45–47, 49, 55–57, 61, 77, 88
– open   4, 36, 51, 313, 394, 415, 416, 419
Szilard, Leo   63, 313

t

Tasmanian Seaway   149, 157, 158
taste   366, 376
tautology   45, 55, 240
taxonomy   208, 209
technology   4, 17, 39, 70, 81, 122, 159, 165, 195, 196, 200, 202, 205, 206, 208, 221, 223, 224, 234, 308–311, 318, 322, 365, 384–386, 411, 453
teleconnection   129, 156
temperature
– definition   41, 42, 53
– scale   41–43
– thermodynamic   41, 76
tentacles   398
terabyte   380
terminator   99
tertiary   147, 157, 158
tetrachromacy   375
tetrapods   147
tetrapyrrole   437
Teviornis gobiensis   147
Thaler   412
Theia   153
thermal inertia   98
thermal noise   42
thermal radiation   63, 72, 73, 85, 91, 94, 98, 109, 110, 119, 120, 122, 148
thermodynamic flux   67, 79, 251, 314
thermodynamic force   66, 79, 115, 251, 314
thermodynamic limit   190, 257, 260, 263, 401, 403
thermodynamic potential   35, 48, 49, 52, 58, 59, 78, 96, 104
thermodynamic probability   61, 76

thermodynamic temperature   41, 76
thermohaline convection   138
thermometer   42, 43
thermometry
– noise   42
– radiation   42
thermoscope   40
third law of thermodynamics   51, 377
thymine   214
tidal friction   145
tides   6, 146
Tierra del Fuego   149
time arrow   35
time reversal   82, 83
tissue   394–397, 399
Titan   155
tomography   141
topology   200, 205, 268, 346–348, 394, 398
tornado   86, 126, 134, 135, 149
trade wind   110, 125, 127, 129, 135
transcription   30, 400
transfer RNA (tRNA)   387, 391, 392, 431, 432, 435
transition probability   180–183, 197–199, 203, 222, 225, 226, 229, 232, 255, 264, 265, 299, 310
transitivity   43, 166–168, 315, 357, 359, 362, 448
translation   367, 384, 406, 425, 432, 434–436
transmission   120, 236, 370, 373, 381, 385, 400
transmitter   368, 370, 383, 387, 397, 399–403
travelling dune   87, 145
tree   6, 157, 160, 161, 188, 192, 193, 296, 320, 360–362, 375, 399, 437, 438, 452, 453
Triassic   147
triple point   41, 52, 116, 155
troll   375
trophic level   292
tropics   106, 119, 120, 134, 140, 157, 452
tropopause   131–134, 153, 155
troposphere   53, 86, 125, 126, 130–134, 155
tsunami   145, 148
turbulence   53, 60, 104, 114, 117, 118, 314, 404, 419, 430, 443, 445, 446
Turing, Alan   2, 3, 394
typhoon   134

u

uncertainty   21, 38, 39, 41, 42, 49, 56, 61, 89, 90, 118, 120, 177, 179, 184, 185, 209, 234, 235, 445, 450
universal common ancestor (UCA)   139

universe 6, 7, 9, 10, 14, 53, 150, 177, 343, 374, 401
upwelling 23, 127, 128, 137, 138, 140, 144, 145, 439
uracil 214
uranium 124, 150
urn model 195, 200, 221, 232

v

vacuum 8, 149, 150
valoric interpretation 60, 63
valuation 5, 31, 32, 240, 241, 316, 321–323, 393
valuation landscape 316, 321, 322
value
– of exchange 31, 33, 63, 308, 310, 318, 412, 448
– of information 31, 33, 235, 237, 365, 448
– law, in economy 308
– selective 240, 242, 248, 259, 262, 274, 275, 277, 284, 290, 300, 303, 306–308, 315–320, 322, 323, 325–330, 332–335, 362, 393, 422, 426, 427, 429, 430, 448
Vancouver 159
van der Pol, Balthasar 25, 163
van der Waals, Johannes Diderik 11
vapor instability 86, 123, 134
vapor pressure 37, 38, 114, 115, 117, 150
variable
– dependent 48
– extensive 38, 40, 45, 46, 48, 50, 51, 56, 57, 60
– external 37, 39
– independent 20, 37, 58, 59, 81, 346
– intensive 37, 38, 40, 42, 43, 53
– internal 39
– state 20, 37, 47–49, 81, 111, 273, 317, 404
ventilation 138, 439
Venus 86, 123, 134, 141, 142, 145, 152, 153, 420, 450
Verhulst-Pearl equation 214, 216, 221, 238
Vienna 41, 51, 378, 379
Vienna Standard Mean Ocean Water (VSMOW) 41
violin 87, 148
virial theorem 53, 446
virus 213
vis vitalis 447
volcano 5, 146, 149
Volkenstein, Mikhail Vladimirovich 28, 33, 60, 61, 234, 235, 335, 365, 380, 416, 417

Volterra, Vito 25, 163, 253, 291, 295, 339
Volvox carteri 394
Volvox globator 394
Vostok 121, 186
vulnerability 392

w

waiting time 197, 317
Warnemünde 112, 184
water cycle 111
water flea 399
waterspout 134
water world 92, 152
weathering 139, 140, 149
web sites 201, 202, 205, 206
Wegener, Alfred 143
Weinberg-Salam theory 9
welfare 235, 236
West Antarctic 157
west-wind belt 102, 125
whitecap 110, 374
Wisconsin ice age 159
Witt, Alexander Adolfovich 25, 163, 173, 339
WMAP 9
wobble hypothesis 389
wood 410, 452, 453
work 2, 5, 16, 19, 21, 30, 32, 36, 46, 49, 50, 54, 58, 59, 61, 62, 65, 66, 68, 76, 81, 82, 92, 119, 148, 163, 190, 191, 202–204, 211, 240, 312, 313, 399, 403, 424, 450, 453
World Meteorological Organisation (WMO) 114
World Wide Web (WWW) 201, 206
Wright, Sewall Green 31, 71, 91, 116, 133, 163, 240, 241, 321
written language 159, 367, 384, 386, 408, 449
Würm ice age 159

y

yeast 201
Yellowstone 419
Yucatan 151

z

Zanclean Flood 158
Zermelo, Ernst 167
zero-point entropy 51, 377, 379, 380
zeroth law of thermodynamics 43, 53
Zloty 413
zooplankton 292, 398
Zwanzig projection 179